STUDENT'S SOLUTIONS MANUAL

C A L C U L U S

One and Several Variables
6th Edition

S. L. SALAS   EINAR HILLE

WILEY

John Wiley and Sons/New York/Chichester/Brisbane/Toronto/Singapore

ISBN 0-471-61198-0

Printed in the United States of America

Printed and bound by the Hamilton Printing Company.

10 9 8 7 6 5 4 3

Computer Typeset by **Laura Salas Hawks**
NORTHWEST COMPUTER SERVICE, INC.
Lakeville, CT 06039

ANSWERS / SOLUTIONS
to All Odd-Numbered Exercises

CONTENTS

Chapter 1   INTRODUCTION -- 1

Chapter 2   LIMITS AND CONTINUITY ---------------------------------- 12

Chapter 3   DIFFERENTIATION--- 24

Chapter 4   THE MEAN-VALUE THEOREM AND APPLICATIONS ---- 57

Chapter 5   INTEGRATION -- 97

Chapter 6   SOME APPLICATIONS OF THE INTEGRAL ---------------- 121

Chapter 7   THE TRANSCENDENTAL FUNCTIONS ---------------------- 138

Chapter 8   TECHNIQUES OF INTEGRATION ------------------------------ 176

Chapter 9   THE CONIC SECTIONS   ------------------------------------ 205

Chapter 10   POLAR COORDINATES; PARAMETRIC EQUATIONS ---- 215

Chapter 11   SEQUENCES; INDETERMINATE FORMS;
             IMPROPER INTEGRALS   --------------------------------------- 239

Chapter 12   INFINITE SERIES   -- 255

Chapter 13   VECTORS --- 276

Chapter 14   VECTOR CALCULUS -- 291

Chapter 15   FUNCTIONS OF SEVERAL VARIABLES ---------------------- 305

Chapter 16   GRADIENTS; EXTREME VALUES; DIFFERENTIALS ---- 315

Chapter 17   DOUBLE AND TRIPLE INTEGRALS --------------------------- 343

Chapter 18   LINE INTEGRALS AND SURFACE INTEGRALS ----------- 365

Appendix A   SOME ELEMENTARY TOPICS ---------------------------------- 387

CHAPTER 1

SECTION 1.3

1. $m = \dfrac{5-1}{(-2)-4} = \dfrac{4}{-6} = -\dfrac{2}{3}$

3. $m = \dfrac{2-2}{4-(-3)} = \dfrac{0}{7} = 0$

5. $m = \dfrac{b-a}{a-b} = -1$

7. $m = \dfrac{0-y_0}{x_0-0} = -\dfrac{y_0}{x_0}$

9. Equation is in the form $y = mx + b$. Slope is 2; y-intercept is -4.

11. Write equation as $y = \frac{1}{3}x + 2$. Slope is $\frac{1}{3}$; y-intercept is 2.

13. Line is vertical: $x = \frac{1}{4}$. Thus slope is undefined and there is no y-intercept.

15. Write equation as $y = \frac{7}{3}x + \frac{4}{3}$. Slope is $\frac{7}{3}$; y-intercept is $\frac{4}{3}$.

17. Line is horizontal: $y = \frac{5}{7}$. Slope is 0; y-intercept is $\frac{5}{7}$.

19. $y = 5x + 2$ 21. $y = -5x + 2$ 23. $y = 3$ 25. $x = -3$

27. Every line parallel to the x-axis has an equation of the form $y = a$ constant. In this case $y = 7$.

29. The line $3y - 2x + 6 = 0$ has slope $\frac{2}{3}$. Every line parallel to it has that same slope. The line through $P(2, 7)$ with slope $\frac{2}{3}$ has equation $y - 7 = \frac{2}{3}(x - 2)$, which reduces to $3y - 2x - 17 = 0$.

31. The line $3y - 2x + 6 = 0$ has slope $\frac{2}{3}$. Every line perpendicular to it has slope $-\frac{3}{2}$.
The line through $P(2, 7)$ with slope $-\frac{3}{2}$ has equation $y - 7 = -\frac{3}{2}(x - 2)$, which reduces to
$2y + 3x - 20 = 0$.

33. $m = 1 = \tan\theta;$ $\theta = 45°$

35. line is vertical: $x = \frac{3}{2};$ $\theta = 90°$

37. $m = -\frac{3}{4} = \tan\theta;$ $\theta \cong 143°$

39. $m = \tan 30° = \frac{1}{3}\sqrt{3}$; line: $y = \frac{1}{3}\sqrt{3}\,x + 2$

41. $m = \tan 120° = -\sqrt{3}$;   line: $y = -\sqrt{3}\,x + 3$

43. $\left(\frac{1}{2}\sqrt{2}, \frac{1}{2}\sqrt{2}\right), \left(-\frac{1}{2}\sqrt{2}, -\frac{1}{2}\sqrt{2}\right)$ [Substitute $y = x$ into $x^2 + y^2 = 1$.]

45. $(3, 4)$ [Write $4x + 3y = 24$ as $y = \frac{4}{3}(6 - x)$ and substitute into $x^2 + y^2 = 25$.]

47. $(1, 1)$; $\alpha \cong 39°$ [$m_1 = 4 = \tan\theta_1$, $\theta_1 \cong 76°$; $m_2 = \frac{3}{4} = \tan\theta_2$, $\theta_2 \cong 37°$]

49. $\left(-\frac{2}{23}, \frac{38}{23}\right)$; $\alpha \cong 17°$ [$m_1 = 4 = \tan\theta_1$, $\theta_1 \cong 76°$; $m_2 = 19 = \tan\theta_2$, $\theta_2 \cong 87°$]

51. Substitute $y = m(x - 5) + 12$ into $x^2 + y^2 = 169$ and you get a quadratic in x that involves m.
That quadratic has a unique solution iff $m = -\frac{5}{12}$. (A quadratic $ax^2 + bx + c = 0$ has a unique solution iff $b^2 - 4ac = 0$. This is clear from the general quadratic formula.)

SECTION 1.4

1. $2 + 3x < 5$

$3x < 3$

$x < 1$

Ans: $(-\infty, 1)$

3. $16x + 64 \leq 16$

$16x \leq -48$

$x \leq -3$

Ans: $(-\infty, -3]$

5. $\frac{1}{2}(1 + x) < \frac{1}{3}(1 - x)$

$3(1 + x) < 2(1 - x)$

$3 + 3x < 2 - 2x$

$5x < -1$

$x < -\frac{1}{5}$

Ans: $(-\infty, -\frac{1}{5})$

7. $x^2 - 1 < 0$

$(x + 1)(x - 1) < 0$

$$+ + + \;\underset{-1}{0}\; - - - - - \;\underset{1}{0}\; + + + + +$$

Ans: $(-1, 1)$

9. $4(x^2 - 3x + 2) > 0$

$4(x - 1)(x - 2) > 0$

$$+ + + + \;\underset{1}{0}\; - - - - - \;\underset{2}{0}\; + + + +$$

Ans: $(-\infty, 1) \cup (2, \infty)$

11. $x(x - 1)(x - 2) > 0$

$$- - \;\underset{0}{0}\; + + \;\underset{1}{0}\; - - \;\underset{2}{0}\; + +$$

Ans: $(0, 1) \cup (2, \infty)$

13. $x^3 - 2x^2 + x \geq 0$

$x(x - 1)^2 \geq 0$

$$- - - - \;\underset{0}{0}\; + + + + \;\underset{1}{0}\; + + + +$$

Ans: $[0, \infty)$

15. $x^2 + 1 > 4x$

$x^2 - 4x + 1 > 0$

$x^2 - 4x + 4 > 3$

$\sqrt{(x - 2)^2} > 3$

$x - 2 > \sqrt{3}$ (or) $x - 2 < -\sqrt{3}$

Ans: $(-\infty, 2 - \sqrt{3}) \cup (2 + \sqrt{3}, \infty)$

17. $\frac{1}{2}(1 + x)^2 < \frac{1}{3}(1 - x)^2$

$3(1 + x)^2 < 2(1 - x)^2$

$3 + 6x + 3x^2 < 2 - 4x + 2x^2$

$x^2 + 10x + 1 < 0$

$x^2 + 10x + 25 < 24$

$(x + 5)^2 < 24$

$-\sqrt{24} < x + 5 < \sqrt{24}$

Ans: $(-5 - 2\sqrt{6}, -5 + 2\sqrt{6})$

19. $1 - 3x^2 < \frac{1}{2}(2 - x^2)$

$2 - 6x^2 < 2 - x^2$

$0 < 5x^2$

True if $x \neq 0$.

Ans: $(-\infty, 0) \cup (0, \infty)$

21. $\frac{1}{x} < x$

$x - \frac{1}{x} > 0$

$\frac{x^2 - 1}{x} > 0$

$x(x - 1)(x + 1) > 0$ (by 1.4.1)

$(x + 1) x (x - 1) > 0$

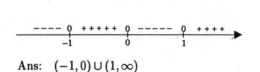

$$- - - - \;\underset{-1}{0}\; + + + + + \;\underset{0}{0}\; - - - - - \;\underset{1}{0}\; + + + +$$

Ans: $(-1, 0) \cup (1, \infty)$

23. $\dfrac{x}{x-5} \geq 0$

$x(x-5) > 0$ or $x = 0$ (by 1.4.1)

Ans: $(-\infty, 0] \cup (5, \infty)$

25. $\dfrac{x}{x-5} > \dfrac{1}{4}$

$\dfrac{x}{x-5} - \dfrac{1}{4} > 0$

$\dfrac{4x - (x-5)}{4(x-5)} > 0$

$\dfrac{3x+5}{4(x-5)} > 0$

$4(x-5)(3x+5) > 0$ (by 1.4.1)

$(3x+5)(x-5) > 0$

Ans: $\left(-\infty, -\dfrac{5}{3}\right) \cup (5, \infty)$

27. $\dfrac{x^2 - 9}{x+1} > 0$

$(x+1)(x-3)(x+3) > 0$ (by 1.4.1)

$(x+3)(x+1)(x-3) > 0$

Ans: $(-3, -1) \cup (3, \infty)$

29. $x^3(x-2)(x+3)^2 < 0$

$(x+3)^2 x (x-2) < 0$

Ans: $(0, 2)$

31. $x^2(x-2)(x+6) > 0$

$(x+6) x^2 (x-2) > 0$

Ans: $(-\infty, -6) \cup (2, \infty)$

33. $5x(x-3)^2 < 0$

Ans: $(-\infty, 0)$

35. $\dfrac{2x}{x^2 - 4} > 0$

$2x(x+2)(x-2) > 0$

$(x+2)\, x\, (x-2) > 0$

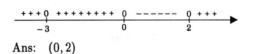

Ans: $(-2, 0) \cup (2, \infty)$

37.
$$\frac{x-1}{9-x^2} < 0$$
$$(x-1)(3-x)(3+x) < 0$$
$$(x+3)(x-1)(x-3) > 0$$

$$--- \; \underset{-3}{0} \; +++++++ \; \underset{1}{0} \; ----- \; \underset{3}{0} \; +++ \longrightarrow$$

Ans: $(-3,1) \cup (3,\infty)$

39.
$$\frac{1}{x-1} + \frac{4}{x-6} > 0$$
$$\frac{x-6+4(x-1)}{(x-1)(x-6)} > 0$$
$$\frac{5x-10}{(x-1)(x-6)} > 0$$
$$5(x-2)(x-1)(x-6) > 0$$
$$(x-1)(x-2)(x-6) > 0$$

$$--- \; \underset{1}{0} \; +++ \; \underset{2}{0} \; ----------- \; \underset{6}{0} +++ \longrightarrow$$

Ans: $(1,2) \cup (6,\infty)$

41.
$$\frac{2x-6}{x^2-6x+5} < 0$$
$$2(x-3)(x-1)(x-5) < 0$$
$$(x-1)(x-3)(x-5) < 0$$

$$---- \; \underset{1}{0} \; +++++ \; \underset{3}{0} \; ----- \; \underset{5}{0} +++ \longrightarrow$$

Ans: $(-\infty,1) \cup (3,5)$

43.
$$\frac{x+3}{x^2(x-5)} < 0$$
$$x^2(x-5)(x+3) < 0$$
$$(x+3)\,x^2(x-5) < 0$$

$$+++ \underset{-3}{0} \; ----- \; \underset{0}{0} \; --------- \; \underset{5}{0} +++ \longrightarrow$$

Ans: $(-3,0) \cup (0,5)$

45.
$$\frac{x^2-4x+3}{x^2} > 0$$
$$x^2(x-1)(x-3) > 0$$

$$+++ \underset{0}{0} \; ++++ \; \underset{1}{0} \; --------- \; \underset{3}{0} ++++ \longrightarrow$$

Ans: $(-\infty,0) \cup (0,1) \cup (3,\infty)$

47. $x < \sqrt{x} < 1 < \dfrac{1}{\sqrt{x}} < \dfrac{1}{x}$

49. $(a-b)^2 > 0 \implies a^2 - 2ab + b^2 > 0 \implies a^2 + b^2 \geq 2ab$

51. With $a \geq 0$ and $b \geq 0$

$$b \geq a \implies b - a = (\sqrt{b} + \sqrt{a})(\sqrt{b} - \sqrt{a}) \geq 0 \implies \sqrt{b} - \sqrt{a} \geq 0 \implies \sqrt{b} \geq \sqrt{a}.$$

53. With $0 \le a \le b$

$$a(1+b) = a + ab \le b + ab = b(1+a).$$

Division by $(1+a)(1+b)$ gives

$$\frac{a}{1+a} \le \frac{b}{1+b}.$$

SECTION 1.5

1. $|x| < 2$

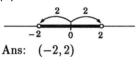

Ans: $(-2, 2)$

3. $|x| > 3$

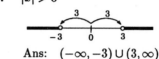

Ans: $(-\infty, -3) \cup (3, \infty)$

5. $|x - 2| < \frac{1}{2}$

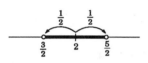

Ans: $\left(\frac{3}{2}, \frac{5}{2}\right)$

7. $|x + 2| < \frac{1}{4}$

$|x - (-2)| < \frac{1}{4}$

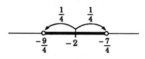

Ans: $\left(-\frac{9}{4}, -\frac{7}{4}\right)$

9. $0 < |x| < 1$

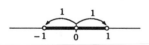

Ans: $(-1, 0) \cup (0, 1)$

11. $0 < |x - 2| < \frac{1}{2}$

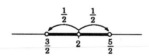

Ans: $\left(\frac{3}{2}, 2\right) \cup \left(2, \frac{5}{2}\right)$

13. $0 < |x - 3| < 8$

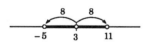

Ans: $(-5, 3) \cup (3, 11)$

15. $|2x - 3| < 1$

$|x - \frac{3}{2}| < \frac{1}{2}$

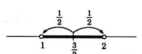

Ans: $(1, 2)$

17. $|2x + 1| < \frac{1}{4}$

$|x - \left(-\frac{1}{2}\right)| < \frac{1}{8}$

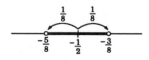

Ans: $\left(-\frac{5}{8}, -\frac{3}{8}\right)$

19. $|2x + 5| > 3$

$|x - \left(-\frac{5}{2}\right)| > \frac{3}{2}$

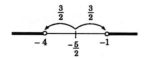

Ans: $(-\infty, -4) \cup (-1, \infty)$

21. $|5x - 1| > 9$

$|x - \frac{1}{5}| > \frac{9}{5}$

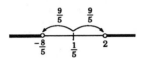

Ans: $\left(-\infty, -\frac{8}{5}\right) \cup (2, \infty)$

23. $(-3, 3)$

$|x - 0| < 3$

Ans: $|x| < 3$

25. $(-3, 7)$

Ans: $|x - 2| < 5$

27. $(-7, 3)$

$|x - (-2)| < 5$

Ans: $|x + 2| < 5$

29. $|x - 2| < A \implies 2|x - 2| = |2x - 4| < 2A \implies |2x - 4| < 3$

provided that $0 < A \leq \frac{3}{2}$

31. $|x + 1| < 2 \implies 3|x + 1| = |3x + 3| < 6 \implies |3x + 3| < A$

provided that $A \geq 6$

33.   By the hint

$$\big|\, |a| - |b|\, \big|^2 = (|a| - |b|)^2 = |a|^2 - 2|a|\,|b| + |b|^2 = a^2 - 2|ab| + b^2$$

$$\leq a^2 - 2ab + b^2 = (a - b)^2.$$

$$(ab \leq |ab|)$$

Taking the square root of the extremes, we have

$$\big|\, |a| - |b|\, \big| \leq \sqrt{(a - b)^2} = |a - b|.$$

SECTION 1.6

1. $x = 1, 3$

3. $x = -2$

5. $x = n\pi$, n an integer

7. $x = n\pi$, n an integer

9. $\text{dom}\,(f) = (-\infty, \infty)$; $\text{ran}\,(f) = [0, \infty)$

11. $\text{dom}\,(f) = (-\infty, \infty)$; $\text{ran}\,(f) = (-\infty, \infty)$

13. $\text{dom}\,(f) = (-\infty, 0) \cup (0, \infty)$; $\text{ran}\,(f) = (0, \infty)$

15. $\text{dom}\,(f) = (-\infty, 1]$; $\text{ran}\,(f) = [0, \infty)$

17. $\text{dom}\,(f) = (-\infty, 1]$; $\text{ran}\,(f) = [-1, \infty)$

19. $\text{dom}\,(f) = (-\infty, 1)$; $\text{ran}\,(f) = (0, \infty)$

21. $\text{dom}\,(f) = (-\infty, \infty)$; $\text{ran}\,(f) = [0, 1]$

23. $\text{dom}\,(f) = (-\infty, \infty)$; $\text{ran}\,(f) = [-2, 2]$

25. $\quad \text{dom}\,(f) = \left\{x \mid x \neq \dfrac{2n+1}{2}\pi, \ n \text{ an integer}\right\}; \quad \text{ran}\,(f) = [1,\infty)$

27.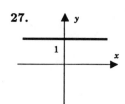

29.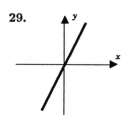

31.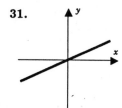

33.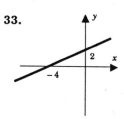

35.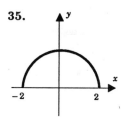

37.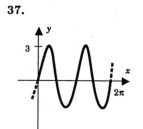

39.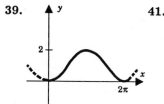

41.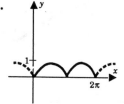

43.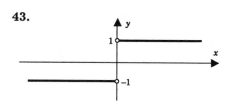

45.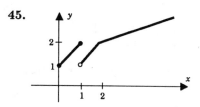

$\text{dom}\,(f) = (-\infty,0) \cup (0,\infty)$
$\text{ran}\,(f) = \{-1,1\}$

$\text{dom}\,(f) = [0,\infty); \quad \text{ran}\,(f) = [1,\infty)$

47.   (a) $(6f + 3g)(x) = 9\sqrt{x}, \quad x > 0$ (b) $(fg)(x) = x - \dfrac{2}{x} - 1, \quad x > 0$

      (c) $(f/g)(x) = \dfrac{x+1}{x-2}, \qquad 0 < x < 2 \ \ \text{and} \ \ x > 2$

49.   (a) $(f+g)(x) = \begin{cases} 1-x, & x \leq 1 \\ 2x-1, & 1 < x < 2 \\ 2x-2, & x \geq 2 \end{cases}$

      (b) $(f-g)(x) = \begin{cases} 1-x, & x \leq 1 \\ 2x-1, & 1 < x < 2 \\ 2x, & x \geq 2 \end{cases}$ (c) $(fg)(x) = \begin{cases} 0, & x < 2 \\ 1-2x, & x \geq 2 \end{cases}$

51.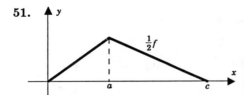

53.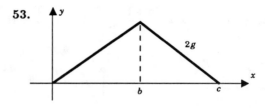

55.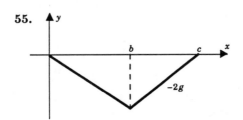

57.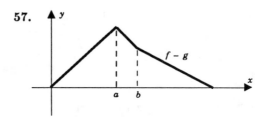

59. odd   **61.** neither   **63.** even   **65.** odd   **67.** even

69. even; $(fg)(-x) = f(-x)g(-x) = f(x)g(x) = (fg)(x)$

71. (a)   (b)

$$f(x) = \left\{ \begin{matrix} 1, & x < -1 \\ -x, & -1 \le x < 0 \end{matrix} \right] \qquad\qquad f(x) = \left\{ \begin{matrix} -1, & x < -1 \\ x, & -1 \le x \le 0 \end{matrix} \right]$$

73. $g(-x) = f(-x) + f(-(-x)) = f(-x) + f(x) = g(x)$

SECTION 1.7

1. $(f \circ g)(x) = 2x^2 + 5$ **3.** $(f \circ g)(x) = \sqrt{x^2 + 5}$

5. $(f \circ g)(x) = \dfrac{1}{(1/x)} = x, \quad x \neq 0$ **7.** $(f \circ g)(x) = \dfrac{1}{x} - 1$

9. $(f \circ g)(x) = \dfrac{1}{\sqrt{(x^2 + 2)^2} - 1} = \dfrac{1}{(x^2 + 2) - 1} = \dfrac{1}{x^2 + 1}$

                           $\sqrt{(x^2 + 2)^2} = |x^2 + 2| = x^2 + 2$

11. $(f \circ g \circ h)(x) = 4\,[g(h(x))] = 4\,[h(x) - 1] = 4(x^2 - 1)$

13. $(f \circ g \circ h)(x) = [g(h(x))]^2 = [h(x) - 1]^2 = (x^4 - 1)^2$

15. $(f \circ g \circ h)(x) = \dfrac{1}{g(h(x))} = \dfrac{1}{1/(2h(x) + 1)} = 2h(x) + 1 = 2x^2 + 1$

17.   Take $f(x) = \dfrac{1}{x}$ since $\dfrac{1 + x^4}{1 + x^2} = F(x) = f(g(x)) = f\left(\dfrac{1 + x^2}{1 + x^4} \right).$

19.   Take $f(x) = 2\sin x$ since $2\sin 3x = F(x) = f(g(x)) = f(3x).$

21. Take $\quad g(x) = \left(1 - \dfrac{1}{x^4}\right)^{2/3} \quad$ since $\quad \left(1 - \dfrac{1}{x^4}\right)^2 = F(x) = f(g(x)) = [g(x)]^3.$

23. Take $\quad g(x) = 2x^3 - 1$ (or $-(2x^3 - 1)$) $\quad$ since $\quad (2x^3 - 1)^2 + 1 = F(x) = f(g(x)) = [g(x)]^2 + 1.$

25. $(f \circ g)(x) = f(g(x)) = \sqrt{g(x)} = \sqrt{x^2} = |x|;$

$(g \circ f)(x) = g(f(x)) = [f(x)]^2 = [\sqrt{x}]^2 = x, \quad x \geq 0$

27. $(f \circ g)(x) = f(g(x)) = 2g(x) = 2\sin x;$

$(g \circ f)(x) = g(f(x)) = \sin f(x) = \sin 2x$

29. $(f \circ g)(x) = \begin{cases} x^2, & x < 0 \\ 1+x, & 0 \leq x < 1 \\ (1+x)^2, & x \geq 1 \end{cases};\qquad (g \circ f)(x) = \begin{cases} 2-x, & x \leq 0 \\ -x^2, & 0 < x < 1 \\ 1+x^2, & x \geq 1 \end{cases}$

31.

	f_1	f_2	f_3	f_4	f_5	f_6
f_1	f_1	f_2	f_3	f_4	f_5	f_6
f_2	f_2	f_1	f_4	f_3	f_6	f_5
f_3	f_3	f_5	f_1	f_6	f_2	f_4
f_4	f_4	f_6	f_2	f_5	f_1	f_3
f_5	f_5	f_3	f_6	f_1	f_4	f_2
f_6	f_6	f_4	f_5	f_2	f_3	f_1

SECTION 1.8

1.
$$f(t) = x$$
$$5t + 3 = x$$
$$5t = x - 3$$
$$t = \tfrac{1}{5}(x - 3)$$
$$f^{-1}(x) = \tfrac{1}{5}(x - 3)$$

3.
$$f(t) = x$$
$$4t - 7 = x$$
$$4t = x + 7$$
$$t = \tfrac{1}{4}(x + 7)$$
$$f^{-1}(x) = \tfrac{1}{4}(x + 7)$$

5. f is not one-to-one; for instance, $f(1) = f(-1)$

7.
$$f(t) = x$$
$$t^5 + 1 = x$$
$$t^5 = x - 1$$
$$t = (x-1)^{1/5}$$
$$f^{-1}(x) = (x-1)^{1/5}$$

9.
$$f(t) = x$$
$$1 + 3t^3 = x$$
$$t^3 = \tfrac{1}{3}(x-1)$$
$$t = \left[\tfrac{1}{3}(x-1)\right]^{1/3}$$
$$f^{-1}(x) = \left[\tfrac{1}{3}(x-1)\right]^{1/3}$$

11.
$$f(t) = x$$
$$(1-t)^3 = x$$
$$1 - t = x^{1/3}$$
$$t = 1 - x^{1/3}$$
$$f^{-1}(x) = 1 - x^{1/3}$$

13.
$$f(t) = x$$
$$(t+1)^3 + 2 = x$$
$$(t+1)^3 = x - 2$$
$$t + 1 = (x-2)^{1/3}$$
$$t = (x-2)^{1/3} - 1$$
$$f^{-1}(x) = (x-2)^{1/3} - 1$$

15.
$$f(t) = x$$
$$t^{3/5} = x$$
$$t = x^{5/3}$$
$$f^{-1}(x) = x^{5/3}$$

17.
$$f(t) = x$$
$$(2-3t)^3 = x$$
$$2 - 3t = x^{1/3}$$
$$3t = 2 - x^{1/3}$$
$$t = \tfrac{1}{3}(2 - x^{1/3})$$
$$f^{-1}(x) = \tfrac{1}{3}(2 - x^{1/3})$$

19.
$$f(t) = x$$
$$\frac{1}{t} = x$$
$$t = \frac{1}{x}$$
$$f^{-1}(x) = \frac{1}{x}$$

21. f is not one-to-one;   for instance, $f\left(\tfrac{1}{2}\right) = f(2)$

23.

$$f(t) = x$$

$$\frac{1}{t^3 + 1} = x$$

$$t^3 + 1 = \frac{1}{x}$$

$$t^3 = \frac{1}{x} - 1$$

$$t = \left(\frac{1}{x} - 1\right)^{1/3}$$

$$f^{-1}(x) = \left(\frac{1}{x} - 1\right)^{1/3}$$

25.

$$f(t) = x$$

$$\frac{t + 2}{t + 1} = x$$

$$t + 2 = xt + x$$

$$t(1 - x) = x - 2$$

$$t = \frac{x - 2}{1 - x}$$

$$f^{-1}(x) = \frac{x - 2}{1 - x}$$

27.   they are equal

29.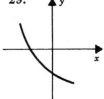

31.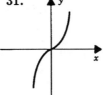

33.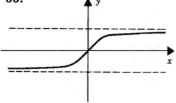

CHAPTER 2

SECTION 2.1

1. (a) 2   (b) −1   (c) does not exist   (d) −3

3. (a) does not exist   (b) −3   (c) does not exist   (d) −3

5. (a) does not exist   (b) does not exist   (c) does not exist   (d) 1

7. (a) 2   (b) 2   (c) 2   (d) −1

9. (a) −1   (b) −1   (c) −1   (d) undefined

11. (a) 0   (b) 0   (c) 0   (d) 0

13. $c = 0, 6$ 15. −1   17. 4   19. 1

21. $\frac{3}{2}$ 23. does not exist   25. $\displaystyle\lim_{x\to3}\frac{2x-6}{x-3} = \lim_{x\to3}2 = 2$

27. $\displaystyle\lim_{x\to3}\frac{x-3}{x^2-6x+9} = \lim_{x\to3}\frac{x-3}{(x-3)^2} = \lim_{x\to3}\frac{1}{x-3};$ does not exist

29. $\displaystyle\lim_{x\to2}\frac{x-2}{x^2-3x+2} = \lim_{x\to2}\frac{x-2}{(x-1)(x-2)} = \lim_{x\to2}\frac{1}{x-1} = 1$

31. does not exist   33. $\displaystyle\lim_{x\to0}\frac{2x-5x^2}{x} = \lim_{x\to0}(2-5x) = 2$

35. $\displaystyle\lim_{x\to1}\frac{x^2-1}{x-1} = \lim_{x\to1}\frac{(x-1)(x+1)}{x-1} = \lim_{x\to1}(x+1) = 2$

37. 0   39. 1   41. 16

43. does not exist   45. does not exist   47. does not exist

49.

$$\lim_{x\to1}\frac{\sqrt{x^2+1}-\sqrt{2}}{x-1} = \lim_{x\to1}\frac{(\sqrt{x^2+1}-\sqrt{2})(\sqrt{x^2+1}+\sqrt{2})}{(x-1)(\sqrt{x^2+1}+\sqrt{2})}$$

$$= \lim_{x\to1}\frac{x^2-1}{(x-1)(\sqrt{x^2+1}+\sqrt{2})} = \lim_{x\to1}\frac{x+1}{\sqrt{x^2+1}+\sqrt{2}} = \frac{2}{2\sqrt{2}} = \frac{1}{\sqrt{2}}$$

51. 1   53. $\frac{3}{2}$

SECTION 2.2

1. $\dfrac{1}{2}$ 3. $\displaystyle\lim_{x\to0}\frac{x(1+x)}{2x^2} = \lim_{x\to0}\frac{1+x}{2x};$ does not exist

5. $\dfrac{4}{\sqrt{5}} = \dfrac{4}{5}\sqrt{5}$ 7. $\displaystyle\lim_{x\to1}\frac{x^4-1}{x-1} = \lim_{x\to1}(x^3+x^2+x+1) = 4$

9. does not exist   11. −1   13. 1   15. 1

17. δ_1 and δ_2 19. $\frac{1}{2}\epsilon$ 21. 2ϵ

23. Since
$$|(2x - 5) - 3| = |2x - 8| = 2|x - 4|,$$
we can take $\delta = \frac{1}{2}\epsilon$:
$$\text{if}\quad 0 < |x - 4| < \tfrac{1}{2}\epsilon \quad \text{then}\quad |(2x - 5) - 3| = 2|x - 4| < \epsilon.$$

25. Since
$$|(6x - 7) - 11| = |6x - 18| = 6|x - 3|,$$
we can take $\delta = \frac{1}{6}\epsilon$:
$$\text{if}\quad 0 < |x - 3| < \tfrac{1}{6}\epsilon \quad \text{then}\quad |(6x - 7) - 11| = 6|x - 3| < \epsilon.$$

27. Since
$$\big||1 - 3x| - 5\big| = \big||3x - 1| - 5\big| \le |3x - 6| = 3|x - 2|,$$
we can take $\delta = \frac{1}{3}\epsilon$:
$$\text{if}\quad 0 < |x - 2| < \tfrac{1}{3}\epsilon \quad \text{then}\quad \big||1 - 3x| - 5\big| \le 3|x - 2| < \epsilon.$$

29. Statements (b), (e), (g), and (i) are necessarily true.

31. (i) $\displaystyle \lim_{x \to 3} \frac{1}{x - 1} = \frac{1}{2}$ (ii) $\displaystyle \lim_{x \to 3} \left(\frac{1}{x - 1} - \frac{1}{2} \right) = 0$

(iii) $\displaystyle \lim_{x \to 3} \left| \frac{1}{x - 1} - \frac{1}{2} \right| = 0$ (iv) $\displaystyle \lim_{h \to 0} \frac{1}{(3 + h) - 1} = \frac{1}{2}$

33. By (2.2.5) parts (i) and (iii) with $l = 0$

35. Let $\epsilon > 0$. If
$$\lim_{x \to c} f(x) = l,$$
then there must exist $\delta > 0$ such that

(*) $\text{if}\quad 0 < |x - c| < \delta \quad \text{then}\quad |f(x) - l| < \epsilon.$

Suppose now that
$$0 < |h| < \delta.$$
Then
$$0 < |(c + h) - c| < \delta$$
and thus by (*)
$$|f(c + h) - l| < \epsilon.$$
This proves that
$$\text{if}\quad \lim_{x \to c} f(x) = l \quad \text{then}\quad \lim_{h \to 0} f(c + h) = l.$$
If, on the other hand,
$$\lim_{h \to 0} f(c + h) = l,$$
then there must exist $\delta > 0$ such that

(**) $\text{if}\quad 0 < |h| < \delta \quad \text{then}\quad |f(c + h) - l| < \epsilon.$

Suppose now that

$$0 < |x - c| < \delta.$$

Then by (**)

$$|f(c + (x - c)) - l| < \epsilon.$$

More simply stated,

$$|f(x) - l| < \epsilon.$$

This proves that

$$\text{if} \quad \lim_{h \to 0} f(c + h) = l \quad \text{then} \quad \lim_{x \to c} f(x) = l.$$

37. (a) Set $\delta = \epsilon\sqrt{c}$. By the hint,

$$\text{if} \quad 0 < |x - c| < \epsilon\sqrt{c} \quad \text{then} \quad |\sqrt{x} - \sqrt{c}| < \frac{1}{\sqrt{c}}|x - c| < \epsilon.$$

(b) Set $\delta = \epsilon^2$. If $0 < x < \epsilon^2$, then $|\sqrt{x} - 0| = \sqrt{x} < \epsilon$.

39. Take $\delta = $ minimum of 1 and $\epsilon/7$. If $0 < |x - 1| < \delta$, then $0 < x < 2$
and $|x - 1| < \epsilon/7$. Therefore

$$|x^3 - 1| = |x^2 + x + 1||x - 1| < 7|x - 1| < 7(\epsilon/7) = \epsilon.$$

41. Set $\delta = \epsilon^2$. If $3 - \epsilon^2 < x < 3$, then $-\epsilon^2 < x - 3$, $0 < 3 - x < \epsilon^2$
and therefore $|\sqrt{3 - x} - 0| < \epsilon$.

43. Suppose, on the contrary, that $\lim_{x \to c} f(x) = l$ for some particular c. Taking $\epsilon = \frac{1}{2}$, there must exist $\delta > 0$ such that

$$\text{if} \quad 0 < |x - c| < \delta, \quad \text{then} \quad |f(x) - l| < \frac{1}{2}.$$

Let x_1 be a rational number satisfying $0 < |x_1 - c| < \delta$ and x_2 an irrational number satisfying $0 < |x_2 - c| < \delta$. (That such numbers exist follows from the fact that every interval contains both rational and irrational numbers.) Now $f(x_1) = l$ and $f(x_2) = 0$. Thus we must have both

$$|1 - l| < \frac{1}{2} \quad \text{and} \quad |0 - l| < \frac{1}{2}.$$

From the first inequality we conclude that $l > \frac{1}{2}$. From the second, we conclude that $l < \frac{1}{2}$. Clearly no such number l exists.

45. We begin by assuming that $\lim_{x \to c^+} f(x) = l$ and showing that

$$\lim_{h \to 0} f(c + |h|) = l.$$

Let $\epsilon > 0$. Since $\lim_{x \to c^+} f(x) = l$, there exists $\delta > 0$ such that

(*) $$\text{if} \quad c < x < c + \delta \quad \text{then} \quad |f(x) - l| < \epsilon.$$

Suppose now that $0 < |h| < \delta$. Then $c < c + |h| < c + \delta$ and, by (*),

$$|f(c + |h|) - l| < \epsilon.$$

Thus $\lim_{h \to 0} f(c + |h|) = l$.

Conversely we now assume that $\lim_{h \to 0} f(c + |h|) = l$. Then for $\epsilon > 0$ there exists $\delta > 0$ such that

(**) $$\text{if} \quad 0 < |h| < \delta \quad \text{then} \quad |f(c + |h|) - l| < \epsilon.$$

Suppose now that $c < x < c + \delta$. Then $0 < x - c < \delta$ so that, by $(**)$,

$$|f(c + (x - c)) - l| = |f(x) - l| < \epsilon.$$

Thus $\displaystyle\lim_{x \to c^+} f(x) = l$.

SECTION 2.3

1.   (a)   3   (b)   4   (c) -2 (d)   0   (e)   does not exist   (f) $\frac{1}{3}$

3. $\displaystyle\lim_{x \to 4} \left(\frac{1}{x} - \frac{1}{4} \right) \left(\frac{1}{x - 4} \right) = \lim_{x \to 4} \left(\frac{4 - x}{4x} \right) \left(\frac{1}{x - 4} \right) = \lim_{x \to 4} \frac{-1}{4x} = \frac{-1}{16};$ Theorem 2.3.2 does not apply

    since $\displaystyle\lim_{x \to 4} \frac{1}{x - 4}$ does not exist.

5.   3   7. -3 9.   5   11.   2   13.   does not exist

15. $-23/20$ 17.   0   19. $\displaystyle\lim_{h \to 0} h \left(1 - \frac{1}{h} \right) = \lim_{h \to 0} (h - 1) = -1$

21. $\displaystyle\lim_{x \to 2} \frac{x - 2}{x^2 - 4} = \lim_{x \to 2} \frac{1}{x + 2} = \frac{1}{4}$

23. $\displaystyle\lim_{x \to -2} \frac{(x^2 - x - 6)^2}{x + 2} = \lim_{x \to -2} \frac{(x - 3)^2 (x + 2)^2}{x + 2} = \lim_{x \to -2} (x - 3)^2 (x + 2) = 0$

25.   0   27. $\displaystyle\lim_{x \to -2} \frac{x^2 - x - 6}{(x + 2)^2} = \lim_{x \to -2} \frac{(x + 2)(x - 3)}{(x + 2)^2} = \lim_{x \to -2} \frac{x - 3}{x + 2};$ does not exist

29. $\displaystyle\lim_{h \to 0} \frac{1 - 1/h^2}{1 - 1/h} = \lim_{h \to 0} \frac{h^2 - 1}{h^2 - h} = \lim_{h \to 0} \frac{(h + 1)(h - 1)}{h(h - 1)} = \lim_{h \to 0} \frac{h + 1}{h};$ does not exist

31. $\displaystyle\lim_{h \to 0} \frac{1 - 1/h}{1 + 1/h} = \lim_{h \to 0} \frac{h - 1}{h + 1} = -1$

33. $\displaystyle\lim_{t \to -1} \frac{t^2 + 6t + 5}{t^2 + 3t + 2} = \lim_{t \to -1} \frac{(t + 1)(t + 5)}{(t + 1)(t + 2)} = \lim_{t \to -1} \frac{t + 5}{t + 2} = 4$

35. $\displaystyle\lim_{t \to 0} \frac{t + a/t}{t + b/t} = \lim_{t \to 0} \frac{t^2 + a}{t^2 + b} = \frac{a}{b}$

37. $\displaystyle\lim_{x \to 1} \frac{x^5 - 1}{x^4 - 1} = \lim_{x \to 1} \frac{(x - 1)(x^4 + x^3 + x^2 + x + 1)}{(x - 1)(x^3 + x^2 + x + 1)} = \lim_{x \to 1} \frac{x^4 + x^3 + x^2 + x + 1}{x^3 + x^2 + x + 1} = \frac{5}{4}$

39. $\displaystyle\lim_{h \to 0} h \left(1 + \frac{1}{h^2} \right) = \lim_{h \to 0} \frac{h^2 + 1}{h};$ does not exist

41. $\displaystyle\lim_{x \to -2} \frac{(x^2 - x - 6)^2}{(x + 2)^2} = \lim_{x \to -2} \frac{(x - 3)^2 (x + 2)^2}{(x + 2)^2} = \lim_{x \to -2} (x - 3)^2 = 25$

43. $\displaystyle\lim_{t \to 1} \frac{t^2 - 2t + 1}{t^3 - 3t^2 + 3t - 1} = \lim_{t \to 1} \frac{(t - 1)^2}{(t - 1)^3} = \lim_{t \to 1} \frac{1}{t - 1};$ does not exist

45. $\displaystyle\lim_{x\to-4}\left(\frac{2x}{x+4}+\frac{8}{x+4}\right)=\lim_{x\to-4}\frac{2x+8}{x+4}=\lim_{x\to-4}2=2$

47. (a) $\displaystyle\lim_{x\to4}\left(\frac{1}{x}-\frac{1}{4}\right)=\lim_{x\to4}\frac{4-x}{4x}=0$

(b) $\displaystyle\lim_{x\to4}\left[\left(\frac{1}{x}-\frac{1}{4}\right)\left(\frac{1}{x-4}\right)\right]=\lim_{x\to4}\left[\left(\frac{4-x}{4x}\right)\left(\frac{1}{x-4}\right)\right]=\lim_{x\to4}\left(-\frac{1}{4x}\right)=-\frac{1}{16}$

(c) $\displaystyle\lim_{x\to4}\left[\left(\frac{1}{x}-\frac{1}{4}\right)(x-2)\right]=\lim_{x\to4}\frac{(4-x)(x-2)}{4x}=0$

(d) $\displaystyle\lim_{x\to4}\left[\left(\frac{1}{x}-\frac{1}{4}\right)\left(\frac{1}{x-4}\right)^2\right]=\lim_{x\to4}\frac{4-x}{4x(x-4)^2}=\lim_{x\to4}\frac{1}{4x(4-x)};$ does not exist

49. (a) $\displaystyle\lim_{x\to4}\frac{f(x)-f(4)}{x-4}=\lim_{x\to4}\frac{(x^2-4x)-(0)}{x-4}=\lim_{x\to4}x=4$

(b) $\displaystyle\lim_{x\to1}\frac{f(x)-f(1)}{x-1}=\lim_{x\to1}\frac{x^2-4x+3}{x-1}=\lim_{x\to1}\frac{(x-1)(x-3)}{x-1}=\lim_{x\to1}(x-3)=-2$

(c) $\displaystyle\lim_{x\to3}\frac{f(x)-f(1)}{x-3}=\lim_{x\to3}\frac{x^2-4x+3}{x-3}=\lim_{x\to3}\frac{(x-1)(x-3)}{x-3}=\lim_{x\to3}(x-1)=2$

(d) $\displaystyle\lim_{x\to3}\frac{f(x)-f(2)}{x-3}=\lim_{x\to3}\frac{x^2-4x+4}{x-3};$ does not exist

51. $f(x)=1/x,\quad g(x)=-1/x$ **53.**   true   **55.**   true

57. false;   set $f(x)=x$ and $c=0$ **59.**   false;   neither limit need exist

61. If $\displaystyle\lim_{x\to c}f(x)=l$ and $\displaystyle\lim_{x\to c}g(x)=l,$ then

$$\lim_{x\to c}h(x)=\lim_{x\to c}\tfrac{1}{2}\{[f(x)+g(x)]-|f(x)-g(x)|\}$$
$$=\lim_{x\to c}\tfrac{1}{2}[f(x)+g(x)]-\lim_{x\to c}|f(x)-g(x)|$$
$$=\tfrac{1}{2}(l+l)-\tfrac{1}{2}(l-l)=l.$$

A similar argument works for H.

63. Suppose on the contrary that the limit exists and is some number l:

$$\lim_{x\to c}f(x)=l.$$

Since

$$\lim_{x\to c}g(x)=0$$

we have

$$\lim_{x\to c}[f(x)g(x)]=\left(\lim_{x\to c}f(x)\right)\left(\lim_{x\to c}g(x)\right)=l\cdot0=0.$$

But since

$$f(x)g(x)=1\quad\text{for all real }x,$$

we know that

$$\lim_{x\to c}[f(x)g(x)]=1.$$

This contradicts the uniqueness of limit.

SECTION 2.4

1.   continuous   **3.**   continuous   **5.**   continuous

7.   removable discontinuity   **9.**   essential discontinuity

11.   continuous   **13.**   essential discontinuity

15.

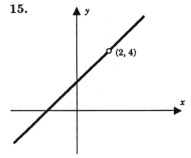

removable discontinuity at 2

17.

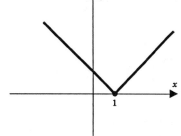

no discontinuities

19.

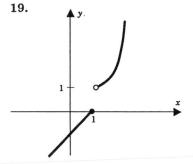

essential discontinuity at 1

21.

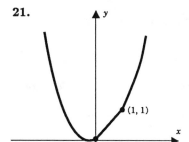

no discontinuities

23.

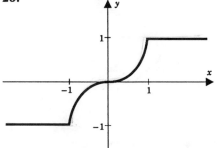

no discontinuities

25.

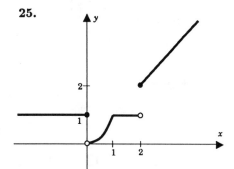

essential discontinuities at 0 and 2

27.

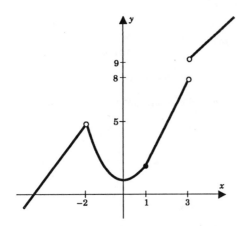

removable discontinuity at -2,
essential discontinuity at 3

29. $f(1) = 2$ **31.** impossible; $\lim\limits_{x \to 1^-} f(x) = -1$ and $\lim\limits_{x \to 1^+} f(x) = 1$

33. Since $\lim\limits_{x \to 1^-} f(x) = 1$ and $\lim\limits_{x \to 1^+} f(x) = A - 3 = f(1)$, take $A = 4$.

35. The function f is continuous at $x = 1$ iff

$$f(1) = \lim\limits_{x \to 1^-} f(x) = A - B \quad \text{and} \quad \lim\limits_{x \to 1^+} f(x) = 3$$

are equal;   that is, $A - B = 3$. The function f is discontinuous at $x = 2$ iff

$$\lim\limits_{x \to 2^-} f(x) = 6 \quad \text{and} \quad \lim\limits_{x \to 2^+} f(x) = f(2) = 4B - A$$

are unequal;   that is, iff $4B - A \neq 6$. More simply we have $A - B = 3$ with $B \neq 3$:

$$A - B = 3, \ 4B - A \neq 6 \implies A - B = 3, \ 3B - 3 \neq 6 \implies A - B = 3, \ B \neq 3.$$

37. Set for instance

$$f(x) = \begin{cases} 0, & x \leq \frac{1}{2} \\ 1, & x > \frac{1}{2} \end{cases}.$$

39. $f(5) = \frac{1}{6}$ **41.** $f(5) = \frac{1}{3}$

43. nowhere;   see Figure 2.1.8

45. $x = 0$, $x = 2$,   and all nonintegral values of x

47. Refer to (2.2.5). Use the equivalence of (i) and (iv) setting $l = f(c)$.

49. Let $\epsilon > 0$ be arbitrary and let A be the billion points on which we have changed the value of f. In $A - \{c\}$ there is one point closest to c. Call it d and set $\delta_1 = |c - d|$. Note that if $0 < |x - c| < \delta_1$, then $f(x) = g(x)$.

Suppose now that g is continuous at c. Then there exists a positive number δ less than δ_1 such that

$$\text{if} \quad 0 < |x - c| < \delta \quad \text{then} \quad |g(x) - g(c)| < \epsilon.$$

But for such x, $f(x) = g(x)$. Therefore we see that

$$\text{if} \quad 0 < |x - c| < \delta, \quad \text{then} \quad |f(x) - g(c)| < \epsilon.$$

This means that

$$\lim_{x \to c} f(x) = g(c)$$

and contradicts the assumption that f has an essential discontinuity at c.

SECTION 2.5

1. $\lim\limits_{x \to 0} \dfrac{\sin 3x}{x} = \lim\limits_{x \to 0} 3\left(\dfrac{\sin 3x}{3x}\right) = 3(1) = 3$

3. $\lim\limits_{x \to 0} \dfrac{3x}{\sin 5x} = \lim\limits_{x \to 0} \dfrac{3}{5}\left(\dfrac{5x}{\sin 5x}\right) = \dfrac{3}{5}(1) = \dfrac{3}{5}$

5. $\lim\limits_{x \to 0} \dfrac{\sin x^2}{x} = \lim\limits_{x \to 0} x\left(\dfrac{\sin x^2}{x^2}\right) = \lim\limits_{x \to 0} x \cdot \lim\limits_{x \to 0} \dfrac{\sin x^2}{x^2} = 0(1) = 0$

7. $\lim\limits_{x \to 0} \dfrac{\sin x}{x^2} = \lim\limits_{x \to 0} \dfrac{(\sin x)/x}{x}$;   does not exist

9. $\lim\limits_{x \to 0} \dfrac{\sin^2 3x}{5x^2} = \lim\limits_{x \to 0} \dfrac{9}{5}\left(\dfrac{\sin 3x}{3x}\right)^2 = \dfrac{9}{5}(1) = \dfrac{9}{5}$

11. $\lim\limits_{x \to 0} \dfrac{2x}{\tan 3x} = \lim\limits_{x \to 0} \dfrac{2x \cos 3x}{\sin 3x} = \lim\limits_{x \to 0} \dfrac{2}{3}\left(\dfrac{3x}{\sin 3x}\right)\cos 3x = \dfrac{2}{3}(1)(1) = \dfrac{2}{3}$

13. $\lim\limits_{x \to 0} x \csc x = \lim\limits_{x \to 0} \dfrac{x}{\sin x} = 1$

15. $\lim\limits_{x \to 0} \dfrac{x^2}{1 - \cos 2x} = \lim\limits_{x \to 0} \dfrac{x^2}{1 - \cos 2x} \cdot \left(\dfrac{1 + \cos 2x}{1 + \cos 2x}\right) = \lim\limits_{x \to 0} \dfrac{x^2(1 + \cos 2x)}{\sin^2 2x}$

$$= \lim\limits_{x \to 0} \dfrac{1}{4}\left(\dfrac{2x}{\sin 2x}\right)^2 (1 + \cos 2x) = \dfrac{1}{4}(1)(2) = \dfrac{1}{2}$$

17. $\lim\limits_{x \to 0} \dfrac{1 - \sec^2 2x}{x^2} = \lim\limits_{x \to 0} \dfrac{-\tan^2 2x}{x^2} = \lim\limits_{x \to 0} \dfrac{-\sin^2 2x}{x^2 \cos^2 2x} = \lim\limits_{x \to 0}\left[-4\left(\dfrac{\sin 2x}{2x}\right)^2 \dfrac{1}{\cos^2 2x}\right] = -4$

19. $\lim\limits_{x \to 0} \dfrac{2x^2 + x}{\sin x} = \lim\limits_{x \to 0}(2x + 1)\dfrac{x}{\sin x} = 1$

21. $\lim\limits_{x \to 0} \dfrac{\tan 3x}{2x^2 + 5x} = \lim\limits_{x \to 0} \dfrac{1}{x(2x + 5)} \dfrac{\sin 3x}{\cos 3x} = \lim\limits_{x \to 0} \dfrac{3}{2x + 5}\left(\dfrac{\sin 3x}{3x}\right)\dfrac{1}{\cos 3x} = \dfrac{3}{5}(1)(1) = \dfrac{3}{5}$

23. $\lim\limits_{x \to 0} \dfrac{\sec x - 1}{x \sec x} = \lim\limits_{x \to 0} \dfrac{\dfrac{1}{\cos x} - 1}{x\left(\dfrac{1}{\cos x}\right)} = \lim\limits_{x \to 0} \dfrac{1 - \cos x}{x} = 0$

25. $\dfrac{2\sqrt{2}}{\pi}$

27. $\lim\limits_{x \to \pi/2} \dfrac{\cos x}{x - \pi/2} \underset{h \to 0}{=} \lim\limits_{h \to 0} \dfrac{\cos (h + \pi/2)}{h} \underset{h \to 0}{=} \lim\limits_{h \to 0} \dfrac{-\sin h}{h} = -1$

$$\raisebox{0pt}{$\llcorner$} \; h = x - \pi/2 \qquad \raisebox{0pt}{$\llcorner$} \; \cos (h + \pi/2) = \cos h \cos \pi/2 - \sin h \sin \pi/2$$

29. $\displaystyle\lim_{x\to\pi/4}\frac{\sin(x+\pi/4)-1}{x-\pi/4}=\lim_{h\to0}\frac{\sin(h+\pi/2)-1}{h}=\lim_{h\to0}\frac{\cos h-1}{h}=0$

$\qquad\qquad h=x-\pi/4$

31. Equivalently we will show that $\displaystyle\lim_{h\to0}\cos(c+h)=\cos c$. The identity

$$\cos(c+h)=\cos c\cos h-\sin c\sin h$$

gives

$$\lim_{h\to0}\cos(c+h)=\cos c\left(\lim_{h\to0}\cos h\right)-\sin c\left(\lim_{h\to0}\sin h\right)$$
$$=(\cos c)(1)-(\sin c)(0)=\cos c.$$

33. $0\le|xf(x)|\le M|x|$ for $x\ne0$ **35.** $0\le|f(x)-l|\le M|x-c|$ for $x\ne c$

37. $0\le|x\sin(1/x)|\le|x|$ for $x\ne0$ **39.** $0\le|\sqrt{x}-\sqrt{c}|\le\dfrac{1}{\sqrt{c}}|x-c|$ for $x\ge0$

SECTION 2.6

1.

3.

5.

7.

9.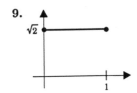

11.

13. impossible by intermediate-value theorem

15.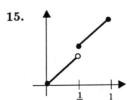

17. Set $g(x)=x-f(x)$. Since g is continuous on $[0,1]$ and $g(0)\le0\le g(1)$, there exists c in $[0,1]$ such that $g(c)=c-f(c)=0$.

19. Set $h(x)=f(x)-g(x)$. Since h is continuous on $[0,1]$ and $h(0)<0<h(1)$, there exists c between 0 and 1 such that $h(c)=f(c)-g(c)=0$.

21. $m_{10}=1.7314$ **23.** $m_{10}=0.6181$ **25.** $m_{12}=1.9997$ **27.** $m_{10}=0.7392$

SECTION 2.7

1. does not exist   **3.** 0

5. $\displaystyle\lim_{x\to3}\frac{x^2-9}{x^2-5x+6}=\lim_{x\to3}\frac{(x-3)(x+3)}{(x-3)(x-2)}=\lim_{x\to3}\frac{x+3}{x-2}=6$

7. $\displaystyle\lim_{x\to 3^-}\frac{x-3}{|x-3|}=\lim_{x\to 3^-}\frac{x-3}{-(x-3)}=\lim_{x\to 3^-}(-1)=-1$

9. $\displaystyle\lim_{x\to 0}\left(\frac{1}{x}-\frac{1-x}{x}\right)=\lim_{x\to 0}\frac{1-(1-x)}{x}=\lim_{x\to 0}1=1$ **11.**   0

13.   1   **15.**   does not exist   **17.**   does not exist   **19.**   0

21. $\displaystyle\lim_{x\to 1}\frac{1-1/x}{1-x^2}=\lim_{x\to 1}\frac{x-1}{x(1-x)(1+x)}=\lim_{x\to 1}\frac{-1}{x(1+x)}=-\frac{1}{2}$ **23.**   1

25. $\displaystyle\lim_{x\to 4}\left[\left(\frac{1}{x}-\frac{1}{4}\right)\left(\frac{1}{x-4}\right)\right]=\lim_{x\to 4}\left[\left(\frac{4-x}{4x}\right)\left(\frac{1}{x-4}\right)\right]=\lim_{x\to 4}\frac{-1}{4x}=-\frac{1}{16}$

27.   does not exist

29. $\displaystyle\lim_{x\to 3^+}\frac{x^2-2x-3}{\sqrt{x-3}}=\lim_{x\to 3^+}\frac{(x+1)(x-3)}{\sqrt{x-3}}=\lim_{x\to 3^+}(x-1)\sqrt{x-3}=0$

31. $\displaystyle\lim_{x\to 2^+}\left(\frac{1}{x-2}-\frac{1}{|x-2|}\right)=\lim_{x\to 2^+}\left(\frac{1}{x-2}-\frac{1}{x-2}\right)=\lim_{x\to 2^+}0=0$

33. $\displaystyle\lim_{x\to 4}\frac{\sqrt{x+5}-3}{x-4}=\lim_{x\to 4}\frac{\sqrt{x+5}-3}{x-4}\left(\frac{\sqrt{x+5}+3}{\sqrt{x+5}+3}\right)=\lim_{x\to 4}\frac{x-4}{(x-4)(\sqrt{x+5}+3)}$

$$=\lim_{x\to 4}\frac{1}{\sqrt{x+5}+3}=\frac{1}{6}$$

35. $\displaystyle\lim_{x\to 0}\frac{5x}{\sin 2x}=\frac{5}{2}\lim_{x\to 0}\frac{2x}{\sin 2x}=\frac{5}{2}$

37. $\displaystyle\lim_{x\to 0}4x^2\cot^2 3x=\lim_{x\to 0}\frac{4}{9}\cos^2 3x\left(\frac{3x}{\sin 3x}\right)^2=\frac{4}{9}$

39. $\displaystyle\lim_{x\to 0}\frac{x^2-3x}{\tan x}=\lim_{x\to 0}(x-3)\left(\frac{x}{\sin x}\right)\cos x=-3$

41. $\displaystyle\lim_{x\to \pi/2}\frac{\cos x}{2x-\pi}=\lim_{h\to 0}\frac{\cos(h+\pi/2)}{2h}=\frac{1}{2}\lim_{h\to 0}\frac{-\sin h}{h}=-\frac{1}{2}$

$$h=x-\frac{\pi}{2}$$

43. $\displaystyle\lim_{x\to 0}\frac{5x^2}{1-\cos 2x}=\lim_{x\to 0}\frac{5x^2}{1-\cos 2x}\left(\frac{1+\cos 2x}{1+\cos 2x}\right)=\lim_{x\to 0}\frac{5x^2(1+\cos 2x)}{\sin^2 2x}$

$$=\frac{5}{4}\lim_{x\to 0}\left(\frac{2x}{\sin 2x}\right)^2(1+\cos 2x)=\frac{5}{4}(1)(2)=\frac{5}{2}$$

45. $\displaystyle\lim_{x\to 2}\frac{x^2-4}{x^3-8}=\lim_{x\to 2}\frac{(x-2)(x+2)}{(x-2)(x^2+2x+4)}=\lim_{x\to 2}\frac{x+2}{x^2+2x+4}=\frac{1}{3}$

47. $\displaystyle\lim_{x\to 2^+}\frac{\sqrt{x-2}}{|x-2|}=\lim_{x\to 2^+}\frac{\sqrt{x-2}}{x-2}=\lim_{x\to 2^+}\frac{1}{\sqrt{x-2}};$ does not exist

49. $\displaystyle\lim_{x\to 2}\left(\frac{6}{x-2}-\frac{3x}{x-2}\right)=\lim_{x\to 2}\frac{6-3x}{x-2}=\lim_{x\to 2}(-3)=-3$

51. $\displaystyle\lim_{x\to -1^+}\frac{|x^2-1|}{x+1}=\lim_{x\to -1^+}\frac{-(x^2-1)}{x+1}=\lim_{x\to -1^+}(1-x)=2$

53. $\displaystyle\lim_{x\to 2}\frac{1-2/x}{1-4/x^2}=\lim_{x\to 2}\frac{x(x-2)}{x^2-4}=\lim_{x\to 2}\frac{x}{x+2}=\frac{1}{2}$

55. $\displaystyle\lim_{x\to 2^+}\frac{\sqrt{x^2-3x+2}}{x-2}=\lim_{x\to 2^+}\frac{\sqrt{(x-1)(x-2)}}{x-2}=\lim_{x\to 2^+}\sqrt{\frac{x-1}{x-2}};$ does not exist

57. $\displaystyle\lim_{x\to 2}\left(1-\frac{2}{x}\right)\left(\frac{3}{4-x^2}\right)=\lim_{x\to 2}\left(\frac{x-2}{x}\right)\left[\frac{3}{(2-x)(2+x)}\right]=\lim_{x\to 2}\frac{-3}{x(2+x)}=-\frac{3}{8}$

59. $\displaystyle\lim_{x\to 7}\frac{x-7}{\sqrt{x+2}-3}=\lim_{x\to 7}\frac{x-7}{\sqrt{x+2}-3}\left(\frac{\sqrt{x+2}+3}{\sqrt{x+2}+3}\right)=\lim_{x\to 7}\frac{(x-7)(\sqrt{x+2}+3)}{x-7}$

$$=\lim_{x\to 7}(\sqrt{x+2}+3)=6$$

61. 6

63. 1

65. (a)

(b) (i) 1   (ii) 1   (iii) 1
     (iv) 5   (v) 5   (vi) 5
     (vii) 9   (viii) 4   (ix) does
                                  not
                                  exist

(c) (i) at 1 and 2   (ii) at 1
     (iii) at 1   (iv) at 2

67. 0   (by the pinching theorem)

69. Since
$$|(5x-4)-6|=|5x-10|=5|x-2|,$$
we can take $\delta=\frac{1}{5}\epsilon$:
$$\text{if}\ \ 0<|x-2|<\tfrac{1}{5}\epsilon\ \ \text{then}\ \ |(5x-4)-6|=5|x-2|<\epsilon.$$

71. Since
$$\big||2x+5|-3\big| \le |2x+8| = 2|x+4|,$$
we can take $\delta = \frac{1}{2}\epsilon$:
$$\text{if} \quad 0 < |x+4| < \tfrac{1}{2}\epsilon \quad \text{then} \quad \big||2x+5|-3\big| \le 2|x+4| < \epsilon.$$

73. Take $\delta =$ minimum of 4 and ϵ.   If $0 < |x-9| < \delta$, then $x-5 > 0$ and $|x-9| < \epsilon$. It follows that
$$|\sqrt{x-5}-2|\,|\sqrt{x-5}+2| = |(x-5)-4| = |x-9| < \epsilon$$
and, since $|\sqrt{x-5}+2| > 1,$ that $|\sqrt{x-5}-2| < \epsilon.$

75. The polynomial
$$P(x) = x^5 - 4x + 1$$
is continuous on $[0, 2]$. (Polynomials are everywhere continuous.) Since
$$P(0) < 7.21 < P(2),$$
we know from the intermediate-value theorem that there is a number x_0 between 0 and 2 such that
$$P(x_0) = 7.21.$$

CHAPTER 3

SECTION 3.1

1. $f'(x) = \lim\limits_{h\to 0} \dfrac{f(x+h)-f(x)}{h} = \lim\limits_{h\to 0} \dfrac{4-4}{h} = \lim\limits_{h\to 0} 0 = 0$

3. $f'(x) = \lim\limits_{h\to 0} \dfrac{f(x+h)-f(x)}{h} = \lim\limits_{h\to 0} \dfrac{[2-3(x+h)]-[2-3x]}{h}$

$\qquad = \lim\limits_{h\to 0} \dfrac{-3h}{h} = \lim\limits_{h\to 0} -3 = -3$

5. $f'(x) = \lim\limits_{h\to 0} \dfrac{f(x+h)-f(x)}{h} = \lim\limits_{h\to 0} \dfrac{[5(x+h)-(x+h)^2]-(5x-x^2)}{h}$

$\qquad = \lim\limits_{h\to 0} \dfrac{5h-2xh-h^2}{h} = \lim\limits_{h\to 0}(5-2x-h) = 5-2x$

7. $f'(x) = \lim\limits_{h\to 0} \dfrac{f(x+h)-f(x)}{h} = \lim\limits_{h\to 0} \dfrac{(x+h)^4-x^4}{h}$

$\qquad = \lim\limits_{h\to 0} \dfrac{(x^4+4x^3h+6x^2h^2+4xh^3+h^4)-x^4}{h}$

$\qquad = \lim\limits_{h\to 0}(4x^3+6x^2h+4xh^2+h^3) = 4x^3$

9. $f'(x) = \lim\limits_{h\to 0} \dfrac{f(x+h)-f(x)}{h} = \lim\limits_{h\to 0} \dfrac{\sqrt{x+h-1}-\sqrt{x-1}}{h}$

$\qquad = \lim\limits_{h\to 0} \dfrac{(x+h-1)-(x-1)}{h(\sqrt{x+h-1}+\sqrt{x-1})} = \lim\limits_{h\to 0} \dfrac{1}{\sqrt{x+h-1}+\sqrt{x-1}} = \dfrac{1}{2\sqrt{x-1}}$

11. $f'(x) = \lim\limits_{h\to 0} \dfrac{f(x+h)-f(x)}{h} = \lim\limits_{h\to 0} \dfrac{\dfrac{1}{(x+h)^2}-\dfrac{1}{x^2}}{h}$

$\qquad = \lim\limits_{h\to 0} \dfrac{x^2-(x^2+2hx+h^2)}{hx^2(x+h)^2} = \lim\limits_{h\to 0} \dfrac{-2x-h}{x^2(x+h)^2} = -\dfrac{2}{x^3}$

13. $f'(2) = \lim\limits_{h\to 0} \dfrac{f(2+h)-f(2)}{h} = \lim\limits_{h\to 0} \dfrac{(3h-1)^2-1}{h}$

$\qquad = \lim\limits_{h\to 0} \dfrac{9h^2-6h}{h} = \lim\limits_{h\to 0}(9h-6) = -6$

15. $f'(2) = \lim\limits_{h\to 0} \dfrac{f(2+h)-f(2)}{h} = \lim\limits_{h\to 0} \dfrac{\dfrac{9}{6+h}-\dfrac{3}{2}}{h}$

$\qquad = \lim\limits_{h\to 0} \dfrac{18-3(6+h)}{2h(6+h)} = \lim\limits_{h\to 0} \dfrac{-3}{2(6+h)} = -\dfrac{1}{4}$

17. $f'(2) = \lim\limits_{h \to 0} \dfrac{f(2+h) - f(2)}{h} = \lim\limits_{h \to 0} \dfrac{(2 + h + \sqrt{4 + 2h}) - 4}{h}$

$\qquad\quad = \lim\limits_{h \to 0} \left(1 + \dfrac{\sqrt{4 + 2h} - 2}{h}\right) = \lim\limits_{h \to 0} 1 + \dfrac{(4 + 2h) - 4}{h(\sqrt{4 + 2h} + 2)}$

$\qquad\qquad\quad = \lim\limits_{h \to 0} 1 + \dfrac{2}{\sqrt{4 + 2h} + 2} = \dfrac{3}{2}$

19. Slope of tangent at $(2, 4)$ is $f'(2) = 4$. Tangent $y - 4 = 4(x - 2)$;
normal $y - 4 = -\frac{1}{4}(x - 2)$.

21. Slope of tangent at $(4, 4)$ is $f'(4) = -3$. Tangent $y - 4 = -3(x - 4)$;
normal $y - 4 = \frac{1}{3}(x - 4)$.

23. Slope of tangent at $(-4, 4)$ is $f'(-4) = -1$. Tangent $y - 4 = -(x + 4)$;
normal $y - 4 = x + 4$.

25. Slope of tangent at $(-3, -1)$ is -1. Tangent $y + 1 = -(x + 3)$;
normal $y + 1 = x + 3$.

27. at $x = -1$ **29.** at $x = 0$ **31.** at $x = 1$

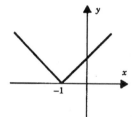 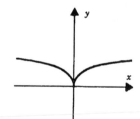 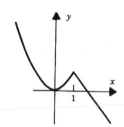

33. $f'(1) = 4$

$\lim\limits_{h \to 0^-} \dfrac{f(1+h) - f(1)}{h} = \lim\limits_{h \to 0^-} \dfrac{4(1+h) - 4}{h} = 4$

$\lim\limits_{h \to 0^+} \dfrac{f(1+h) - f(1)}{h} = \lim\limits_{h \to 0^+} \dfrac{2(1+h)^2 + 2 - 4}{h} = 4$

35. $f'(-1)$ does not exist

$\lim\limits_{h \to 0^-} \dfrac{f(-1+h) - f(-1)}{h} = \lim\limits_{h \to 0^-} \dfrac{h - 0}{h} = 1$

$\lim\limits_{h \to 0^+} \dfrac{f(-1+h) - f(-1)}{h} = \lim\limits_{h \to 0^+} \dfrac{h^2 - 0}{h} = 0$

37. 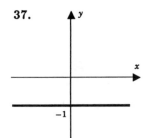   **39.** 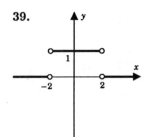   **41.**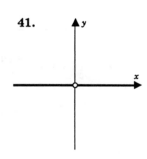

43. Since $f(1) = 1$ and $\lim\limits_{x \to 1+} f(x) = 2$, f is not continuous at 1 and thus, by (3.1.4), is not differentiable at 1.

45. (a) $f'(x) = \begin{cases} 2(x+1), & x < 0 \\ 2(x-1), & x > 0 \end{cases}$

(b) $\lim\limits_{h \to 0^-} \dfrac{f(0+h) - f(0)}{h} = \lim\limits_{h \to 0^-} \dfrac{(h+1)^2 - 1}{h} = \lim\limits_{h \to 0^-} (h+2) = 2,$

$\lim\limits_{h \to 0^+} \dfrac{f(0+h) - f(0)}{h} = \lim\limits_{h \to 0^+} \dfrac{(h-1)^2 - 1}{h} = \lim\limits_{h \to 0^+} (h-2) = -2.$

47. $f(x) = c$, c any constant

49. $f(x) = |x+1|$; $f(x) = \begin{cases} 0, & x \neq -1 \\ 1, & x = -1 \end{cases}$

51. $f(x) = 2x + 5$

53. $f(x) = \begin{cases} 0, & x \text{ rational} \\ 1, & x \text{ irrational} \end{cases}$

SECTION 3.2

1. $F'(x) = -1$ **3.** $F'(x) = 55x^4 - 18x^2$ **5.** $F'(x) = 2ax + b$

7. $F'(x) = 2x^{-3}$ **9.** $G'(x) = (x^2 - 1)(1) + (x - 3)(2x) = 3x^2 - 6x - 1$

11. $G'(x) = \dfrac{(1-x)(3x^2) - x^3(-1)}{(1-x)^2} = \dfrac{3x^2 - 2x^3}{(1-x)^2}$

13. $G'(x) = \dfrac{(2x+3)(2x) - (x^2-1)(2)}{(2x+3)^2} = \dfrac{2(x^2 + 3x + 1)}{(2x+3)^2}$

15. $G'(x) = (x-1)(1) + (x-2)(1) = 2x - 3$

17. $G'(x) = \dfrac{(x-2)(1/x^2) - (6 - 1/x)(1)}{(x-2)^2} = \dfrac{-2(3x^2 - x + 1)}{x^2(x-2)^2}$

19. $G'(x) = (9x^8 - 8x^9)\left(1 - \dfrac{1}{x^2}\right) + \left(x + \dfrac{1}{x}\right)(72x^7 - 72x^8)$

$\qquad = -80x^9 + 81x^8 - 64x^7 + 63x^6$

21. $f'(x) = -x(x-2)^{-2}$, $f'(0) = -\frac{1}{4}$, $f'(1) = -1$

23. $f'(x) = \dfrac{(1+x^2)(-2x) - (1-x^2)(2x)}{(1+x^2)^2} = \dfrac{-4x}{(1+x^2)^2}$, $f'(0) = 0$, $f'(1) = -1$

25. $f'(x) = \dfrac{(cx+d)a - (ax+b)c}{(cx+d)^2} = \dfrac{ad-bc}{(cx+d)^2}$, $f'(0) = \dfrac{ad-bc}{d^2}$, $f'(1) = \dfrac{ad-bc}{(c+d)^2}$

27. $f'(x) = xh'(x) + h(x)$, $f'(0) = 0h'(0) + h(0) = 0(2) + 3 = 3$

29. $f'(x) = h'(x) + \dfrac{h'(x)}{[h(x)]^2}$, $f'(0) = h'(0) + \dfrac{h'(0)}{[h(0)]^2} = 2 + \dfrac{2}{3^2} = \dfrac{20}{9}$

31. $f'(x) = \dfrac{(x+2)(1) - x(1)}{(x+2)^2} = \dfrac{2}{(x+2)^2}$,

slope of tangent at $(-4, 2)$ is $f'(-4) = 1/2$,

equation for tangent $y - 2 = \frac{1}{2}(x+4)$

33. $f'(x) = (x^2 - 3)(5 - 3x^2) + (5x - x^3)(2x)$,

slope of tangent at $(1, -8)$ is $f'(1) = (-2)(2) + (4)(2) = 4$,

equation for tangent $y + 8 = 4(x-1)$

35. $f'(x) = -12x^{-3}$, slope of tangent at $(3, \frac{2}{3})$ is $f'(3) = -\frac{4}{9}$,

equation for tangent $y - \frac{2}{3} = -\frac{4}{9}(x-3)$

37. $f'(x) = (x-2)(2x-1) + (x^2 - x - 11)(1) = 3(x-3)(x+1)$,

$f'(x) = 0$ at $x = -1, 3$

39. $f'(x) = \dfrac{(x^2+1)(5) - 5x(2x)}{(x^2+1)^2} = \dfrac{5(1-x^2)}{(x^2+1)^2}$, $f'(x) = 0$ at $x = \pm 1$

41. $f'(x) = 1 - 8/x^3$, $f'(x) = 0$ at $x = 2$

43. slope of line 4,

slope of tangent $-2x$,

$-2x = 4$ at $x = -2$

45. slope of line $-1/5$,

slope of tangent $3x^2 - 2x$,

$3x^2 - 2x = 5$ at $x = -1, 5/3$

47.

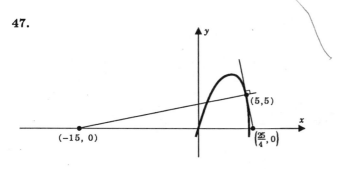

slope of tangent at $(5, 5)$ is $f'(5) = -4$

tangent $y - 5 = -4(x-5)$ intersects

x-axis at $\left(\frac{25}{4}, 0\right)$

normal $y - 5 = \frac{1}{4}(x-5)$ intersects

x-axis at $(-15, 0)$

area of triangle is

$$\frac{1}{2}(5)\left(15 + \frac{25}{4}\right) = \frac{425}{8}$$

49. If the point $(1,3)$ lies on the graph, we have $f(1) = 3$ and thus

(*) $A + B + C = 3.$

If the line $4x + y = 8$ (slope -4) is tangent to the graph at $(2,0)$, then

$f(2) = 0$ and $f'(2) = -4$. Thus,

(**) $4A + 2B + C = 0$ and $4A + B = -4.$

Solving the equations in (*) and (**), we find that $A = -1,$ $B = 0,$ $C = 4.$

51. Since

$$\left(\frac{f}{g}\right)(x) = \frac{f(x)}{g(x)} = f(x) \cdot \frac{1}{g(x)},$$

it follows from the product and reciprocal rules that

$$\left(\frac{f}{g}\right)'(x) = \left(f \cdot \frac{1}{g}\right)'(x) = f(x)\left(-\frac{g'(x)}{[g(x)]^2}\right) + f'(x) \cdot \frac{1}{g(x)} = \frac{g(x)f'(x) - f(x)g'(x)}{[g(x)]^2}.$$

53. $g(x) = [f(x)]^2 = (ff)(x)$

$g'(x) = f(x)f'(x) + f(x)f'(x) = 2f(x)f'(x)$

55. We want f to be continuous at $x = 2$. That is, we want

$$\lim_{x \to 2^-} f(x) = f(2) = \lim_{x \to 2^+} f(x).$$

This gives

(1) $8A + 2B + 2 = 4B - A.$

We also want

$$\lim_{x \to 2^-} f'(x) = \lim_{x \to 2^+} f'(x).$$

This gives

(2) $12A + B = 4B.$

Equations (1) and (2) together imply that $A = -2$ and $B = -8$.

SECTION 3.3

1. $\dfrac{dy}{dx} = 12x^3 - 2x$ **3.** $\dfrac{dy}{dx} = 1 + \dfrac{1}{x^2}$

5. $\dfrac{dy}{dx} = \dfrac{(1+x^2)(1) - x(2x)}{(1+x^2)^2} = \dfrac{1 - x^2}{(1+x^2)^2}$ **7.** $\dfrac{dy}{dx} = \dfrac{(1-x)2x - x^2(-1)}{(1-x)^2} = \dfrac{x(2-x)}{(1-x)^2}$

9. $\dfrac{dy}{dx} = \dfrac{(x^3-1)3x^2 - (x^3+1)3x^2}{(x^3-1)^2} = \dfrac{-6x^2}{(x^3-1)^2}$ **11.** $\dfrac{d}{dx}(2x - 5) = 2$

13. $\dfrac{d}{dx}[(3x^2 - x^{-1})(2x + 5)] = (3x^2 - x^{-1})2 + (2x + 5)(6x + x^{-2})$

$$= 18x^2 + 30x + 5x^{-2}$$

15. $\dfrac{d}{dt}\left(\dfrac{t^2+1}{t^2-1}\right) = \dfrac{(t^2-1)2t - (t^2+1)(2t)}{(t^2-1)^2} = \dfrac{-4t}{(t^2-1)^2}$

17. $\dfrac{d}{dt}\left(\dfrac{t^4}{2t^3-1}\right) = \dfrac{(2t^3-1)4t^3 - t^4(6t^2)}{(2t^3-1)^2} = \dfrac{2t^3(t^3-2)}{(2t^3-1)^2}$

19. $\dfrac{d}{du}\left(\dfrac{2u}{1-2u}\right) = \dfrac{(1-2u)2 - 2u(-2)}{(1-2u)^2} = \dfrac{2}{(1-2u)^2}$

21. $\dfrac{d}{du}\left(\dfrac{u}{u-1} - \dfrac{u}{u+1}\right) = \dfrac{(u-1)(1) - u(1)}{(u-1)^2} - \dfrac{(u+1)(1) - u(1)}{(u+1)^2}$

$$= \dfrac{-1}{(u-1)^2} - \dfrac{1}{(u+1)^2} = \dfrac{-2(1+u^2)}{(u^2-1)^2}$$

23. $\dfrac{d}{dx}\left(\dfrac{x^2}{1-x^2} - \dfrac{1-x^2}{x^2}\right) = \dfrac{(1-x^2)2x - x^2(-2x)}{(1-x^2)^2} - \dfrac{x^2(-2x) - (1-x^2)2x}{x^4}$

$$= \dfrac{2x}{(1-x^2)^2} + \dfrac{2}{x^3}$$

25. $\dfrac{d}{dx}\left(\dfrac{x^3+x^2+x+1}{x^3-x^2+x-1}\right)$

$$= \dfrac{(x^3-x^2+x-1)(3x^2+2x+1) - (x^3+x^2+x+1)(3x^2-2x+1)}{(x^3-x^2+x-1)^2}$$

$$= \dfrac{-2(x^4+2x^2+1)}{(x^2+1)^2(x-1)^2} = \dfrac{-2}{(x-1)^2}$$

27. $\dfrac{dy}{dx} = (x+1)\dfrac{d}{dx}[(x+2)(x+3)] + (x+2)(x+3)\dfrac{d}{dx}(x+1)$

$$= (x+1)(2x+5) + (x+2)(x+3)$$

At $x = 2$, $\dfrac{dy}{dx} = (3)(9) + (4)(5) = 47$.

29. $\dfrac{dy}{dx} = \dfrac{(x+2)\dfrac{d}{dx}[(x-1)(x-2)] - (x-1)(x-2)(1)}{(x+2)^2}$

$$= \dfrac{(x+2)(2x-3) - (x-1)(x-2)}{(x+2)^2}$$

At $x = 2$, $\quad \dfrac{dy}{dx} = \dfrac{4(1) - 1(0)}{16} = \dfrac{1}{4}$.

31. $f'(x) = 21x^2 - 30x^4$ **33.** $f'(x) = 1 + 3x^{-2}$ **35.** $f'(x) = 4x + 4x^{-3}$

$f''(x) = 42x - 120x^3$ $f''(x) = -6x^{-3}$ $f''(x) = 4 - 12x^{-4}$

37. $\dfrac{dy}{dx} = x^2 + x + 1$ **39.** $\dfrac{dy}{dx} = 8x - 20$ **41.** $\dfrac{dy}{dx} = 3x^2 + 3x^{-4}$

$\dfrac{d^2y}{dx^2} = 2x + 1$ $\dfrac{d^2y}{dx^2} = 8$ $\dfrac{d^2y}{dx^2} = 6x - 12x^{-5}$

$\dfrac{d^3y}{dx^3} = 2$ $\dfrac{d^3y}{dx^3} = 0$ $\dfrac{d^3y}{dx^3} = 6 + 60x^{-6}$

43. $\dfrac{d}{dx}\left[x\dfrac{d}{dx}(x - x^2)\right] = \dfrac{d}{dx}[x(1 - 2x)] = \dfrac{d}{dx}[x - 2x^2] = 1 - 4x$

45. $\dfrac{d^4}{dx^4}[3x - x^4] = \dfrac{d^3}{dx^3}[3 - 4x^3] = \dfrac{d^2}{dx^2}[-12x^2] = \dfrac{d}{dx}[-24x] = -24$

47. $\dfrac{d^2}{dx^2}\left[(1 + 2x)\dfrac{d^2}{dx^2}(5 - x^3)\right] = \dfrac{d^2}{dx^2}[(1 + 2x)(-6x)] = \dfrac{d^2}{dx^2}[-6x - 12x^2] = -24$

49. It suffices to give a single counterexample. For instance, if

$f(x) = g(x) = x$, then $(fg)(x) = x^2$ so that $(fg)''(x) = 2$ but

$f(x)g''(x) + f''(x)g(x) = x \cdot 0 + 0 \cdot x = 0.$

51. $f''(x) = 6x$;   (a) $x = 0$ (b) $x > 0$ (c) $x < 0$

53. $f''(x) = 12x^2 + 12x - 24$;   (a) $x = -2, 1$ (b) $x < -2,$ $x > 1$ (c) $-2 < x < 1$

55. The result is true for $n = 1$:

$$\dfrac{d^1y}{dx^1} = \dfrac{dy}{dx} = -x^{-2} = (-1)^1 1! \, x^{-1-1}.$$

If the result is true for $n = k$:

$$\dfrac{d^ky}{dx^k} = (-1)^k k! \, x^{-k-1}$$

then the result is true for $n = k + 1$:

$$\dfrac{d^{k+1}y}{dx^{k+1}} = (-1)^k k!(-k - 1)x^{-k-2} = (-1)^{k+1}(k + 1)! \, x^{-(k+1)-1}.$$

SECTION 3.4

1. $A = \pi r^2$, $\dfrac{dA}{dr} = 2\pi r$. When $r = 2$, $\dfrac{dA}{dr} = 4\pi$.

3. $A = \dfrac{1}{2}z^2$, $\dfrac{dA}{dz} = z$. When $z = 4$, $\dfrac{dA}{dz} = 4$.

5. $y = \dfrac{1}{x(1 + x)}$, $\dfrac{dy}{dx} = \dfrac{-(2x + 1)}{x^2(1 + x)^2}$.   At $x = 2$, $\dfrac{dy}{dx} = -\dfrac{5}{36}$.

7. $V = \dfrac{4}{3}\pi r^3$, $\dfrac{dV}{dr} = 4\pi r^2 =$ the surface area of the ball.

9. $y = 2x^2 + x - 1$, $\dfrac{dy}{dx} = 4x + 1$. $\dfrac{dy}{dx} = 4$ at $x = \dfrac{3}{4}$. Therefore $x_0 = \dfrac{3}{4}$.

11. (a) $w = s\sqrt{2}$, $V = s^3 = \left(\dfrac{w}{\sqrt{2}}\right)^3 = \dfrac{\sqrt{2}}{4}w^3$, $\dfrac{dV}{dw} = \dfrac{3\sqrt{2}}{4}w^2$.

    (b) $z^2 = s^2 + w^2 = 3s^2$, $z = s\sqrt{3}$. $V = s^3 = \left(\dfrac{z}{\sqrt{3}}\right)^3 = \dfrac{\sqrt{3}}{9}z^3$, $\dfrac{dV}{dz} = \dfrac{\sqrt{3}}{3}z^2$.

13. (a) $\dfrac{dA}{d\theta} = \dfrac{1}{2}r^2$ (b) $\dfrac{dA}{dr} = r\theta$

    (c) $\theta = \dfrac{2A}{r^2}$ so $\dfrac{d\theta}{dr} = \dfrac{-4A}{r^3} = \dfrac{-4}{r^3}\left(\dfrac{1}{2}r^2\theta\right) = \dfrac{-2\theta}{r}$

15. $V = s^3$, $dV/ds = 3s^2$. Since s is measured in centimeters, V is measured in cubic centimeters. At $s = 5$, $dV/ds = 75$. Thus, numerically, V changes 75 times as fast as s. Since s decreases at the rate of 3 centimeters per second, V decreases at the rate of 225 centimeters per second.

17. $y = ax^2 + bx + c$, $z = bx^2 + ax + c$.

    $\dfrac{dy}{dx} = 2ax + b$, $\dfrac{dz}{dx} = 2bx + a$.

    $\dfrac{dy}{dx} = \dfrac{dz}{dx}$ iff $2ax + b = 2bx + a$. With $a \neq b$, this occurs only at $x = \dfrac{1}{2}$.

SECTION 3.5

1. $x(5) = -6$, $v(t) = 3 - 2t$ so $v(5) = -7$ and speed $= 7$, $a(t) = -2$ so $a(5) = -2$.

3. $x(2) = -4$, $v(t) = 3t^2 - 6$ so $v(2) = 6$ and speed $= 6$, $a(t) = 6t$ so $a(2) = 12$.

5. $x(1) = 6$, $v(t) = -18/(t+2)^2$ so $v(1) = -2$ and speed $= 2$,

    $a(t) = 36/(t+2)^3$ so $a(1) = 4/3$.

7. $x(1) = 0$, $vt = 4t^3 + 18t^2 + 6t - 10$ so $v(1) = 18$ and speed $= 18$,

    $a(t) = 12t^2 + 36t + 6$ so $a(1) = 54$.

9. $v(t) = 3t^2 - 6t + 3 = 3(t-1)^2 \geq 0$; the object never changes direction.

11. $v(t) = 1 - \dfrac{5}{(t+2)^2}$; the object changes direction (from left to right) at $t = -2 + \sqrt{5}$.

13. $v(t) = \dfrac{8 - t^2}{(t^2 + 8)^2}$; the object changes direction (from right to left) at $t = 2\sqrt{2}$.

15. A 17. A 19. A and B 21. A 23. A and C

25. The object is moving right when $v(t) > 0$. Here,

    $v(t) = 4t^3 - 36t^2 + 56t = 4t(t-2)(t-7)$ and $v(t) > 0$ when $0 < t < 2$ and $7 < t$.

27. The object is speeding up when $v(t)$ and $a(t)$ have the same sign.

$v(t) = 5t^3(4 - t)$ sign of $v(t)$:

$a(t) = 20t^2(3 - t)$ sign of $a(t)$:

Thus, $0 < t < 3$ and $4 < t$.

29. The object is moving left and slowing down when $v(t) < 0$ and $a(t) > 0$.

$v(t) = 3(t - 5)(t + 1)$ sign of $v(t)$:

$a(t) = 6(t - 2)$ sign of $a(t)$:

Thus, $2 < t < 5$.

31. The object is moving right and speeding up when $v(t) > 0$ and $a(t) > 0$.

$v(t) = 4t(t - 2)(t - 4)$ sign of $v(t)$:

$a(t) = 4(3t^2 - 12t + 8)$ sign of $a(t)$:

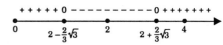

Thus, $0 < t < 2 - \frac{2}{3}\sqrt{3}$ and $4 < t$.

33. Since $v_0 = 0$ the equation of motion is

$$y(t) = -16t^2 + y_0.$$

We want to find y_0 so that $y(6) = 0$. From

$$0 = -16(6)^2 + y_0$$

we get $y_0 = 576$ feet.

35. In general, $mgy + \frac{1}{2}mv^2 = C$. Since $y = 0$ when $v = v_0$, it is clear that $C = \frac{1}{2}v_0^2$. We therefore have

$$mgy + \tfrac{1}{2}mv^2 = \tfrac{1}{2}mv_0^2.$$

At maximum height y_1, $v = 0$ and therefore

$$mgy_1 = \tfrac{1}{2}mv_0^2.$$

This gives $y_1 = v_0^2/2g$.

37. clear from the energy equation

39. In the equation

$$y(t) = -16t^2 + v_0 t + y_0$$

we take $v_0 = -80$ and $y_0 = 224$. The ball first strikes the ground when

$$-16t^2 - 80t + 224 = 0;$$

that is, at $t = 2$. Since

$$v(t) = y\prime(t) = -32t - 80,$$

we have $v(2) = -144$ so that the speed of the ball the first time it strikes the ground is 144 ft/sec. Thus, the speed of the ball the third time it strikes the ground is $\frac{1}{4}\left[\frac{1}{4}(144)\right] = 9$ ft/sec.

41. The equation is $y(t) = -16t^2 + 32t$.   (Here $y_0 = 0$ and $v_0 = 32$.)

(a)   We solve $y(t) = 0$ to find that the stone strikes the ground at $t = 2$ seconds.

(b)   The stone attains its maximum height when $v(t) = 0$. Solving

$$v(t) = -32t + 32 = 0, \quad \text{we get } t = 1 \quad \text{and, thus, the maximum height is } y(1) = 16 \text{ feet.}$$

(This result can be obtained from the energy equation.)

(c)   We want to choose v_0 in

$$y(t) = -16t^2 + v_0 t$$

so that $y(t_0) = 36$ when $v(t_0) = 0$ for some time t_0.

From $v(t) = -32t + v_0 = 0$ we get $t_0 = v_0/32$ so that

$$-16\left(\frac{v_0}{32}\right)^2 + v_0\left(\frac{v_0}{32}\right) = 36, \quad \text{or} \quad \frac{v_0{}^2}{64} = 36.$$

Thus, $v_0 = 48$ ft/sec. (This result can be obtained from the energy equation.)

43.   For all three parts of the problem the basic equation is

$$y(t) = -16t^2 + v_0 t + y_0$$

with

(∗) $y(t_0) = 100$ and $y(t_0 + 2) = 16$

for some time $t_0 > 0$.

We are asked to find y_0 for a given value of v_0.

From (∗) we get

$$16 - 100 = y(t_0 + 2) - y(t_0)$$

$$= [-16(t_0 + 2)^2 + v_0(t_0 + 2) + y_0] - [-16t_0{}^2 + v_0 t_0 + y_0]$$

$$= -64t_0 - 64 + 2v_0$$

so that

$$t_0 = \tfrac{1}{32}(v_0 + 10).$$

Substituting this result in the basic equation and noting that $y(t_0) = 100$, we have

$$-16\left(\frac{v_0 + 10}{32}\right)^2 + v_0\left(\frac{v_0 + 10}{32}\right) + y_0 = 100$$

and therefore

(**) $$y_0 = 100 - \frac{v_0^2}{64} + \frac{25}{16}.$$

We use (**) to find the answer to each part of the problem.

(a) $v_0 = 0$ so $y_0 = \frac{1625}{16}$ ft   (b) $v_0 = -5$ so $y_0 = \frac{6475}{64}$ ft   (c) $v_0 = 10$ so $y_0 = 100$ ft

SECTION 3.6

1. $f(x) = x^4 + 2x^2 + 1, \quad f'(x) = 4x^3 + 4x = 4x(x^2 + 1)$

   $f(x) = (x^2 + 1)^2, \quad f'(x) = 2(x^2 + 1)(2x) = 4x(x^2 + 1)$

3. $f(x) = 8x^3 + 12x^2 + 6x + 1, \quad f'(x) = 24x^2 + 24x + 6 = 6(2x + 1)^2$

   $f(x) = (2x + 1)^3, \quad f'(x) = 3(2x + 1)^2(2) = 6(2x + 1)^2$

5. $f(x) = x^2 + 2 + x^{-2}, \quad f'(x) = 2x - 2x^{-3} = 2x(1 - x^{-4})$

   $f(x) = (x + x^{-1})^2, \quad f'(x) = 2(x + x^{-1})(1 - x^{-2}) = 2x(1 + x^{-2})(1 - x^{-2}) = 2x(1 - x^{-4})$

7. $f'(x) = -1(1 - 2x)^{-2} \dfrac{d}{dx}(1 - 2x) = 2(1 - 2x)^{-2}$

9. $f'(x) = 20(x^5 - x^{10})^{19} \dfrac{d}{dx}(x^5 - x^{10}) = 20(x^5 - x^{10})^{19}(5x^4 - 10x^9)$

11. $f'(x) = 4\left(x - \dfrac{1}{x}\right)^3 \dfrac{d}{dx}\left(x - \dfrac{1}{x}\right) = 4\left(x - \dfrac{1}{x}\right)^3 \left(1 + \dfrac{1}{x^2}\right)$

13. $f'(x) = 4(x - x^3 - x^5)^3 \dfrac{d}{dx}(x - x^3 - x^5) = 4(x - x^3 - x^5)^3(1 - 3x^2 - 5x^4)$

15. $f'(t) = 100(t^2 - 1)^{99} \dfrac{d}{dt}(t^2 - 1) = 200t(t^2 - 1)^{99}$

17. $f'(t) = 4(t^{-1} + t^{-2})^3 \dfrac{d}{dt}(t^{-1} + t^{-2}) = 4(t^{-1} + t^{-2})^3(-t^{-2} - 2t^{-3})$

19. $f'(x) = 4\left(\dfrac{3x}{x^2 + 1}\right)^3 \dfrac{d}{dx}\left(\dfrac{3x}{x^2 + 1}\right) = 4\left(\dfrac{3x}{x^2 + 1}\right)^3 \left[\dfrac{(x^2 + 1)3 - 3x(2x)}{(x^2 + 1)^2}\right] = \dfrac{324x^3(1 - x^2)}{(x^2 + 1)^5}$

21. $f'(x) = 2(x^4 + x^2 + x)^1 \dfrac{d}{dx}(x^4 + x^2 + x) = 2(x^4 + x^2 + x)(4x^3 + 2x + 1)$

23. $f'(x) = -\left(\dfrac{x^3}{3} + \dfrac{x^2}{2} + \dfrac{x}{1}\right)^{-2} \dfrac{d}{dx}\left(\dfrac{x^3}{3} + \dfrac{x^2}{2} + \dfrac{x}{1}\right) = -\left(\dfrac{x^3}{3} + \dfrac{x^2}{2} + x\right)^{-2}(x^2 + x + 1)$

25. $f'(x) = 3\left(\dfrac{1}{x + 2} - \dfrac{1}{x - 2}\right)^2 \dfrac{d}{dx}\left(\dfrac{1}{x + 2} - \dfrac{1}{x - 2}\right)$

   $= 3\left(\dfrac{1}{x + 2} - \dfrac{1}{x - 2}\right)^2 \left(\dfrac{-1}{(x + 2)^2} + \dfrac{1}{(x - 2)^2}\right) = \dfrac{384x}{(x + 2)^4(x - 2)^4}$

27. $\dfrac{dy}{dx} = \dfrac{dy}{du}\dfrac{du}{dx} = \dfrac{-2u}{(1+u^2)^2} \cdot (2)$

At $x = 0$, we have $u = 1$ and thus $\dfrac{dy}{dx} = \dfrac{-4}{4} = -1.$

29. $\dfrac{dy}{dx} = \dfrac{dy}{du}\dfrac{du}{dx} = \dfrac{(1-4u)2 - 2u(-4)}{(1-4u)^2} \cdot 4(5x^2+1)^3(10x) = \dfrac{2}{(1-4u)^2} \cdot 40x(5x^2+1)^3$

At $x = 0$, we have $u = 1$ and thus $\dfrac{dy}{dx} = \dfrac{2}{9}(0) = 0.$

31. $\dfrac{dy}{dt} = \dfrac{dy}{du}\dfrac{du}{dx}\dfrac{dx}{dt} = \dfrac{(1+u^2)(-7) - (1-7u)(2u)}{(1+u^2)^2}(2x)(2)$

$= \dfrac{7u^2 - 2u - 7}{(1+u^2)^2}(4x) = \dfrac{4x(7x^4 + 12x - 2)}{(x^4 + 2x^2 + 2)^2} = \dfrac{4(2t-5)[7(2t-5)^4 + 12(2t-5)^2 - 2]}{[(2t-5)^4 + 2(2t-5)^2 + 2]^2}$

33. $\dfrac{dy}{dx} = \dfrac{dy}{ds}\dfrac{ds}{dt}\dfrac{dt}{dx} = 2(s+3) \cdot \dfrac{1}{2\sqrt{t-3}} \cdot (2x)$

At $x = 2$, we have $t = 4$ so that $s = 1$ and thus $\dfrac{dy}{dx} = 2(4)\dfrac{1}{2 \cdot 1}(4) = 16.$

35. $(f \circ g)'(0) = f'(g(0))g'(0) = f'(2)g'(0) = (1)(1) = 1$

37. $(f \circ g)'(2) = f'(g(2))g'(2) = f'(2)g'(2) = (1)(1) = 1$

39. $(g \circ f)'(1) = g'(f(1))f'(1) = g'(0)f'(1) = (1)(1) = 1$

41. $(f \circ h)'(0) = f'(h(0))h'(0) = f'(1)h'(0) = (1)(2) = 2$

43. $(h \circ f)'(0) = h'(f(0))f'(0) = h'(1)f'(0) = (1)(2) = 2$

45. $(g \circ f \circ h)'(2) = g'(f(h(2)))\, f'(h(2))h'(2) = g'(1)f'(0)h'(2) = (0)(2)(2) = 0$

47. $2xf'(x^2 + 1)$ 49. $2f(x)f'(x)$

51. $f'(x) = -4x(1+x^2)^{-3};$ (a) $x = 0$ (b) $x < 0$ (c) $x > 0$

53. $f'(x) = \dfrac{1 - x^2}{(1+x^2)^2};$ (a) $x = \pm 1$ (b) $-1 < x < 1$ (c) $x < -1, \ x > 1$

55. $v(t) = 5(t+1)(t-9)^2(t-3);$ the object changes direction (from left to right) at $t = 3.$

57. $v(t) = 12t^3(t^2 - 12)^3(t^2 - 4);$ the object changes direction (from right to left) at $t = 2$

and (from left to right) at $t = 2\sqrt{3}.$

59. If $p(x) = (x-a)^2 g(x),$ then $p'(x) = (x-a)[2g(x) + (x-a)g'(x)]$

61. $f'(x) = -[(7x - x^{-1})^{-2} + 3x^2]^{-2} \left\{ \dfrac{d}{dx}(7x - x^{-1})^{-2} + \dfrac{d}{dx}(3x^2) \right\}$

$= -[(7x - x^{-1})^{-2} + 3x^2]^{-2} \left\{ -2(7x - x^{-1})^{-3}(7 + x^{-2}) + 6x \right\}$

$= 2[(7x - x^{-1})^{-2} + 3x^2]^{-2}[(7x - x^{-1})^{-3}(7 + x^{-2}) - 3x]$

63. $f'(x) = 3[(x + x^{-1})^2 - (x^2 + x^{-2})^{-1}]^2 \left\{ \dfrac{d}{dx}(x + x^{-1})^2 - \dfrac{d}{dx}(x^2 + x^{-2})^{-1} \right\}$

$= 3\left[(x + x^{-1})^2 - (x^2 + x^{-2})^{-1}\right]^2 [2(x + x^{-1})(1 - x^{-2}) + (x^2 + x^{-2})^{-2}(2x - 2x^{-3})]$

65. By numerical work $f'(1) \cong -20$; by the chain rule

$$f'(x) = 20(x^2 - 2)^9, \quad f'(1) = -20$$

SECTION 3.7

1. $\dfrac{dy}{dx} = -3\sin x - 4\sec x \tan x$

3. $\dfrac{dy}{dx} = 3x^2 \csc x - x^3 \csc x \cot x$

5. $\dfrac{dy}{dt} = -2\cos t \sin t$

7. $\dfrac{dy}{du} = 4\sin^3 \sqrt{u}\, \dfrac{d}{du}(\sin \sqrt{u}) = 4\sin^3 \sqrt{u}\, \cos \sqrt{u}\, \dfrac{d}{du}(\sqrt{u}) = 2u^{-1/2}\sin^3 \sqrt{u}\, \cos \sqrt{u}$

9. $\dfrac{dy}{dx} = \sec^2 x^2\, \dfrac{d}{dx}(x^2) = 2x \sec^2 x^2$

11. $\dfrac{dy}{dx} = 4[x + \cot \pi x]^3[1 - \pi \csc^2 \pi x]$

13. $\dfrac{dy}{dx} = \cos x, \quad \dfrac{d^2 y}{dx^2} = -\sin x$

15. $\dfrac{dy}{dx} = \dfrac{(1 + \sin x)(-\sin x) - \cos x\,(\cos x)}{(1 + \sin x)^2} = \dfrac{-\sin x - (\sin^2 x + \cos^2 x)}{(1 + \sin x)^2} = -(1 + \sin x)^{-1}$

$\dfrac{d^2 y}{dx^2} = (1 + \sin x)^{-2}\dfrac{d}{dx}(1 + \sin x) = \cos x\,(1 + \sin x)^{-2}$

17. $\dfrac{dy}{du} = 3\cos^2 2u\, \dfrac{d}{du}(\cos 2u) = -6\cos^2 2u \sin 2u$

$\dfrac{d^2 y}{du^2} = -6[\cos^2 2u\, \dfrac{d}{du}(\sin 2u) + \sin 2u\, \dfrac{d}{du}(\cos^2 2u)]$

$= -6[2\cos^3 2u + \sin 2u\,(-4\cos 2u \sin 2u)] = 12\cos 2u\,[2\sin^2 2u - \cos^2 2u]$

19. $\dfrac{dy}{dt} = 2\sec^2 2t, \quad \dfrac{d^2 y}{dt^2} = 4\sec 2t\, \dfrac{d}{dt}(\sec 2t) = 8\sec^2 2t \tan 2t$

21. $\dfrac{dy}{dx} = x^2(3\cos 3x) + 2x\sin 3x$

$\dfrac{d^2y}{dx^2} = [x^2(-9\sin 3x) + 2x(3\cos 3x)] + [2x(3\cos 3x) + 2(\sin 3x)]$

$\qquad = (2 - 9x^2)\sin 3x + 12x\cos 3x$

23. $y = \sin^2 x + \cos^2 x = 1$ so $\dfrac{dy}{dx} = \dfrac{d^2y}{dx^2} = 0$

25. $\dfrac{d^4}{dx^4}(\sin x) = \dfrac{d^3}{dx^3}(\cos x) = \dfrac{d^2}{dx^2}(-\sin x) = \dfrac{d}{dx}(-\cos x) = \sin x$

27. $\dfrac{d}{dt}\left[t^2\dfrac{d^2}{dt^2}(t\cos 3t)\right] = \dfrac{d}{dt}\left[t^2\dfrac{d}{dt}(\cos 3t - 3t\sin 3t)\right]$

$\qquad\qquad = \dfrac{d}{dt}[t^2(-3\sin 3t - 3\sin 3t - 9t\cos 3t)]$

$\qquad\qquad = \dfrac{d}{dt}[-6t^2\sin 3t - 9t^3\cos 3t]$

$\qquad\qquad = (-18t^2\cos 3t - 12t\sin 3t) + (27t^3\sin 3t - 27t^2\cos 3t)$

$\qquad\qquad = (27t^3 - 12t)\sin 3t - 45t^2\cos 3t$

29. $\dfrac{d}{dx}[f(\sin 3x)] = f'(\sin 3x)\dfrac{d}{dx}(\sin 3x) = 3\cos 3x\,f'(\sin 3x)$

31. $\dfrac{dy}{dx} = \cos x;$ slope of tangent at $(0,0)$ is 1, an equation for tangent is $y = x$.

33. $\dfrac{dy}{dx} = -\csc^2 x;$ slope of tangent at $\left(\dfrac{\pi}{6}, \sqrt{3}\right)$ is -4, an equation for tangent is $y - \sqrt{3} = -4\left(x - \dfrac{\pi}{6}\right)$.

35. $\dfrac{dy}{dx} = \sec x\tan x,$ slope of tangent at $\left(\dfrac{\pi}{4}, \sqrt{2}\right)$ is $\sqrt{2}$, an equation for tangent is $y - \sqrt{2} = \sqrt{2}\left(x - \dfrac{\pi}{4}\right)$.

37. $\dfrac{dy}{dx} = -\sin x;$ $x = \pi$

39. $\dfrac{dy}{dx} = \cos x - \sqrt{3}\sin x;$ $\dfrac{dy}{dx} = 0$ gives $\tan x = \dfrac{1}{\sqrt{3}};$ $x = \dfrac{\pi}{6}, \dfrac{7\pi}{6}$

41. $\dfrac{dy}{dx} = 2\sin x\cos x = \sin 2x;$ $x = \dfrac{\pi}{2}, \pi, \dfrac{3\pi}{2}$

43. $\dfrac{dy}{dx} = \sec^2 x - 2;$ $\dfrac{dy}{dx} = 0$ gives $\sec x = \pm\sqrt{2};$ $x = \dfrac{\pi}{4}, \dfrac{3\pi}{4}, \dfrac{5\pi}{4}, \dfrac{7\pi}{4}$

45. $\dfrac{dy}{dx} = 2\sec x \tan x + \sec^2 x;$ since $\sec x$ is never zero, $\dfrac{dy}{dx} = 0$ gives

$2\tan x + \sec x = 0$ so that $\sin x = -1/2;$ $x = \dfrac{7\pi}{6}, \dfrac{11\pi}{6}$

47. $\dfrac{dy}{dx} = \dfrac{(1+\cot x)(-\csc c \cot x) - \csc x\,(-\csc^2 x)}{(1+\cot x)^2}$

$= \dfrac{\csc x\,(-\cot x - \cot^2 x + \csc^2 x)}{(1+\cot x)^2} = \dfrac{\csc x\,(1-\cot x)}{(1+\cot x)^2};$

since $\csc x$ is never zero, $\dfrac{dy}{dx} = 0$ gives $\cot x = 1;$ $x = \dfrac{\pi}{4}, \dfrac{5\pi}{4}$

49. We want $v(t) > 0$ and $a(t) > 0.$

$v(t) = 3\cos 3t$

$a(t) = -9\sin 3t$

sign of $v(t)$:

sign of $a(t)$:

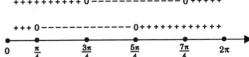

Thus, $\pi < t < \dfrac{2\pi}{3},$ $\dfrac{7\pi}{6} < t < \dfrac{4\pi}{3},$ $\dfrac{11\pi}{6} < t < 2\pi.$

51. We want $v(t) > 0$ and $a(t) > 0.$

$v(t) = \cos t + \sin t$

$a(t) = -\sin t + \cos t$

sign of $v(t)$:

sign of $a(t)$:

Thus, $0 < t < \dfrac{\pi}{4}$ and $\dfrac{7\pi}{4} < t < 2\pi.$

53. We want $v(t) > 0$ and $a(t) > 0.$

$v(t) = 1 - 2\sin t$

$a(t) = -2\cos t$

sign of $v(t)$:

sign of $a(t)$:

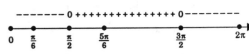

Thus, $\dfrac{5\pi}{6} < t < \dfrac{3\pi}{2}.$

55. (a) $\dfrac{dy}{dt} = \dfrac{dy}{du}\dfrac{du}{dx}\dfrac{dx}{dt} = (2u)(\sec x \tan x)\pi = 2\pi \sec^2 \pi t \tan \pi t$

(b) $y = \sec^2 \pi t - 1,$ $\dfrac{dy}{dt} = 2\sec \pi t\,(\sec \pi t \tan \pi t)\pi = 2\pi \sec^2 \pi t \tan \pi t$

57. (a) $\dfrac{dy}{dt} = \dfrac{dy}{du}\dfrac{du}{dx}\dfrac{dx}{dt} = 4\left[\dfrac{1}{2}(1-u)\right]^3\left(-\dfrac{1}{2}\right)(-\sin x)(2) = 4\left[\dfrac{1}{2}(1-\cos 2t)\right]^3 \sin 2t$

$= 4\sin^6 t\,(2\sin t \cos t) = 8\sin^7 t \cos t$

(b) $y = \left[\dfrac{1}{2}(1-\cos 2t)\right]^4 = \sin^8 t,$ $\dfrac{dy}{dt} = 8\sin^7 t \cos t$

59. $\dfrac{d^n}{dx^n}(\cos x) = \left\{ \begin{array}{l} (-1)^{(n+1)/2}\sin x, \ \ n \text{ odd} \\ (-1)^{n/2}\cos x, \ \ n \text{ even} \end{array} \right]$

61. By numerical work $f'(0) \cong 0$;   by the chain rule $f'(x) = -2x \sin x^2$, $f'(0) = 0$.

SECTION 3.8

1. $\dfrac{dy}{dx} = \dfrac{1}{2}(x^3+1)^{-1/2}\dfrac{d}{dx}(x^3+1) = \dfrac{3}{2}x^2(x^3+1)^{-1/2}$

3. $\dfrac{dy}{dx} = x\left(\dfrac{1}{2}(x^2+1)^{-1/2}(2x)\right) + (x^2+1)^{1/2} = (1+2x^2)(x^2+1)^{-1/2}$

5. $\dfrac{dy}{dx} = \dfrac{1}{4}(2x^2+1)^{-3/4}\dfrac{d}{dx}(2x^2+1) = x(2x^2+1)^{-3/4}$

7. $\dfrac{dy}{dx} = \sqrt{2-x^2}\left[\dfrac{-x}{\sqrt{3-x^2}}\right] + \sqrt{3-x^2}\left[\dfrac{-x}{\sqrt{2-x^2}}\right] = \dfrac{x(2x^2-5)}{\sqrt{2-x^2}\sqrt{3-x^2}}$

9. $\dfrac{d}{dx}\left(\sqrt{x} + \dfrac{1}{\sqrt{x}}\right) = \dfrac{d}{dx}(x^{1/2} + x^{-1/2}) = \dfrac{1}{2}x^{-1/2} - \dfrac{1}{2}x^{-3/2} = \dfrac{1}{2}x^{-3/2}(x-1)$

11. $\dfrac{d}{dx}\left(\dfrac{x}{\sqrt{x^2+1}}\right) = \dfrac{d}{dx}\left(x(x^2+1)^{-1/2}\right)$

$$= x\left(-\dfrac{1}{2}(x^2+1)^{-3/2}(2x)\right) + (x^2+1)^{-1/2} = (x^2+1)^{-3/2}$$

13. $\dfrac{d}{dx}(x^{1/3} + x^{-1/3}) = \dfrac{1}{3}x^{-2/3} - \dfrac{1}{3}x^{-4/3} = \dfrac{1}{3}x^{-4/3}(x^{2/3}-1)$

15. (a) 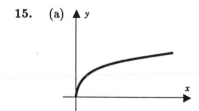 (b) 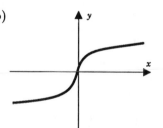 (c)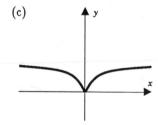

17. $y = (a+bx)^{1/3}$; $\dfrac{dy}{dx} = \dfrac{b}{3}(a+bx)^{-2/3}$; $\dfrac{d^2y}{dx^2} = \dfrac{-2b^2}{9}(a+bx)^{-5/3}$

19. $y = (a^2-x^2)^{1/2}$; $\dfrac{dy}{dx} = -x(a^2-x^2)^{-1/2}$;

$$\dfrac{d^2y}{dx^2} = -(a^2-x^2)^{-1/2} - x(a^2-x^2)^{-3/2}(x) = -a^2(a^2-x^2)^{-3/2}$$

21. $\dfrac{dy}{dx} = \sqrt{x}\left(\cos\sqrt{x}\left(\dfrac{1}{2\sqrt{x}}\right)\right) + \dfrac{1}{2\sqrt{x}}(\sin\sqrt{x}) = \dfrac{1}{2}\cos\sqrt{x} + \dfrac{1}{2}x^{-1/2}\sin\sqrt{x}$

$$\dfrac{d^2y}{dx^2} = -\dfrac{1}{2}\sin\sqrt{x}\left(\dfrac{1}{2\sqrt{x}}\right) + \dfrac{1}{2}x^{-1/2}\cos\sqrt{x}\left(\dfrac{1}{2\sqrt{x}}\right) - \dfrac{1}{4}x^{-3/2}\sin\sqrt{x}$$

$$= \dfrac{\sqrt{x}\cos\sqrt{x} - (x+1)\sin\sqrt{x}}{4x\sqrt{x}}$$

23. $\dfrac{d}{dx}[f(\sqrt{x}+1)] = f'(\sqrt{x}+1)\dfrac{d}{dx}(\sqrt{x}+1) = \dfrac{1}{2\sqrt{x}}f'(\sqrt{x}+1)$

25. $\dfrac{d}{dx}\left[\sqrt{f(x^2+1)}\right] = \dfrac{1}{2}[f(x^2+1)]^{-1/2}\dfrac{d}{dx}[f(x^2+1)]$

$$= \dfrac{1}{2}[f(x^2+1)]^{-1/2}f'(x^2+1)(2x) = \dfrac{xf'(x^2+1)}{\sqrt{f(x^2+1)}}$$

27. $(f^{-1})'(x) = \dfrac{1}{f'(f^{-1}(x))} = \dfrac{1}{f(f^{-1}(x))} = \dfrac{1}{x}$

29. $(f^{-1})'(x) = \dfrac{1}{f'(f^{-1}(x))} = \dfrac{1}{\sqrt{1-[f(f^{-1}(x))]^2}} = \dfrac{1}{\sqrt{1-x^2}}$

31. $\epsilon = \dfrac{P}{Q}\left|\dfrac{dQ}{dP}\right| = \dfrac{P}{\sqrt{300-P}}\left|-\dfrac{1}{2\sqrt{300-P}}\right| = \dfrac{P}{2(300-P)}.$

With $0 < P < 300$,

$$\epsilon < 1 \implies P < 600 - 2P \implies 3P < 600 \implies P < 200$$

$$\epsilon = 1 \implies P = 600 - 2P \implies 3P = 600 \implies P = 200$$

$$\epsilon > 1 \implies P > 600 - 2P \implies 3P > 600 \implies P > 200.$$

Answer: inelastic for $0 < P < 200$,   unitary at $P = 200$,   elastic for $200 < P < 300$.

33. $(f^{-1})'(4) = \dfrac{1}{f'(f^{-1}(4))} = \dfrac{1}{f'(2)} = \dfrac{1}{7}$ **35.** $(f^{-1})'(5) = \dfrac{1}{f'(f^{-1}(5))} = \dfrac{1}{f'(3)} = \dfrac{1}{8}$

37. $g'(2) = f'(f(2))f'(2) = f'(4)f'(2) = 9 \cdot 7 = 63$

39. $(g^{-1})'(12) = \dfrac{1}{g'(g^{-1}(12))} = \dfrac{1}{g'(2)} = \dfrac{1}{63}$ by Ex. 37

41.   By numerical work $f'(16) \cong 0.375$;   from (3.8.4)

$$f'(x) = \tfrac{3}{4}x^{-1/4}, \quad f'(16) = \tfrac{3}{8} = 0.375.$$

43.   By numerical work $l \cong 0.125$;

$$\lim_{x \to 32}\dfrac{x^{2/5}-4}{x^{4/5}-16} = \lim_{x \to 32}\dfrac{1}{x^{2/5}+4} = \dfrac{1}{8} = 0.125.$$

SECTION 3.9

1. $\dfrac{dy}{dx} = \dfrac{2}{3}x^{-1/3}$ **3.** $y = ax^{-1} + b + cx, \quad \dfrac{dy}{dx} = -ax^{-2} + c$

5. $y = \sqrt{a}\,(x^{1/2} + x^{-1/2}), \quad \dfrac{dy}{dx} = \dfrac{\sqrt{a}}{2}\,(x^{-1/2} - x^{-3/2})$

7. $\dfrac{dr}{d\theta} = \dfrac{1}{2}(1-2\theta)^{-1/2}(-2) = -(1-2\theta)^{-1/2}$

9. $f'(x) = -\frac{1}{2}(a^2 - x^2)^{-3/2}(-2x) = x(a^2 - x^2)^{-3/2}$

11. $\dfrac{dy}{dx} = 3\left(a + \dfrac{b}{x^2}\right)^2\left(-\dfrac{2b}{x^3}\right) = -\dfrac{6b}{x^3}\left(a + \dfrac{b}{x^2}\right)^2$

13. $\dfrac{ds}{dt} = t\left[\dfrac{1}{2}(a^2 + t^2)^{-1/2}(2t)\right] + (a^2 + t^2)^{1/2} = \dfrac{a^2 + 2t^2}{\sqrt{a^2 + t^2}}$

15. $\dfrac{dy}{dx} = \dfrac{(a^2 - x^2)2x - (a^2 + x^2)(-2x)}{(a^2 - x^2)^2} = \dfrac{4a^2 x}{(a^2 - x^2)^2}$

17. $\dfrac{dy}{dx} = \dfrac{(1 + 2x^2)(-1) - (2 - x)4x}{(1 + 2x^2)^2} = \dfrac{2x^2 - 8x - 1}{(1 + 2x^2)^2}$

19. $\dfrac{dy}{dx} = \dfrac{\sqrt{a - bx} - x(\frac{1}{2}(a - bx)^{-1/2}(-b))}{(\sqrt{a - bx})^2} = \dfrac{2a - bx}{2(a - bx)^{3/2}}$

21. $\dfrac{dy}{dx} = \dfrac{1}{2}\left(\dfrac{1 - cx}{1 + cx}\right)^{-1/2}\left[\dfrac{(1 + cx)(-c) - (1 - cx)c}{(1 + cx)^2}\right] = \dfrac{-c}{(1 + cx)^2}\sqrt{\dfrac{1 + cx}{1 - cx}}$

23. $\dfrac{dy}{dx} = \dfrac{1}{2}\left(\dfrac{a^2 + x^2}{a^2 - x^2}\right)^{-1/2}\left[\dfrac{(a^2 - x^2)2x - (a^2 + x^2)(-2x)}{(a^2 - x^2)^2}\right] = \dfrac{2a^2 x}{(a^2 - x^2)^2}\sqrt{\dfrac{a^2 - x^2}{a^2 + x^2}}$

25. $\dfrac{dr}{d\theta} = \dfrac{\theta\left[\frac{1}{3}(a + b\theta)^{-2/3}b\right] - (a + b\theta)^{1/3}}{\theta^2} = -\dfrac{3a + 2b\theta}{3\theta^2(a + b\theta)^{2/3}}$

27. $\dfrac{dy}{dx} = \dfrac{b}{a} \cdot \dfrac{1}{2}(a^2 - x^2)^{-1/2}(-2x) = -\dfrac{bx}{a\sqrt{a^2 - x^2}}$

29. $\dfrac{dy}{dx} = \dfrac{3}{2}(a^{2/3} - x^{2/3})^{1/2}\left(-\dfrac{2}{3}x^{-1/3}\right) = -x^{-1/3}(a^{2/3} - x^{2/3})^{1/2}$

31. $\dfrac{dy}{dx} = 2x \sec x^2 \tan x^2$ 33. $\dfrac{dy}{dx} = -6\cot^2 2x \csc^2 2x$

35. $\dfrac{dy}{dx} = -\sin x \cos(\cos x)$

37. $\dfrac{dy}{dx} = 3(x^2 - x)^2(2x - 1);$ at $x = 3,$ $\dfrac{dy}{dx} = 3(6)^2(5) = 540$

39. $\dfrac{dy}{dx} = \dfrac{1}{3}x^{-2/3} + \dfrac{1}{2}x^{-1/2};$ at $x = 64,$ $\dfrac{dy}{dx} = \dfrac{1}{3}\left(\dfrac{1}{16}\right) + \dfrac{1}{2}\left(\dfrac{1}{8}\right) = \dfrac{1}{12}$

41. $\dfrac{dy}{dx} = \dfrac{1}{3}(2x)^{-2/3}(2) + \dfrac{2}{3}(2x)^{-1/3}(2);$ at $x = 4,$ $\dfrac{dy}{dx} = \dfrac{2}{3}\left(\dfrac{1}{4}\right) + \dfrac{4}{3}\left(\dfrac{1}{2}\right) = \dfrac{5}{6}$

43. $\dfrac{dy}{dx} = -\dfrac{1}{2}(25 - x^2)^{-3/2}(-2x);$ at $x = 3,$ $\dfrac{dy}{dx} = \dfrac{3}{64}$

45. $\dfrac{dy}{dx} = \dfrac{x[\frac{1}{2}(16 + 3x)^{-1/2} \cdot 3] - \sqrt{16 + 3x}}{x^2};$ at $x = 3,$ $\dfrac{dy}{dx} = \dfrac{3\left(\frac{3}{2}\right)\left(\frac{1}{5}\right) - 5}{9} = -\dfrac{41}{90}$

47. $\dfrac{dy}{dx} = x\left[\dfrac{1}{2}(8-x^2)^{-1/2}(-2x)\right] + \sqrt{8-x^2};$ at $x = 2$, $\dfrac{dy}{dx} = -2 + 2 = 0$

49. $\dfrac{dy}{dx} = x^2\left[\dfrac{1}{2}(1+x^3)^{-1/2}(3x^2)\right] + 2x\sqrt{1+x^3};$ at $x = 2$, $\dfrac{dy}{dx} = 4\left[\dfrac{1}{2}\cdot\dfrac{1}{3}(12)\right] + 4(3) = 20$

51. $\dfrac{dy}{dx} = 2\sec^2 2x;$ at $x = \dfrac{\pi}{6}$, $\dfrac{dy}{dx} = 8$

53. $\dfrac{dy}{dx} = -12\cos^2 4x \sin 4x;$ at $x = \dfrac{\pi}{12}$, $\dfrac{dy}{dx} = -12\left(\dfrac{1}{2}\right)^2\left(\dfrac{\sqrt{3}}{2}\right) = -\dfrac{3}{2}\sqrt{3}$

55. $\dfrac{dy}{dx} = \csc \pi x - \pi x \csc \pi x \cot \pi x;$ at $x = \dfrac{1}{4}$, $\dfrac{dy}{dx} = \sqrt{2} - \dfrac{\pi}{4}\sqrt{2}(1) = \left(1 - \dfrac{\pi}{4}\right)\sqrt{2}$

SECTION 3.10

1.
$$x^2 + y^2 = 4$$
$$2x + 2y\frac{dy}{dx} = 0$$
$$\frac{dy}{dx} = \frac{-x}{y}$$

3.
$$4x^2 + 9y^2 = 36$$
$$8x + 18y\frac{dy}{dx} = 0$$
$$\frac{dy}{dx} = \frac{-4x}{9y}$$

5.
$$x^4 + 4x^3 y + y^4 = 1$$
$$4x^3 + 12x^2 y + 4x^3\frac{dy}{dx} + 4y^3\frac{dy}{dx} = 0$$
$$\frac{dy}{dx} = -\frac{x^3 + 3x^2 y}{x^3 + y^3}$$

7.
$$(x - y)^2 - y = 0$$
$$2(x - y)\left(1 - \frac{dy}{dx}\right) - \frac{dy}{dx} = 0$$
$$\frac{dy}{dx} = \frac{2(x - y)}{2(x - y) + 1}$$

9.
$$\sin(x + y) = xy$$
$$\cos(x + y)\left(1 + \frac{dy}{dx}\right) = x\frac{dy}{dx} + y$$
$$\frac{dy}{dx} = \frac{y - \cos(x + y)}{\cos(x + y) - x}$$

11.
$$y^2 + 2xy = 16$$

$$2y\frac{dy}{dx} + 2x\frac{dy}{dx} + 2y = 0$$

$$(x+y)\frac{dy}{dx} + y = 0.$$

Differentiating a second time, we have

$$(x+y)\frac{d^2y}{dx^2} + \frac{dy}{dx}\left(2 + \frac{dy}{dx}\right) = 0.$$

Substituting $\dfrac{dy}{dx} = \dfrac{-y}{x+y}$,   we have

$$(x+y)\frac{d^2y}{dx^2} - \frac{y}{(x+y)}\left(\frac{2x+y}{x+y}\right) = 0, \quad \frac{d^2y}{dx^2} = \frac{2xy + y^2}{(x+y)^3} = \frac{16}{(x+y)^3}.$$

13.
$$y^2 + xy - x^2 = 9$$

$$2y\frac{dy}{dx} + x\frac{dy}{dx} + y - 2x = 0.$$

Differentiating a second time, we have

$$\left[2\left(\frac{dy}{dx}\right)^2 + 2y\frac{d^2y}{dx^2}\right] + \left[x\frac{d^2y}{dx^2} + \frac{dy}{dx}\right] + \frac{dy}{dx} - 2 = 0$$

$$(2y+x)\frac{d^2y}{dx^2} + 2\left[\left(\frac{dy}{dx}\right)^2 + \frac{dy}{dx} - 1\right] = 0.$$

Substituting $\dfrac{dy}{dx} = \dfrac{2x-y}{2y+x}$,   we have

$$(2y+x)\frac{d^2y}{dx^2} + 2\left[\frac{(2x-y)^2 + (2x-y)(2y+x) - (2y+x)^2}{(2y+x)^2}\right] = 0$$

$$\frac{d^2y}{dx^2} = \frac{10(y^2 + xy - x^2)}{(2y+x)^3} = \frac{90}{(2y+x)^3}.$$

15. $x^2 - 4y^2 = 9, \quad 2x - 8y\frac{dy}{dx} = 0.$

At $(5,2)$, we get $\dfrac{dy}{dx} = \dfrac{5}{8}.$ Then,

$$2 - 8\left[y\frac{d^2y}{dx^2} + \left(\frac{dy}{dx}\right)^2\right] = 0.$$

At $(5,2)$ we get

$$2 - 8\left[2\frac{d^2y}{dx^2} + \frac{25}{64}\right] = 0 \quad \text{so that} \quad \frac{d^2y}{dx^2} = -\frac{9}{128}.$$

17. $\cos(x + 2y) = 0 \qquad -\sin(x + 2y)\left(1 + 2\dfrac{dy}{dx}\right) = 0.$

At $(\pi/6, \pi/6)$, we get $\dfrac{dy}{dx} = -1/2$. Then,

$$-\cos(x + 2y)\left(1 + 2\dfrac{dy}{dx}\right)^2 - \sin(x + 2y)\left(2\dfrac{d^2y}{dx^2}\right) = 0.$$

At $(\pi/6, \pi/6)$, we get

$$-\cos\dfrac{\pi}{2}(0)^2 - \sin\dfrac{\pi}{2}\left(2\dfrac{d^2y}{dx^2}\right) = 0 \quad \text{so that} \quad \dfrac{d^2y}{dx^2} = 0.$$

19. $\qquad\qquad 2x + 3y = 5$

$\qquad\qquad 2 + 3\dfrac{dy}{dx} = 0$

slope of tangent at $(-2, 3)$: $\quad -2/3$

tangent: $\quad y - 3 = -\frac{2}{3}(x + 2)$

normal: $\quad y - 3 = \frac{3}{2}(x + 2)$

21. $\qquad\qquad x^2 + xy + 2y^2 = 28$

$\qquad\qquad 2x + x\dfrac{dy}{dx} + y + 4y\dfrac{dy}{dx} = 0$

slope of tangent at $(-2, -3)$: $\quad -\frac{1}{2}$

tangent: $\quad y + 3 = -\frac{1}{2}(x + 2)$

normal: $\quad y + 3 = 2(x + 2)$

23. Differentiation of $x^2 + y^2 = r^2$ gives us $2x + 2y\dfrac{dy}{dx} = 0$ so that the slope of the normal line is

$$\dfrac{-1}{dy/dx} = \dfrac{y}{x} \quad (x \neq 0).$$

Let (x_0, y_0) be a point on the circle. Clearly, if $x_0 = 0$, the normal line, $x = 0$, passes through the origin. If $x_0 \neq 0$, the normal line is

$$y - y_0 = \dfrac{y_0}{x_0}(x - x_0), \quad \text{which simplifies to} \quad y = \dfrac{y_0}{x_0}x,$$

a line through the origin.

25. For the parabola $y^2 = 2px + p^2$ we have $2y\dfrac{dy}{dx} = 2p$ and the slope of a tangent is given by $m_1 = p/y$.

For the parabola $y^2 = p^2 - 2px$ we obtain $m_2 = -p/y$ as the slope of a tangent. The parabolas intersect at the points $(0, \pm p)$. At each of these points $m_1 m_2 = -1$; the parabolas intersect at right angles.

27. The hyperbola and the ellipse intersect at the four points $(\pm 3, \pm 2)$. For the hyperbola, $\dfrac{dy}{dx} = \dfrac{x}{y}$.

For the ellipse, $\dfrac{dy}{dx} = -\dfrac{4x}{9y}$. The product of these slopes is therefore $-\dfrac{4x^2}{9y^2}$. At each of the points of intersection this product is -1.

29. The line $2y + x + 3 = 0$ has slope $-1/2$. Thus we want the tangents to the ellipse with slope 2. Let (x_0, y_0) be a point of tangency. From $4x^2 + y^2 = 72$ we obtain $8x + 2y\dfrac{dy}{dx} = 0$. Requiring that $\dfrac{dy}{dx} = 2$ at (x_0, y_0), we have $8x_0 + 4y_0 = 0$ and thus $y_0 = -2x_0$. Substituting this result in the equation $4x_0^2 + y_0^2 = 72$, we find that the points of tangency (x_0, y_0) are $(3, -6)$ and $(-3, 6)$. The tangent lines have equations

$$y + 6 = 2(x - 3) \quad \text{and} \quad y - 6 = 2(x + 3).$$

SECTION 3.11

1. $x + 2y = 2,$ $\dfrac{dx}{dt} + 2\dfrac{dy}{dt} = 0$

    (a)   If $\dfrac{dx}{dt} = 4,$ then $\dfrac{dy}{dt} = -2$ units/sec.   (b)   If $\dfrac{dy}{dt} = -2,$ then $\dfrac{dx}{dt} = 4$ units/sec.

3.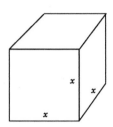

       Find $\dfrac{dx}{dt}$ and $\dfrac{dS}{dt}$ when $V = 27\text{m}^3$

       given that $\dfrac{dV}{dt} = -2\text{m}^3/\text{min}.$

       (*) $V = x^3,$ $S = 6x^2$

Differentiation of equations (*) gives

$$\frac{dV}{dt} = 3x^2\frac{dx}{dt} \quad \text{and} \quad \frac{dS}{dt} = 12x\frac{dx}{dt}.$$

When $V = 27,$ $x = 3.$ Substituting $x = 3$ and $dV/dt = -2,$ we get

$$-2 = 27\frac{dx}{dt} \quad \text{so that} \quad \frac{dx}{dt} = \frac{-2}{27} \quad \text{and} \quad \frac{dS}{dt} = 12(3)\left(\frac{-2}{27}\right) = -8/3.$$

The rate of change of an edge is $-2/27$ m/min;   the rate of change of the surface area is $-8/3$ m²/min.

5.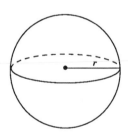

       Find $\dfrac{dr}{dt}$ and $\dfrac{dS}{dt}$ when $r = 10$ ft

       given that $\dfrac{dV}{dt} = 8$ ft³/min.

       (*) $V = \frac{4}{3}\pi r^3,$ $S = 4\pi r^2$

Differentiation of equations (*) with respect to t gives

$$\frac{dV}{dt} = 4\pi r^2\frac{dr}{dt} \quad \text{and} \quad \frac{dS}{dt} = 8\pi r\frac{dr}{dt}.$$

Substituting $r = 10$ and $dV/dt = 8,$ we get

$$8 = 4\pi(10)^2\frac{dr}{dt} \quad \text{so that} \quad \frac{dr}{dt} = \frac{1}{50\pi} \quad \text{and} \quad \frac{dS}{dt} = 8\pi(10)\frac{1}{50\pi} = \frac{8}{5}.$$

The radius is increasing $\dfrac{1}{50\pi}$ ft/min; the surface area is increasing $\dfrac{8}{5}$ ft²/min.

7.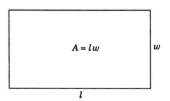

       We will find the values of l for which $\dfrac{dA}{dt} < 0$

       given that $\dfrac{dl}{dt} = 1$ cm/sec and

       $P = 2(l + w) = 24.$

We combine $A = lw$ and $l + w = 12$ to write $A = 12l - l^2$. Differentiating with respect to t, we have

$$\frac{dA}{dt} = 12\frac{dl}{dt} - 2l\frac{dl}{dt} = 2(6 - l)\frac{dl}{dt}.$$

Since $dl/dt = 1$, $\frac{dA}{dt} < 0$ for $l > 6$. The area of the rectangle starts to decrease when the length is 6 cm.

9.

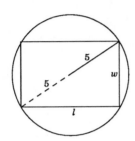

Find $\dfrac{dA}{dt}$ when $l = 6$ in.

given that $\dfrac{dl}{dt} = -2$ in./sec.

By the Pythagorean theorem

$$l^2 + w^2 = 100.$$

Also, $A = lw$. Thus, $A = l\sqrt{100 - l^2}$. Differentiation with respect to t gives

$$\frac{dA}{dt} = l\left(\frac{-l}{\sqrt{100 - l^2}}\right)\frac{dl}{dt} + \sqrt{100 - l^2}\,\frac{dl}{dt}.$$

Substituting $l = 6$ and $dl/dt = -2$, we get

$$\frac{dA}{dt} = 6\left(\frac{-6}{8}\right)(-2) + (8)(-2) = -7.$$

The area is decreasing at the rate of 7 in.^{2}/sec.

11.

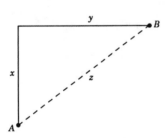

Compare $\dfrac{dy}{dt}$ to $\dfrac{dx}{dt} = -13$ mph

given that $z = 16$ and $\dfrac{dz}{dt} = -17$

when $x = y$.

By the Pythagorean theorem $x^2 + y^2 = z^2$. Thus,

$$2x\frac{dx}{dt} + 2y\frac{dy}{dt} = 2z\frac{dz}{dt}.$$

Since $x = y$ when $z = 16$, we have $x = y = 8\sqrt{2}$ and

$$2(8\sqrt{2})(-13) + 2(8\sqrt{2})\frac{dy}{dt} = 2(16)(-17).$$

Solving for dy/dt, we get

$$-13\sqrt{2} + \sqrt{2}\,\frac{dy}{dt} = -34 \quad \text{or} \quad \frac{dy}{dt} = \frac{1}{\sqrt{2}}(13\sqrt{2} - 34) \cong -11.$$

Thus, boat A wins the race.

13. We want to find dV/dt when $V = 1000$ ft^3 and $P = 5$ lb/in.^{2} given that $dP/dt = -0.05$ lb/in.^{2}/hr.
Differentiating $PV = C$ with respect to t, we get

$$P\frac{dV}{dt} + V\frac{dP}{dt} = 0 \quad \text{so that} \quad 5\frac{dV}{dt} + 1000(-0.05) = 0. \quad \text{Thus,} \quad \frac{dV}{dt} = 10.$$

The volume increases at the rate of 10 ft^3/hr.

15. length of arc $= r\theta$, speed $= \dfrac{d}{dt}[r\theta] = r\dfrac{d\theta}{dt} = r\omega$

17. We know that $d\theta/dt = \omega$ and, at time t, $\theta = \theta_0$. Therefore $\theta = \omega t + \theta_0$. It follows that

$$x(t) = r\cos(\omega t + \theta_0) \quad \text{and} \quad y(t) = r\sin(\omega t + \theta_0).$$

19. For the sector $\quad A = \dfrac{1}{2}r^2\theta, \quad \dfrac{dA}{dt} = \dfrac{1}{2}r^2\dfrac{d\theta}{dt} = \dfrac{1}{2}r^2\omega \quad$ is constant.

For triangle T

$$A = \tfrac{1}{2}(2r\sin\tfrac{1}{2}\theta)(r\cos\tfrac{1}{2}\theta)$$

$$= \tfrac{1}{2}r^2(2\sin\tfrac{1}{2}\theta\cos\tfrac{1}{2}\theta) = \tfrac{1}{2}r^2\sin\theta,$$

$$\frac{dA}{dt} = \frac{1}{2}r^2\cos\theta\frac{d\theta}{dt} = \frac{1}{2}r^2\omega\cos\theta \quad \text{varies with } \theta.$$

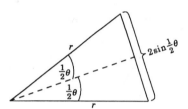

For segment S

$$A = \tfrac{1}{2}r^2\theta - \tfrac{1}{2}r^2\sin\theta = \tfrac{1}{2}r^2(\theta - \sin\theta),$$

$$\frac{dA}{dt} = \frac{1}{2}r^2\left(\frac{d\theta}{dt} - \cos\frac{d\theta}{dt}\right) = \frac{1}{2}r^2\omega(1 - \cos\theta) \quad \text{varies with } \theta.$$

21.

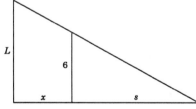

Find $\dfrac{ds}{dt}$ when $x = 3$ ft (and $s = 4$ ft)

given that $\dfrac{dx}{dt} = 400$ ft/min.

By similar triangles

$$\frac{L}{x+s} = \frac{6}{s}.$$

Substitution of $x = 3$ and $s = 4$ gives us $\dfrac{L}{7} = \dfrac{6}{4}$ so that the lamp post is

$L = 10.5$ ft tall. Rewriting

$$\frac{10.5}{x+s} = \frac{6}{s} \quad \text{as} \quad s = \frac{4}{3}x$$

and differentiating with respect to t, we find that

$$\frac{ds}{dt} = \frac{4}{3}\frac{dx}{dt} = \frac{1600}{3}.$$

The shadow lengthens at the rate of 1600/3 ft/min.

23.

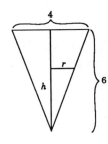

Find $\dfrac{dh}{dt}$ when $h = 3$ in.

given that $\dfrac{dV}{dt} = -\dfrac{1}{2}$ cu in./min.

By similar triangles

$$r = \tfrac{1}{3}h.$$

Thus $V = \tfrac{1}{3}\pi r^2 h = \tfrac{1}{27}\pi h^3$. Differentiating with respect to t , we get

$$\frac{dV}{dt} = \frac{1}{9}\pi h^2 \frac{dh}{dt}.$$

When $h = 3$,

$$-\frac{1}{2} = \frac{1}{9}\pi(9)\frac{dh}{dt} \quad \text{and} \quad \frac{dh}{dt} = -\frac{1}{2\pi}.$$

The water level is dropping at the rate of $1/2\pi$ inches per minute.

25.

Find $\dfrac{d\theta}{dt}$ when $x = 4$ ft

given that $\dfrac{dx}{dt} = 2$ in./min.

$$(*) \qquad \tan\frac{\theta}{2} = \frac{3}{x}$$

Differentiation of $(*)$ with respect to t gives

$$\frac{1}{2}\sec^2\frac{\theta}{2}\frac{d\theta}{dt} = -\frac{3}{x^2}\frac{dx}{dt} \quad \text{or} \quad \frac{d\theta}{dt} = -\frac{6}{x^2}\cos^2\frac{\theta}{2}\frac{dx}{dt}.$$

Note that $dx/dt = 2$ in./min$=1/6$ ft/min. When $x = 4$, we have $\cos\theta/2 = 4/5$ and thus

$$\frac{d\theta}{dt} = -\frac{6}{16}\left(\frac{4}{5}\right)^2\left(\frac{1}{6}\right) = -\frac{1}{25}.$$

The vertex angle decreases at the rate of 0.04 rad/min.

27.

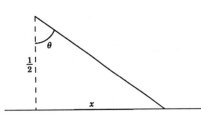

Find $\dfrac{dx}{dt}$ when $x = 1$ mi

given that $\dfrac{d\theta}{dt} = 2\pi$ rad/min.

$$(*) \qquad \tan\theta = \frac{x}{1/2} = 2x$$

Differentiation of $(*)$ with respect to t gives

$$\sec^2\theta\,\frac{d\theta}{dt} = 2\frac{dx}{dt}.$$

When $x = 1$, we get $\sec\theta = \sqrt{5}$ and thus $\dfrac{dx}{dt} = 5\pi$. The light is traveling at 5π mi/min.

29.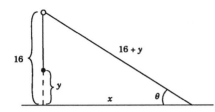

Find $\dfrac{d\theta}{dt}$ when $y = 4$ ft

given that $\dfrac{dx}{dt} = 3$ ft/sec.

$\tan \theta = \dfrac{16}{x}$, $\quad x^2 + (16)^2 = (16 + y)^2$

Differentiating $\tan \theta = 16/x$ with respect to t, we obtain

$$\sec^2 \theta \frac{d\theta}{dt} = \frac{-16}{x^2}\frac{dx}{dt} \quad \text{and thus} \quad \frac{d\theta}{dt} = \frac{-16}{x^2}\cos^2 \theta \frac{dx}{dt}.$$

From $x^2 + (16)^2 = (16 + y)^2$ we conclude that $x = 12$, when $y = 4$. Thus

$$\cos \theta = \frac{x}{16 + y} = \frac{12}{20} = \frac{3}{5} \quad \text{and} \quad \frac{d\theta}{dt} = \frac{-16}{(12)^2}\left(\frac{3}{5}\right)^2 (3) = \frac{-3}{25}.$$

The angle decreases at the rate of 0.12 rad/sec.

31.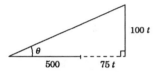

Find $\dfrac{d\theta}{dt}$ when $t = 6$ min.

$$\tan \theta = \frac{100t}{500 + 75t} = \frac{4t}{20 + 3t}$$

Differentiation with respect to t gives

$$\sec^2 \theta \frac{d\theta}{dt} = \frac{(20 + 3t)4 - 4t(3)}{(20 + 3t)^2} = \frac{80}{(20 + 3t)^2}.$$

When $t = 6$

$$\tan \theta = \frac{24}{38} = \frac{12}{19} \quad \text{and} \quad \sec^2 \theta = 1 + \left(\frac{12}{19}\right)^2 = \frac{505}{361}$$

so that

$$\frac{d\theta}{dt} = \frac{80}{(20 + 3t)^2} \cdot \frac{1}{\sec^2 \theta} = \frac{80}{(38)^2} \cdot \frac{361}{505} = \frac{4}{101}.$$

The angle increases at the rate of 4/101 rad/min.

SECTION 3.12

1.
$$\Delta V = (x + h)^3 - x^3$$

$$= (x^3 + 3x^2h + 3xh^2 + h^3) - x^3$$

$$= 3x^2h + 3xh^2 + h^3,$$

$$dV = 3x^2h,$$

$$\Delta V - dV = 3xh^2 + h^3 \quad \text{(see figure)}$$

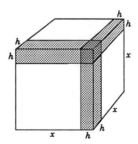

3. $f(x) = x^{1/3}, \quad x = 1000, \quad h = 10, \quad f'(x) = \frac{1}{3}x^{-2/3}$

$\sqrt[3]{1010} = f(x+h) \cong f(x) + hf'(x) = \sqrt[3]{1000} + 10\left(\frac{1}{3}(1000)^{-2/3}\right) = 10\frac{1}{30}$

5. $f(x) = x^{1/4}, \quad x = 16, \quad h = -1, \quad f'(x) = \frac{1}{4}x^{-3/4}$

$(15)^{1/4} = f(x+h) \cong f(x) + hf'(x) = (16)^{1/4} + (-1)\left(\frac{1}{4}(16)^{-3/4}\right) = 1\frac{31}{32}$

7. $f(x) = x^{1/5}, \quad x = 32, \quad h = -2, \quad f'(x) = \frac{1}{5}x^{-4/5}$

$(30)^{1/5} = f(x+h) \cong f(x) + hf'(x) = (32)^{1/5} + (-2)\left(\frac{1}{5}(32)^{-4/5}\right) = 1.975$

9. $f(x) = x^{3/5}, \quad x = 32, \quad h = 1, \quad f'(x) = \frac{3}{5}x^{-2/5}$

$(33)^{3/5} = f(x+h) \cong f(x) + hf'(x) = (32)^{3/5} + (1)\left(\frac{3}{5}(32)^{-2/5}\right) = 8.15$

11. $f(x) = \sin x, \quad x = \frac{\pi}{4}, \quad h = \frac{\pi}{180}, \quad f'(x) = \cos x$

$\sin 46° = f(x+h) \cong f(x) + hf'(x) = \sin\frac{\pi}{4} + \frac{\pi}{180}\cos\frac{\pi}{4} = \frac{\sqrt{2}}{2}\left(1 + \frac{\pi}{180}\right) \cong 0.719$

13. $f(x) = \tan x, \quad x = \frac{\pi}{6}, \quad h = \frac{-\pi}{90}, \quad f'(x) = \sec^2 x$

$\tan 28° = f(x+h) \cong f(x) + hf'(x) = \tan\frac{\pi}{6} + \left(\frac{-\pi}{90}\right)\left(\frac{4}{3}\right) = \frac{\sqrt{3}}{3} - \frac{2\pi}{135} \cong 0.531$

15. $f(2.8) \cong f(3) + (-0.2)f'(3) = 2 + (-0.2)(2) = 1.6$

17. $V(x) = \pi x^2 h; \quad$ volume $= V(r+t) - V(r) \cong tV'(r) = 2\pi r h t$

19. $V(x) = x^3, \quad V'(x) = 3x^2, \quad \Delta V \cong dV = V'(10)h = 300h$

$|dV| \le 3 \quad \Longrightarrow \quad |300h| \le 3 \quad \Longrightarrow \quad |h| \le 0.01, \quad$ error ≤ 0.01 feet

21. $t(l) = \pi\sqrt{\frac{l}{g}}, \quad t'(l) = \frac{\pi}{2\sqrt{gl}}$

$t(l+0.01) - t(l) \cong dt = t'(l)(0.01) = \frac{(3.14)(0.01)}{2\sqrt{(32)(3.26)}} \cong 0.00153$

about 0.00153 seconds

23. $A(x) = \frac{1}{4}\pi x^2, \quad dA = \frac{1}{2}\pi x h, \quad \frac{dA}{A} = 2\frac{h}{x}$

$\frac{dA}{A} \le 0.01 \quad \Longleftrightarrow \quad 2\frac{h}{x} \le 0.01 \quad \Longleftrightarrow \quad \frac{h}{x} \le 0.005 \quad$ within $\frac{1}{2}\%$

25. (a) $x_{n+1} = \frac{1}{2}x_n + 12\left(\frac{1}{x_n}\right)$ (b) $x_4 \cong 4.89898$

27. (a) $x_{n+1} = \frac{2}{3}x_n + \frac{25}{3}\left(\frac{1}{x_n}\right)^2$ (b) $x_4 \cong 2.92402$

29. (a) $x_{n+1} = \dfrac{x_n \sin x_n + \cos x_n}{\sin x_n + 1}$ (b) $x_4 \cong 0.73909$

31. (a) $x_{n+1} = \dfrac{2x_n \cos x_n - 2 \sin x_n}{2 \cos x_n - 1}$ (b) $x_4 \cong 1.89549$

33. (a) and (b)

35. $\displaystyle\lim_{h \to 0} \frac{g_1(h) + g_2(h)}{h} = \lim_{h \to 0} \frac{g_1(h)}{h} + \lim_{h \to 0} \frac{g_2(h)}{h} = 0 + 0 = 0$

$\displaystyle\lim_{h \to 0} \frac{g_1(h)g_2(h)}{h} = \lim_{h \to 0} h \,\frac{g_1(h)g_2(h)}{h^2} = \left(\lim_{h \to 0} h\right)\left(\lim_{h \to 0} \frac{g_1(h)}{h}\right)\left(\lim_{h \to 0} \frac{g_2(h)}{h}\right) = (0)(0)(0) = 0$

SECTION 3.13

1. (a) $f'(x_0) = 2f'(1),\quad 2x_0 = 4,\quad x_0 = 2$ (b) $f'(x_0) = 2f'(1),\quad 6x_0{}^3 012,\quad x_0 = \pm\sqrt{2}$

(c) $f'(x_0) = 2f'(1),\quad \dfrac{1}{2\sqrt{x_0}} = 1,\quad x_0 = \dfrac{1}{4}$

3. $y = \dfrac{2}{3}x^{3/2},\quad \dfrac{dy}{dx} = x^{1/2} = m$

(a) $m = \tan 45° \implies x^{1/2} = 1$ $\left(1, \frac{2}{3}\right)$ (b) $m = \tan 60° \implies x^{1/2} = \sqrt{3}$ $(3, 2\sqrt{3})$

(c) $m = \tan 30° \implies x^{1/2} = \dfrac{1}{\sqrt{3}}$ $\left(\dfrac{1}{3}, \dfrac{2}{27}\sqrt{3}\right)$

5. $y - a = \dfrac{1}{m}(x - b)$ since $(f^{-1})'(b) = \dfrac{1}{f'(f^{-1}(b))} = \dfrac{1}{f'(a)} = \dfrac{1}{m}$ **7.** $\dfrac{d^n y}{dx^n} = b^n n!$

9. $m = \dfrac{dy}{dx} = 4 - 3x^2$ and $\dfrac{dx}{dt} = \dfrac{1}{3}$. Thus, at $x = 2$, $\dfrac{dm}{dt} = -6t\,\dfrac{dx}{dt} = -6(2)\left(\dfrac{1}{3}\right) = -4$.

The slope is decreasing at the rate of 4 units/sec.

11. $f'(2) = \displaystyle\lim_{h \to 0} \frac{(2+h)^2 - 2^2}{h} = \lim_{h \to 0} \frac{4h + h^2}{h} = \lim_{h \to 0} (4 + h) = 4$ **13.** $\frac{1}{5}$

$f'(2) = \displaystyle\lim_{x \to 2} \frac{x^2 - 2^2}{x - 2} = \lim_{x \to 2} (x + 2) = 4$

15. $f(x) = \sin x$ has derivative $f'(x) = \cos x$:

$$\lim_{x \to \pi} \frac{\sin x}{x - \pi} = \lim_{x \to \pi} \frac{\sin x - \sin \pi}{x - \pi} = f'(\pi) = \cos \pi = -1.$$

17. Let (a, b) be the point of tangency. The slope of the tangent is given by $f'(a)$ and by $(b-2)/(a-(-2))$. Thus,

(1)
$$3a^2 - 1 = \frac{b-2}{a+2}.$$

Since (a, b) lies on the curve,

(2)
$$b = a^3 - a.$$

Substituting equation (2) in equation (1), we have

$$3a^2 - 1 = \frac{a^3 - a - 2}{a+2}, \quad 3a^3 + 6a^2 - a - 2 = a^3 - a - 2, \quad a^2(a+3) = 0.$$

Thus, $a = 0$ or $a = -3$. The points of tangency are $(0, 0)$ and $(-3, -24)$. The tangent line at the first point is $y = -x$; the tangent line at the second point is $y + 24 = 26(x + 3)$.

19. 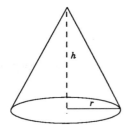 Find $\dfrac{dh}{dt}$ when $r = 4$ in. and $h = 15$ in.

given that $\dfrac{dr}{dt} = 0.3$ in./min and $\dfrac{dV}{dt} = 0$.

$$V = \frac{1}{3} = \pi r^2 h, \qquad \frac{dV}{dt} = \frac{\pi}{3}\left(r^2 \frac{dh}{dt} + 2rh \frac{dr}{dt}\right).$$

Thus, $\quad 0 = \dfrac{\pi}{3}\left[16\dfrac{dh}{dt} + 2(4)(15)(0.3)\right]\quad$ so that $\quad \dfrac{dh}{dt} = -\dfrac{9}{4}.\quad$ The height is decreasing

at the rate of 2.25 in./min.

21.  Find $\dfrac{dA}{dt}$ when $x = 7$ cm and $\dfrac{dx}{dt} = -0.4$ cm/min

given that $\quad P = 2(x + y) = 20$.

$$A = xy, \qquad \frac{dA}{dt} = x\frac{dy}{dt} + y\frac{dx}{dt}.$$

Since $\quad x + y = 10,\quad$ we know that $\quad \dfrac{dy}{dt} = -\dfrac{dx}{dt}.\quad$ Thus,

$$\frac{dA}{dt} = 7(0.4) + 3(-0.4) = 1.6.$$

The area increases at the rate of 1.6 cm^2/min.

23. $V = \dfrac{4}{3}\pi r^3, \quad S = 4\pi r^2.$ Find $\dfrac{dV}{dt}$ when $\dfrac{dS}{dt} = 4$ in.^{2}/min and $\dfrac{dr}{dt} = 0.1$ in./min.

Differentiating $S = 4\pi r^2$ with respect to t, we have

$$\frac{dS}{dt} = 8\pi r\frac{dr}{dt} \quad \text{and thus} \quad 4 = 8\pi r(0.1) \quad \text{or} \quad r = \frac{5}{\pi}.$$

Then, when $r = 5/\pi$,

$$\frac{dV}{dt} = 4\pi r^2 \frac{dr}{dt} = 4\pi \left(\frac{5}{\pi}\right)^2 (0.1) = 10/\pi.$$

The volume is increasing at the rate of $10/\pi$ in.³/min.

25. We need to determine those times t between 0 and 2π when the velocity $v(t)$ and the acceleration $a(t)$ have opposite signs.

$v(t) = 1 - 2\sin t$ sign of $v(t)$:

$a(t) = -2\cos t$ sign of $a(t)$:

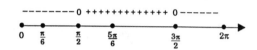

Thus, $0 < t < \dfrac{\pi}{6},$ $\dfrac{\pi}{2} < t < \dfrac{5\pi}{6},$ $\dfrac{3\pi}{2} < t < 2\pi.$

27. $S(r) = 4\pi r^2,$ $dS = 8\pi rh,$ $\dfrac{dS}{S} = \dfrac{8\pi rh}{4\pi r^2} = 2\dfrac{h}{r} = 0.002$

$V(r) = \dfrac{4}{3}\pi r^3,$ $dV = 4\pi r^2 h,$ $\dfrac{dV}{V} = \dfrac{4\pi r^2 h}{\frac{4}{3}\pi r^3} = 3\dfrac{h}{r} = 0.003$

29.

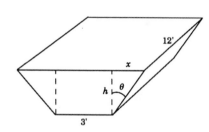

Find $\dfrac{dh}{dt}$ when $h = 2$ ft

given that $\dfrac{dV}{dt} = 10$ ft³/min.

$V = \frac{1}{2}h[3 + (2x + 3)] \cdot 12 = 6h(2x + 6).$

Since $\sin\theta = \dfrac{4}{5},$ we know that $\tan\theta = \dfrac{4}{3} = \dfrac{x}{h}.$ Thus, $x = \dfrac{4h}{3}$ and

$$V = 6h\left(\tfrac{8}{3}h + 6\right) = 36h + 16h^2.$$

Then,

$$\frac{dV}{dt} = 36\frac{dh}{dt} + 32h\frac{dh}{dt}$$

and substituting $dV/dt = 10$ and $h = 2$, we obtain

$$10 = 36\frac{dh}{dt} + 64\frac{dh}{dt}   \text{ so that }   \frac{dh}{dt} = 0.1.$$

The water level is rising at the rate of 0.1 ft/min.

31.

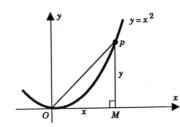

Find $\dfrac{dA}{dt}$ when $x = a$

given that $\dfrac{dx}{dt} = k$ units/sec.

$$A = \frac{1}{2}xy = \frac{1}{2}x^3, \quad \frac{dA}{dt} = \frac{3}{2}x^2\frac{dx}{dt} = \frac{3k}{2}x^2.$$

When $x = a$, $\dfrac{dA}{dt} = \dfrac{3}{2}a^2k$. The area is increasing at the rate of $\dfrac{3}{2}a^2k$ units^2/sec.

33. (a) $\dfrac{dP}{dl} = 0.16l^{-1/2}$. When $l = 9$, $\dfrac{dP}{dl} = \dfrac{16}{300} \cong 0.0533$ sec.

(b) $dP = 0.2\dfrac{dP}{dl} \cong 0.0106$ sec.

35. Since $v_0 = 5$ ft/sec (velocity imparted by rising balloon), the equation of motion is

$$y(t) = -16t^2 + 5t + y_0.$$

The ballast strikes the ground in 8 seconds: $x(8) = 0$. Thus,

$$0 = -16(8)^2 + 5(8) + y_0 \quad \text{so that} \quad y_0 = 984.$$

The ballast was dropped from an altitude of 984 ft.

37. The slope of the isocost line $y = -\dfrac{A}{B}x + \dfrac{1}{B}$ is $-A/B$. The slope of the tangent to the indifference curve is $dy/dx = -1/x^2$. Thus,

$$\frac{A}{B} = \frac{1}{x^2} \quad \text{so that} \quad x = \sqrt{\frac{B}{A}} \quad (x > 0).$$

Then,

$$y = -\frac{A}{B}\sqrt{\frac{B}{A}} + \frac{1}{B} = \frac{1 - \sqrt{AB}}{B}.$$

Finally,

$$C = y - \frac{1}{x} = \frac{1 - \sqrt{AB}}{B} - \sqrt{\frac{A}{B}} = \frac{1 - 2\sqrt{AB}}{B}.$$

At the equilibrium point $x = \sqrt{\dfrac{B}{A}} = \dfrac{\sqrt{AB}}{A}$ and $y = \dfrac{1 - \sqrt{AB}}{B}$.

39. Since the points $(1, 3)$ and $(2, 3)$ lie on the curve,

$$(*) \qquad\qquad 3 = A + B + C \quad \text{and} \quad 3 = 4A + 2B + C.$$

At (2, 3) the derivative $\dfrac{dy}{dx} = 2Ax + B$ is the slope of the line $x - y + 1 = 0$. Therefore

(**) $1 = 4A + B.$

Equations (*) and (**) together give $A = 1,\quad B = -3,\quad C = 5.$

41. $v(t) = -2\sin 2t + 2\sin t = -4\sin t \cos t + 2\sin t = 2\sin t\,(1 - 2\cos t).$
The object changes direction at $t = \pi/3,\ \pi,\ 5\pi/3.$

43. There are two cases: $A = \frac{1}{3}\pi,\quad A = \frac{2}{3}\pi.$

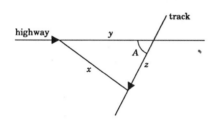

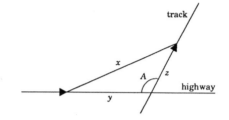

Case 1: $A = \frac{1}{3}\pi$ Case 2: $A = \frac{2}{3}\pi$

We want to find dx/dt when $y = z = \dfrac{500}{5280}$ mi given that $\dfrac{dy}{dt} = -30$ mph and $\dfrac{dz}{dt} = 60$ mph.

By the law of cosines

$$x^2 = y^2 + z^2 - 2yz\cos A,$$

$$2x\frac{dx}{dt} = 2y\frac{dy}{dt} + 2z\frac{dz}{dt} - 2\left(y\frac{dz}{dt} + z\frac{dy}{dt}\right)\cos A.$$

At the given instant

$$x\frac{dx}{dt} = \frac{500}{5280}(-30) + \frac{500}{5280}(60) - \left[\frac{500}{5280}(60) + \frac{500}{5280}(-30)\right]\cos A,$$

$$x\frac{dx}{dt} = \frac{15000}{5280}(1 - \cos A).$$

In case 1

$$\cos A = \frac{1}{2},\qquad\qquad x = \frac{500}{5280},\qquad\qquad \frac{dx}{dt} = 15.$$

In case 2

$$\cos A = -\frac{1}{2},\qquad\qquad x = \frac{500}{5280}\sqrt{3},\qquad\qquad \frac{dx}{dt} = 15\sqrt{3}.$$

Thus, the distance increases 15 mph or $15\sqrt{3}$ mph.

45. The speed of the projection of P onto the x-axis is $|dx/dt|$; the speed of the projection of P
onto the y-axis is $|dy/dt|$. By Exercise 17 of Section 3.11

$$x = r\cos(\omega t + \theta_0),\quad y = r\sin(\omega t + \theta_0).\qquad (\omega,\ \theta_0\quad\text{constant})$$

Differentiation with respect to t gives

$$\frac{dx}{dt} = -r\omega \sin{(\omega t + \theta_0)}, \quad \frac{dy}{dt} = r\omega \cos{(\omega t + \theta_0)}.$$

$|dx/dt|$ is maximal when $\sin{(\omega t + \theta_0)} = \pm 1$ and thus $\cos{(\omega t + \theta_0)} = 0$. Then (x, y) is $(0, r)$ or $(0, -r)$.

$|dy/dt|$ is maximal when $\cos{(\omega t + \theta_0)} = \pm 1$ and thus $\sin{(\omega t + \theta_0)} = 0$. Then (x, y) is $(r, 0)$ or $(-r, 0)$.

CHAPTER 4

SECTION 4.1

1. $f'(c) = 2c$, $\dfrac{f(b) - f(a)}{b - a} = \dfrac{4 - 1}{2 - 1} = 3$; $2c = 3 \implies c = 3/2$

3. $f'(c) = 3c^2$, $\dfrac{f(b) - f(a)}{b - a} = \dfrac{27 - 1}{3 - 1} = 13$; $3c^2 = 13 \implies c = \dfrac{1}{3}\sqrt{39}$ $\left(-\dfrac{1}{3}\sqrt{39} \text{ is not in } [a, b]\right)$

5. $f'(c) = \dfrac{-c}{\sqrt{1 - c^2}}$, $\dfrac{f(b) - f(a)}{b - a} = \dfrac{0 - 1}{1 - 0} = -1$; $\dfrac{-c}{\sqrt{1 - c^2}} = -1 \implies c = \dfrac{1}{2}\sqrt{2}$

    $(-\dfrac{1}{2}\sqrt{2}$ is not in $[a, b])$

7. f is continuous on $[-1, 1]$, differentiable on $(-1, 1)$ and $f(-1) = f(1) = 0$.

    $f'(x) = \dfrac{-x(5 - x^2)}{(3 + x^2)^2\sqrt{1 - x^2}}$, $f'(c) = 0$ for c in $(-1, 1)$ implies $c = 0$.

9. f is everywhere continuous and everywhere differentiable except possibly at $x = -1$.

    f is continuous at $x = -1$: as you can check,

    $$\lim_{x \to -1^-} f(x) = 0, \quad \lim_{x \to -1^+} f(x) = 0, \quad \text{and} \quad f(-1) = 0.$$

    f is differentiable at $x = -1$ and $f'(-1) = 2$: as you can check,

    $$\lim_{h \to 0^-} \dfrac{f(-1 + h) - f(-1)}{h} = 2 \quad \text{and} \quad \lim_{h \to 0^+} \dfrac{f(-1 + h) - f(-1)}{h} = 2.$$

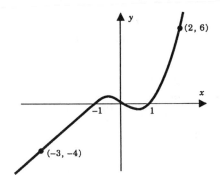

Thus f satisfies the conditions of the mean-value theorem on every closed interval $[a, b]$.

$$f'(x) = \left\{ \begin{array}{ll} 2, & x \le -1 \\ 3x^2 - 1, & x > -1 \end{array} \right\};$$

$$\dfrac{f(2) - f(-3)}{2 - (-3)} = \dfrac{6 - (-4)}{2 - (-3)} = 2.$$

$f'(c) = 2$ with $c \in (-3, 2)$ iff $c = 1$ or $-3 < c \le -1$.

11. $\dfrac{f(1) - f(-1)}{1 - (-1)} = 0$ and $f'(x)$ is never zero. This result does not violate the mean-value theorem

    since f is not differentiable at 0; the theorem does not apply.

13.   Set $P(x) = 6x^4 - 7x + 1$. If there existed three numbers $a < b < c$ at which $P(x) = 0$, then by Rolle's theorem $P'(x)$ would have to be zero for some x in (a, b) and also for some x in (b, c). This is not the case: $P'(x) = 24x^3 - 7$ is zero only at $x = (7/24)^{1/3}$.

15.   Set $P(x) = x^3 + 9x^2 + 33x - 8$. Note that $P(0) < 0$ and $P(1) > 0$. Thus by the intermediate-value theorem there exists some number c between 0 and 1 at which $P(x) = 0$. If the equation $P(x) = 0$ had an additional real root, then by Rolle's theorem there would have to be some real number at which $P'(x) = 0$. This is not the case: $P'(x) = 3x^2 + 18x + 33$ is never zero since the discriminant $b^2 - 4ac = (18)^2 - 12(33) < 0$.

17.   Let c and d be two consecutive roots of the equation $P'(x) = 0$. The equation $P(x) = 0$ cannot have two or more roots between c and d for then by Rolle's theorem $P'(x)$ would have to be zero somewhere between these two roots and thus between c and d. In this case c and d would no longer be consecutive roots of $P'(x) = 0$.

19.   If $x_1 = x_2$, then $|f(x_1) - f(x_2)|$ and $|x_1 - x_2|$ are both 0 and the inequality holds. If $x_1 \neq x_2$, then you know by the mean-value theorem that

$$\frac{f(x_1) - f(x_2)}{x_1 - x_2} = f'(c)$$

for some number c between x_1 and x_2. Since $|f'(c)| \leq 1$, you can conclude that

$$\left| \frac{f(x_1) - f(x_2)}{x_1 - x_2} \right| \leq 1 \quad \text{and thus that} \quad |f(x_1) - f(x_2)| \leq |x_1 - x_2|.$$

21.   Set, for instance, $f(x) = \begin{cases} 1, & a < x < b \\ 0, & x = a, b \end{cases}$.

23.   (a)   Between any two times that the object is at the origin there is at least one instant when the velocity is zero.

   (b)   During any time interval there is at least one instant when the instantaneous velocity equals the average velocity over that interval.

25.
$$f'(x_0) = \lim_{y \to 0} \frac{f(x_0 + y) - f(x_0)}{y} = \lim_{y \to 0} \frac{f'(x_0 + \theta y)y}{y} = \lim_{y \to 0} f'(x_0 + \theta y)$$
$$\underset{\text{(by the hint)}}{}$$

$$= \lim_{x \to x_0} f'(x) = L$$
$$\underset{\text{(by 2.2.5)}}{}$$

27.   2.99

SECTION 4.2

1. $f'(x) = 3x^2 - 3 = 3\left(x^2 - 1\right) = 3(x + 1)(x - 1)$

   f increases on $(-\infty, -1]$ and $[1, \infty)$, decreases on $[-1, 1]$

3. $f'(x) = 1 - \dfrac{1}{x^2} = \dfrac{x^2 - 1}{x^2} = \dfrac{(x+1)(x-1)}{x^2}$

    f increases on $(-\infty, -1]$ and $[1, \infty)$, decreases on $[-1, 0)$ and $(0, 1]$ (f is not defined at 0)

5. $f'(x) = 3x^2 + 4x^3 = x^2(3 + 4x)$

    f increases on $\left[-\frac{3}{4}, \infty\right)$,   decreases on $\left(-\infty, -\frac{3}{4}\right]$

7. $f'(x) = 4(x+1)^3$

    f increases on $[-1, \infty)$,   decreases on $(-\infty, -1]$

9. $f(x) = \begin{cases} \dfrac{1}{2-x}, & x < 2 \\ \dfrac{1}{x-2}, & x > 2 \end{cases}$ $f'(x) = \begin{cases} \dfrac{1}{(2-x)^2}, & x < 2 \\ \dfrac{-1}{(x-2)^2}, & x > 2 \end{cases}$

    f increases on $(-\infty, 2)$,   decreases on $(2, \infty)$ (f is not defined at 2)

11. $f'(x) = -\dfrac{4x}{(x^2 - 1)^2}$

    f increases on $(-\infty, -1)$ and $(-1, 0]$, decreases on $[0, 1)$ and $(1, \infty)$ (f is not defined at ± 1)

13. $f(x) = \begin{cases} x^2 - 5, & x < -\sqrt{5} \\ -(x^2 - 5), & -\sqrt{5} \le x \le \sqrt{5} \\ x^2 - 5, & \sqrt{5} < x \end{cases}$, $f'(x) = \begin{cases} 2x, & x < -\sqrt{5} \\ -2x, & -\sqrt{5} < x < \sqrt{5} \\ 2x, & \sqrt{5} < x \end{cases}$

    f increases on $[-\sqrt{5}, 0]$ and $[\sqrt{5}, \infty)$, decreases on $(-\infty, -\sqrt{5}]$ and $[0, \sqrt{5}]$

15. $f'(x) = \dfrac{2}{(x+1)^2}$

    f increases on $(-\infty, -1)$ and $(-1, \infty)$ (f is not defined at -1)

17. $f'(x) = -\dfrac{7(1 - \sqrt{x})^6}{\sqrt{x}(1 + \sqrt{x})^8}$ 19. $f'(x) = \dfrac{x}{(2 + x^2)^2} \sqrt{\dfrac{2 + x^2}{1 + x^2}}$

    f decreases on $[0, \infty)$ f increases on $[0, \infty)$

    decreases on $(-\infty, 0]$

21. $f'(x) = \dfrac{-3}{2x^2} \sqrt{\dfrac{x}{3-x}}$ 23. $f'(x) = 1 + \sin x$

    f decreases on $(0, 3]$ f increases on $[0, 2\pi]$

25. $f'(x) = -2 \sin 2x - 2 \sin x = -2 \sin x (2 \cos x + 1)$

    f increases on $\left[\frac{2}{3}\pi, \pi\right]$,   decreases on $\left[0, \frac{2}{3}\pi\right]$

27. $f'(x) = \sqrt{3} + 2 \sin 2x$

    f increases on $\left[0, \frac{2}{3}\pi\right]$ and $\left[\frac{5}{6}\pi, \pi\right]$, decreases on $\left[\frac{2}{3}\pi, \frac{5}{6}\pi\right]$

29. $\dfrac{d}{dx}\left(\dfrac{x^3}{3}-x\right)=f'(x) \quad\Longrightarrow\quad f(x)=\dfrac{x^3}{3}-x+C$

$f(1)=2 \quad\Longrightarrow\quad 2=\frac{1}{3}-1+C,$ so $C=\frac{8}{3}.$ Thus, $f(x)=\frac{1}{3}x^3-x+\frac{8}{3}.$

31. $\dfrac{d}{dx}\left(x^5+x^4+x^3+x^2+x\right)=f'(x) \quad\Longrightarrow\quad f(x)=x^5+x^4+x^3+x+C$

$f(0)=5 \quad\Longrightarrow\quad 5=0+C,$ so $C=5.$ Thus, $f(x)=x^5+x^4+x^3+x^2+x+5.$

33. $\dfrac{d}{dx}\left(\dfrac{3}{4}x^{4/3}-\dfrac{2}{3}x^{3/2}\right)=f'(x) \quad\Longrightarrow\quad f(x)=\dfrac{3}{4}x^{4/3}-\dfrac{2}{3}x^{3/2}+C$

$f(0)=1 \quad\Longrightarrow\quad 1=0+C,$ so $C=1.$ Thus, $f(x)=\frac{3}{4}x^{4/3}-\frac{2}{3}x^{3/2}+1,\ x\ge 0.$

35. $\dfrac{d}{dx}\left(2x-\cos x\right)=f'(x) \quad\Longrightarrow\quad f(x)=2x-\cos x+C$

$f(0)=3 \quad\Longrightarrow\quad 3=0-1+C,$ so $C=4.$ Thus, $f(x)=2x-\cos x+4.$

37. $f'(x)=\begin{cases} 1, & x<-3 \\ -1, & -3<x<-1 \\ 1, & -1<x<1 \\ -2, & 1<x \end{cases}$

f increases on $(-\infty,-3)$ and $[-1,1],$

decreases on $[-3,-1]$ and $[1,\infty)$

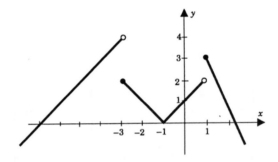

39. $f'(x)=\begin{cases} -2x, & x<1 \\ -2, & 1<x<3 \\ 3, & 3<x \end{cases}$

f increases on $(-\infty,0]$ and $[3,\infty),$

decreases on $[0,1)$ and $[1,3]$

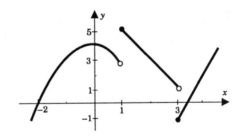

41. (a) $M\le L\le N$ (b) none   (c) $M=L=N$

43. Set, for instance, $f(x)=\begin{cases} 1, & x\ \text{rational} \\ 0, & x\ \text{irrational} \end{cases}.$

45. Set $h(x)=f(x)-g(x).$ Since $h'(x)>0$ for all $x,$ h increases everywhere. Since $h(0)=0,$

$$h(x)<0 \text{ on } (-\infty,0) \quad \text{and} \quad h(x)>0 \text{ on } (0,\infty).$$

It follows that

(a) $f(x) < g(x)$ on $(-\infty, 0)$ and   (b) $f(x) > g(x)$ on $(0, \infty)$.

SECTION 4.3

1. $f'(x) = 3x^2 + 3 > 0;$ no critical pts, no local extreme values

3. $f'(x) = 1 - \dfrac{1}{x^2}$

critical pts $-1, 1$

$f''(x) = \dfrac{2}{x^3},\ f''(-1) = -2,\ f''(1) = 2$

$f(-1) = -2$ local max, $f(1) = 2$ local min

5. $f'(x) = 2x - 3x^2 = x(2 - 3x)$

critical pts $0, \frac{2}{3}$

$f''(x) = 2 - 6x,\ f''(0) = 2,$

$f''\left(\frac{2}{3}\right) = -2$

$f(0) = 0$ local min,

$f\left(\frac{2}{3}\right) = \frac{4}{27}$ local max

7. $f'(x) = \dfrac{2}{(1-x)^2};$ no critical pts, no local extreme values

9. $f'(x) = -\dfrac{2(2x+1)}{x^2(x+1)^2};$ critical pt $-\dfrac{1}{2}$

$f\left(-\frac{1}{2}\right) = -8$ local max

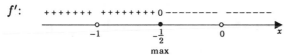

11. $f'(x) = x^2(5x - 3)(x - 1);$ critical pts $0, \frac{3}{5}, 1$

$f\left(\dfrac{3}{5}\right) = \dfrac{2^2 3^3}{5^5}$ local max

$f(1) = 0$ local min

no local extreme at 0

13. $f'(x) = (5 - 8x)(x - 1)^2;$ critical pts $\frac{5}{8}, 1$

$f\left(\frac{5}{8}\right) = \frac{27}{2048}$ local max

no local extreme at 1

15. $f'(x) = \dfrac{x(2+x)}{(1+x)^2}$;   critical pts $-2, 0$

$f(-2) = -4$ local max

$f(0) = 0$ local min

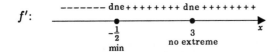

$f':$ $+++++++0 \ --------- \ --------- \ 0+++++++$ x

-2 max -1 0 min

17. $f'(x) = \begin{cases} 2x+1, & x < -2, x > 1 \\ -(2x+1), & -2 < x < 1 \end{cases}$;   critical pts $-2, -\tfrac{1}{2}, 1$

$f(-2) = 0$ local min

$f\left(-\tfrac{1}{2}\right) = \tfrac{9}{4}$ local max

$f(1) = 0$ local min

$f':$ $----- \ dne ++++++0 \ ------- \ dne +++++$ x

-2 min $-\tfrac{1}{2}$ max 1 min

19. $f'(x) = \tfrac{1}{3}x(7x+12)(x+2)^{-2/3}$;   critical pts $-2, -\tfrac{12}{7}, 0$

$f\left(-\tfrac{12}{7}\right) = \tfrac{144}{49}\left(\tfrac{2}{7}\right)^{1/3}$ local max

$f(0) = 0$ local min

$f':$ $++++++++ \ dne +++ 0 \ ----------- \ 0++++$ x

-2 no extreme $-\tfrac{12}{7}$ max 0 min

21. $f(x) = \begin{cases} 2-3x, & x \le -\tfrac{1}{2} \\ x+4, & -\tfrac{1}{2} < x < 3 \\ 3x-2, & 3 \le x \end{cases}$, $f'(x) = \begin{cases} -3, & x < -\tfrac{1}{2} \\ 1, & -\tfrac{1}{2} < x < 3 \\ 3, & 3 < x \end{cases}$;

critical pts $-\tfrac{1}{2}, 3$

$f\left(-\tfrac{1}{2}\right) = \tfrac{7}{2}$ local min

no local extreme at 3

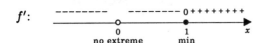

$f':$ $------- \ dne +++++++++ \ dne ++++++++$ x

$-\tfrac{1}{2}$ min 3 no extreme

23. $f'(x) = \tfrac{2}{3}x^{-4/3}(x-1)$;   critical pts $0, -1$

$f(1) = 3$ local min

no local extreme at 0

$f':$ $-------- \ -------- \ 0++++++++$ x

0 no extreme 1 min

25. $f'(x) = \cos x - \sin x$;   critical pts $\tfrac{1}{4}\pi, \tfrac{5}{4}\pi$

$f''(x) = -\sin x - \cos x$, $f''\left(\tfrac{1}{4}\pi\right) = -\sqrt{2}$, $f''\left(\tfrac{5}{4}\pi\right) = \sqrt{2}$

$f\left(\tfrac{1}{4}\pi\right) = \sqrt{2}$ local max, $f\left(\tfrac{5}{4}\pi\right) = -\sqrt{2}$ local min

27. $f'(x) = \cos x \left(2 \sin x - \sqrt{3}\right)$;   critical pts $\frac{1}{2}\pi$, $\frac{1}{3}\pi$, $\frac{2}{3}\pi$

$f\left(\frac{1}{3}\pi\right) = f\left(\frac{2}{3}\pi\right) = -\frac{3}{4}$ local mins

f':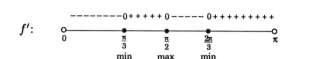

$f\left(\frac{1}{2}\pi\right) = 1 - \sqrt{3}$ local max

29. $f'(x) = \cos^2 x - \sin^2 x - 3 \cos x + 2 = (2\cos x - 1)(\cos x - 1)$ critical pts $\frac{1}{3}\pi$, $\frac{5}{3}\pi$

$f\left(\frac{1}{3}\pi\right) = \frac{2}{3}\pi - \frac{5}{4}\sqrt{3}$ local min

f':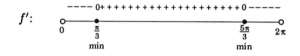

$f\left(\frac{5}{3}\pi\right) = \frac{10}{3}\pi + \frac{5}{4}\sqrt{3}$ local max

31.   (i) f increases on $(c - \delta, c]$ and decreases on $[c, c + \delta)$.

      (ii) f decreases on $(c - \delta, c]$ and increases on $[c, c + \delta)$.

      (iii)   If $f'(x) > 0$ on $(c - \delta, c) \cup (c, c + \delta)$, then, since f is continuous at c, f increases on $(c - \delta, c]$ and also on $[c, c + \delta)$. Therefore, in this case, f increases on $(c - \delta, c + \delta)$. A similar argument shows that, if $f'(x) < 0$ on $(c - \delta, c) \cup (c, c + \delta)$, then f decreases on $(c - \delta, c + \delta)$.

33.

$$P(x) = x^4 - 8x^3 + 22x^2 - 24x + 4$$

$$P'(x) = 4x^3 - 24x^2 + 44x - 24$$

$$P''(x) = 12x^2 - 48x + 44$$

Since obviously $P'(1) = 0$, $P'(x)$ is divisible by $x - 1$. Division by $x - 1$ gives

$$P'(x) = (x - 1)\left(4x^2 - 20x + 24\right) = 4(x - 1)(x - 2)(x - 3).$$

The critical pts are 1, 2, 3. Since

$$P''(1) > 0, \quad P''(2) < 0, \quad P''(3) > 0,$$

$P(1) = -5$ is a local min, $P(2) = -4$ is a local max, and $P(3) = -5$ is a local min.

Since $P'(x) < 0$ for $x < 0$, P decreases on $(-\infty, 0]$. Since $P(0) > 0$, P does not take on the value 0 on $(-\infty, 0]$.

Since $P(0) > 0$ and $P(1) < 0$, P takes on the value 0 at least once on $(0, 1)$. Since $P'(x) < 0$ on $(0, 1)$, P decreases on $[0, 1]$. It follows that P takes on the value zero only once on $[0, 1]$.

Since $P'(x) > 0$ on $(1, 2)$ and $P'(x) < 0$ on $(2, 3)$, P increases on $[1, 2]$ and decreases on $[2, 3]$. Since $P(1)$, $P(2)$, $P(3)$ are all negative, P cannot take on the value 0 between 1 and 3.

Since $P(3) < 0$ and $P(100) > 0$, P takes on the value 0 at least once on $(3, 100)$. Since $P'(x) > 0$ on $(3, 100)$, P increases on $[3, 100]$. It follows that P takes on the value zero only once on $[3, 100]$.

Since $P'(x) > 0$ on $(100, \infty)$, P increases on $[100, \infty)$. Since $P(100) > 0$, P does not take on the value 0 on $[100, \infty)$.

SECTION 4.4

1. $f'(x) = \frac{1}{2}(x+2)^{-1/2}, \ x > -2;$ $f':$

    critical pt $-2;$

    $f(-2) = 0$ endpt and abs min;   as $x \to \infty, \ f(x) \to \infty;$ so no abs max

3. $f'(x) = 2x - 4, \ x \in (0,3);$ $f':$

    critical pts $0, 2, 3;$

    $f(0) = 1$ endpt and abs max, $f(2) = -3$ local and abs min, $f(3) = -2$ endpt max

5. $f'(x) = 2x - \dfrac{1}{x^2} = \dfrac{2x^3 - 1}{x^2}, \ x \ne 0; \quad f'(x) = 0$ at $x = 2^{-1/3}$

    critical pt $2^{-1/3}; \quad f''(x) = 2 + \dfrac{2}{x^3}, \ f''(2^{-1/3}) = 6$

    $f(2^{-1/3}) = 2^{-2/3} + 2^{1/3} = 2^{-2/3} + 2 \cdot 2^{-2/3} = 3 \cdot 2^{-2/3}$ local min

7. $f'(x) = \dfrac{2x^3 - 1}{x^2}, \ x \in \left(\dfrac{1}{10}, 2\right);$ $f':$

    critical pts $\frac{1}{10}, 2^{-1/3}, 2;$

    $f(\frac{1}{10}) = 10\frac{1}{100}$ endpt and abs max, $f(2^{-1/3}) = 3 \cdot 2^{-2/3}$ local and abs min,

    $f(2) = 4\frac{1}{2}$ endpt max

9. $f'(x) = 2x - 3, \ x \in (0,2);$ $f':$

    critical pts $0, \frac{3}{2}, 2;$

    $f(0) = 2$ endpt and abs max, $f(\frac{3}{2}) = -\frac{1}{4}$ local and abs min,

    $f(2) = 0$ endpt max

11. $f'(x) = \dfrac{(2-x)(2+x)}{(4+x^2)^2}, \ x \in (-3,1);$ $f':$

    critical pts $-3, -2, 1;$

    $f(-3) = -\frac{3}{13}$ endpt max, $f(-2) = -\frac{1}{4}$ local and abs min,

    $f(1) = \frac{1}{5}$ endpt and abs max

13. $f'(x) = 2(x - \sqrt{x})\left(1 - \dfrac{1}{2\sqrt{x}}\right), \ x > 0;$ $f':$

    critical pts $0, \frac{1}{4}, 1;$

$f(0) = 0$ endpt and abs min, $f\left(\frac{1}{4}\right) = \frac{1}{16}$ local max, $f(1) = 0$ local and abs min;

as $x \to \infty$, $f(x) \to \infty$;   so no abs max

15. $f'(x) = \dfrac{3(2 - x)}{2\sqrt{3 - x}},\quad x < 3$ f' :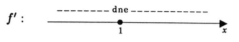

critical pts 2, 3 ;

$f(2) = 2$ local and abs max, $f(3) = 0$ endpt min;

as $x \to -\infty$, $f(x) \to -\infty$;   so no abs min

17. $f'(x) = -\frac{1}{3}(x - 1)^{-2/3},\quad x \neq 1$; f' :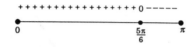

critical pt 1 ;

no local extremes; $\left.\begin{array}{l} \text{as} \quad x \to \infty, \quad f(x) \to -\infty \\ \text{as} \quad x \to -\infty, \quad f(x) \to \infty \end{array}\right\}$ no abs extremes

19. $f'(x) = \sin x \left(2 \cos x + \sqrt{3}\right), x \in (0, \pi)$; f' :

critical pts 0, $\frac{5}{6}\pi$, π ;

$f(0) = -\sqrt{3}$ endpt and abs min, $f\left(\frac{5}{6}\pi\right) = \frac{7}{4}$ local and abs max,

$f(\pi) = \sqrt{3}$ endpt min

21. $f'(x) = -3 \sin x \left(2 \cos^2 x + 1\right) < 0,\quad x \in (0, \pi)$;   critical pts 0, π ;

$f(0) = 5$ endpt and abs max, $f(\pi) = -5$ endpt and abs min

23. $f'(x) = \sec^2 x - 1 \geq 0,\quad x \in \left(-\frac{1}{3}\pi, \frac{1}{2}\pi\right)$;   critical pts $-\frac{1}{3}\pi$, 0;

$f\left(-\frac{1}{3}\pi\right) = \frac{1}{3}\pi - \sqrt{3}$ endpt and abs min, no abs max

25.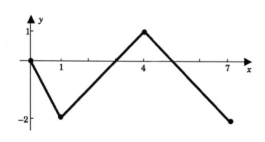

$f'(x) = \begin{cases} -2, \, 0 < x < 1 \\ 1, \, 1 < x < 4 \\ -1, \, 4 < x < 7 \end{cases}$

critical pts 0, 1, 4, 7

$f(0) = 0$ endpt max, $f(1) = -2$ local and abs min,

$f(4) = 1$ local and absolute max, $f(7) = -2$ endpt and abs min

27.

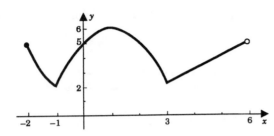

$$f'(x) = \begin{cases} 2x, & -2 < x < -1 \\ 2 - 2x, & -1 < x < 3 \\ 1, & 3 < x < 6 \end{cases}$$

critical pts $-2, -1, 1, 3$

$f(-2) = 5$ endpt max, $f(-1) = 2$ local and abs min,

$f(1) = 6$ local and abs max, $f(3) = 2$ local and abs min

29.

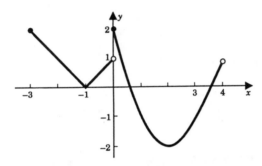

$$f'(x) = \begin{cases} -1, & -3 < x < -1 \\ 1, & -1 < x < 0 \\ 2x - 4, & 0 < x < 3 \\ 2, & 3 \le x < 4 \end{cases}$$

critical pts $-3, -1, 0, 2$

$f(-3) = 2$ endpt and abs max, $f(-1) = 0$ local min,

$f(0) = 2$ local and abs max, $f(2) = -2$ local and abs min

31.   By contradiction. If f is continuous at c, then, by the first-derivative test (4.3.3), $f(c)$ is not a local maximum.

33.   If f is not differentiable on (a, b), then f has a critical point at each point c in (a, b) where $f'(c)$ does not exist. If f is differentiable on (a, b), then by the mean-value theorem there exists c in (a, b) where $f'(c) = [f(b) - f(a)]/(b - a) = 0$. This means c is a critical point of f.

35.
$$P(x) - M \ge a_0 x^n - \left(|a_1| x^{n-1} + \cdots + |a_{n-1}| x + |a_n| + M\right)$$
for $x > 0$

$$\ge a_0 x^n - \left(|a_1| + \cdots + |a_{n-1}| + |a_n| + M\right)$$
for $x > 1$

$$\ge 0 \quad \text{for} \quad x \ge \left(\frac{|a_1| + \cdots + |a_{n-1}| + |a_n| + M}{a_0}\right)^{1/n} + 1.$$

SECTION 4.5

1. Set $P = xy$ and $y = 40 - x$. We want to maximize

$$P(x) = x(40 - x), \quad 0 < x < 40. \quad \text{(key step completed)}$$

$$P'(x) = 40 - 2x, \quad P'(x) = 0 \implies x = 20.$$

Since P increases on $(0, 20]$ and decreases on $[20, 40)$, the abs max of P occurs when $x = 20$. Then, $y = 20$ and $xy = 400$.

The maximal value of xy is 400.

3. 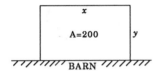   <u>Minimize P</u>

$$P = x + 2y, \quad 200 = xy, \quad y = 200/x$$

$$P(x) = x + \frac{400}{x}, \quad 0 < x. \quad \text{(key step completed)}$$

$$P'(x) = 1 - \frac{400}{x^2}, \quad P'(x) = 0 \implies x = 20.$$

Since P decreases on $(0, 20]$ and increases on $[20, \infty)$, the abs min of P occurs when $x = 20$.

To minimize the fencing, make the garden 20 ft (parallel to barn) by 10 ft.

5. 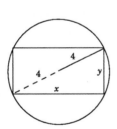   <u>Maximize A</u>

$$A = xy, \quad x^2 + y^2 = 8^2, \quad y = \sqrt{64 - x^2}$$

$$A(x) = x\sqrt{64 - x^2}, \quad 0 < x < 8. \quad \text{(key step completed)}$$

$$A'(x) = \sqrt{64 - x^2} + x\left(\frac{-x}{\sqrt{64 - x^2}}\right) = \frac{64 - 2x^2}{\sqrt{64 - x^2}}, \quad A'(x) = 0 \implies x = 4\sqrt{2}.$$

Since A increases on $(0, 4\sqrt{2}]$ and decreases on $[4\sqrt{2}, 8)$, the abs max of A occurs when $x = 4\sqrt{2}$. Then, $y = 4\sqrt{2}$ and $xy = 32$.

The maximal area is 32.

7.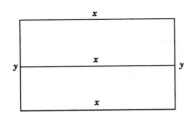

Maximize A

$$A = xy, \quad 2y + 3x = 600, \quad y = \frac{600 - 3x}{2}$$

$A(x) = x\left(300 - \frac{3}{2}x\right), \quad 0 < x < 200.$ (key step completed)

$A'(x) = 300 - 3x, \quad A'(x) = 0 \implies x = 100.$

Since A increases on $(0, 100]$ and decreases on $[100, 200)$, the abs max of A occurs when $x = 100$. Then, $y = 150$.

The playground of greatest area measures 100 ft by 150 ft. (The fence divider is 100 ft long.)

9.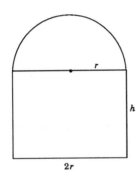

Maximize L

To account for the semi-circular portion admitting less light per square foot, we multiply its area by 1/3.

$$L = 2rh + \frac{1}{3}\left(\frac{\pi r^2}{2}\right),$$

$$2r + 2h + \pi r = 30, \quad h = \frac{1}{2}(30 - 2r - \pi r)$$

$$L = 2r\left(\frac{30 - 2r - \pi r}{2}\right) + \frac{1}{6}\pi r^2$$

$L(r) = 30r - \left(2 + \frac{5}{6}\pi\right)r^2, \quad 0 < r < \frac{30}{2 + \pi}.$ (key step completed)

$L'(r) = 30 - \left(4 + \frac{5}{3}\pi\right)r, \quad L'(r) = 0 \implies r = \frac{90}{12 + 5\pi}.$

Since $L''(r) < 0$ for all r in the domain of L, the local max at $r = 90/(12 + 5\pi)$ is the abs max.

For the window that admits the most light, take the radius of the semicircle as $\frac{90}{12 + 5\pi} \cong 3.25$ ft and the height of the rectangular portion as $\frac{90 + 30\pi}{12 + 5\pi} \cong 6.65$ ft.

11.

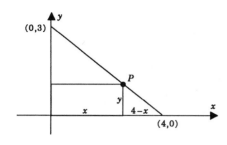

Maximize A

$$A = xy, \quad \frac{3}{4} = \frac{y}{4-x} \quad \text{(similar triangles)}$$

$$y = \tfrac{3}{4}(4-x)$$

$$A(x) = \frac{3x}{4}(4-x), \quad 0 < x < 4. \qquad \text{(key step completed)}$$

$$A'(x) = 3 - \frac{3x}{2}, \quad A'(x) = 0 \implies x = 2.$$

Since A increases on $(0, 2]$ and decreases on $[2, 4)$, the abs max of A occurs when $x = 2$.

To maximize the area of the rectangle, take P as the point $\left(2, \frac{3}{2}\right)$.

13.

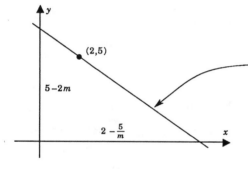

Minimize A

$$A = \tfrac{1}{2}(x\text{-intercept})\,(y\text{-intercept})$$

Equation of line: $\quad y - 5 = m(x - 2)$

x-intercept: $\quad 2 - \dfrac{5}{m}$

y-intercept: $\quad 5 - 2m$

$$A = \frac{1}{2}\left(2 - \frac{5}{m}\right)(5 - 2m) = 10 - 2m - \frac{25}{2m}$$

$$A(m) = 10 - 2m - \frac{25}{2m}, \quad m < 0. \qquad \text{(key step completed)}$$

$$A'(m) = -2 + \frac{25}{2m^2}, \quad A'(m) = 0 \implies m = -\frac{5}{2}.$$

Since $A''(m) = -25/m^3 > 0$ for $m < 0$, the local min at $m = -5/2$ is the abs min.

The triangle of minimal area is formed by the line of slope $-5/2$.

15.

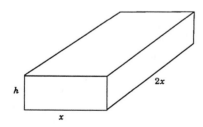

Maximize V

$$V = 2x^2 h, \quad 2\left(2x^2 + xh + 2xh\right) = 100, \quad h = \frac{50 - 2x^2}{3x}$$

$$V = 2x^2 \left(\frac{50 - 2x^2}{3x}\right)$$

$$V(x) = \tfrac{100}{3}x - \tfrac{4}{3}x^3, \quad 0 < x < 5. \qquad \text{(key step completed)}$$

$V'(x) = \frac{100}{3} - 4x^2$, $V'(x) = 0 \implies x = \frac{5}{3}\sqrt{3}$.

Since $V''(x) = -8x < 0$ on $(0,5)$, the local max at $x = \frac{5}{3}\sqrt{3}$ is the abs max.

The base of the box of greatest volume measures $\frac{5}{3}\sqrt{3}$ in. by $\frac{10}{3}\sqrt{3}$ in.

17.

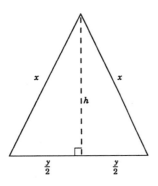

Maximize A

$A = \frac{1}{2}hy$

$2x + y = 12 \implies y = 12 - 2x$

Pythagorean Theorem:

$$h^2 + \left(\frac{y}{2}\right)^2 = x^2 \implies h = \sqrt{x^2 - \left(\frac{y}{2}\right)^2}$$

Thus, $h = \sqrt{x^2 - (6-x)^2} = \sqrt{12x - 36}$.

$A(x) = (6-x)\sqrt{12x - 36}$, $3 < x < 6$.   (key step completed)

$A'(x) = -\sqrt{12x - 36} + (6-x)\left(\dfrac{6}{\sqrt{12x-36}}\right) = \dfrac{72 - 18x}{\sqrt{12x-36}}$,

$A'(x) = 0 \implies x = 4$.

Since A increases on $(3,4]$ and decreases on $[4,6)$, the abs max of A occurs at $x = 4$.

The triangle of maximal area is equilateral with side of length 4.

19.

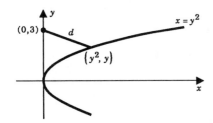

Minimize d

$$d = \sqrt{(y^2 - 0)^2 + (y-3)^2}$$

The square-root function is increasing;

d is minimal when $D = d^2$ is minimal.

$D(y) = y^4 + (y-3)^2$, y real.   (key step completed)

$D'(y) = 4y^3 + 2(y-3) = (y-1)\left(4y^2 + 4y + 6\right)$, $D'(y) = 0$ at $y = 1$.

Since $D''(y) = 12y^2 + 2 > 0$, the local min at $y = 1$ is the abs min.

The point $(1,1)$ is the point on the parabola closest to $(0,3)$.

21.

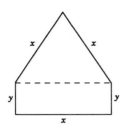

<u>Maximize A</u>

$$A = xy + \frac{\sqrt{3}}{4}x^2, \quad 30 = 3x + 2y, \quad y = \frac{30 - 3x}{2}$$

$$A(x) = 15x - \frac{3}{2}x^2 + \frac{\sqrt{3}}{4}x^2, \quad 0 < x < 10. \qquad \text{(key step completed)}$$

$$A'(x) = 15 - 3x + \frac{\sqrt{3}}{2}x, \quad A'(x) = 0 \quad \Longrightarrow \quad x = \frac{30}{6 - \sqrt{3}} = \frac{10}{11}\left(6 + \sqrt{3}\right).$$

Since $A''(x) = -3 + \frac{\sqrt{3}}{2} < 0$ on $(0, 10)$, the local max at $x = \frac{10}{11}\left(6 + \sqrt{3}\right)$ is the abs max.

The pentagon of greatest area is composed of an equilateral triangle with side $\frac{10}{11}\left(6 + \sqrt{3}\right) \cong 7.03$ in. and rectangle with height $\frac{15}{11}\left(5 - \sqrt{3}\right) \cong 4.46$ in.

23.

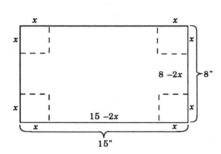

<u>Maximize V</u>

$$V = x(8 - 2x)(15 - 2x)$$

$$\begin{bmatrix} x > 0 \\ 8 - 2x > 0 \\ 15 - 2x > 0 \end{bmatrix} \Longrightarrow \quad 0 < x < 4$$

$$V(x) = 120x - 46x^2 + 4x^3, \quad 0 < x < 4. \qquad \text{(key step completed)}$$

$$V'(x) = 120 - 92x + 12x^2 = 4(3x - 5)(x - 6), \quad V'(x) = 0 \text{ at } x = \tfrac{5}{3}.$$

Since V increases on $\left(0, \frac{5}{3}\right]$ and decreases on $\left[\frac{5}{3}, 4\right)$, the abs max of V occurs when $x = \frac{5}{3}$.

The box of maximal volume is made by cutting out squares 5/3 inches on a side.

25.

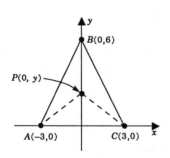

<u>Minimize $\overline{AP} + \overline{BP} + \overline{CP} = S$</u>

length $AP = \sqrt{9 + y^2}$

length $BP = 6 - y$

length $CP = \sqrt{9 + y^2}$

$$S(y) = 6 - y + 2\sqrt{9 + y^2}, \quad 0 \le y \le 6. \qquad \text{(key step completed)}$$

$$S'(y) = -1 + \frac{2y}{\sqrt{9 + y^2}}, \quad S'(y) = 0 \quad \Longrightarrow \quad y = \sqrt{3}.$$

Since

$$S(0) = 12, \quad S\left(\sqrt{3}\right) = 6 + 3\sqrt{3} \cong 11.2, \quad \text{and} \quad S(6) = 6\sqrt{5} \cong 13.4,$$

the abs min of S occurs when $y = \sqrt{3}$.

To minimize the sum of the distances, take P as the point $\left(0, \sqrt{3}\right)$.

27.

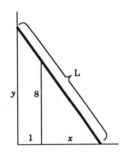

Minimize L

$L^2 = y^2 + (x+1)^2$.

By similar triangles $\dfrac{y}{x+1} = \dfrac{8}{x}, \quad y = \dfrac{8}{x}(x+1)$.

$L^2 = \left[\left(\dfrac{8}{x}\right)(x+1)\right]^2 + (x+1)^2 = (x+1)^2\left(\dfrac{64}{x^2}+1\right)$

Since L is minimal when L^2 is minimal, we consider the function

$$f(x) = (x+1)^2\left(\frac{64}{x^2}+1\right), \quad x > 0. \qquad \text{(key step completed)}$$

$$f'(x) = 2(x+1)\left(\frac{64}{x^2}+1\right) + (x+1)^2\left(\frac{-128}{x^3}\right)$$

$$= \frac{2(x+1)}{x^3}\left[x^3 - 64\right], \quad f'(x) = 0 \implies x = 4.$$

Since f decreases on $(0, 4]$ and increases on $[4, \infty)$, the abs min of f occurs when $x = 4$.

The shortest ladder is $5\sqrt{5}$ ft long.

29.

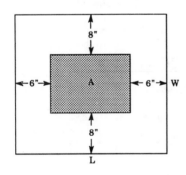

Maximize A

(We use feet rather than inches to reduce arithmetic.)

$A = (L-1)\left(W - \frac{4}{3}\right)$

$LW = 27 \implies W = \dfrac{27}{L}$

$A = (L-1)\left(\dfrac{27}{L} - \dfrac{4}{3}\right) = \dfrac{85}{3} - \dfrac{27}{L} - \dfrac{4}{3}L$

$$A(L) = \frac{85}{3} - \frac{27}{L} - \frac{4}{3}L, \quad 1 < L < \frac{81}{4}. \qquad \text{(key step completed)}$$

$$A'(L) = \frac{27}{L^2} - \frac{4}{3}, \quad A'(L) = 0 \implies L = \frac{9}{2}.$$

Since $A'(L) = -54/L^3 < 0$ for $1 < L < \frac{81}{4}$, the max at $L = \frac{9}{2}$ is the abs max.

The banner has length $9/2$ ft $= 54$ in. and height 6 ft $= 72$ in.

31.

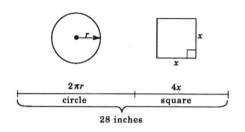

Find the extreme values of A

$$A = \pi r^2 + x^2$$

$$2\pi r + 4x = 28 \quad \Longrightarrow \quad x = 7 - \tfrac{1}{2}\pi r.$$

$$A(r) = \pi r^2 + \left(7 - \frac{1}{2}\pi r\right)^2, \quad 0 \le r \le \frac{14}{\pi}. \qquad \text{(key step completed)}$$

Note: the endpoints of the domain correspond to the instances when the string is not cut: $r = 0$ when no circle is formed, $r = 14/\pi$ when no square is formed.

$$A'(r) = 2\pi r - \pi \left(7 - \frac{1}{2}\pi r\right), \quad A'(r) = 0 \quad \Longrightarrow \quad r = \frac{14}{4 + \pi}.$$

Since $A''(r) = 2\pi + \pi^2/2 > 0$ on $(0, 14/\pi)$, the abs min of A occurs when $r = 14/(4 + \pi)$ and the abs max of A occurs at one of the endpts: $A(0) = 49$, $A(14/\pi) = 196/\pi > 49$.

(a) To maximize the sum of the two areas, use all of the string to form the circle.

(b) To minimize the sum of the two areas, use $2\pi r = 28\pi/(4 + \pi) \cong 12.32$ inches of string for the circle.

33.

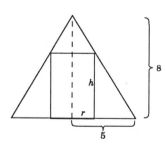

Maximize V

$$V = \pi r^2 h$$

By similar triangles

$$\frac{8}{5} = \frac{h}{5 - r} \quad \text{or} \quad h = \frac{8}{5}(5 - r).$$

$$V(r) = \frac{8\pi}{5} r^2 (5 - r), \quad 0 < r < 5. \qquad \text{(key step completed)}$$

$$V'(r) = \frac{8\pi}{5}\left(10r - 3r^2\right), \quad V'(r) = 0 \quad \Longrightarrow \quad r = 10/3.$$

Since V increases on $(0, 10/3]$ and decreases on $[10/3, 5)$, the abs max of V occurs when $r = 10/3$.

The cylinder with maximal volume has radius $10/3$ and height $8/3$.

35.

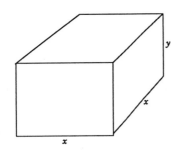

<u>Minimize C</u>

In dollars,

$C =$ cost base $+$ cost top $+$ cost sides

$\quad = .35\left(x^2\right) + .15\left(x^2\right) + .20(4xy)$

$\quad = \frac{1}{2}x^2 + \frac{4}{5}xy$

Volume $= x^2 y = 1250 \quad y = \dfrac{1250}{x^2}$

$C(x) = \dfrac{1}{2}x^2 + \dfrac{1000}{x}, \quad x > 0.$ (key step completed)

$C'(x) = x - \dfrac{1000}{x^2}, \quad C'(x) = 0 \quad \Longrightarrow \quad x = 10.$

Since $C''(x) = 1 + 2000/x^3 > 0$ for $x > 0$, the local min of C at $x = 10$ is the abs min.

The least expensive box is 12.5 ft tall with a square base 10 ft on a side.

37.

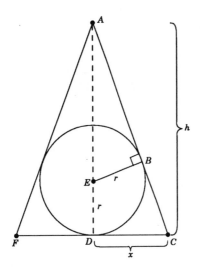

<u>Minimize A</u>

$A = \frac{1}{2}(h)(2x) = hx$

Triangles ADC and ABE are similar:

$\dfrac{AD}{DC} = \dfrac{AB}{BE} \quad$ or $\quad \dfrac{h}{x} = \dfrac{AB}{r}.$

Pythagorean Theorem:

$r^2 + (AB)^2 = (h - r)^2.$

Thus

$r^2 + \left(\dfrac{hr}{x}\right)^2 = (h - r)^2.$

Solving this equation for h we find that

$h = \dfrac{2x^2 r}{x^2 - r^2}.$

$A(x) = \dfrac{2x^3 r}{x^2 - r^2}, \quad x > r.$ (key step completed)

$A'(x) = \dfrac{\left(x^2 - r^2\right)\left(6x^2 r\right) - 2x^3 r(2x)}{\left(x^2 - r^2\right)^2} = \dfrac{2x^2 r\left(x^2 - 3r^2\right)}{\left(x^2 - r^2\right)^2},$

$A'(x) = 0 \quad \Longrightarrow \quad x = r\sqrt{3}.$

Since A decreases on $\left(r, r\sqrt{3}\right]$ and increases on $\left[r\sqrt{3}, \infty\right)$, the local min at $x = r\sqrt{3}$ is the abs min of A. When $x = r\sqrt{3}$, we get $h = 3r$ so that $FC = 2r\sqrt{3}$ and $AF = FC = \sqrt{h^2 + x^2} = 2r\sqrt{3}$.

The triangle of least area is equilateral with side of length $2r\sqrt{3}$.

39.

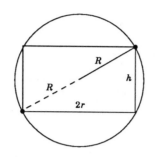

<u>Maximize V</u>

$V = \pi r^2 h$

By the Pythagorean Theorem,

$(2r)^2 + h^2 = (2R)^2$

so

$h = 2\sqrt{R^2 - r^2}.$

$V(r) = 2\pi r^2 \sqrt{R^2 - r^2}, \quad 0 < r < R.$ (key step completed)

$V'(r) = 2\pi \left[2r\sqrt{R^2 - r^2} - \dfrac{r^3}{\sqrt{R^2 - r^2}} \right] = \dfrac{2\pi r \left(2R^2 - 3r^2 \right)}{\sqrt{R^2 - r^2}}$

$V'(r) = 0 \implies r = \frac{1}{3} R\sqrt{6}.$

Since V increases on $\left(0, \frac{1}{3} R\sqrt{6} \right]$ and decreases on $\left[\frac{1}{3} R\sqrt{6}, R \right)$, the local max at $r = \frac{1}{3} R\sqrt{6}$ is the abs max.

The cylinder of maximal volume has base radius $\frac{1}{3} R\sqrt{6}$ and height $\frac{2}{3} R\sqrt{3}$.

41.

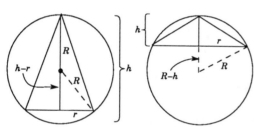

<u>Maximize V</u>

$V = \frac{1}{3} \pi r^2 h$

Pythagorean Theorem

Case 1 : $(h - R)^2 + r^2 = R^2$

Case 2 : $(R - h)^2 + r^2 = R^2$

Case 1 : $h \geq R$ Case 2 : $h \leq R$ In both cases

$$r^2 = R^2 - (R - h)^2 = 2hR - h^2.$$

$V(h) = \frac{1}{3} \pi \left(2h^2 R - h^3 \right), \quad 0 < h < 2R.$ (key step completed)

$V'(h) = \frac{1}{3} \pi \left(4hR - 3h^2 \right), \quad V'(h) = 0 \quad \text{at} \quad h = \dfrac{4R}{3}.$

Since V increases on $\left(0, \frac{4}{3} R \right]$ and decreases on $\left[\frac{4}{3} R, 2R \right)$, the local max at $h = \frac{4}{3} R$ is the abs max.

The cone of maximal volume has height $\frac{4}{3} R$ and radius $\frac{2}{3} R\sqrt{2}$.

43.

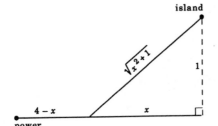

Minimize C

In units of $10,000$,

$$C = \frac{\text{cost of cable}}{\text{underground}} + \frac{\text{cost of cable}}{\text{under water}}$$

$$= 3(4-x) + 5\sqrt{x^2+1}.$$

Clearly, the cost is unnecessarily high if

$$x > 4 \quad \text{or} \quad x < 0.$$

$$C(x) = 12 - 3x + 5\sqrt{x^2+1}, \quad 0 \le x \le 4. \qquad \text{(key step completed)}$$

$$C'(x) = -3 + \frac{5x}{\sqrt{x^2+1}}, \quad C'(x) = 0 \implies x = 3/4.$$

Since the domain of C is closed, the abs min can be identified by evaluating C at each critical point:

$$C(0) = 17, \quad C\left(\tfrac{3}{4}\right) = 16, \quad C(4) = 5\sqrt{17} \cong 20.6.$$

The minimum cost is $160,000$.

45. $P'(\theta) = \dfrac{-mW(m\cos\theta - \sin\theta)}{(m\sin\theta + \cos\theta)^2}; \quad P$ is minimized when $\tan\theta = m$.

47.

Maximize θ

Since the tangent function increases on

$[0, \pi/2)$, we can maximize θ by maximizing $\tan\theta$.

$$\tan\theta = \tan(B - A)$$

$$= \frac{\tan B - \tan A}{1 + \tan B \tan A}$$

$$= \frac{36/x - 6/x}{1 + (36/x)(6/x)} = \frac{30x}{x^2 + 216}.$$

Thus, we consider

$$f(x) = \frac{30x}{x^2 + 216}, \quad x \ge 0. \qquad \text{(key step completed)}$$

$$f'(x) = \frac{(x^2 + 216)\,30 - 30x(2x)}{(x^2 + 216)^2} = \frac{30\left(216 - x^2\right)}{(x^2 + 216)^2},$$

$$f'(x) = 0 \implies x = 6\sqrt{6}.$$

Since f increases on $[0, 6\sqrt{6}]$ and decreases on $[6\sqrt{6}, \infty)$, the local max at $x = 6\sqrt{6}$ is the abs max. The observer should sit $6\sqrt{6}$ ft from the screen.

SECTION 4.6

1. $f'(x) = -x^{-2}, \quad f''(x) = 2x^{-3};$

    concave down on $(-\infty, 0)$, concave up on $(0, \infty)$;   no pts of inflection

3. $f'(x) = 3x^2 - 3, \quad f''(x) = 6x;$

    concave down on $(-\infty, 0)$, concave up on $(0, \infty)$;   pt of inflection $(0, 2)$

5. $f'(x) = x^3 - x, \quad f''(x) = 3x^2 - 1;$

    concave up on $\left(-\infty, -\frac{1}{3}\sqrt{3}\right)$ and on $\left(\frac{1}{3}\sqrt{3}, \infty\right)$, concave down on $\left(-\frac{1}{3}\sqrt{3}, \frac{1}{3}\sqrt{3}\right)$;

    pts of inflection $\left(-\frac{1}{3}\sqrt{3}, -\frac{5}{36}\right)$ and $\left(\frac{1}{3}\sqrt{3}, -\frac{5}{36}\right)$

7. $f'(x) = -\dfrac{x^2 + 1}{\left(x^2 - 1\right)^2}, \quad f''(x) = \dfrac{2x\left(x^2 + 3\right)}{\left(x^2 - 1\right)^3};$

    concave down on $(-\infty, -1)$ and on $(0, 1)$, concave up on $(-1, 0)$ and on $(1, \infty)$;

    pt of inflection $(0, 0)$

9. $f'(x) = 4x^3 - 4x, \quad f''(x) = 12x^2 - 4;$

    concave up on $\left(-\infty, -\frac{1}{3}\sqrt{3}\right)$ and on $\left(\frac{1}{3}\sqrt{3}, \infty\right)$, concave down on $\left(-\frac{1}{3}\sqrt{3}, \frac{1}{3}\sqrt{3}\right)$;

    pts of inflection $\left(-\frac{1}{3}\sqrt{3}, \frac{4}{9}\right)$ and $\left(\frac{1}{3}\sqrt{3}, \frac{4}{9}\right)$

11. $f'(x) = \dfrac{-1}{\sqrt{x}\left(1 + \sqrt{x}\right)^2}, \quad f''(x) = \dfrac{1 + 3\sqrt{x}}{2x\sqrt{x}\left(1 + \sqrt{x}\right)^3};$

    concave up on $(0, \infty)$;   no pts of inflection

13. $f'(x) = \frac{5}{3}(x + 2)^{2/3}, \quad f''(x) = \frac{10}{9}(x + 2)^{-1/3};$

    concave down on $(-\infty, -2)$, concave up on $(-2, \infty)$;   pt of inflection $(-2, 0)$

15. $f'(x) = 2 \sin x \cos x = \sin 2x, \quad f''(x) = 2 \cos 2x;$

    concave up on $\left(0, \frac{1}{4}\pi\right)$ and $\left(\frac{3}{4}\pi, \pi\right)$, concave down on $\left(\frac{1}{4}\pi, \frac{3}{4}\pi\right)$;

    pts of inflection $\left(\frac{1}{4}\pi, \frac{1}{2}\right)$ and $\left(\frac{3}{4}\pi, \frac{1}{2}\right)$

17. $f'(x) = 2x + 2 \cos 2x, \quad f''(x) = 2 - 4 \sin 2x;$

    concave up on $\left(0, \frac{1}{12}\pi\right)$ and on $\left(\frac{5}{12}\pi, \pi\right)$, concave down on $\left(\frac{1}{12}\pi, \frac{5}{12}\pi\right)$;

    pts of inflection $\left(\dfrac{1}{12}\pi, \dfrac{72 + \pi^2}{144}\right)$ and $\left(\dfrac{5}{12}\pi, \dfrac{72 + 25\pi^2}{144}\right)$

19. Since $f''(x) = 6x - 2(a + b + c)$, set $d = \frac{1}{3}(a + b + c)$. Note that $f''(d) = 0$ and that f is concave down on $(-\infty, d)$ and concave up on (d, ∞); $(d, f(d))$ is a point of inflection.

21.   Since $(-1, 1)$ lies on the graph, $1 = -a + b$.

Since $f''(x)$ exists for all x and there is a pt of inflection at $x = \frac{1}{3}$, we must have $f''\left(\frac{1}{3}\right) = 0$. Therefore

$$0 = 2a + 2b.$$

Solving these two equations, we find $a = -\frac{1}{2}$ and $b = \frac{1}{2}$.

Verification: the function

$$f(x) = -\tfrac{1}{2}x^3 + \tfrac{1}{2}x^2$$

has second derivative $f''(x) = -3x + 1$. This does change sign at $x = \frac{1}{3}$.

23.   First, we require that $\left(\frac{1}{6}\pi, 5\right)$ lie on the curve:

$$5 = \tfrac{1}{2}A + B.$$

Next we require that $\dfrac{d^2y}{dx^2} = -4A\cos 2x - 9B\sin 3x$ be zero (and change sign) at $x = \dfrac{1}{6}\pi$:

$$0 = -2A - 9B.$$

Solving these two equations, we find $A = 18$, $B = -4$.

Verification: the function

$$f(x) = 18\cos 2x - 4\sin 3x$$

has second derivative $f''(x) = -72\cos 2x + 36\sin 3x$. This does change sign at $x = \frac{1}{6}\pi$.

SECTION 4.7

[Rough sketches; not scale drawings]

1. $f(x) = (x - 2)^2$

$f'(x) = 2(x - 2)$

$f''(x) = 2$

f':

f'':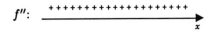

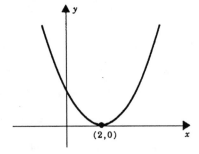

3. $f(x) = x^3 - 2x^2 + x + 1$

$f'(x) = (3x - 1)(x - 1)$

$f''(x) = 6x - 4$

f':

f'':

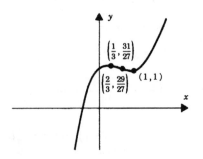

5. $f(x) = x^3 + 6x^2, \quad x \in [-4, 4]$

$f'(x) = 3x(x+4)$

$f''(x) = 6x + 12$

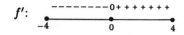

$f':$

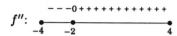

$f'':$

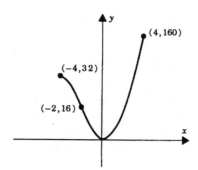

7. $f(x) = \frac{2}{3}x^3 - \frac{1}{2}x^2 - 10x - 1$

$f'(x) = (2x - 5)(x + 2)$

$f''(x) = 4x - 1$

$f':$

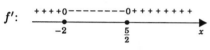

$f'':$

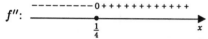

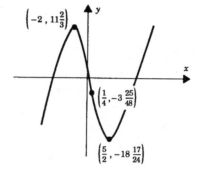

9. $f(x) = 3x^4 - 4x^3 + 1$

$f'(x) = 12x^2(x - 1)$

$f''(x) = 12x(3x - 2)$

$f':$

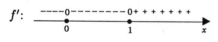

$f'':$

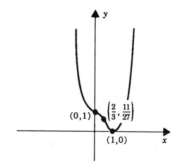

11. $f(x) = 2x^{1/2} - x, \quad x \in [0, 4]$

$f'(x) = x^{-1/2}\left(1 - x^{1/2}\right)$

$f''(x) = -\frac{1}{2}x^{-3/2}$

$f':$

$f'':$

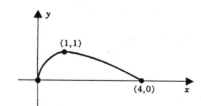

13. $f(x) = 2 + (x+1)^{6/5}$

$f'(x) = \frac{6}{5}(x+1)^{1/5}$

$f''(x) = \frac{6}{25}(x+1)^{-4/5}$

f':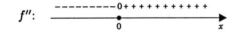

f'':

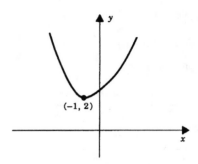

15. $f(x) = 3x^5 + 5x^3$

$f'(x) = 15x^2\left(x^2+1\right)$

$f''(x) = 30x\left(2x^2+1\right)$

f':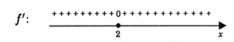

f'':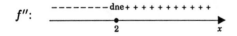

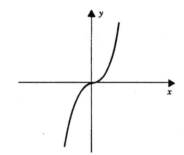

17. $f(x) = 1 + (x-2)^{5/3}$

$f'(x) = \frac{5}{3}(x-2)^{2/3}$

$f''(x) = \frac{10}{9}(x-2)^{-1/3}$

f':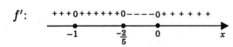

f'':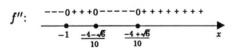

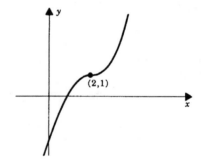

19. $f(x) = x^2(1+x)^3$

$f'(x) = x(1+x)^2(5x+2)$

$f''(x) = 2(1+x)\left(10x^2+8x+1\right)$

f':

f'':

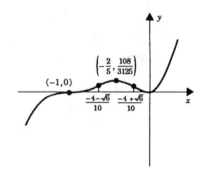

21. $f(x) = x(1-x)^{1/2}$

$f'(x) = \frac{1}{2}(1-x)^{-1/2}(2-3x)$

$f''(x) = \frac{1}{4}(1-x)^{-3/2}(3x-4)$

f':

f'':

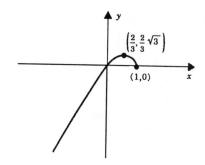

23. $f(x) = x + \sin 2x, \quad x \in [0, \pi]$

$f'(x) = 1 + 2\cos 2x$

$f''(x) = -4\sin 2x$

f':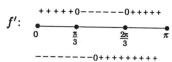

f'':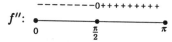

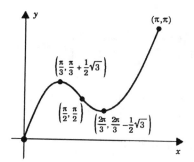

25. $f(x) = \cos^4 x, \quad x \in [0, \pi]$

$f'(x) = -4\cos^3 x \sin x$

$f''(x) = 4\cos^2 x \left(3\sin^2 x - \cos^2 x\right)$

f':

f'':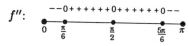

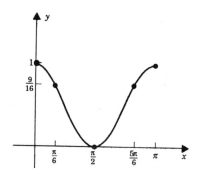

27. $f(x) = 2\sin^3 x + 3\sin x, \quad x \in [0, \pi]$

$f'(x) = 3\cos x \left(2\sin^2 x + 1\right)$

$f''(x) = 9\sin x \left(1 - 2\sin^2 x\right)$

f':

f'':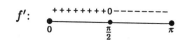

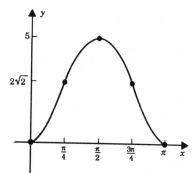

SECTION 4.8

1.   vertical: $x = \frac{1}{3}$;   horizontal: $y = \frac{1}{3}$

3.   vertical: $x = 2$;   horizontal: none

5.   vertical: $x = \pm 3$;   horizontal: $y = 0$

7.   vertical: $x = -\frac{4}{3}$;   horizontal: $y = \frac{4}{9}$

9.   vertical: $x = \frac{5}{2}$;   horizontal: $y = 0$

11.   vertical: none;   horizontal: $y = \pm\frac{3}{2}$

13.   vertical: $x = 1$;   horizontal: $y = 0$

15.   vertical: none;   horizontal: $y = 0$

17.   vertical: $x = \left(2n + \frac{1}{2}\right)\pi$;   horizontal: none

19. $f'(x) = \frac{4}{3}(x + 3)^{1/3}$; neither

21. $f'(x) = -\frac{4}{5}(2 - x)^{-1/5}$;   cusp

23. $f'(x) = \frac{6}{5}x^{-2/5}\left(1 - x^{3/5}\right)$;   tangent

25. $f(-2)$ undefined;   neither

27. $f'(x) = \begin{cases} \frac{1}{2}(x - 1)^{-1/2}, & x > 1 \\ -\frac{1}{2}(1 - x)^{-1/2}, & x < 1 \end{cases}$;   cusp

29. $f'(x) = \begin{cases} \frac{1}{3}(x + 8)^{-23}, & x > -8 \\ -\frac{1}{3}(x + 8)^{-2/3}, & x < -8 \end{cases}$;   cusp

31. $f'(x) = \begin{cases} \frac{1}{3}x^{-2/3}, & x > 1 \\ -\frac{1}{3}x^{-2/3}, & x < 1 \end{cases}$;   tangent

33. f not continuous at 0;   neither

35.   Given a positive number ϵ, there exists a negative number K such that, if $x \leq K$, then $|f(x) - l| < \epsilon$.

SECTION 4.9

[Rough sketches; not scale drawings]

1. $f(x) = [(x + 1) - 1]^3 + 1$

$f'(x) = 3x^2$

$f''(x) = 6x$

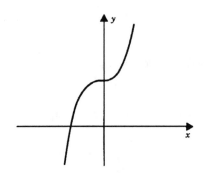

3. $f(x) = x^2(5 - x)^3$

$f'(x) = 5x(2 - x)(5 - x)^2$

$f''(x) = 10(5 - x)\left(2x^2 - 8x + 5\right)$

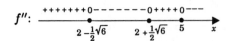

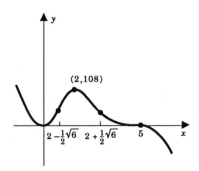

5. $f(x) = x^2 + 2x^{-1}$

$f'(x) = 2x - 2x^{-2} = 2\left(x^3 - 1\right)/x^2$

$f''(x) = 2 + 4x^{-3}$

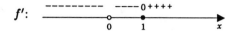

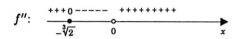

f':

f'':

asymptote: $x = 0$

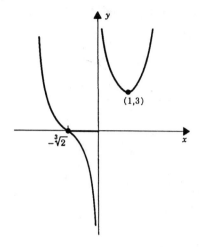

$(1,3)$

$-\sqrt[3]{2}$

7. $f(x) = (x-4)/x^2$

$f'(x) = (8-x)/x^3$

$f''(x) = (2x-24)/x^4$

f':

f'':

asymptotes: $x = 0, \; y = 0$

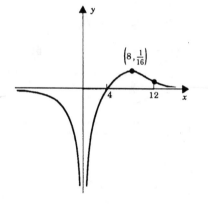

$\left(8, \frac{1}{16}\right)$

9. $f(x) = \begin{cases} 4 - x^2, & |x| > 1 \\ x^2 + 2, & -1 \le x \le 1 \end{cases}$

$f'(x) = \begin{cases} -2x, & |x| > 1 \\ 2x, & -1 < x < 1 \end{cases}$

$f''(x) = \begin{cases} -2, & |x| > 1 \\ 2, & -1 < x < 1 \end{cases}$

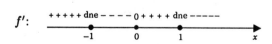

f':

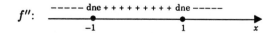

f'':

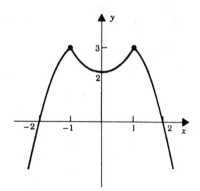

11. $f(x) = x^2 - 6x^{1/3}$

$f'(x) = 2x^{-2/3}\left(x^{5/3} - 1\right)$

$f''(x) = \frac{2}{3}x^{-5/3}\left(3x^{5/3} + 2\right)$

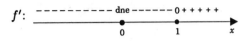

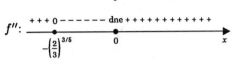

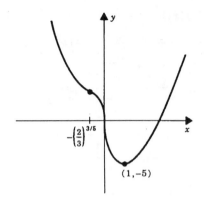

vertical tangent at $(0,0)$

13. $f(x) = x(x-1)^{1/5}$

$f'(x) = \frac{1}{5}(x-1)^{-4/5}(6x-5)$

$f''(x) = \frac{2}{25}(x-1)^{-9/5}(3x-5)$

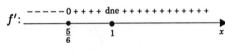

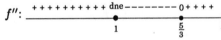

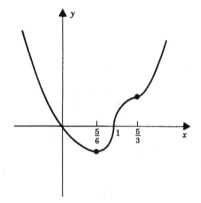

vertical tangent at $(1,0)$

15. $f(x) = \dfrac{2x}{4x - 3}$

$f'(x) = -6(4x-3)^{-2}$

$f''(x) = 48(4x-3)^{-3}$

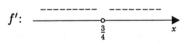

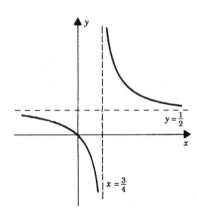

asymptotes: $x = 3/4,\ y = 1/2$

17. $f(x) = \dfrac{x}{(x+3)^2}$

$f'(x) = \dfrac{3-x}{(x+3)^3}$

$f''(x) = \dfrac{2x-12}{(x+3)^4}$

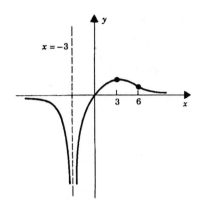

f':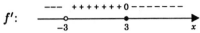

f'':

asymptotes: $x = -3,\ y = 0$

19. $f(x) = \dfrac{x^2}{x^2 - 4}$

$f'(x) = \dfrac{-8x}{\left(x^2 - 4\right)^2}$

$f''(x) = \dfrac{8\left(3x^2 + 4\right)}{\left(x^2 - 4\right)^3}$

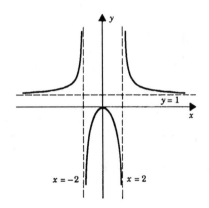

f':

f'':

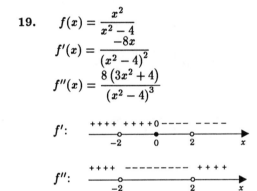

asymptotes: $x = -2,\ x = 2,\ y = 1$

21. $f(x) = \left(\dfrac{x}{x-2}\right)^{1/2}; \quad x \le 0,\ x > 2$

$f'(x) = -\left(\dfrac{x}{x-2}\right)^{-1/2} (x-2)^{-2}$

$f''(x) = (2x-1)\left(\dfrac{x}{x-2}\right)^{-3/2} (x-2)^{-4}$

f':

f'':

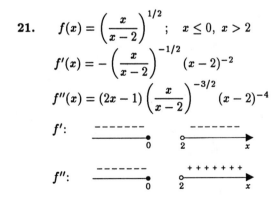

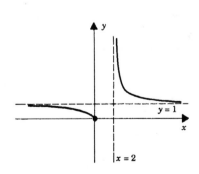

asymptotes: $x = 2,\ y = 1$

23. $f(x) = \dfrac{x}{\sqrt{4x^2 + 1}}$

$f'(x) = \left(4x^2 + 1\right)^{-3/2}$

$f''(x) = -12x \left(4x^2 + 1\right)^{-5/2}$

f':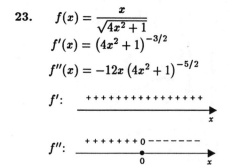

f'':

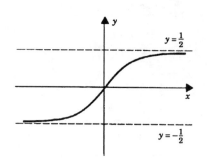

asymptotes: $y = -1/2,\ y = 1/2$

25. $f(x) = x^2 \left(x^2 - 2\right)^{-1/2}, \quad |x| > \sqrt{2}$

$f'(x) = x \left(x^2 - 4\right) \left(x^2 - 2\right)^{-3/2}$

$f''(x) = 2 \left(x^2 + 4\right) \left(x^2 - 2\right)^{-5/2}$

f':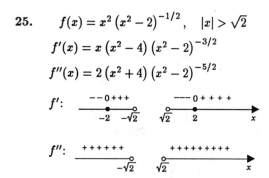

f'':

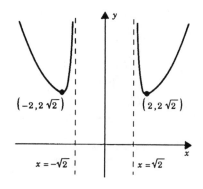

asymptotes: $x = -\sqrt{2},\ x = \sqrt{2}$

27. $f(x) = 3\sin 2x, \quad x \in [0, \pi]$

$f'(x) = 6\cos 2x$

$f''(x) = -12\sin 2x$

f':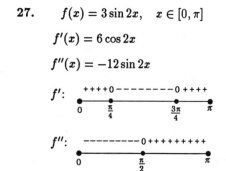

f'':

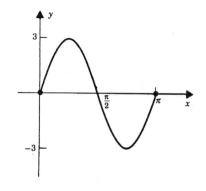

29. $f(x) = 2\sin 3x, \quad x \in [0, \pi]$

$f'(x) = 6\cos 3x$

$f''(x) = -18\sin 3x$

f':

f'':

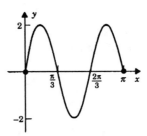

31. $f(x) = (\sin x - \cos x)^2 = \sin^2 x + \cos^2 x - 2\sin x \cos x$

$\qquad = 1 - \sin 2x, \quad x \in [0, \pi]$

$f'(x) = -2\cos 2x$

$f''(x) = 4\sin 2x$

f':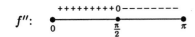

f'':

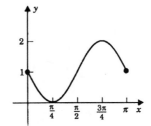

33. $f(x) = 2\tan x - \sec^2 x, \quad x \in (0, \pi/2)$

$\qquad = -(1 - \tan x)^2$

$f'(x) = 2\sec^2 x\,(1 - \tan x)$

$f''(x) = -2\sec^2 x\,(3\tan^2 x - 2\tan x + 1)$

f':

f'':

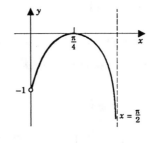

asymptote: $x = \tfrac{1}{2}\pi$

SECTION 4.10

1. Let x be the number of customers and P the profit in dollars. We assume the profit is non-negative.

$$P(x) = \begin{cases} 15x, & 0 \le x \le 1000 \\ [15 - .01(x - 1000)]x, & 1000 < x \le 2500 \end{cases};$$

$$P'(x) = \begin{cases} 15, & 0 < x < 1000 \\ -.02x + 25, & 1000 < x < 2500 \end{cases}.$$

The critical points are $x = 0, 1000, 1250, 2500$. From $P(0) = 0$, $P(1000) = 15000$, $P(1250) = 15625$, and $P(2500) = 0$ we conclude that the maximum profit occurs when there are 1250 customers.

3.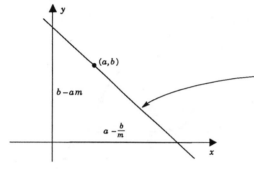

    Minimize A

    $A = \frac{1}{2}(x\text{-intercept})(y\text{-intercept})$

    equation of line: $y - b = m(x - a)$

    x-intercept: $a - \dfrac{b}{m}$

    y-intercept: $b - am$

    $$A = \frac{1}{2}\left(a - \frac{b}{m}\right)(b - am) = ab - \frac{b^2}{2m} - \frac{a^2 m}{2}$$

    $A(m) = ab - \dfrac{b^2}{2m} - \dfrac{a^2 m}{2}, \quad m < 0.$ (key step completed)

    $A'(m) = \dfrac{b^2}{2m^2} - \dfrac{a^2}{2}, \quad A'(m) = 0 \quad \Longrightarrow \quad m = -b/a.$

    Since $A''(m) = -b^2/m^3 > 0$ for $m < 0$, the local min at $m = -b/a$ is the abs min.

    The triangle of minimal area is formed by the line of slope $-b/a$.

5. Set $S = x^n + y^n$ and $y = 100 - x$. We want to minimize

    $$S(x) = x^n + (100 - x)^n, \quad 0 < x < 100.$$

    $$S'(x) = nx^{n-1} + n(100 - x)^{n-1}(-1) = n\left[x^{n-1} - (100 - x)^{n-1}\right]$$

    $S'(x) = 0$ implies $x = 100 - x$; that is, $x = y = 50$. Since S decreases on $(0, 50]$ and increases on $[50, 100)$, the abs min of S occurs when $x = 50$.

    To minimize $x^n + y^n$ take $x = y = 50$.

7.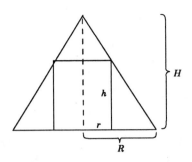

Maximize V

$$V = \pi r^2 h$$

By similar triangles

$$\frac{H}{R} = \frac{h}{R-r}, \quad h = \frac{H}{R}(R-r).$$

$$V(r) = \frac{\pi H}{R} r^2(R-r), \quad 0 < r < R. \qquad \text{(key step completed)}$$

$$V'(r) = \frac{\pi H}{R}\left(2rR - 3r^2\right), \quad V'(r) = 0 \implies r = \frac{2}{3}R.$$

Since V increases on $\left(0, \frac{2R}{3}\right]$ and decreases on $\left[\frac{2R}{3}, R\right)$, the abs max of V occurs when $r = \frac{2R}{3}$.

Then, $h = \frac{H}{R}\left(R - \frac{2R}{c}\right) = \frac{H}{3}.$

The cylinder with maximal volume has radius $\frac{2R}{3}$ and height $\frac{H}{3}$.

9. $P(Q) = (\beta - \alpha Q)Q - \left(aQ^2 + bQ + c\right), \quad Q > 0.$

$P'(Q) = (\beta - 2\alpha Q) - (2aQ + b).$

The profit is maximized by an output of $\dfrac{\beta - b}{2(a + \alpha)}$ tons.

11. Assuming the price $p = (375 - 5x)/3$ is positive, we analyze the profit function P:

$$P(x) = \left(\frac{375 - 5x}{3}\right)x - \left(500 + 15x + \frac{1}{5}x^2\right), \quad 0 < x < 75.$$

$$P'(x) = \left(\frac{375}{3} - \frac{10}{3}x\right) - \left(15 + \frac{2}{5}x\right) = 110 - \frac{56}{15}x.$$

Since $P'(x) = 0$ when $x = 825/28 \cong 29.46$ and $P''(x) < 0$ for $0 < x < 75$, the maximum value of P (for integral x) occurs when $x = 29$ or when $x = 30$.

As $P(29) \cong 1120.13$ and $P(30) = 1120$, the maximum profit is obtained by setting production at 29 lamps per week.

13. $P(x) = \left(\dfrac{2500 - x^2}{20}\right)x - \left(500 + 15x + \dfrac{1}{5}x^2\right), \quad 0 < x < 50.$

$P'(x) = \left(125 - \frac{3}{20}x^2\right) - \left(15 + \frac{2}{5}x\right) = 110 - \frac{2}{5}x - \frac{3}{20}x^2$

$P''(x) = -\frac{2}{5} - \frac{3}{10}x.$

$P'(x) = 0$ on $(0, 50)$ only at $x \cong 25.78$. Since $P'(x) < 0$ on $(0, 50)$ this point gives the mathematical maximum.

As $P(25) = 1343.75$ and $P(26) = 1346$, the optimal number of lamps is 26.

15. The total receipts is given by the function f:

$$f(Q_1) = Q_1 + 2\left(\frac{40 - 5Q_1}{10 - Q_1}\right) = Q_1 + 2\left[5 - \frac{10}{10 - Q_1}\right], \quad 0 \le Q_1 \le 8.$$

$$f'(Q_1) = 1 - \frac{20}{(10 - Q_1)^2}, \quad f'(Q_1) = 0 \implies Q_1 = 10 - \sqrt{20}.$$

The abs max of f on $[0, 8]$ occurs at a critical point. Since $f(0) = 8$, $f(8) = 8$ and $f\left(10 - \sqrt{20}\right) = \left(10 - \sqrt{20}\right) + 2\left(5 - \sqrt{5}\right) = 20 - 4\sqrt{5} > 8$, the abs max occurs when $Q_1 = 10 - \sqrt{20} \cong 5.5$.

17. The government wants to set the excise rate r so as to maximize its revenues rQ_0, where Q_0 is the output set by the industry to maximize its own profit P with this tax in effect.

Thus, we first find Q_0 in terms of r to maximize the profit P:

$$\text{profit} = P(Q) = \text{revenue} - \text{cost} - \text{tax} = R(Q) - C(Q) - rQ$$

$$= (\beta Q - \alpha Q^2) - (aQ^2 + bQ + c) - rQ$$

$$P'(Q) = (\beta - 2\alpha Q) - (2aQ + b) - r.$$

Since $P'(Q) = 0$ implies $Q = \dfrac{\beta - b - r}{2(a + \alpha)}$ and since $P''(Q) = -2(\alpha + a) < 0$ for all Q, the industry can maximize its profit P by taking

$$Q_0 = \frac{\beta - b - r}{2(a + \alpha)}.$$

Next, we determine the value of r that maximizes the tax revenue T:

$$T(r) = rQ_0 = \frac{(\beta - b - r)r}{2(a + \alpha)}.$$

As you can check, T is maximized by taking $r = \frac{1}{2}(\beta - b)$.

19.

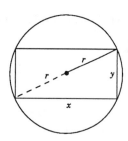

Maximize A

$$A = xy, \quad x^2 + y^2 = (2r)^2, \quad y = \sqrt{4r^2 - x^2}$$

$A(x) = x\sqrt{4r^2 - x^2}, \quad 0 < x < 2r.$ (key step completed)

$$A'(x) = \sqrt{4r^2 - x^2} + x\left(\frac{-x}{\sqrt{4r^2 - x^2}}\right) = \frac{4r^2 - 2x^2}{\sqrt{4r^2 - x^2}}, \quad A'(x) = 0 \implies x = r\sqrt{2}.$$

Since A increases on $(0, r\sqrt{2}]$ and decreases on $[r\sqrt{2}, 2r)$, the abs max of A occurs when $x = r\sqrt{2}$. Then, $y = r\sqrt{2}$ and $xy = 2r^2$.

The maximal area is $2r^2$.

21. Let x be the number of passengers and R the revenue in dollars.

$$R(x) = \begin{cases} 37x, & 16 \leq x \leq 35 \\ [37 - \frac{1}{2}(x - 35)]\, x, & 35 < x \leq 48 \end{cases};$$

$$R'(x) = \begin{cases} 37, & 16 < x < 35 \\ \frac{109}{2} - x, & 35 < x < 48 \end{cases}.$$

The critical points are $x = 16, 35,$ and 48. From $R(16) = 592$, $R(35) = 1295$, and $R(48) = 1464$ we conclude that the revenue is maximized by taking a full load of 48 passengers.

23.

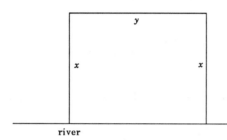
river

Maximize A

$A = xy, \quad 2x + y = 800, \quad y = 800 - 2x$

From $y = 800 - 2x$ and $x, y \geq 220$, we conclude that $200 \leq x \leq 290$.

$A(x) = x(800 - 2x), \quad 220 \leq x \leq 290.$ (key step completed)

$A'(x) = 800 - 4x = 4(200 - x).$

Since $A'(x) < 0$ on $(220, 290)$ and A is continuous at the endpoints, A decreases on $[220, 290]$. The maximal value of A occurs at $x = 220$. Then $y = 800 - 440 = 360$.

The largest field measures 360 ft (parallel to the river) by 220 ft.

25.

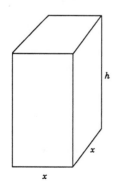

Find extremes of S

$S = 2\left(x^2\right) + 4(xh)$

Specifications:

$$V = x^2 h = 27 \quad \Longrightarrow \quad h = \frac{27}{x^2}$$

and thus $S = 2x^2 + 4x\left(\dfrac{27}{x^2}\right).$

$x^2 \leq 18 \quad \Longrightarrow \quad x \leq 3\sqrt{2}.$

$h \leq 2 \quad \Longrightarrow \quad \dfrac{27}{x^2} \leq 2 \quad \Longrightarrow \quad x \geq \dfrac{3}{2}\sqrt{6}.$

$$S(x) = 2x^2 + \frac{108}{x}, \quad \frac{3}{2}\sqrt{6} \leq x \leq 3\sqrt{2}. \qquad \text{(key step completed)}$$

$$S'(x) = 4x - \frac{108}{x^2} = \frac{4(x^3 - 27)}{x^2}.$$

Since $S'(x) > 0$ on $\left(\frac{3}{2}\sqrt{6}, 3\sqrt{2}\right)$, S increases on $\left[\frac{3}{2}\sqrt{6}, 3\sqrt{2}\right]$.

(a)   The box with minimal surface area: $\frac{3}{2}\sqrt{6}$ by $\frac{3}{2}\sqrt{6}$ by 2 ft.

(b)   The box with maximal surface area: $3\sqrt{2}$ by $3\sqrt{2}$ by $\frac{3}{2}$ ft.

27.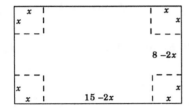

<u>Maximize V</u>

$$V = x(8 - 2x)(15 - 2x)$$

$$\begin{bmatrix} x \geq 2 \\ 8 - 2x \geq 2 \\ 15 - 2x \geq 2 \end{bmatrix} 2 \leq x \leq 3$$

$$V(x) = 120x - 46x^2 + 4x^3, \quad 2 \leq x \leq 3. \qquad \text{(key step completed)}$$

$$V'(x) = 120 - 92x + 12x^2, \quad V'(x) = 0 \text{ has no solutions.}$$

Since the domain of V is closed, the abs max can be identified by evaluating V at the endpoints:

$$V(2) = 88, \quad V(3) = 54.$$

The box of maximal volume: 2 by 4 by 11 inches.

29.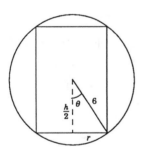

<u>Maximize V</u>

$$V = \pi r^2 h$$

Since

$$\sin\theta = \frac{r}{6} \quad \text{and} \quad \cos\theta = \frac{h/2}{6},$$

$$V = \pi(6\sin\theta)^2(12\cos\theta).$$

$$V(\theta) = 432\pi \sin^2\theta \cos\theta, \quad 0 < \theta < \pi/2. \qquad \text{(key step completed)}$$

$$V'(\theta) = 432\pi \left[2\sin\theta\cos^2\theta - \sin^3\theta\right]$$

$$= 432\pi \sin\theta \left[2\cos^2\theta - \sin^2\theta\right]$$

$$V'(\theta) = 0 \implies 2\cos^2\theta - \sin^2\theta = 0 \implies \tan\theta = \sqrt{2}$$

Let $0 < \alpha < \pi/2$ be the angle for which $\tan \alpha = \sqrt{2}$. Since $V'(\theta) > 0$ on $(0, \alpha)$ and $V'(\theta) < 0$ on $(\alpha, \pi/2)$, the abs max of V occurs when $\theta = \alpha$. Then

$$h = 12 \cos \alpha = 12 \left(\frac{1}{\sqrt{3}} \right) = 4\sqrt{3} \quad \text{and} \quad r = 6 \sin \theta = 6 \left(\frac{\sqrt{2}}{\sqrt{3}} \right) = 2\sqrt{6}.$$

The cylinder of maximal volume has base radius $2\sqrt{6}$ inches and height $4\sqrt{3}$ inches.

31. The ball strikes the ground $(y = 0)$ when $x = \dfrac{800m}{m^2 + 1}$. Thus, we want to find the maximum of the function

$$f(m) = \frac{800m}{m^2 + 1}, \quad m > 0. \qquad \text{(key step completed)}$$

$$f'(m) = \frac{800 \left(1 - m^2 \right)}{\left(m^2 + 1 \right)^2}, \quad f'(m) = 0 \quad \Longrightarrow \quad m = 1.$$

Since $f'(x) > 0$ on $(0, 1)$ and $f'(x) < 0$ on $(1, \infty)$ and f is continuous, f increases on $[0, 1]$ and decreases on $[1, \infty)$. The abs max of f occurs when $m = 1$.

33. We want to maximize the ratio

$$\frac{\text{income}}{\text{cost}} = \frac{200,000n}{1,000,000n + 100,000[1 + 2 + \cdots + (n-1)] + 5,000,000}$$

where n is the number of stories. Since

$$[1 + 2 + \cdots + (n-1)] = \tfrac{1}{2}(n-1)n,$$

the ratio can be written

$$\frac{2n}{10n + \tfrac{1}{2}(n-1)n + 50} = \frac{4n}{n^2 + 19n + 100}.$$

Thus, we consider the function

$$f(x) = \frac{4x}{x^2 + 19x + 100}, \quad x \geq 0. \qquad \text{(key step completed)}$$

$$f'(x) = \frac{4 \left(100 - x^2 \right)}{\left(x^2 + 19x + 100 \right)^2}, \quad f'(x) = 0 \quad \Longrightarrow \quad x = 10.$$

Since $f'(x) > 0$ on $(0, 10)$ and decreases on $(10, \infty)$ and f is continuous, f increases on $[0, 10]$ and decreases on $[10, \infty)$. The abs max of f occurs when $x = 10$.

A ten story building will provide the greatest rate of return on investment.

35. (a)   All points on the line $y = x$ have the form (x, x). The distance between (x, x) and (x_1, y_1) is

$$f(x) = \sqrt{(x - x_1)^2 + (x - y_1)^2}.$$

This function has its minimum at $x = \frac{1}{2}(x_1 + y_1)$. Therefore the point on the line closest to (x_1, y_1) is $\left(\frac{1}{2}[x_1 + y_1], \frac{1}{2}[x_1 + y_1]\right)$.

(b)   As you can check, $f\left(\frac{1}{2}[x_1 + y_1]\right) = \frac{1}{2}\sqrt{2}|x_1 - y_1|$. This is the distance from (x_1, y_1) to the line $y = x$.

37. All points on the curve $y = x^{3/2}$ have the form $(x, x^{3/2})$, $x \geq 0$. The distance between $(x, x^{3/2})$ and $(\frac{1}{2}, 0)$ is

$$f(x) = \sqrt{\left(x - \frac{1}{2}\right)^2 + x^3}. \qquad (x \geq 0)$$

This function has its minimum at $x = \frac{1}{3}$. Therefore the point on the curve closest to $(\frac{1}{2}, 0)$ is the point $\left(\frac{1}{3}, \left(\frac{1}{3}\right)^{3/2}\right) = \left(\frac{1}{3}, \frac{1}{9}\sqrt{3}\right)$.

39.
$$I(x) = \frac{a}{x^2} + \frac{b}{(s - x)^2}, \quad 0 \leq x \leq s.$$

$$I'(x) = \frac{-2a}{x^3} + \frac{2b}{(s - x)^3}.$$

$$I'(x) = 0 \implies \left(\frac{s - x}{x}\right)^3 = \frac{2b}{2a} \implies \frac{s - x}{x} = \left(\frac{b}{a}\right)^{1/3} \implies x = \frac{a^{1/3}s}{a^{1/3} + b^{1/3}}.$$

Clearly, this x value lies in $(0, s)$ and minimizes I since $I''(x) > 0$ on $(0, s)$.

41.

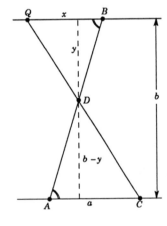

Minimize S

$$S = \frac{1}{2}a(b - y) + \frac{1}{2}xy$$

Triangles DAC and DBQ are similar:

$$\frac{b - y}{a} = \frac{y}{x} \text{ and therefore } y = \frac{bx}{x + a}.$$

Thus,

$$S = \frac{1}{2}a\left(b - \frac{bx}{x + a}\right) + \frac{1}{2}\frac{bx^2}{x + a}$$

$$= \frac{1}{2}\frac{a^2b}{x + a} + \frac{1}{2}\frac{bx^2}{x + a}.$$

$$S(x) = \frac{b}{2}\left(\frac{x^2 + a^2}{x + a}\right), \quad x > 0. \qquad \text{(key step completed)}$$

$$S'(x) = \frac{b}{2}\left[\frac{(x + a)2x - (x^2 + a^2)}{(x + a)^2}\right].$$

$$S'(x) = 0 \implies x^2 + 2ax - a^2 = 0 \implies x = a(\sqrt{2} - 1).$$

To minimize the sum of the areas of the two triangles, place Q exactly $a\left(\sqrt{2}-1\right)$ units to the left of B.

43. Let the solid have base dimensions L, W, H. Since the diagonal of the solid is a diameter of the sphere,

$$(*) \qquad\qquad L^2 + W^2 + H^2 = (2r)^2.$$

We begin by considering all the solids with the same height H. The bases of these solids are rectangles inscribed in a fixed circle. By Exercise 19 the inscribed rectangle of greatest area is a square; thus, the solid of height H with greatest volume has a square base. Thus, $L = W$ and

$$V = LWH = L^2 H.$$

By $(*)$

$$2L^2 + H^2 = (2r)^2 \quad \text{and} \quad L^2 = 2r^2 - \tfrac{1}{2}H^2.$$

Thus we need to find the value of H that maximizes the function

$$V(H) = 2r^2 H - \tfrac{1}{2}H^3, \quad 0 < H < 2r.$$

$$V'(H) = 2r^2 - \tfrac{3}{2}H^2. \qquad V'(H) = 0 \implies H = \tfrac{2}{3}\sqrt{3}\,r.$$

Since $V''(H) = -3H < 0$ on $(0, 2r)$, the local max at $H = \tfrac{2}{3}\sqrt{3}\,r$ is the abs max.

When $H = \tfrac{2}{3}\sqrt{3}\,r$, we get $L = W = \tfrac{2}{3}\sqrt{3}\,r$. The solid of greatest volume is a cube with edge $\tfrac{2}{3}\sqrt{3}\,r$ units long.

45.

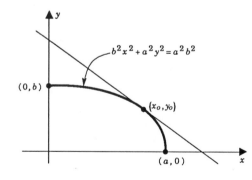

Minimize A

$$A = \tfrac{1}{2}(x\text{-intercept})(y\text{-intercept})$$

$$2b^2 x + 2a^2 y\,\frac{dy}{dx} = 0 \quad \text{so} \quad m = -\frac{b^2 x_0}{a^2 y_0}.$$

Equation of tangent:

$$y - y_0 = -\frac{b^2 x_0}{a^2 y_0}\left(x - x_0\right).$$

x intercept:

$$\frac{b^2 x_0{}^2 + a^2 y_0{}^2}{b^2 x_0} = \frac{a^2 b^2}{b^2 x_0} = \frac{a^2}{x_0}.$$

y intercept:

$$\frac{b^2 x_0{}^2 + a^2 y_0{}^2}{a^2 y_0} = \frac{a^2 b^2}{a^2 y_0} = \frac{b^2}{y_0}.$$

Thus, $A = \dfrac{a^2 b^2}{2x_0 y_0}$. Since (x_0, y_0) lies on the ellipse,

$$y_0 = \frac{b}{a}\sqrt{a^2 - x_0{}^2} \quad \text{so that} \quad A = \frac{a^3 b}{2x_0\sqrt{a^2 - x_0{}^2}}.$$

We consider the function

$$A(x) = \frac{a^3 b}{2}\left(\frac{1}{x\sqrt{a^2 - x^2}}\right), \quad 0 < x < a. \qquad \text{(key step completed)}$$

$$A'(x) = -\frac{a^3 b}{2}\left[\frac{\sqrt{a^2 - x^2} - x^2/\sqrt{a^2 - x^2}}{\left(x\sqrt{a^2 - x^2}\right)^2}\right] = -\frac{a^3 b}{2}\frac{\left(a^2 - 2x^2\right)}{x^2\left(a^2 - x^2\right)^{3/2}}.$$

Solving $A'(x) = 0$, we get $x = \frac{1}{2}a\sqrt{2}$. Since A decreases on $\left(0, \frac{1}{2}a\sqrt{2}\right]$ and increases on $\left[\frac{1}{2}a\sqrt{2}\, a\right)$, the abs min occurs when $x_0 = \frac{1}{2}a\sqrt{2}$. Then, $y_0 = \frac{1}{2}b\sqrt{2}$. The point of tangency is $\left(\frac{1}{2}a\sqrt{2}, \frac{1}{2}b\sqrt{2}\right)$.

CHAPTER 5

SECTION 5.2

1. $L_f(P) = 0(\frac{1}{4}) + \frac{1}{2}(\frac{1}{4}) + 1(\frac{1}{2}) = \frac{5}{8}$, $U_f(P) = \frac{1}{2}(\frac{1}{4}) + 1(\frac{1}{4}) + 2(\frac{1}{2}) = \frac{11}{8}$

3. $L_f(P) = \frac{1}{4}(\frac{1}{2}) + \frac{1}{16}(\frac{1}{4}) + 0(\frac{1}{4}) = \frac{9}{64}$, $U_f(P) = 1(\frac{1}{2}) + \frac{1}{4}(\frac{1}{4}) + \frac{1}{16}(\frac{1}{4}) = \frac{37}{64}$

5. $L_f(P) = 1(\frac{1}{2}) + \frac{9}{8}(\frac{1}{2}) = \frac{17}{16}$, $U_f(P) = \frac{9}{8}(\frac{1}{2}) + 2(\frac{1}{2}) = \frac{25}{16}$

7. $L_f(P) = \frac{1}{2}(\frac{1}{2}) + 0(\frac{1}{2}) + 0(\frac{1}{4}) + \frac{1}{4}(\frac{3}{4}) = \frac{7}{16}$, $U_f(P) = 1(\frac{1}{2}) + \frac{1}{2}(\frac{1}{2}) + \frac{1}{4}(\frac{1}{4}) + 1(\frac{3}{4}) = \frac{25}{16}$

9. $L_f(P) = \frac{1}{16}(\frac{3}{4}) + 0(\frac{1}{2}) + \frac{1}{16}(\frac{1}{4}) + \frac{1}{4}(\frac{1}{2}) = \frac{3}{16}$, $U_f(P) = 1(\frac{3}{4}) + \frac{1}{16}(\frac{1}{2}) + \frac{1}{4}(\frac{1}{4}) + 1(\frac{1}{2}) = \frac{43}{32}$

11. $L_f(P) = 0\left(\frac{\pi}{6}\right) + \frac{1}{2}\left(\frac{\pi}{3}\right) + 0\left(\frac{\pi}{2}\right) = \frac{\pi}{6}$, $U_f(P) = \frac{1}{2}\left(\frac{\pi}{6}\right) + 1\left(\frac{\pi}{3}\right) + 1\left(\frac{\pi}{2}\right) = \frac{11\pi}{12}$

13.   (a) $L_f(P) \le U_f(P)$ but $3 \not\le 2$.

     (b) $L_f(P) \le \displaystyle\int_{-1}^{1} f(x)\,dx \le U_f(P)$ but $3 \not\le 2 \le 6$.

     (c) $L_f(P) \le \displaystyle\int_{-1}^{1} f(x)\,dx \le U_f(P)$ but $3 \le 10 \not\le 6$.

15.   (a) $L_f(P) = -3x_1(x_1 - x_0) - 3x_2(x_2 - x_1) - \cdots - 3x_n(x_n - x_{n-1})$,

         $U_f(P) = -3x_0(x_1 - x_0) - 3x_1(x_2 - x_1) - \cdots - 3x_{n-1}(x_n - x_{n-1})$

     (b)   For each index i

$$-3x_i \le -\tfrac{3}{2}(x_i + x_{i-1}) \le -3x_{i-1}.$$

Multiplication by $\Delta x_i = x_i - x_{i-1}$ gives

$$-3x_i\,\Delta x_i \le -\tfrac{3}{2}(x_i^2 - x_{i-1}^2) \le -3x_{i-1}\,\Delta x_i.$$

Summing from $i = 1$ to $i = n$, we find that

$$L_f(P) \le -\tfrac{3}{2}(x_1^2 - x_0^2) - \cdots - \tfrac{3}{2}(x_n^2 - x_{n-1}^2) \le U_f(P).$$

The middle sum collapses to

$$-\tfrac{3}{2}(x_n^2 - x_0^2) = -\tfrac{3}{2}(b^2 - a^2).$$

Thus

$$L_f(P) \le -\frac{3}{2}(b^2 - a^2) \le U_f(P) \quad \text{so that} \quad \int_a^b -3x\,dx = -\frac{3}{2}(b^2 - a^2).$$

17. $L_f(P) = x_0{}^3(x_1 - x_0) + x_1{}^3(x_2 - x_1) + \cdots + x_{n-1}^3(x_n - x_{n-1})$

     $U_f(P) = x_1{}^3(x_1 - x_0) + x_2{}^3(x_2 - x_1) + \cdots + x_n{}^3(x_n - x_{n-1})$

For each index i

$$x_{i-1}^3 \le \tfrac{1}{4}\left(x_i{}^3 + x_i{}^2 x_{i-1} + x_i x_{i-1}^2 + x_{i-1}^3\right) \le x_i{}^3$$

and thus by the hint

$$x_{i-1}^3(x_i - x_{i-1}) \le \tfrac{1}{4}\left(x_i^4 - x_{i-1}^4\right) \le x_i^3(x_i - x_{i-1}).$$

Adding up these inequalities, we find that

$$L_f(P) \le \tfrac{1}{4}\left(x_n^4 - x_0^4\right) \le U_f(P).$$

Since $x_n = 1$ and $x_0 = 0$, the middle term is $\dfrac{1}{4}$: $\quad \displaystyle\int_0^1 x^3\,dx = \dfrac{1}{4}$.

19.   Let P be an arbitrary partition of $[0, 4]$. Since each $m_i = 2$ and each $M_i \ge 2$,

$$L_g(P) = 2\Delta x_1 + \cdots + 2\Delta x_n = 2(\Delta x_1 + \cdots + \Delta x_n) = 2 \cdot 4 = 8$$

and

$$U_g(P) \ge 2\Delta x_1 + \cdots + 2\Delta x_n = 2(\Delta x_1 + \cdots + \Delta x_n) = 2 \cdot 4 = 8.$$

Thus

$$L_g(P) \le 8 \le U_g(P) \quad \text{for all partitions } P \text{ of } [0, 4].$$

Uniqueness: Suppose that

$(*)$ $\qquad\qquad L_g(P) \le I \le U_g(P)$ for all partitions P of $[0,4]$.

Since $L_g(P) = 8$ for all P, I is at least 8. Suppose now that $I > 8$ and choose a partition P of $[0,4]$ with max $\Delta x_i < \tfrac{1}{5}(I - 8)$ and

$$0 = x_1 < \cdots < x_{i-1} < 3 < x_i < \cdots < x_n = 4.$$

Then

$$U_g(P) = 2\Delta x_1 + \cdots + 2\Delta x_{i-1} + 7\Delta x_i + 2\Delta x_{i+1} + \cdots + 2\Delta x_n$$

$$= 2(\Delta x_1 + \cdots + \Delta x_n) + 5\Delta x_i$$

$$= 8 + 5\Delta x_i < 8 + \tfrac{5}{5}(I - 8) = I$$

and I does not satisfy $(*)$. This contradiction proves that I is not greater than 8 and therefore $I = 8$.

SECTION 5.3

1.   (a) $\displaystyle\int_0^5 f(x)\,dx = \int_0^2 f(x)\,dx + \int_2^5 f(x)\,dx = 4 + 1 = 5$

(b) $\displaystyle\int_1^2 f(x)\,dx = \int_0^2 f(x)\,dx - \int_0^1 f(x)\,dx = 4 - 6 = -2$

(c) $\displaystyle\int_1^5 f(x)\,dx = \int_0^5 f(x)\,dx - \int_0^1 f(x)\,dx = 5 - 6 = -1$

(d) 0   (e) $\displaystyle\int_2^0 f(x)\,dx = -\int_0^2 f(x)\,dx = -4$

(f) $\displaystyle\int_5^1 f(x)\,dx = -\int_1^5 f(x)\,dx = 1$

3. With $P = \left\{ 1, \frac{3}{2}, 2 \right\}$ and $f(x) = \frac{1}{x}$, we have

$$0.5 < \frac{7}{12} = L_f(P) \leq \int_1^2 \frac{dx}{x} \leq U_f(P) = \frac{5}{6} < 1.$$

5. (a) $F(0) = 0$ (b) $F'(x) = x\sqrt{x+1}$ (c) $F'(2) = 2\sqrt{3}$

    (d) $F(2) = \int_0^2 t\sqrt{t+1}\,dt$ (e) $-F(x) = \int_x^0 t\sqrt{t+1}\,dt$

7. $F'(x) = \frac{1}{x^2+9}$;   (a) $\frac{1}{10}$ (b) $\frac{1}{9}$ (c) $\frac{4}{37}$ (d) $\frac{-2x}{(x^2+9)^2}$

9. $F'(x) = -x\sqrt{x^2+1}$;   (a) $\sqrt{2}$ (b) 0 (c) $-\frac{1}{4}\sqrt{5}$ (d) $-\frac{1+2x^2}{\sqrt{x^2+1}}$

11. $F'(x) = \cos \pi x$;   (a) -1 (b) 1 (c) 0 (d) $-\pi \sin \pi x$

13. (a) Since $P_1 \subseteq P_2$, $U_f(P_2) \leq U_f(P_1)$ but $5 \not\leq 4$.

    (b) Since $P_1 \subseteq P_2$, $L_f(P_1) \leq L_f(P_2)$ but $5 \not\leq 4$.

15. By the hint $\dfrac{F(b) - F(a)}{b - a} = F'(c)$ for some c in (a, b). The desired result follows by observing that

$$F(b) = \int_a^b f(t)\,dt, \quad F(a) = 0, \quad \text{and} \quad F'(c) = f(c).$$

17. Set $G(x) = \int_a^x f(t)\,dt$. Then $F(x) = \int_c^a f(t)\,dt + G(x)$. By (5.3.5) G, and thus F,

is continuous on $[a, b]$, is differentiable on (a, b), and $F'(x) = G'(x) = f(x)$ for all x in (a, b).

SECTION 5.4

1. $\int_0^1 (2x - 3)\,dx = [x^2 - 3x]_0^1 = (-2) - (0) = -2$

3. $\int_{-1}^0 5x^4\,dx = [x^5]_{-1}^0 = (0) - (-1) = 1$

5. $\int_1^4 \sqrt{x}\,dx = \int_1^4 x^{1/2}\,dx = \left[\frac{2}{3}x^{3/2} \right]_1^4 = \frac{2}{3}\left[x^{3/2} \right]_1^4 = \frac{2}{3}(8 - 1) = \frac{14}{3}$

7. $\int_1^5 2\sqrt{x-1}\,dx = \int_1^5 2(x-1)^{1/2}\,dx = \left[\frac{4}{3}(x-1)^{3/2} \right]_1^5 = \frac{4}{3}[4^{3/2} - 0] = \frac{32}{3}$

9. $\int_{-2}^0 (x+1)(x-2)\,dx = \int_{-2}^0 (x^2 - x - 2)\,dx = \left[\frac{x^3}{3} - \frac{x^2}{2} - 2x \right]_{-2}^0 = \left[0 - \left(\frac{-8}{3} - 2 + 4 \right) \right] = \frac{2}{3}$

11. $\displaystyle\int_3^3 \sqrt{x}\,dx = 0$

13. $\displaystyle\int_0^1 \frac{dt}{(t+2)^3} = \int_0^1 (t+2)^{-3}\,dt = \left[-\frac{1}{2}(t+2)^{-2}\right]_0^1 = -\frac{1}{2}[3^{-2} - 2^{-2}] = \frac{5}{72}$

15. $\displaystyle\int_1^2 \left(3t + \frac{4}{t^2}\right)dt = \int_1^2 (3t + 4t^{-2})\,dt = \left[\frac{3}{2}t^2 - 4t^{-1}\right]_1^2 = \left[(6-2) - \left(\frac{3}{2} - 4\right)\right] = \frac{13}{2}$

17. $\displaystyle\int_0^1 (x^{3/2} - x^{1/2})\,dx = \left[\frac{2}{5}x^{5/2} - \frac{2}{3}x^{3/2}\right]_0^1 = \left[\left(\frac{2}{5} - \frac{2}{3}\right) - 0\right] = \frac{-4}{15}$

19. $\displaystyle\int_0^1 (x+1)^{17}\,dx = \left[\frac{1}{18}(x+1)^{18}\right]_0^1 = \frac{1}{18}(2^{18} - 1)$

21. $\displaystyle\int_0^a (\sqrt{a} - \sqrt{x})^2\,dx = \int_0^a (a - 2\sqrt{a}\,x^{1/2} + x)\,dx = \left[ax - \frac{4}{3}\sqrt{a}\,x^{3/2} + \frac{x^2}{2}\right]_0^a = a^2 - \frac{4}{3}a^2 + \frac{a^2}{2} = \frac{1}{6}a^2$

23. $\displaystyle\int_1^2 \frac{6-t}{t^3}\,dt = \int_1^2 (6t^{-3} - t^{-2})\,dt = [-3t^{-2} + t^{-1}]_1^2 = \left[\frac{-3}{4} + \frac{1}{2}\right] - [-3 + 1] = \frac{7}{4}$

25. $\displaystyle\int_0^1 x^2(x-1)\,dx = \int_0^1 (x^3 - x^2)\,dx = \left[\frac{x^4}{4} - \frac{x^3}{3}\right]_0^1 = -\frac{1}{12}$

27. $\displaystyle\int_1^2 2x(x^2 + 1)\,dx = \int_1^2 (2x^3 + 2x)\,dx = \left[\frac{x^4}{2} + x^2\right]_1^2 = 12 - \frac{3}{2} = \frac{21}{2}$

29. $\displaystyle\int_0^{\pi/2} \cos x\,dx = [\sin x]_0^{\pi/2} = 1$ **31.** $\displaystyle\int_0^{\pi/4} \sec^2 x\,dx = [\tan x]_0^{\pi/4} = 1$

33. $\displaystyle\int_{\pi/6}^{\pi/4} \csc x\,\cot x\,dx = [-\csc x]_{\pi/6}^{\pi/4} = -\sqrt{2} - (-2) = 2 - \sqrt{2}$

35. $\displaystyle\int_0^{2\pi} \sin x\,dx = [-\cos x]_0^{2\pi} = -1 - (-1) = 0$

37. (a) $\displaystyle F(x) = \int_2^x \frac{dt}{t}$ (b) $\displaystyle F(x) = -3 + \int_2^x \frac{dt}{t}$

39. $\displaystyle\frac{d}{dx}\left[\int_a^x f(t)\,dt\right] = f(x); \quad \int_a^x \frac{d}{dt}[f(t)]\,dt = f(x) - f(a)$

SECTION 5.5

1. $\displaystyle A = \int_0^1 (2 + x^3)\,dx = \left[2x + \frac{x^4}{4}\right]_0^1 = \frac{9}{6}$ ⟨

3. $\displaystyle A = \int_3^8 \sqrt{x+1}\,dx = \int_3^8 (x+1)^{1/2}\,dx = \left[\frac{2}{3}(x+1)^{3/2}\right]_3^8 = \frac{2}{3}[27 - 8] = \frac{38}{3}$

5. $\displaystyle A = \int_0^1 (2x^2 + 1)^2\,dx = \int_0^1 (4x^4 + 4x^2 + 1)\,dx = \left[\frac{4}{5}x^5 + \frac{4}{3}x^3 + x\right]_0^1 = \frac{47}{15}$

7. $A = \displaystyle\int_1^2 [0 - (x^2 - 4)]\,dx = \int_1^2 (4 - x^2)\,dx = \left[4x - \dfrac{x^3}{3}\right]_1^2 = \left[8 - \dfrac{8}{3}\right] - \left[4 - \dfrac{1}{3}\right] = \dfrac{5}{3}$

9. $A = \displaystyle\int_{\pi/3}^{\pi/2} \sin x\,dx = [-\cos x]_{\pi/3}^{\pi/2} = (0) - \left(-\dfrac{1}{2}\right) = \dfrac{1}{2}$

11.

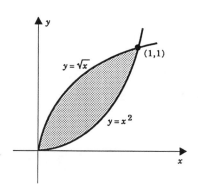

$A = \displaystyle\int_0^1 [x^{1/2} - x^2]\,dx$

$\quad = \left[\tfrac{2}{3}x^{3/2} - \tfrac{1}{3}x^3\right]_0^1 = \tfrac{1}{3}$

13.

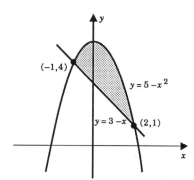

$A = \displaystyle\int_{-1}^2 [(5 - x^2) - (3 - x)]\,dx$

$\quad = \displaystyle\int_{-1}^2 (2 + x - x^2)\,dx$

$\quad = \left[2x + \dfrac{x^2}{2} - \dfrac{x^3}{3}\right]_{-1}^2$

$\quad = [4 + 2 - \tfrac{8}{3}] - [-2 + \tfrac{1}{2} + \tfrac{1}{3}] = \tfrac{9}{2}$

15.

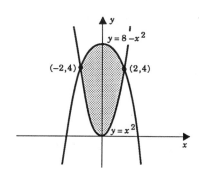

$A = \displaystyle\int_{-2}^2 [(8 - x^2) - (x^2)]\,dx$

$\quad = \displaystyle\int_{-2}^2 (8 - 2x^2)\,dx$

$\quad = \left[8x - \tfrac{2}{3}x^3\right]_{-2}^2$

$\quad = \left[16 - \tfrac{16}{3}\right] - \left[-16 + \tfrac{16}{3}\right] = \tfrac{64}{3}$

17.

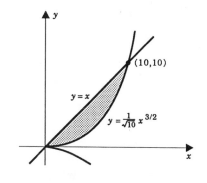

$A = \displaystyle\int_0^{10} \left[x - \dfrac{1}{\sqrt{10}}\,x^{3/2}\right]dx$

$\quad = \left[\dfrac{x^2}{2} - \dfrac{2\sqrt{10}}{50}\,x^{5/2}\right]_0^{10}$

$\quad = 50 - \dfrac{2\sqrt{10}}{50}(10)^{5/2} = 10$

19.

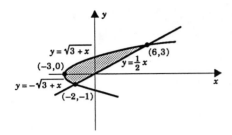

$$A = \int_{-3}^{-2} [(\sqrt{3+x}) - (-\sqrt{3+x})]\, dx + \int_{-2}^{6} \left[(\sqrt{3+x}) - \left(\frac{1}{2}x\right)\right]\, dx$$

$$= \int_{-3}^{-2} 2(3+x)^{1/2}\, dx + \int_{-2}^{6} \left[(3+x)^{1/2} - \frac{1}{2}x\right]\, dx$$

$$= \left[\frac{4}{3}(3+x)^{3/2}\right]_{-3}^{-2} + \left[\frac{2}{3}(3+x)^{3/2} - \frac{x^2}{4}\right]_{-2}^{6} = \left[\frac{4}{3} - 0\right] + \left[(18-9) - \left(\frac{2}{3} - 1\right)\right] = \frac{32}{3}$$

21.

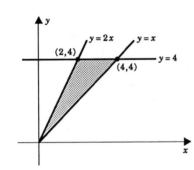

$$A = \int_{0}^{2} [2x - x]\, dx + \int_{2}^{4} [4 - x]\, dx$$

$$= \left[\tfrac{1}{2}x^2\right]_{0}^{2} + \left[4x - \tfrac{1}{2}x^2\right]_{2}^{4}$$

$$= 2 + [8 - 6] = 4$$

23.

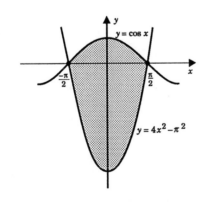

$$A = \int_{-\pi/2}^{\pi/2} [\cos x - (4x^2 - \pi^2)]\, dx$$

$$= \left[\sin x - \tfrac{4}{3}x^3 + \pi^2 x\right]_{-\pi/2}^{\pi/2}$$

$$= \left[1 - \tfrac{1}{6}\pi^3 + \tfrac{1}{2}\pi^3\right] - \left[-1 + \tfrac{1}{6}\pi^3 - \tfrac{1}{2}\pi^3\right]$$

$$= 2 + \tfrac{2}{3}\pi^3$$

25.

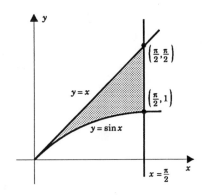

$$A = \int_0^{\pi/2} [x - \sin x]\, dx$$

$$= \left[\frac{x^2}{2} + \cos x \right]_0^{\pi/2}$$

$$= \frac{\pi^2}{8} - 1$$

SECTION 5.6

1. $\displaystyle \int \frac{dx}{x^4} = \int x^{-4}\, dx = -\frac{1}{3}x^{-3} + C$

3. $\displaystyle \int (ax + b)\, dx = \frac{a}{2}x^2 + bx + C$

5. $\displaystyle \int \frac{dx}{\sqrt{1+x}} = \int (1+x)^{-1/2}\, dx = 2(1+x)^{1/2} + C$

7. $\displaystyle \int \left(\frac{x^3 - 1}{x^2} \right) dx = \int (x - x^{-2})\, dx = \frac{x^2}{2} + x^{-1} + C$

9. $\displaystyle \int (t - a)(t - b)\, dt = \int [t^2 - (a+b)t + ab]\, dt = \frac{t^3}{3} - \frac{a+b}{2}t^2 + abt + C$

11. $\displaystyle \int \frac{(t^2 - a)(t^2 - b)}{\sqrt{t}}\, dt = \int [t^{7/2} - (a+b)t^{3/2} + abt^{-1/2}]\, dt$

$$= \tfrac{2}{9}t^{9/2} - \tfrac{2}{5}(a+b)t^{5/2} + 2abt^{1/2} + C$$

13. $\displaystyle \int g(x)g'(x)\, dx = \frac{1}{2}[g(x)]^2 + C$

15. $\displaystyle \int \tan x \sec^2 x\, dx = \int \sec x \frac{d}{dx}[\sec x]\, dx = \frac{1}{2}\sec^2 x + C$

$$\int \tan x \sec^2 x\, dx = \int \tan x \frac{d}{dx}[\tan x]\, dx = \frac{1}{2}\tan^2 x + C$$

17. $\displaystyle \int \frac{4}{(4x+1)^2}\, dx = \int 4(4x+1)^{-2}\, dx = -(4x+1)^{-1} + C$

19. $\displaystyle f(x) = \int f'(x)\, dx = \int (2x - 1)\, dx = x^2 - x + C.$

Since $f(3) = 4$, we get $4 = 9 - 3 + C$ so that $C = -2$ and

$$f(x) = x^2 - x - 2.$$

21. $f(x) = \displaystyle\int f'(x)\,dx = \int (ax + b)\,dx = \frac{1}{2}ax^2 + bx + C.$

Since $f(2) = 0$, we get $0 = 2a + 2b + C$ so that $C = -2a - 2b$ and

$$f(x) = \tfrac{1}{2}ax^2 + bx - 2a - 2b.$$

23. $f(x) = \displaystyle\int f'(x)\,dx = \int \sin x\,dx = -\cos x + C.$

Since $f(0) = 2$, we get $2 = -1 + C$ so that $C = 3$ and

$$f(x) = 3 - \cos x.$$

25.   First,

$$f'(x) = \int f''(x)\,dx = \int (6x - 2)\,dx = 3x^2 - 2x + C.$$

Since $f'(0) = 1$, we get $1 = 0 + C$ so that $C = 1$ and

$$f'(x) = 3x^2 - 2x + 1.$$

Next,

$$f(x) = \int f'(x)\,dx = \int (3x^2 - 2x + 1)\,dx = x^3 - x^2 + x + K.$$

Since $f(0) = 2$, we get $2 = 0 + K$ so that $K = 2$ and

$$f(x) = x^3 - x^2 + x + 2.$$

27.   First,

$$f'(x) = \int f''(x)\,dx = \int (x^2 - x)\,dx = \frac{1}{3}x^3 - \frac{1}{2}x^2 + C.$$

Since $f'(1) = 0$, we get $0 = \tfrac{1}{3} - \tfrac{1}{2} + C$ so that $C = \tfrac{1}{6}$ and

$$f'(x) = \tfrac{1}{3}x^3 - \tfrac{1}{2}x^2 + \tfrac{1}{6}.$$

Next,

$$f(x) = \int f'(x)\,dx = \int \left(\frac{1}{3}x^3 - \frac{1}{2}x^2 + \frac{1}{6}\right)\,dx = \frac{1}{12}x^4 - \frac{1}{6}x^3 + \frac{1}{6}x + K.$$

Since $f(1) = 2$, we get $2 = \tfrac{1}{12} - \tfrac{1}{6} + \tfrac{1}{6} + K$ so that $K = \tfrac{23}{12}$ and

$$f(x) = \frac{x^4}{12} - \frac{x^3}{6} + \frac{x}{6} + \frac{23}{12} = \frac{1}{12}(x^4 - 2x^3 + 2x + 23).$$

29.   First,

$$f'(x) = \int f''(x)\,dx = \int \cos x\,dx = \sin x + C.$$

Since $f'(0) = 1$, we get $1 = 0 + C$ so that $C = 1$ and

$$f'(x) = \sin x + 1.$$

Next,

$$f(x) = \int f'(x)\,dx = \int (\sin x + 1)\,dx = -\cos x + x + K.$$

Since $f(0) = 2$, we get $2 = -1 + 0 + K$ so that $K = 3$ and

$$f(x) = -\cos x + x + 3.$$

31. First,

$$f'(x) = \int f''(x)\,dx = \int (2x - 3)\,dx = x^2 - 3x + C.$$

Then,

$$f(x) = \int f'(x)\,dx = \int (x^2 - 3x + C)\,dx = \frac{1}{3}x^3 - \frac{3}{2}x^2 + Cx + K.$$

Since $f(2) = -1$, we get

(1) $$-1 = \tfrac{8}{3} - 6 + 2C + K;$$

and, from $f(0) = 3$, we conclude that

(2) $$3 = 0 + K.$$

Solving (1) and (2) simultaneously, we get $K = 3$ and $C = -\tfrac{1}{3}$ so that

$$f(x) = \tfrac{1}{3}x^3 - \tfrac{3}{2}x^2 - \tfrac{1}{3}x + 3.$$

33. $\dfrac{d}{dx}\left[\int f(x)\,dx\right] = f(x);\quad \int \dfrac{d}{dx}[f(x)]\,dx = f(x) + C$

SECTION 5.7

1. (a) $x(t) = \int v(t)\,dt = \int (6t^2 - 6)\,dt = 2t^3 - 6t + C.$

Since $x(0) = -2$, we get $-2 = 0 + C$ so that $C = -2$ and

$$x(t) = 2t^3 - 6t - 2.\quad \text{Therefore}\quad x(3) = 34.$$

Three seconds later the object is 34 units to the right of the origin .

(b) $s = \displaystyle\int_0^3 |v(t)|\,dt = \int_0^3 |6t^2 - 6|\,dt = \int_0^1 (6 - 6t^2)\,dt + \int_1^3 (6t^2 - 6)\,dt$

$$= [6t - 2t^3]_0^1 + [2t^3 - 6t]_1^3 = 4 + [36 - (-4)] = 44.$$

The object traveled 44 units.

3. (a) $v(t) = \int a(t)\,dt = \int (t+1)^{-1/2}\,dt = 2(t+1)^{1/2} + C.$

Since $v(0) = 1$, we get $1 = 2 + C$ so that $C = -1$ and

$$v(t) = 2(t+1)^{1/2} - 1.$$

(b) We know $v(t)$ by part (a). Therefore,

$$x(t) = \int v(t)\,dt = \int [2(t+1)^{1/2} - 1]\,dt = \frac{4}{3}(t+1)^{3/2} - t + C.$$

Since $x(0) = 0$, we get $0 = \tfrac{4}{3} - 0 + C$ so that

$$C = -\tfrac{4}{3}\quad\text{and}\quad x(t) = \tfrac{4}{3}(t+1)^{3/2} - t - \tfrac{4}{3}.$$

5. (a) $v_0 = 60$ mph $= 88$ feet per second. In general, $v(t) = at + v_0$. Here, in feet and seconds, $v(t) = -20t + 88$. Thus $v(t) = 0$ at $t = 4.4$ seconds.

(b) In general, $x(t) = \frac{1}{2}at^2 + v_0 t + x_0$. Here we take $x_0 = 0$. In feet and seconds

$$x(t) = -10(4.4)^2 + 88(4.4) = 10(4.4)^2 = 193.6 \text{ ft.}$$

7.
$$[v(t)]^2 = (at + v_0)^2 = a^2 t^2 + 2av_0 t + v_0{}^2$$
$$= v_0{}^2 + a(at^2 + 2v_0 t)$$
$$= v_0{}^2 + 2a(\tfrac{1}{2}at^2 + v_0 t)$$

$$x(t) = \tfrac{1}{2}at^2 + v_0 t + x_0 \Big\downarrow$$
$$= v_0{}^2 + 2a\,[x(t) - x_0]$$

9. The car can accelerate to 60 mph (88 ft/sec) in 20 seconds thereby covering a distance of 880 ft. It can decelerate from 88 ft/sec to 0 ft/sec in 4 seconds thereby covering a distance of 176 ft. At full speed, 88 ft/sec, it must cover a distance of

$$\tfrac{5280}{2} - 880 - 176 = 1584 \text{ ft.}$$

This takes $\frac{1584}{88} = 18$ seconds. The run takes at least $20 + 18 + 4 = 42$ seconds.

11. $v(t) = \displaystyle\int a(t)\,dt = \int (2A + 6Bt)\,dt = 2At + 3Bt^2 + C.$

Since $v(0) = v_0$, we have $v_0 = 0 + C$ so that $v(t) = 2At + 3Bt^2 + v_0.$

$$x(t) = \int v(t)\,dt = \int (2At + 3Bt^2 + v_0)\,dt = At^2 + Bt^3 + v_0 t + K.$$

Since $x(0) = x_0$, we have $x_0 = 0 + K$ so that $K = x_0$ and

$$x(t) = x_0 + v_0 t + At^2 + Bt^3.$$

13.
$$x'(t) = t^2 - 5, \qquad\qquad\qquad y'(t) = 3t,$$
$$x(t) = \tfrac{1}{3}t^3 - 5t + C. \qquad\qquad y(t) = \tfrac{3}{2}t^2 + K.$$

When $t = 2$, the particle is at $(4, 2)$. Thus, $x(2) = 4$ and $y(2) = 2$.

$$4 = \tfrac{8}{3} - 10 + C \implies C = \tfrac{34}{3}. \qquad\qquad 2 = 6 + K \implies K = -4.$$
$$x(t) = \tfrac{1}{3}t^3 - 5t + \tfrac{34}{3}, \qquad\qquad\qquad y(t) = \tfrac{3}{2}t^2 - 4.$$

Four seconds later the particle is at $(x(6),\, y(6)) = (\tfrac{160}{3},\, 50).$

15. Since $v(0) = 2$, we have $2 = A \cdot 0 + B$ so that $B = 2$. Therefore

$$x(t) = \int v(t)\,dt = \int (At + 2)\,dt = \frac{1}{2}At^2 + 2t + C.$$

Since $x(2) = x(0) - 1$, we have

$$2A + 4 + C = C - 1 \quad \text{so that} \quad A = -\tfrac{5}{2}.$$

17. $x(t) = \displaystyle\int v(t)\,dt = \int \sin t\,dt = -\cos t + C$

Since $x(\pi/6) = 0$, we have $\quad 0 = -\dfrac{\sqrt{3}}{2} + C \quad$ so that $\quad C = \dfrac{\sqrt{3}}{2} \quad$ and $\quad x(t) = \dfrac{\sqrt{3}}{2} - \cos t.$

(a) At $t = 11\pi/6$ sec.

(b) We want to find the smallest $t_0 > \pi/6$

for which $x(t_0) = 0$ and $v(t_0) > 0$. We get

$$t_0 = 13\pi/6 \text{ seconds.}$$

19. The mean-value theorem. With obvious notation

$$\frac{x(1/2) - x(0)}{1/12} = \frac{4}{1/12} = 48.$$

By the mean-value theorem there exists some time t_0 at which

$$x'(t_0) = \frac{x(1/12) - x(0)}{1/12}.$$

21. $\dfrac{v'(t)}{[v(t)]^2} = 2 \implies -[v(t)]^{-1} = 2t - v_0^{-1}.$

$$\implies [v(t)]^{-1} = v_0^{-1} - 2t \implies v(t) = \frac{1}{v_0^{-1} - 2t} = \frac{v_0}{1 - 2tv_0}.$$

SECTION 5.8

1. $\left\{ \begin{array}{l} u = 2 - 3x \\ du = -3\,dx \end{array} \right\}$; $\displaystyle\int \frac{dx}{(2-3x)^2} = \int (2-3x)^{-2}\,dx = -\frac{1}{3}\int u^{-2}\,du = \frac{1}{3}u^{-1} + C$

$$= \tfrac{1}{3}(2-3x)^{-1} + C$$

3. $\left\{ \begin{array}{l} u = 2x + 1 \\ du = 2\,dx \end{array} \right\}$; $\displaystyle\int \sqrt{2x+1}\,dx = \int (2x+1)^{1/2}\,dx = \frac{1}{2}\int u^{1/2}\,du = \frac{1}{3}u^{3/2} + C$

$$= \tfrac{1}{3}(2x+1)^{3/2} + C$$

5. $\left\{ \begin{array}{l} u = ax + b \\ du = a\,dx \end{array} \right\}$; $\displaystyle\int (ax+b)^{3/4}\,dx = \frac{1}{a}\int u^{3/4}\,du = \frac{4}{7a}u^{7/4} + C$

$$= \frac{4}{7a}(ax+b)^{7/4} + C$$

7. $\left\{ \begin{array}{l} u = 4t^2 + 9 \\ du = 8t\,dt \end{array} \right\}$; $\displaystyle\int \frac{t}{(4t^2+9)^2}\,dt = \frac{1}{8}\int \frac{du}{u^2} = -\frac{1}{8}u^{-1} + C = -\frac{1}{8}(4t^2+9)^{-1} + C$

9. $\left\{ \begin{array}{l} u = 5t^3 + 9 \\ du = 15t^2\,dt \end{array} \right\}$; $\displaystyle\int t^2\left(5t^3+9\right)^4\,dt = \frac{1}{15}\int u^4\,du = \frac{1}{75}u^5 + C$

$$= \tfrac{1}{75}(5t^3+9)^5 + C$$

11. $\left\{\begin{array}{l} u = 1 + x^3 \\ du = 3x^2\,dx \end{array}\right\}$; $\displaystyle\int x^2\left(1+x^3\right)^{1/4}dx = \frac{1}{3}\int u^{1/4}\,du = \frac{4}{15}u^{5/4} + C$

$$= \tfrac{4}{15}(1+x^3)^{5/4} + C$$

13. $\left\{\begin{array}{l} u = 1 + s^2 \\ du = 2s\,ds \end{array}\right\}$; $\displaystyle\int \frac{s}{(1+s^2)^3}\,ds = \frac{1}{2}\int \frac{du}{u^3} = -\frac{1}{4}u^{-2} + C = -\frac{1}{4}(1+s^2)^{-2} + C$

15. $\left\{\begin{array}{l} u = x^2 + 1 \\ du = 2x\,dx \end{array}\right\}$; $\displaystyle\int \frac{x}{\sqrt{x^2+1}}\,dx = \int \left(x^2+1\right)^{-1/2}x\,dx = \frac{1}{2}\int u^{-1/2}\,du$

$$= u^{1/2} + C = \sqrt{x^2+1} + C$$

17. $\left\{\begin{array}{l} u = 1 - x^3 \\ du = -3x^2\,dx \end{array}\right\}$; $\displaystyle\int x^2\left(1-x^3\right)^{2/3}dx = -\frac{1}{3}\int u^{2/3}\,du$

$$= -\tfrac{1}{5}u^{5/3} + C = -\tfrac{1}{5}(1-x^3)^{5/3} + C$$

19. $\left\{\begin{array}{l} u = x^2 + 1 \\ du = 2x \end{array}\right\}$; $\displaystyle\int 5x\left(x^2+1\right)^{-3}dx = \frac{5}{2}\int u^{-3}\,du = -\frac{5}{4}u^{-2} + C$

$$= -\tfrac{5}{4}(x^2+1)^{-2} + C$$

21. $\left\{\begin{array}{l} u = x^{1/4} + 1 \\ du = \frac{1}{4}x^{-3/4}\,dx \end{array}\right\}$; $\displaystyle\int x^{-3/4}\left(x^{1/4}+1\right)^{-2}dx = 4\int u^{-2}\,du = -4u^{-1} + C$

$$= -4(x^{1/4}+1)^{-1} + C$$

23. $\left\{\begin{array}{l} u = 1 - a^4x^4 \\ du = -4a^4x^3\,dx \end{array}\right\}$; $\displaystyle\int \frac{b^3x^3}{\sqrt{1-a^4x^4}}\,dx = -\frac{b^3}{4a^4}\int u^{-1/2}\,du = -\frac{b^3}{2a^4}u^{1/2} + C$

$$= -\frac{b^3}{2a^4}\sqrt{1-a^4x^4} + C$$

25. $\left\{\begin{array}{l|l} u = x^2 + 1 & x = 0 \implies u = 1 \\ du = 2x\,dx & x = 1 \implies u = 2 \end{array}\right\}$; $\displaystyle\int_0^1 x\left(x^2+1\right)^3 dx = \frac{1}{2}\int_1^2 u^3\,du$

$$= \tfrac{1}{8}\left[u^4\right]_1^2 = \tfrac{15}{8}$$

27. $\left\{\begin{array}{l|l} u = 1 + x^2 & x = 0 \implies u = 1 \\ du = 2x\,dx & x = 1 \implies u = 2 \end{array}\right\}$; $\displaystyle\int_0^1 5x\left(1+x^2\right)^4 dx = \frac{5}{2}\int_1^2 u^4\,du$

$$= \tfrac{1}{2}\left[u^5\right]_1^2 = \tfrac{31}{2}$$

29. 0; integrand is an odd function

31.
$$\left\{\begin{array}{l} u = a^2 - y^2 \\ \\ du = -2y\,dy \end{array}\right. \left|\begin{array}{l} y = 0 \implies u = a^2 \\ \\ y = a \implies u = 0 \end{array}\right\};\qquad \int_0^a y\sqrt{a^2 - y^2}\;dy = -\frac{1}{2}\int_{a^2}^0 u^{1/2}\,du$$

$$= -\tfrac{1}{3}\left[u^{3/2}\right]_{a^2}^0 = \tfrac{1}{3}\left(a^2\right)^{3/2} = \tfrac{1}{3}|a|^3$$

33.
$$\left\{\begin{array}{l} u = 1 - y^3/a^3 \\ \\ du = -\left(3y^2/a^3\right)dy \end{array}\right. \left|\begin{array}{l} y = -a \implies u = 2 \\ \\ y = 0 \implies u = 1 \end{array}\right\};$$

$$\int_{-a}^0 y^2\left(1 - \frac{y^3}{a^3}\right)^{-2} dy = -\frac{a^3}{3}\int_2^1 u^{-2}\,du = -\frac{a^3}{3}\left[-u^{-1}\right]_2^1 = \frac{a^3}{6}$$

35.
$$\left\{\begin{array}{l} u = x + 1 \\ \\ du = dx \end{array}\right\};\qquad \int x\sqrt{x+1}\;dx = \int (u-1)\sqrt{u}\;du = \int \left(u^{3/2} - u^{1/2}\right)du$$

$$= \tfrac{2}{5}u^{5/2} - \tfrac{2}{3}u^{3/2} + C = \tfrac{2}{5}(x+1)^{5/2} - \tfrac{2}{3}(x+1)^{3/2} + C$$

37.
$$\left\{\begin{array}{l} u = 2x - 1 \\ \\ du = 2\,dx \end{array}\right\};\qquad \int x\sqrt{2x-1}\;dx = \frac{1}{2}\int \frac{u+1}{2}\sqrt{u}\;du = \frac{1}{4}\int \left(u^{3/2} + u^{1/2}\right)du$$

$$= \tfrac{1}{10}u^{5/2} + \tfrac{1}{6}u^{3/2} + C = \tfrac{1}{10}(2x-1)^{5/2} + \tfrac{1}{6}(2x-1)^{3/2} + C$$

39.
$$\left\{\begin{array}{l} u = y + 1 \\ \\ du = dy \end{array}\right\};\qquad \int y(y+1)^{12}\;dy = \int (u-1)u^{12}\;du = \int \left(u^{13} - u^{12}\right)du$$

$$= \tfrac{1}{14}u^{14} - \tfrac{1}{13}u^{13} + C = \tfrac{1}{14}(y+1)^{14} - \tfrac{1}{13}(y+1)^{13} + C$$

41.
$$\left\{\begin{array}{l} u = t - 2 \\ \\ du = dt \end{array}\right\};\qquad \int t^2(t-2)^5\;dt = \int (u+2)^2 u^5\;du = \int \left(u^7 + 4u^6 + 4u^5\right)du$$

$$= \tfrac{1}{8}u^8 + \tfrac{4}{7}u^7 + \tfrac{2}{3}u^6 + C = \tfrac{1}{8}(t-2)^8 + \tfrac{4}{7}(t-2)^7 + \tfrac{2}{3}(t-2)^6 + C$$

43.
$$\left\{\begin{array}{l} u = t - 2 \\ \\ du = dt \end{array}\right\};\qquad \int t^2(t-2)^{-5}\;dt = \int (u+2)^2 u^{-5}\;du = \int \left(u^2 + 4u + 4\right)u^{-5}\;du$$

$$= \int \left(u^{-3} + 4u^{-4} + 4u^{-5}\right)du = -\tfrac{1}{2}u^{-2} - \tfrac{4}{3}u^{-3} - u^{-4} + C$$

$$= -\tfrac{1}{2}(t-2)^{-2} - \tfrac{4}{3}(t-2)^{-3} - (t-2)^{-4} + C$$

45.
$$\left\{\begin{array}{l} u = x + 1 \\ \\ du = dx \end{array}\right\};\qquad \int x(x+1)^{-1/5}\;dx = \int (u-1)u^{-1/5}\;du = \int \left(u^{4/5} - u^{-1/5}\right)du$$

$$= \tfrac{5}{9}u^{9/5} - \tfrac{5}{4}u^{4/5} + C = \tfrac{5}{9}(x+1)^{9/5} - \tfrac{5}{4}(x+1)^{4/5} + C$$

47.
$$\left\{\begin{array}{l} u = x+1 \\ du = dx \end{array}\right. \left|\begin{array}{l} x=0 \implies u=1 \\ x=1 \implies u=2 \end{array}\right\};\quad \int_0^1 \frac{x+3}{\sqrt{x+1}}\,dx = \int_1^2 \frac{u+2}{\sqrt{u}}\,du$$

$$= \int_1^2 \left(u^{1/2} + 2u^{-1/2}\right)\,du$$

$$= \left[\tfrac{2}{3}u^{3/2} + 4u^{1/2}\right]_1^2 = \tfrac{2}{3}\sqrt{8} + 4\sqrt{2} - \tfrac{2}{3} - 4 = \tfrac{16}{3}\sqrt{2} - \tfrac{14}{3}$$

49.
$$\left\{\begin{array}{l} u = x^2+1 \\ du = 2x\,dx \end{array}\right. \left|\begin{array}{l} x=-1 \implies u=2 \\ x=0 \implies u=1 \end{array}\right\};\quad \int_{-1}^0 x^3\left(x^2+1\right)^6\,dx = \frac{1}{2}\int_2^1 (u-1)u^6\,du$$

$$= \frac{1}{2}\int_2^1 (u^7 - u^6)\,du = \left[\frac{1}{16}u^8 - \frac{1}{14}u^7\right]_2^1 = -\frac{255}{16} + \frac{127}{14} = -\frac{769}{112}$$

SECTION 5.9

1. $\displaystyle\int \cos(3x-1)\,dx = \frac{1}{3}\sin(3x-1) + C$ **3.** $\displaystyle\int \csc^2 \pi x\,dx = -\frac{1}{\pi}\cot \pi x + C$

5. $\left\{\begin{array}{l} u = 3-2x \\ du = -2\,dx \end{array}\right\};\quad \displaystyle\int \sin(3-2x)\,dx = \int -\frac{1}{2}\sin u\,du = \frac{1}{2}\cos u + C = \frac{1}{2}\cos(3-2x) + C$

7. $\left\{\begin{array}{l} u = \cos x \\ du = -\sin x\,dx \end{array}\right\};\quad \displaystyle\int \cos^4 x \sin x\,dx = \int -u^4\,du = -\frac{1}{5}u^5 + C = -\frac{1}{5}\cos^5 x + C$

9. $\left\{\begin{array}{l} u = x^{1/2} \\ du = \frac{1}{2}x^{-1/2}\,dx \end{array}\right\};\quad \displaystyle\int x^{-1/2}\sin x^{1/2}\,dx = \int 2\sin u\,du = -2\cos u + C$

$$= -2\cos x^{1/2} + C$$

11. $\left\{\begin{array}{l} u = 1+\sin x \\ du = \cos x\,dx \end{array}\right\};\quad \displaystyle\int \sqrt{1+\sin x}\ \cos x\,dx = \int u^{1/2}\,du = \frac{2}{3}u^{3/2} + C$

$$= \tfrac{2}{3}(1+\sin x)^{3/2} + C$$

13. $\displaystyle\int \frac{1}{\cos^2 x}\,dx = \int \sec^2 x\,dx = \tan x + C$

15. $\left\{\begin{array}{l} u = x^2 \\ du = 2x\,dx \end{array}\right\};\quad \displaystyle\int x\sin^3 x^2 \cos x^2\,dx = \int \frac{1}{2}\sin^3 u \cos u\,du = \frac{1}{8}\sin^4 u + C$

$$= \tfrac{1}{8}\sin^4 x^2 + C$$

17. $\left\{\begin{array}{l} u = \cot x \\ du = -\csc^2 x\, dx \end{array}\right\};$ $\displaystyle\int (1 + \cot^2 x)\csc^2 x\, dx = \int -(1+u^2)\, du = -u - \frac{1}{3}u^3 + C$

$$= -\cot x - \tfrac{1}{3}\cot^3 x + C$$

19. $\displaystyle\int \sin^2 3x\, dx = \int \left(\frac{1}{2} - \frac{1}{2}\cos 6x\right) dx = \frac{x}{2} - \frac{1}{12}\sin 6x + C$

21. $\left\{\begin{array}{l} u = 1 + \tan x \\ du = \sec^2 x\, dx \end{array}\right\};$ $\displaystyle\int \frac{\sec^2 x}{\sqrt{1 + \tan x}}\, dx = \int u^{-1/2}\, du = 2u^{1/2} + C$

$$= 2(1 + \tan x)^{1/2} + C$$

23.   0;   the sine is an odd function   **25.** $\displaystyle\int_{1/4}^{1/3} \sec^2 \pi x\, dx = \frac{1}{\pi}[\tan \pi x]_{1/4}^{1/3} = \frac{1}{\pi}(\sqrt{3} - 1)$

27. $\displaystyle\int_0^{\pi/2} \sin^3 x \cos x\, dx = \frac{1}{4}[\sin^4 x]_0^{\pi/2} = \frac{1}{4}$ **29.** $\displaystyle\int_{\pi/6}^{\pi/4} \csc x \cot x\, dx = [-\csc x]_{\pi/6}^{\pi/4} = 2 - \sqrt{2}$

31. $\displaystyle\int_0^{2\pi} \cos^2 x\, dx = \int_0^{2\pi} \left(\frac{1}{2} + \frac{1}{2}\cos 2x\right) dx = \left[\frac{x}{2} + \frac{\sin 2x}{4}\right]_0^{2\pi} = \pi$

33. $\displaystyle A = \int_0^{1/4} (\cos^2 \pi x - \sin^2 \pi x)\, dx = \int_0^{1/4} \cos 2\pi x\, dx$

$$= \frac{1}{2\pi}[\sin 2\pi x]_0^{1/4} = \frac{1}{2\pi}$$

35. $\displaystyle A = \int_{1/6}^{1/4} (\csc^2 \pi x - \sec^2 \pi x)\, dx = \frac{1}{\pi}[-\cot \pi x - \tan \pi x]_{1/6}^{1/4}$

$$= \frac{1}{\pi}\left(-2 + \cot \frac{\pi}{6} + \tan \frac{\pi}{6}\right)$$

$$= \frac{1}{\pi}\left(-2 + \sqrt{3} + \frac{1}{\sqrt{3}}\right) = \frac{1}{3\pi}(4\sqrt{3} - 6)$$

37. (a) $\left\{\begin{array}{l} u = \sec x \\ du = \sec x \tan x\, dx \end{array}\right\};$ $\displaystyle\int \sec^2 x \tan x\, dx = \int u\, du = \frac{1}{2}u + C$

$$= \tfrac{1}{2}\sec^2 x + C$$

(b) $\begin{cases} u = \tan x \\ du = \sec^2 x \, dx \end{cases}$; $\displaystyle\int \sec^2 x \tan x \, dx = \int u \, du = \frac{1}{2}u^2 + C'$

$$= \tfrac{1}{2}\tan^2 x + C'$$

(c) $C' = C + \frac{1}{2}$

39. $\displaystyle A = \frac{4b}{a}\int_0^a \sqrt{a^2 - x^2}\, dx = \frac{4b}{a}\left(\frac{\text{area of circle of radius } a}{4}\right)$

$$= \frac{4b}{a}\left(\frac{\pi a^2}{4}\right) = \pi ab$$

SECTION 5.10

1.   Yes; $\displaystyle\int_a^b [f(x) - g(x)]\, dx = \int_a^b f(x)\, dx - \int_a^b g(x)\, dx > 0.$

3.   Yes;   otherwise we would have $f(x) \le g(x)$ for all $x \in [a, b]$ and it would follow that

$$\int_a^b f(x)\, dx \le \int_a^b g(x)\, dx.$$

5.   No;   take $f(x) = 0$, $g(x) = -1$ on $[0,1]$.

7.   No;   take, for example, any odd function on an interval of the form $[-c, c]$.

9.   No; $\displaystyle\int_{-1}^1 x \, dx = 0$ but $\displaystyle\int_{-1}^1 |x|\, dx \ne 0.$

11.   Yes; $\displaystyle U_f(P) \ge \int_a^b f(x)\, dx = 0.$ **13.**   No; $\displaystyle L_f(P) \le \int_a^b f(x)\, dx = 0.$

15.   Yes; $\displaystyle\int_a^b [f(x) + 1]\, dx = \int_a^b f(x)\, dx + \int_a^b 1\, dx = 0 + b - a = b - a.$

17. $\displaystyle\frac{d}{dx}\left[\int_0^{1+x^2} \frac{dt}{\sqrt{2t + 5}}\right] = \frac{1}{\sqrt{2(1 + x^2) + 5}}\,\frac{d}{dx}(1 + x^2) = \frac{2x}{\sqrt{2x^2 + 7}}$

19. $\displaystyle\frac{d}{dx}\left[\int_x^a f(t)\, dt\right] = \frac{d}{dx}\left[-\int_a^x f(t)\, dt\right] = -f(x)$

21. $\displaystyle\frac{d}{dx}\left[\int_x^{x^2} \frac{dt}{t}\right] = \frac{1}{x^2}\frac{d}{dx}(x^2) - \frac{1}{x}\frac{d}{dx}(x) = \frac{2x}{x^2} - \frac{1}{x} = \frac{1}{x}$

23.
$$\frac{d}{dx}\left[\int_{x^{1/3}}^{2+3x}\frac{dt}{1+t^{3/2}}\right] = \frac{1}{1+(2+3x)^{3/2}}\frac{d}{dx}(2+3x) - \frac{1}{1+(x^{1/3})^{3/2}}\frac{d}{dx}\left(x^{1/3}\right)$$

$$= \frac{3}{1+(2+3x)^{3/2}} - \frac{1}{3x^{2/3}(1+x^{1/2})}$$

25. (a) With P a partition of $[a, b]$

$$L_f(P) \le \int_a^b f(x)\,dx.$$

If f is nonnegative on $[a, b]$, then $L_f(P)$ is nonnegative and, consequently, so is the integral. If f is positive on $[a, b]$, then $L_f(P)$ is positive and, consequently, so is the integral.

(b) Take F as an antiderivative of f on $[a, b]$. Observe that

$$F'(x) = f(x) \text{ on } (a, b) \quad \text{and} \quad \int_a^b f(x)\,dx = F(b) - F(a).$$

If $f(x) \ge 0$ on $[a, b]$, then F is nondecreasing on $[a, b]$ and $F(b) - F(a) \ge 0$. If $f(x) > 0$ on $[a, b]$, then F is increasing on $[a, b]$ and $F(b) - F(a) > 0$.

27. For all $x \in [a, b]$

$$-f(x) \le |f(x)| \quad \text{and} \quad f(x) \le |f(x)|.$$

It follows from II that

$$-\int_a^b f(x)\,dx = \int_a^b -f(x)\,dx \le \int_a^b |f(x)|\,dx \quad \text{and} \quad \int_a^b f(x)\,dx \le \int_a^b |f(x)|\,dx,$$

and thus

$$\left|\int_a^b f(x)\,dx\right| \le \int_a^b |f(x)|\,dx.$$

29.
$$H(x) = \int_{2x}^{x^3-4}\frac{x\,dt}{1+\sqrt{t}} = x\int_{2x}^{x^3-4}\frac{dt}{1+\sqrt{t}},$$

$$H'(x) = x\cdot\left[\frac{3x^2}{1+\sqrt{x^3-4}} - \frac{2}{1+\sqrt{2x}}\right] + 1\cdot\int_{2x}^{x^3-4}\frac{dt}{1+\sqrt{t}},$$

$$H'(2) = 2\left[\frac{12}{3} - \frac{2}{3}\right] + \underbrace{\int_4^4\frac{dt}{1+\sqrt{t}}}_{=0} = \frac{20}{3}$$

SECTION 5.11

1.   A.V. $= \dfrac{1}{c} \displaystyle\int_0^c (mx + b)\, dx = \dfrac{1}{c} \left[\dfrac{m}{2}x^2 + bx\right]_0^c = \dfrac{mc}{2} + b;$ at $x = c/2$

3.   A.V. $= \dfrac{1}{2} \displaystyle\int_{-1}^1 x^3\, dx = 0$ since the integrand is odd;   at $x = 0$

5.   A.V. $= \dfrac{1}{4} \displaystyle\int_{-2}^2 |x|\, dx = \dfrac{1}{2}\displaystyle\int_0^2 |x|\, dx = \dfrac{1}{2}\displaystyle\int_0^2 x\, dx = \dfrac{1}{2}\left[\dfrac{x^2}{2}\right]_0^2 = 1;$ at $x = \pm 1$

7.   A.V. $= \dfrac{1}{2} \displaystyle\int_0^2 (2x - x^2)\, dx = \dfrac{1}{2}\left[x^2 - \dfrac{x^3}{3}\right]_0^2 = \dfrac{2}{3};$ at $x = 1 \pm \dfrac{1}{3}\sqrt{3}$

9.   A.V. $= \dfrac{1}{9} \displaystyle\int_0^9 \sqrt{x}\, dx = \dfrac{1}{9}\left[\dfrac{2}{3}x^{3/2}\right]_0^9 = 2;$ at $x = 4$

11.   A.V. $= \dfrac{1}{2\pi} \displaystyle\int_0^{2\pi} \sin x\, dx = \dfrac{1}{2\pi}[-\cos x]_0^{2\pi} = 0;$ at $x = 0,\, \pi,\, 2\pi$

13. $0 = \displaystyle\int_a^b [f(x) - A]\, dx = \displaystyle\int_a^b f(x)\, dx - A\displaystyle\int_a^b dx = \displaystyle\int_a^b f(x)\, dx - A\,(b - a).$

   Thus, $A = \dfrac{1}{b - a} \displaystyle\int_a^b f(x)\, dx = $ average value of f on $[a, b]$.

15.   Average of f' on $[a, b] = \dfrac{1}{b - a} \displaystyle\int_a^b f'(x)\, dx = \dfrac{1}{b - a}\,[f(x)]_a^b = \dfrac{f(b) - f(a)}{b - a}.$

17.   (a)   A.V. $= \dfrac{1}{\sqrt{3}} \displaystyle\int_0^{\sqrt{3}} y\, dx = \dfrac{1}{\sqrt{3}}\displaystyle\int_0^{\sqrt{3}} x^2\, dx = \dfrac{1}{\sqrt{3}}\left[\dfrac{x^3}{3}\right]_0^{\sqrt{3}} = 1$

   (b)   A.V. $= \dfrac{1}{3} \displaystyle\int_0^3 x\, dy = \dfrac{1}{3}\displaystyle\int_0^3 \sqrt{y}\, dy = \dfrac{1}{3}\left[\dfrac{2}{3}y^{3/2}\right]_0^3 = \dfrac{2}{3}\sqrt{3}$

   (c)   A.V. $= \dfrac{1}{\sqrt{3}} \displaystyle\int_0^{\sqrt{3}} \sqrt{(x - 0)^2 + (x^2 - 0)^2}\, dx = \dfrac{1}{\sqrt{3}}\displaystyle\int_0^{\sqrt{3}} x\sqrt{1 + x^2}\, dx$

   $= \dfrac{1}{\sqrt{3}} \left[\dfrac{1}{3}(1 + x^2)^{3/2}\right]_0^{\sqrt{3}} = \dfrac{7}{9}\sqrt{3}$

19.   The distance the stone has fallen after t seconds is given by $s(t) = 16t^2$.

   (a)   The terminal velocity after x seconds is $s'(x) = 32x$. The average velocity
   is $\dfrac{s(x) - s(0)}{x - 0} = 16x$. Thus the terminal velocity is twice the average velocity.

   (b)   For the first $\frac{1}{2}x$ seconds, aver. vel. $= \dfrac{s\left(\frac{1}{2}x\right) - s(0)}{\frac{1}{2}x - 0} = 8x.$

For the next $\frac{1}{2}x$ seconds, aver. vel. $= \dfrac{s(x) - s(\frac{1}{2}x)}{x - \frac{1}{2}x} = 24x.$

Thus, for the first $\frac{1}{2}x$ seconds the average velocity is one-third of the average velocity during the next $\frac{1}{2}x$ seconds.

21.   (a)
$$M = \int_0^L k\sqrt{x}\, dx = k\left[\frac{2}{3}x^{3/2}\right]_0^L = \frac{2}{3}kL^{3/2}$$

$$x_M M = \int_0^L x\left(k\sqrt{x}\right) dx = \int_0^L kx^{3/2}\, dx = \left[\frac{2}{5}kx^{5/2}\right]_0^L = \frac{2}{5}kL^{5/2}$$

$$x_M = \left(\tfrac{2}{5}kL^{5/2}\right) / \left(\tfrac{2}{3}kL^{3/2}\right) = \tfrac{3}{5}L$$

(b)
$$M = \int_0^L k\left(L - x\right)^2 dx = \left[-\frac{1}{3}k\left(L - x\right)^3\right]_0^L = \frac{1}{3}kL^3$$

$$x_M M = \int_0^L x\left[k\left(L - x\right)^2\right] dx = \int_0^L k\left(L^2 x - 2Lx^2 + x^3\right) dx$$

$$= k\left[\tfrac{1}{2}L^2 x^2 - \tfrac{2}{3}Lx^3 + \tfrac{1}{4}x^4\right]_0^L = \tfrac{1}{12}kL^4$$

$$x_M = \left(\tfrac{1}{12}kL^4\right) / \left(\tfrac{1}{3}kL^3\right) = \tfrac{1}{4}L$$

23.
$$\tfrac{1}{4}LM = \tfrac{1}{8}LM_1 + x_{M_2}M_2$$

$$x_{M_2} = \frac{1}{M_2}\left(\frac{1}{4}LM - \frac{1}{8}LM_1\right) = \frac{L}{8M_2}(2M - M_1)$$

25.   If f is continuous on $[a, b]$, then, by Theorem 5.3.5, F satisfies the conditions of the mean-value theorem of differential calculus (Theorem 4.1.1). Therefore, by that theorem, there is at least one number c in (a, b) for which
$$F'(c) = \frac{F(b) - F(a)}{b - a}.$$
Then
$$\int_a^b f(x)\, dx = F(b) - F(a) = F'(c)(b - a) = f(c)(b - a).$$

27.   If f and g take on the same average value on every interval $[a, x]$, then
$$\frac{1}{x - a}\int_a^x f(t)\, dt = \frac{1}{x - a}\int_a^x g(t)\, dt.$$

Multiplication by $(x - a)$ gives

$$\int_a^x f(t)\, dt = \int_a^x g(t)\, dt.$$

Differentiation with respect to x gives $f(x) = g(x)$. This shows that, if the averages are everywhere the same, then the functions are everywhere the same.

SECTION 5.12

1. $\displaystyle\int (\sqrt{x-a} - \sqrt{x-b})\, dx \quad = \int \left[(x-a)^{1/2} - (x-b)^{1/2} \right] dx$

$$= \tfrac{2}{3} \left[(x-a)^{3/2} - (x-b)^{3/2} \right] + C$$

3. $\left\{ \begin{array}{l} u = t^{2/3} - 1 \\ du = \frac{2}{3} t^{-1/3}\, dt \end{array} \right\}; \quad \displaystyle\int t^{-1/3}(t^{2/3} - 1)^2\, dt = \int \frac{3}{2} u^2\, du = \frac{1}{2} u^3 + C$

$$= \tfrac{1}{2}(t^{2/3} - 1)^3 + C$$

5. $\displaystyle\int (1 + 2\sqrt{x})^2\, dx = \int (1 + 4x^{1/2} + 4x)\, dx = x + \frac{8}{3} x^{3/2} + 2x^2 + C$

7. $\displaystyle\int (x^{1/5} - x^{-1/5})^2\, dx = \int (x^{2/5} - 2 + x^{-2/5})\, dx = \frac{5}{7} x^{7/5} - 2x + \frac{5}{3} x^{3/5} + C$

9. $\left\{ \begin{array}{l} u = 2 - x \\ du = -dx \end{array} \right\}; \quad \displaystyle\int x\sqrt{2-x}\, dx = -\int (2-u)u^{1/2}\, du = \int (u^{3/2} - 2u^{1/2})\, du$

$$= \tfrac{2}{5} u^{5/2} - \tfrac{4}{3} u^{3/2} + C = \tfrac{2}{5}(2-x)^{5/2} - \tfrac{4}{3}(2-x)^{3/2} + C$$

11. $\left\{ \begin{array}{l} u = a + b\sqrt{y+1} \\ du = \dfrac{b\, dy}{2\sqrt{y+1}} \end{array} \right\}; \quad \displaystyle\int \frac{(a + b\sqrt{y+1})^2}{\sqrt{y+1}}\, dy = \int \frac{2}{b} u^2\, du = \frac{2}{3b} u^3 + C$

$$= \frac{2}{3b} \left(a + b\sqrt{y+1} \right)^3 + C$$

13. $\left\{ \begin{array}{l} u = g(x) \\ du = g'(x)\, dx \end{array} \right\}; \quad \displaystyle\int \frac{g'(x)}{[g(x)]^3}\, dx = \int \frac{du}{u^3} = \int u^{-3}\, du = -\frac{1}{2} u^{-2} + C$

$$= -\frac{1}{2[g(x)]^2} + C$$

15. $\left\{ \begin{array}{l} u = g(x) \\ du = g'(x)\, dx \end{array} \right\}; \quad \displaystyle\int \frac{g(x)g'(x)}{\sqrt{1 + [g(x)]^2}}\, dx = \int \frac{u}{\sqrt{1 + u^2}}\, du = \sqrt{1 + u^2} + C$

$$= \sqrt{1 + [g(x)]^2} + C$$

17. $\displaystyle\int (\sec\theta - \tan\theta)^2\, d\theta = \int (\sec^2\theta - 2\sec\theta\tan\theta + \tan^2\theta)\, d\theta$

$$= \int (2\sec^2\theta - 2\sec\theta\tan\theta - 1)\, d\theta = 2\tan\theta - 2\sec\theta - \theta + C$$

19. $\displaystyle\int \frac{dx}{1 + \cos 2x}\quad = \int \left(\frac{1 - \cos 2x}{1 - \cos^2 2x}\right) dx = \int \left(\frac{1 - \cos 2x}{\sin^2 2x}\right) dx$

$$= \int \left(\sec^2 2x - \frac{\cos 2x}{\sin^2 2x}\right) dx = -\frac{1}{2}\cot 2x + \frac{1}{2}\left(\frac{1}{\sin 2x}\right) + C$$

$$= \tfrac{1}{2}(\csc 2x - \cot 2x) + C$$

21. $\displaystyle\int \sec^4\pi x\tan\pi x\, dx = \int \sec^3\pi x\, (\sec\pi x\tan\pi x)\, dx$

$$= \frac{1}{4\pi}\sec^4\pi x + C$$

23. $\displaystyle\int (1 + \sin^2\pi x)^{-3}\sin 2\pi x\, dx = \int (1 + \sin^2\pi x)^{-3}\cdot 2\sin\pi x\cos\pi x\, dx$

$$= -\frac{1}{2\pi}\left(1 + \sin^2\pi x\right)^{-2} + C$$

25. $\displaystyle A = \int_0^2 x\sqrt{2x^2 + 1}\, dx \qquad \left\{\begin{array}{l|l} u = 2x^2 + 1 & x = 0 \implies u = 1 \\ du = 4x\, dx & x = 2 \implies u = 9 \end{array}\right\}$

$$= \int_1^9 \frac{1}{4}u^{1/2}\, du = \left[\frac{1}{6}u^{3/2}\right]_1^9 = \frac{13}{3}$$

27. $\displaystyle A = \int_1^2 x^{-3}(1 + x^{-2})^{-3}\, dx \qquad \left\{\begin{array}{l|l} u = 1 + x^{-2} & x = 1 \implies u = 2 \\ du = -2x^{-3}\, dx & x = 2 \implies u = 5/4 \end{array}\right\}$

$$= \int_2^{5/4} \left(-\frac{1}{2}u^{-3}\right) du = \left[\frac{1}{4}u^{-2}\right]_2^{5/4} = \frac{39}{400}$$

29. $\displaystyle\left\{\begin{array}{l} u = x^2 + 1 \\ du = 2x\, dx \end{array}\right\};\qquad y = \int x\sqrt{x^2 + 1}\, dx = \frac{1}{2}\int u^{1/2}\, du = \frac{1}{3}u^{3/2} + C$

$$= \tfrac{1}{3}\left(x^2 + 1\right)^{3/2} + C$$

Since $(0, 1)$ lies on the curve, we get $1 = \frac{1}{3}(1) + C$ so $C = 2/3$. The equation for the curve can be written $y = \frac{1}{3}(x^2 + 1)^{3/2} + \frac{2}{3}$.

31. (a) $\displaystyle\int_0^{x_0} [g(x) - y_0]\, dx$ \qquad\qquad (b) $\displaystyle\int_0^{x_0} [y_0 - f(x)]\, dx$

(c) $\displaystyle\int_0^{x_0} f(x)\, dx + \int_{x_0}^a g(x)\, dx$

(d) $\displaystyle\int_0^{x_0} \left[\left(\frac{c - b}{a}x + b\right) - g(x)\right] dx + \int_{x_0}^a \left[\left(\frac{c - b}{a}x + b\right) - f(x)\right] dx$

(e) $\displaystyle\int_{x_0}^a [f(x) - g(x)]\, dx$

33.

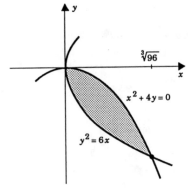

$$A = \int_0^{(96)^{1/3}} \left[\left(-\frac{1}{4}x^2\right) - (-\sqrt{6x})\right]\,dx$$

$$= \int_0^{(96)^{1/3}} \left[(6x)^{1/2} - \frac{1}{4}x^2\right]\,dx$$

$$= \left[\tfrac{1}{9}(6x)^{3/2} - \tfrac{1}{12}x^3\right]_0^{(96)^{1/3}}$$

$$= \left[\tfrac{1}{9}(12)^2 - 8\right] = 8$$

35.

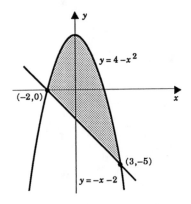

$$A = \int_{-2}^3 \left[(4 - x^2) - (-x - 2)\right]\,dx$$

$$= \int_{-2}^3 (6 + x - x^2)\,dx$$

$$= \left[6x + \frac{x^2}{2} - \frac{x^3}{3}\right]_{-2}^3$$

$$= \left(\tfrac{27}{2}\right) - \left(-\tfrac{22}{3}\right) = \tfrac{125}{6}$$

37.

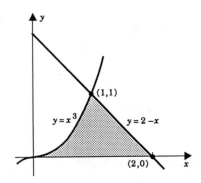

$$A = \int_0^1 x^3\,dx + \int_1^2 (2 - x)\,dx$$

$$= \left[\frac{x^4}{4}\right]_0^1 + \left[2x - \frac{x^2}{2}\right]_1^2$$

$$= \tfrac{1}{4} + \left(2 - \tfrac{3}{2}\right) = \tfrac{3}{4}$$

39.

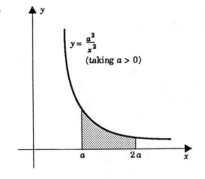

$$A = \int_a^{2a} \frac{a^2}{x^2}\,dx = \left[-a^2 x^{-1}\right]_a^{2a} = \frac{a}{2}$$

41. necessarily holds: $L_g(P) \le \int_a^b g(x)\,dx < \int_a^b f(x)\,dx \le U_f(P)$

43. necessarily holds: $L_g(P) \le \int_a^b g(x)\,dx < \int_a^b f(x)\,dx$

45. necessarily holds: $U_f(P) \ge \int_a^b f(x)\,dx > \int_a^b g(x)\,dx$

47. $a(t) = -\omega^2 A \sin(\omega t + \phi_0)$ gives $v(t) = \omega A \cos(\omega t + \phi_0) + C.$

Since $v_0 = v(0) = \omega A \cos\phi_0 + C,$ we have $C = v_0 - \omega A \cos\phi_0.$ Therefore,

$$v(t) = \omega A \cos(\omega t + \phi_0) + v_0 - \omega A \cos\phi_0.$$

This gives

$$x(t) = A \sin(\omega t + \phi_0) + (v_0 - \omega A \cos\phi_0)t + K.$$

Since $x_0 = x(0) = A \sin\phi_0 + K,$ we have $K = x_0 - A \sin\phi_0.$ Therefore,

$$x(t) = A \sin(\omega t + \phi_0) + (v_0 - \omega A \cos\phi_0)t + x_0 - A \sin\phi_0.$$

49. $\left\{ \begin{array}{l} u = 1 + \sqrt{x} \\ du = \dfrac{dx}{2\sqrt{x}} \end{array} \right\};$ $y = \int -\dfrac{dx}{2\sqrt{x}\,(1 + \sqrt{x})^2} = \int -\dfrac{1}{u^2}\,du = \dfrac{1}{u} + C = \dfrac{1}{1 + \sqrt{x}} + C.$

Since $(4, \tfrac{1}{3})$ lies on the curve, we get $\dfrac{1}{3} = \dfrac{1}{1+2} + C$ so that $C = 0.$

The equation of the curve can be written $y = 1/(1 + \sqrt{x}).$

51. $A \cong 0.615$:

$$L_f(P) = \tfrac{1}{100}\left(\tfrac{1}{1.1} + \tfrac{3}{1.2} + \tfrac{5}{1.3} + \tfrac{7}{1.4} + \tfrac{9}{1.5} + \tfrac{11}{1.6} + \tfrac{13}{1.7} + \tfrac{15}{1.8} + \tfrac{17}{1.9} + \tfrac{19}{2}\right) \cong 0.595580$$

$$U_f(P) = \tfrac{1}{100}\left(\tfrac{1}{1} + \tfrac{3}{1.1} + \tfrac{5}{1.2} + \tfrac{7}{1.3} + \tfrac{9}{1.4} + \tfrac{11}{1.5} + \tfrac{13}{1.6} + \tfrac{15}{1.7} + \tfrac{17}{1.8} + \tfrac{19}{1.9}\right) \cong 0.634334$$

$$\tfrac{1}{2}\left[L_f(P) + U_f(P)\right] \cong \tfrac{1}{2}\left[0.595580 + 0.634334\right] \cong 0.615$$

53. $\dfrac{2x}{1 + x^4}$

55. $\dfrac{d}{dx}\left(\int_0^{\sin x} \dfrac{dt}{1 - t^2}\right) = \dfrac{1}{1 - \sin^2 x}\dfrac{d}{dx}(\sin x) = \dfrac{\cos x}{\cos^2 x} = \sec x$

57. $\dfrac{d}{dx}\left(\int_0^{\tan x} \dfrac{dt}{1 + t^2}\right) = \dfrac{1}{1 + \tan^2 x}\dfrac{d}{dx}(\tan x) = \cos^2 x \sec^2 x = 1$

59. $M = \displaystyle\int_0^a k(2a - x)\,dx = k\left[2ax - \tfrac{1}{2}x^2\right]_0^a = \tfrac{3}{2}ka^2$

$$x_M M = \int_0^a kx(2a - x)\,dx = \int_0^a k(2ax - x^2)\,dx = k\left[ax^2 - \tfrac{1}{3}x^3\right]_0^a = \tfrac{2}{3}ka^3$$

$$x_M = \left(\tfrac{2}{3}ka^3\right) / \left(\tfrac{3}{2}ka^2\right) = \tfrac{4}{9}a$$

61. $f'(x) = \dfrac{x-1}{1+x^2}$ is negative for $x < 1$, zero at $x = 1$, positive for $x > 1$.

Therefore f decreases on $(-\infty, 0]$ and increases on $[0, \infty)$. Thus f takes on its minimum value at $x = 1$.

SECTION 5.13

1.

3. (a) $\Delta x_1 = \Delta x_2 = \frac{1}{8}, \quad \Delta x_3 = \Delta x_4 = \Delta x_5 = \frac{1}{4}$

(b) $\|P\| = \frac{1}{4}$

(c) $m_1 = 0, \quad m_2 = \frac{1}{4}, \quad m_3 = \frac{1}{2}, \quad m_4 = 1, \quad m_5 = \frac{3}{2}$

(d) $f(x_1^*) = \frac{1}{8}, \quad f(x_2^*) = \frac{3}{8}, \quad f(x_3^*) = \frac{3}{4}, \quad f(x_4^*) = \frac{5}{4},$

  $f(x_5^*) = \frac{3}{2}$

(e) $M_1 = \frac{1}{4}, \quad M_2 = \frac{1}{2}, \quad M_3 = 1, \quad M_4 = \frac{3}{2}, \quad M_5 = 2$

(f) $L_f(P) = \frac{25}{32}$ (g) $S^*(P) = \frac{15}{16}$

(h) $U_f(P) = \frac{39}{32}$ (i) $\displaystyle\int_a^b f(x)\,dx = 1$

5. (a) $\dfrac{1}{n^2}(1 + 2 + \cdots + n) = \dfrac{1}{n^2}\left[\dfrac{n(n+1)}{2}\right] = \dfrac{1}{2} + \dfrac{1}{2n}$

(b) $S_n^* = \dfrac{1}{2} + \dfrac{1}{2n}, \quad \displaystyle\int_0^1 x\,dx = \left[\dfrac{1}{2}x^2\right]_0^1 = \dfrac{1}{2}$

$\left| S_n^* - \displaystyle\int_0^1 x\,dx \right| = \dfrac{1}{2n} < \dfrac{1}{n} < \epsilon \quad \text{if} \quad n > \dfrac{1}{\epsilon}$

7. (a) $\dfrac{1}{n^4}\left(1^3 + 2^3 + \cdots + n^3\right) = \dfrac{1}{n^4}\left[\dfrac{n^2(n+1)^2}{4}\right] = \dfrac{1}{4} + \dfrac{1}{2n} + \dfrac{1}{4n^2}$

(b) $S_n^* = \dfrac{1}{4} + \dfrac{1}{2n} + \dfrac{1}{4n^2}, \quad \displaystyle\int_0^1 x^3 dx = \left[\dfrac{1}{4}x^4\right]_0^1 = \dfrac{1}{4}$

$\left| S_n^* - \displaystyle\int_0^1 x^3 dx \right| = \dfrac{1}{2n} + \dfrac{1}{4n^2} < \dfrac{1}{n} < \epsilon \quad \text{if} \quad n > \dfrac{1}{\epsilon}$

9. $S^*(P) = \frac{1}{3}\left[\frac{1}{6}\cos\left(\frac{1}{6}\right)^2 + \frac{3}{6}\cos\left(\frac{3}{6}\right)^2 + \frac{5}{6}\cos\left(\frac{5}{6}\right)^2 + \frac{7}{6}\cos\left(\frac{7}{6}\right)^2 + \frac{9}{6}\cos\left(\frac{9}{6}\right)^2 + \frac{11}{6}\cos\left(\frac{11}{6}\right)^2\right]$

  $\cong -0.3991$

$\displaystyle\int_0^2 x\cos x^2\,dx = \left[\frac{1}{2}\sin x^2\right]_0^2 = \frac{1}{2}\sin 4 \cong -0.3784$

CHAPTER 6

SECTION 6.1

1.

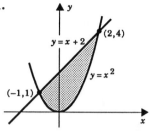

(a) $\displaystyle\int_{-1}^{2} [(x+2) - x^2]\, dx$

(b) $\displaystyle\int_{0}^{1} [(\sqrt{y}) - (-\sqrt{y})]\, dy + \int_{1}^{4} [(\sqrt{y}) - (y-2)]\, dy$

3.

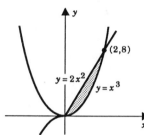

(a) $\displaystyle\int_{0}^{2} [(2x^2) - (x^3)]\, dx$

(b) $\displaystyle\int_{0}^{8} \left[(y^{1/3}) - \left(\frac{1}{2}y\right)^{1/2} \right] dy$

5.

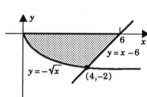

(a) $\displaystyle\int_{0}^{4} [(0) - (-\sqrt{x})]\, dx + \int_{4}^{6} [(0) - (x-6)]\, dx$

(b) $\displaystyle\int_{-2}^{0} [(y+6) - (y^2)]\, dy$

7.

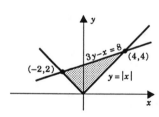

(a) $\displaystyle\int_{-2}^{0} \left[\left(\frac{8+x}{3}\right) - (-x) \right] dx + \int_{0}^{4} \left[\left(\frac{8+x}{3}\right) - (x) \right] dx$

(b) $\displaystyle\int_{0}^{2} [(y) - (-y)]\, dy + \int_{2}^{4} [(y) - (3y - 8)]\, dy$

9.

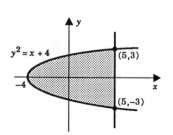

(a) $\displaystyle\int_{-4}^{5} [(\sqrt{4+x}) - (-\sqrt{4+x})]\, dx$

(b) $\displaystyle\int_{-3}^{3} [(5) - (y^2 - 4)]\, dy$

11.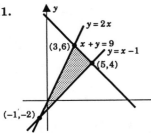

(a) $\displaystyle\int_{-1}^{3} [(2x) - (x-1)]\, dx + \int_{3}^{5} [(9-x) - (x-1)]\, dx$

(b) $\displaystyle\int_{-2}^{4} \left[(y+1) - \left(\frac{1}{2}y\right)\right] dy + \int_{4}^{6} \left[(9-y) - \left(\frac{1}{2}y\right)\right] dy$

13.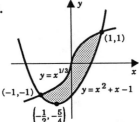

(a) $\displaystyle\int_{-1}^{1} \left[\left(x^{1/3}\right) - (x^2 + x - 1)\right] dx$

(b) $\displaystyle\int_{-5/4}^{-1} \left[\left(-\frac{1}{2} + \frac{1}{2}\sqrt{4y+5}\right) - \left(-\frac{1}{2} - \frac{1}{2}\sqrt{4y+5}\right)\right] dy$

$\displaystyle +\int_{-1}^{1} \left[\left(-\frac{1}{2} + \frac{1}{2}\sqrt{4y+5}\right) - (y^3)\right] dy$

15.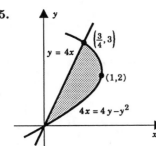

$\displaystyle A = \int_{0}^{3} \left[\left(\frac{4y - y^2}{4}\right) - \left(\frac{y}{4}\right)\right] dy$

$\displaystyle = \int_{0}^{3} \left(\frac{3}{4}y - \frac{1}{4}y^2\right) dy$

$\displaystyle = \left[\frac{3}{8}y^2 - \frac{1}{12}y^3\right]_0^3 = \frac{9}{8}$

17.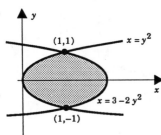

$\displaystyle A = 2\int_{0}^{1} \left[(3 - 2y^2) - (y^2)\right] dy$

$\displaystyle = 2\int_{0}^{1} (3 - 3y^2)\, dy$

$\displaystyle = 2\left[3y - y^3\right]_0^1 = 2\,(2) = 4$

19.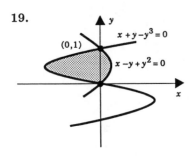

$\displaystyle A = \int_{0}^{1} \left[(y - y^2) - (y^3 - y)\right] dy$

$\displaystyle = \int_{0}^{1} (2y - y^2 - y^3)\, dy$

$\displaystyle = \left[y^2 - \frac{1}{3}y^3 - \frac{1}{4}y^4\right]_0^1 = \frac{5}{12}$

SECTION 6.2

1.

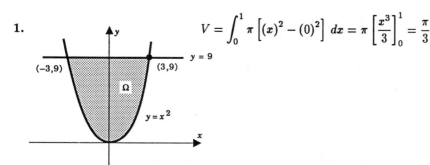

$$V = \int_0^1 \pi \left[(x)^2 - (0)^2 \right] \, dx = \pi \left[\frac{x^3}{3} \right]_0^1 = \frac{\pi}{3}$$

3.

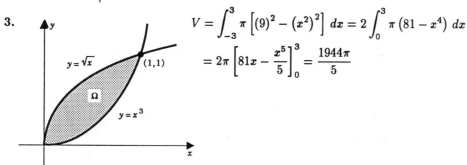

$$V = \int_{-3}^3 \pi \left[(9)^2 - \left(x^2 \right)^2 \right] \, dx = 2 \int_0^3 \pi \left(81 - x^4 \right) \, dx$$

$$= 2\pi \left[81x - \frac{x^5}{5} \right]_0^3 = \frac{1944\pi}{5}$$

5.

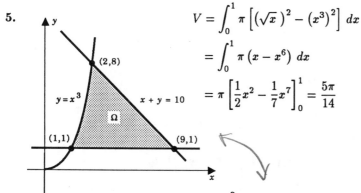

$$V = \int_0^1 \pi \left[\left(\sqrt{x} \right)^2 - \left(x^3 \right)^2 \right] \, dx$$

$$= \int_0^1 \pi \left(x - x^6 \right) \, dx$$

$$= \pi \left[\frac{1}{2} x^2 - \frac{1}{7} x^7 \right]_0^1 = \frac{5\pi}{14}$$

7.

$$V = \int_1^2 \pi \left[\left(x^3 \right)^2 - (1)^2 \right] \, dx + \int_2^9 \pi \left[(10-x)^2 - (1)^2 \right] \, dx$$

$$= \int_1^2 \pi \left(x^6 - 1 \right) \, dx + \int_2^9 \pi \left(99 - 20x + x^2 \right) \, dx$$

$$= \pi \left[\frac{1}{7} x^7 - x \right]_1^2 + \pi \left[99x - 10x^2 + \frac{1}{3} x^3 \right]_2^9 = \frac{3790\pi}{21}$$

9.

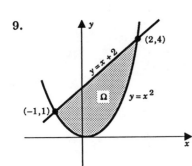

$$V = \int_{-1}^{2} \pi \left[(x+2)^2 - \left(x^2 \right)^2 \right] \, dx$$

$$= \int_{-1}^{2} \pi \left(x^2 + 4x + 4 - x^4 \right) \, dx$$

$$= \pi \left[\tfrac{1}{3} x^3 + 2x^2 + 4x - \tfrac{1}{5} x^5 \right]_{-1}^{2} = \tfrac{72}{5} \pi$$

11.

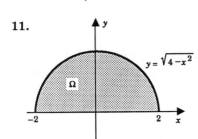

$$V = \int_{-2}^{2} \pi \left[\sqrt{4-x^2} \right]^2 \, dx = 2 \int_{0}^{2} \pi \left(4 - x^2 \right) \, dx$$

$$= 2\pi \left[4x - \frac{x^3}{3} \right]_{0}^{2} = \frac{32}{3} \pi$$

13.

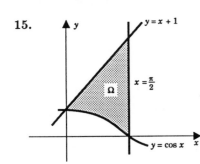

$$V = \int_{0}^{\pi/4} \pi \sec^2 x \, dx = \pi \left[\tan x \right]_{0}^{\pi/4} = \pi$$

15.

$$V = \int_{0}^{\pi/2} \pi \left[(x+1)^2 - (\cos x)^2 \right] \, dx$$

$$= \int_{0}^{\pi/2} \pi \left[(x+1)^2 - \left(\frac{1}{2} + \frac{1}{2} \cos 2x \right) \right] \, dx$$

$$= \pi \left[\frac{1}{3} (x+1)^3 - \frac{1}{2} x - \frac{1}{4} \sin 2x \right]_{0}^{\pi/2} = \frac{\pi^2}{24} \left(\pi^2 + 6\pi + 6 \right)$$

17.

$$V = \int_{0}^{4} \pi \left(\frac{y}{2} \right)^2 \, dy = \frac{\pi}{12} \left[y^3 \right]_{0}^{4} = \frac{16\pi}{3}$$

19.

$$V = \int_0^2 \pi \left[(8)^2 - \left(y^3\right)^2 \right] dy$$

$$= \int_0^2 \pi \left(64 - y^6 \right) dy$$

$$= \pi \left[64y - \tfrac{1}{7}y^7 \right]_0^2 = \tfrac{768}{7}\pi$$

21.

$$V = \int_0^1 \pi \left[\left(y^{1/3}\right)^2 - \left(y^2\right)^2 \right] dy$$

$$= \int_0^1 \pi \left[y^{2/3} - y^4 \right] dy$$

$$= \pi \left[\tfrac{3}{5}y^{5/3} - \tfrac{1}{5}y^5 \right]_0^1 = \tfrac{2}{5}\pi$$

23.

$$V = \int_0^4 \pi \left[y^2 - \left(\tfrac{y}{2}\right)^2 \right] dy + \int_4^8 \pi \left[4^2 - \left(\tfrac{y}{2}\right)^2 \right] dy$$

$$= \int_0^4 \pi \left[\tfrac{3}{4}y^2 \right] dy + \int_4^8 \pi \left[16 - \tfrac{1}{4}y^2 \right] dy$$

$$= \pi \left[\tfrac{1}{4}y^3 \right]_0^4 + \pi \left[16y - \tfrac{1}{12}y^3 \right]_4^8 = \tfrac{128}{3}\pi$$

25.

$$V = \int_{-1}^1 \pi \left[\left(2 - y^2\right)^2 - \left(y^2\right)^2 \right] dy$$

$$= 2 \int_0^1 \pi \left[4 - 4y^2 \right] dy = 2\pi \left[4y - \tfrac{4}{3}y^3 \right]_0^1 = \tfrac{16}{3}\pi$$

27. (a) $V = \int_{-r}^r \left(2\sqrt{r^2 - x^2} \right)^2 dx = 8 \int_0^r \left(r^2 - x^2 \right) dx = 8 \left[r^2 x - \tfrac{1}{3}x^3 \right]_0^r = \tfrac{16}{3}r^3$

(b) $V = \int_{-r}^r \tfrac{\sqrt{3}}{4} \left(2\sqrt{r^2 - x^2} \right)^2 dx = 2\sqrt{3} \int_0^r \left(r^2 - x^2 \right) dx = \tfrac{4\sqrt{3}}{3}r^3$

29. (a) $V = \displaystyle\int_{-2}^{2} \left(4 - x^2\right)^2 \, dx = 2\int_{0}^{2} \left(16 - 8x^2 + x^4\right)\, dx = 2\left[16x - \dfrac{8}{3}x^3 + \dfrac{1}{5}x^5\right]_0^2 = \dfrac{512}{15}$

  (b) $V = \displaystyle\int_{-2}^{2} \dfrac{\pi}{2}\left(\dfrac{4 - x^2}{2}\right)^2 \, dx = \dfrac{\pi}{4}\int_{0}^{2}\left(4 - x^2\right)^2\, dx = \dfrac{\pi}{4}\left(\dfrac{256}{15}\right) = \dfrac{64}{15}\pi$

  (c) $V = \displaystyle\int_{-2}^{2} \dfrac{\sqrt{3}}{4}\left(4 - x^2\right)^2 \, dx = \dfrac{\sqrt{3}}{2}\int_{0}^{2}\left(4 - x^2\right)^2\, dx = \dfrac{\sqrt{3}}{2}\left(\dfrac{256}{15}\right) = \dfrac{128}{15}\sqrt{3}$

31. (a) $V = \displaystyle\int_{0}^{4}\left[(\sqrt{y}) - (-\sqrt{y})\right]^2 \, dy = \int_{0}^{4} 4y\, dy = \left[2y^2\right]_0^4 = 32$

  (b) $V = \displaystyle\int_{0}^{4} \dfrac{\pi}{2}\left(\sqrt{y}\right)^2 \, dy = \dfrac{\pi}{2}\int_{0}^{4} y\, dy = \dfrac{\pi}{2}\left[\dfrac{1}{2}y^2\right]_0^4 = 4\pi$

  (c) $V = \displaystyle\int_{0}^{4} \dfrac{\sqrt{3}}{4}\left[(\sqrt{y}) - (-\sqrt{y})\right]^2 \, dy = \sqrt{3}\int_{0}^{4} y\, dy = 8\sqrt{3}$

33. $V = \displaystyle\int_{-a}^{a} \pi\left(b\sqrt{1 - \dfrac{x^2}{a^2}}\right)^2 \, dx = \dfrac{2b^2}{a^2}\int_{0}^{a} \pi\left(a^2 - x^2\right)^2\, dx = \dfrac{2b^2}{a^2}\pi\left[a^2 x - \dfrac{1}{3}x^3\right]_0^a$

$$= \dfrac{2b^2}{a^2}\pi\left(\dfrac{2}{3}a^3\right) = \dfrac{4}{3}\pi a b^2$$

35.

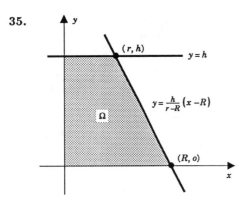

The specified frustum is generated by revolving the region Ω about the y-axis.

$$V = \int_{0}^{h} \pi\left[\dfrac{r - R}{h}y + R\right]^2 \, dy$$

$$= \pi\left[\dfrac{h}{3(r - R)}\left(\dfrac{r - R}{h}y + R\right)^3\right]_0^h$$

$$= \dfrac{\pi h}{3(r - R)}\left(r^3 - R^3\right) = \dfrac{\pi h}{3}\left(r^2 + rR + R^2\right)$$

37. Capacity of basin $= \dfrac{1}{2}\left(\dfrac{4}{3}\pi r^3\right) = \dfrac{2}{3}\pi r^3$.

  (a) Volume of water $= \displaystyle\int_{r/2}^{r} \pi\left[\sqrt{r^2 - x^2}\,\right]^2 \, dx$

$$= \pi\int_{r/2}^{r}\left(r^2 - x^2\right)\, dx = \pi\left[r^2 x - \dfrac{1}{3}x^3\right]_{r/2}^{r} = \dfrac{5}{24}\pi r^3.$$

  The basin is $\left(\dfrac{5}{24}\pi r^3\right)(100)\big/\left(\dfrac{2}{3}\pi r^3\right) = 31\tfrac{1}{4}\%$ full.

  (b) Volume of water $= \displaystyle\int_{2r/3}^{r} \pi\left[\sqrt{r^2 - x^2}\,\right]^2 \, dx = \pi\int_{2r/3}^{r}\left(r^2 - x^2\right)\, dx = \dfrac{8}{81}\pi r^3.$

  The basin is $\left(\dfrac{8}{81}\pi r^3\right)(100)\big/\left(\dfrac{2}{3}\pi r^3\right) = 14\tfrac{22}{27}\%$ full.

39.   (a) $V = \int_0^4 \pi \left(x^{3/2}\right)^2 dx = \pi \int_0^4 x^3 dx = \pi \left[\frac{1}{4}x^4\right]_0^4 = 64\pi$

(b) $V = \int_0^8 \pi \left(4 - y^{3/2}\right)^2 dy = \pi \int_0^8 \left(16 - 8y^{2/3} + y^{4/3}\right) dy$

$= \pi \left[16y - \frac{24}{5}y^{5/3} + \frac{3}{7}y^{7/3}\right]_0^8 = \frac{1024}{35}\pi$

(c) $V = \int_0^4 \pi \left[(8)^2 - \left(8 - x^{3/2}\right)^2\right] dx = \pi \int_0^4 \left(16x^{3/2} - x^3\right) dx$

$= \pi \left[\frac{32}{5}x^{5/2} - \frac{1}{4}x^4\right]_0^4 = \frac{704}{5}\pi$

(d) $V = \int_0^8 \pi \left[(4)^2 - \left(y^{2/3}\right)^2\right] dy = \pi \int_0^8 \left(16 - y^{4/3}\right) dy = \pi \left[16y - \frac{3}{7}y^{7/3}\right]_0^8 = \frac{512}{7}\pi$

SECTION 6.3

1.

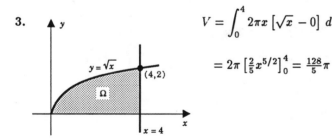

$V = \int_0^1 2\pi x \left[x - 0\right] dx = 2\pi \int_0^1 x^2 dx$

$= 2\pi \left[\frac{1}{3}x^3\right]_0^1 = \frac{2\pi}{3}$

3.

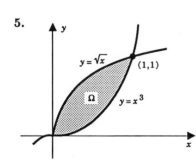

$V = \int_0^4 2\pi x \left[\sqrt{x} - 0\right] dx = 2\pi \int_0^4 x^{3/2} dx$

$= 2\pi \left[\frac{2}{5}x^{5/2}\right]_0^4 = \frac{128}{5}\pi$

5.

$V = \int_0^1 2\pi x \left[\sqrt{x} - x^3\right] dx$

$= 2\pi \int_0^1 \left(x^{3/2} - x^4\right) dx$

$= 2\pi \left[\frac{2}{5}x^{5/2} - \frac{1}{5}x^5\right]_0^1 = \frac{2\pi}{5}$

7.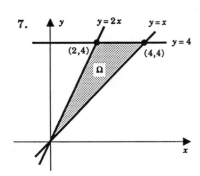

$$V = \int_0^2 2\pi x \left[2x - x\right] dx + \int_2^4 2\pi x \left[4 - x\right] dx$$

$$= 2\pi \int_0^2 x^2 dx + 2\pi \int_2^4 \left(4x - x^2\right) dx$$

$$= 2\pi \left[\tfrac{1}{3}x^3\right]_0^2 + 2\pi \left[2x^2 - \tfrac{1}{3}x^3\right]_2^4 = 16\pi$$

9.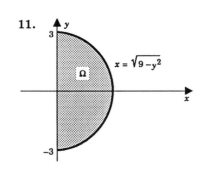

$$V = \int_0^1 2\pi x \left[(\sqrt{x}) - (-\sqrt{x})\right] dx + \int_1^4 2\pi x \left[(\sqrt{x}) - (x - 2)\right] dx$$

$$= 4\pi \int_0^1 x^{3/2} dx + 2\pi \int_1^4 \left(x^{3/2} - x^2 + 2x\right) dx$$

$$= 4\pi \left[\tfrac{2}{5}x^{5/2}\right]_0^1 + 2\pi \left[\tfrac{2}{5}x^{5/2} - \tfrac{1}{3}x^3 + x^2\right]_1^4 = \tfrac{72}{5}\pi$$

11.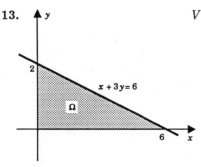

$$V = \int_0^3 2\pi x \left[\sqrt{9 - x^2} - \left(-\sqrt{9 - x^2}\right)\right] dx$$

$$= 4\pi \int_0^3 x \left(9 - x^2\right)^{1/2} dx$$

$$= 4\pi \left[-\tfrac{1}{3}\left(9 - x^2\right)^{3/2}\right]_0^3 = 36\pi$$

13.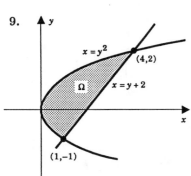

$$V = \int_0^2 2\pi y \left[6 - 3y\right] dy$$

$$= 6\pi \int_0^2 \left(2y - y^2\right) dy$$

$$= 6\pi \left[y^2 - \tfrac{1}{3}y^3\right]_0^2 = 8\pi$$

15.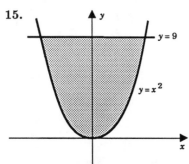

$$V = \int_0^9 2\pi y \left[(\sqrt{y}) - (-\sqrt{y})\right] dy$$

$$= 4\pi \int_0^9 y^{3/2} dy$$

$$= 4\pi \left[\tfrac{2}{5}y^{5/2}\right]_0^9 = \tfrac{1944}{5}\pi$$

17.

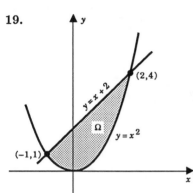

$$V = \int_0^1 2\pi y \left[y^{1/3} - y^2 \right] dy$$

$$= 2\pi \int_0^1 \left(y^{4/3} - y^3 \right) dy$$

$$= 2\pi \left[\tfrac{3}{7} y^{7/3} - \tfrac{1}{4} y^4 \right]_0^1 = \tfrac{5}{14}\pi$$

19.

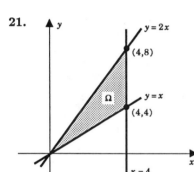

$$V = \int_0^1 2\pi y \left[(\sqrt{y}) - (-\sqrt{y}) \right] dy + \int_1^4 2\pi y \left[(\sqrt{y}) - (y-2) \right] dy$$

$$= 4\pi \int_0^1 y^{3/2} dy + 2\pi \int_1^4 \left(y^{3/2} - y^2 + 2y \right) dy$$

$$= 4\pi \left[\tfrac{2}{5} y^{5/2} \right]_0^1 + 2\pi \left[\tfrac{2}{5} y^{5/2} - \tfrac{1}{3} y^3 + y^2 \right]_1^4 = \tfrac{72}{5}\pi$$

21.

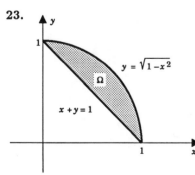

$$V = \int_0^4 2\pi y \left[y - \tfrac{y}{2} \right] dy + \int_4^8 2\pi y \left[4 - \tfrac{y}{2} \right] dy$$

$$= \pi \int_0^4 y^2 dy + \pi \int_4^8 \left(8y - y^2 \right) dy$$

$$= \pi \left[\tfrac{1}{3} y^3 \right]_0^4 + \pi \left[4y^2 - \tfrac{1}{3} y^3 \right]_4^8 = 64\pi$$

23.

$$V = \int_0^1 2\pi y \left[\sqrt{1 - y^2} - (1 - y) \right] dy$$

$$= 2\pi \int_0^1 \left[y \left(1 - y^2 \right)^{1/2} - y + y^2 \right] dy$$

$$= 2\pi \left[-\frac{1}{3} \left(1 - y^2 \right)^{3/2} - \frac{1}{2} y^2 + \frac{1}{3} y^3 \right]_0^1 = \frac{\pi}{3}$$

25. $V = \int_0^a 2\pi x \left[2b\sqrt{1 - \dfrac{x^2}{a^2}} \right] dx = \dfrac{4\pi b}{a} \int_0^a x \left(a^2 - x^2\right)^{1/2} dx = \dfrac{4\pi b}{a} \left[-\dfrac{1}{3} \left(a^2 - x^2\right)^{3/2} \right]_0^a = \dfrac{4}{3}\pi a^2 b$

27.

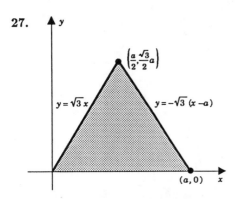

By the shell method

$$V = \int_0^{a/2} 2\pi x \left(\sqrt{3}\, x\right) dx + \int_{a/2}^a 2\pi x \left[\sqrt{3}\,(a - x) \right] dx$$

$$= 2\pi\sqrt{3} \int_0^{a/2} x^2 \, dx + 2\pi\sqrt{3} \int_{a/2}^a \left(ax - x^2\right) dx$$

$$= 2\pi\sqrt{3} \left[\frac{1}{3}x^3 \right]_0^{a/2} + 2\pi\sqrt{3} \left[\frac{a}{2}x^2 - \frac{1}{3}x^3 \right]_{a/2}^a = \frac{\sqrt{3}}{4}a^3\pi$$

29. (a) $V = \int_0^8 2\pi y \left[4 - y^{2/3} \right] dy = 2\pi \int_0^8 \left(4y - y^{5/3}\right) dy = 2\pi \left[2y^2 - \frac{3}{8}y^{8/3} \right]_0^8 = 64\pi$

   (b) $V = \int_0^4 2\pi (4 - x) \left[x^{3/2} \right] dx = 2\pi \int_0^4 \left(4x^{3/2} - x^{5/2}\right) dx$

   $= 2\pi \left[\frac{8}{5}x^{5/2} - \frac{2}{7}x^{7/2} \right]_0^4 = \frac{1024}{35}\pi$

   (c) $V = \int_0^8 2\pi (8 - y) \left[4 - y^{2/3} \right] dy = 2\pi \int_0^8 \left(32 - 4y - 8y^{2/3} + y^{5/3}\right) dy$

   $= 2\pi \left[32y - 2y^2 - \frac{24}{5}y^{5/3} + \frac{3}{8}y^{8/3} \right]_0^8 = \frac{704}{5}\pi$

   (d) $V = \int_0^4 2\pi x \left[x^{3/2} \right] dx = 2\pi \int_0^4 x^{5/2} \, dx = 2\pi \left[\frac{2}{7}x^{7/2} \right]_0^4 = \frac{512}{7}\pi$

SECTION 6.4

1.

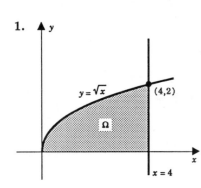

$A = \int_0^4 \sqrt{x} \, dx = \dfrac{16}{3}$

$\bar{x}A = \int_0^4 x\sqrt{x} \, dx = \dfrac{64}{5}, \quad \bar{x} = \dfrac{12}{5}$

$\bar{y}A = \int_0^4 \dfrac{1}{2} \left(\sqrt{x}\right)^2 dx = 4, \quad \bar{y} = \dfrac{3}{4}$

$V_x = 2\pi\bar{y}A = 8\pi, \quad V_y = 2\pi\bar{x}A = \dfrac{128}{5}\pi$

3.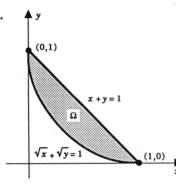

$$A = \int_0^1 \left(x^{1/3} - x^2 \right) dx = \frac{5}{12}$$

$$\overline{x}A = \int_0^1 x \left(x^{1/3} - x^2 \right) dx = \frac{5}{28}, \quad \overline{x} = \frac{3}{7}$$

$$\overline{y}A = \int_0^1 \frac{1}{2} \left[\left(x^{1/3} \right)^2 - \left(x^2 \right)^2 \right] dx = \frac{1}{5}, \quad \overline{y} = \frac{12}{25}$$

$$V_x = 2\pi \overline{y}A = \tfrac{2}{5}\pi, \quad V_y = 2\pi \overline{x}A = \tfrac{5}{14}\pi$$

5.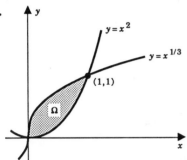

$$A = \int_1^3 (2x - 2) \, dx = 4$$

$$\overline{x}A = \int_1^3 x (2x - 2) \, dx = \frac{28}{3}, \quad \overline{x} = \frac{7}{3}$$

$$\overline{y}A = \int_1^3 \frac{1}{2} \left[(2x)^2 - (2)^2 \right] dx = \frac{40}{3}, \quad \overline{y} = \frac{10}{3}$$

$$V_x = 2\pi \overline{y}A = \tfrac{80}{3}\pi, \quad V_y = 2\pi \overline{x}A = \tfrac{56}{3}\pi$$

7.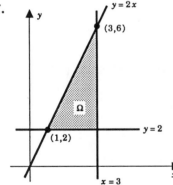

$$A = \int_0^2 \left[6 - \left(x^2 + 2x \right) \right] dx = \frac{16}{3}$$

$$\overline{x}A = \int_0^2 x \left[6 - \left(x^2 + 2 \right) \right] dx = 4, \quad \overline{x} = \frac{3}{4}$$

$$\overline{y}A = \int_0^2 \frac{1}{2} \left[(6)^2 - \left(x^2 + 2 \right)^2 \right] dx = \frac{352}{15}, \quad \overline{y} = \frac{22}{5}$$

$$V_x = 2\pi \overline{y}A = \tfrac{704}{15}\pi, \quad V_y = 2\pi \overline{x}A = 8\pi$$

9.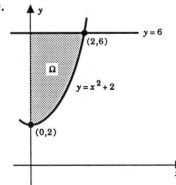

$$A = \int_0^1 \left[(1 - x) - \left(1 - \sqrt{x} \right)^2 \right] dx = \frac{1}{3}$$

$$\overline{x}A = \int_0^1 x \left[(1 - x) - \left(1 - \sqrt{x} \right)^2 \right] dx = \frac{2}{15}, \quad \overline{x} = \frac{2}{5}$$

$$\overline{y} = \tfrac{2}{5} \qquad \text{by symmetry}$$

$$V_x = 2\pi \overline{y}A = \tfrac{4}{15}\pi, \quad V_y = \tfrac{4}{15}\pi \qquad \text{by symmetry}$$

11.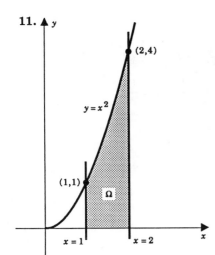

$$A = \int_1^2 x^2\, dx = \frac{7}{3}$$

$$\overline{x}A = \int_1^2 x\left(x^2\right) dx = \frac{15}{4}, \quad \overline{x} = \frac{45}{28}$$

$$\overline{y}A = \int_1^2 \frac{1}{2}\left(x^2\right)^2 dx = \frac{31}{10}, \quad \overline{y} = \frac{93}{70}$$

$$V_x = 2\pi\overline{y}A = \tfrac{31}{5}\pi, \quad V_y = 2\pi\overline{x}A = \tfrac{15}{2}\pi$$

13.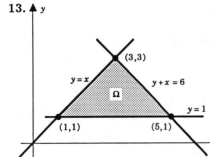

$$A = \tfrac{1}{2}bh = 4; \qquad \text{by symmetry,} \quad \overline{x} = 3$$

$$\overline{y}A = \int_1^3 y\left[(6-y) - y\right] dy = \frac{20}{3}, \quad \overline{y} = \frac{5}{3}$$

$$V_x = 2\pi\overline{y}A = \tfrac{40}{3}\pi, \quad V_y = 2\pi\overline{x}A = 24\pi$$

15. $\left(\frac{5}{2}, 5\right)$ **17.** $\left(1, \frac{8}{5}\right)$ **19.** $\left(\frac{10}{3}, \frac{40}{21}\right)$ **21.** $(2,4)$ **23.** $\left(-\frac{3}{5}, 0\right)$

25. (a) $(0,0)$ by symmetry

(b) Ω_1 smaller quarter disc, Ω_2 the larger quarter disc

$$A_1 = \frac{1}{16}\pi, \quad A_2 = \pi; \quad \overline{x}_1 = \overline{y}_1 = \frac{2}{3\pi}, \quad \overline{x}_2 = \overline{y}_2 = \frac{8}{3\pi} \quad \text{(Problem 1)}$$

$$\overline{x}A = \left(\frac{8}{3\pi}\right)(\pi) - \frac{2}{3\pi}\left(\frac{1}{16}\pi\right)\frac{63}{24}, \quad A = \frac{15}{16}\pi$$

$$\overline{x} = \left(\frac{63}{24}\right)\bigg/\left(\frac{15\pi}{16}\right) = \frac{14}{5\pi}, \quad \overline{y} = \overline{x} = \frac{14}{5\pi} \quad \text{(symmetry)}$$

(c) $\overline{x} = 0, \quad \overline{y} = \frac{14}{5\pi}$

27. Use theorem of Pappus. Centroid of rectangle is located

$$c + \sqrt{\left(\frac{a}{2}\right)^2 + \left(\frac{b}{2}\right)^2} \quad \text{units}$$

from line l. The area of the rectangle is ab. Thus,

$$\text{volume} = 2\pi \left[c + \sqrt{\left(\frac{a}{2}\right)^2 + \left(\frac{b}{2}\right)^2} \right] (ab) = \pi ab \left(2c + \sqrt{a^2 + b^2} \right).$$

29. (a) $\left(\frac{2}{3}a, \frac{1}{3}h\right)$ (b) $\left(\frac{2}{3}a + \frac{1}{3}b, \frac{1}{3}h\right)$ (c) $\left(\frac{1}{3}a + \frac{1}{3}b, \frac{1}{3}h\right)$

31. (a) $V = \frac{2}{3}\pi R^3 \sin^3\theta + \frac{1}{3}\pi R^3 \sin^2\theta \cos\theta = \frac{1}{3}\pi R^3 \sin^2\theta \left(2\sin\theta + \cos\theta\right)$

(b) $\bar{x} = \dfrac{V}{2\pi A} = \dfrac{\frac{1}{3}\pi R^3 \sin^2\theta \left(2\sin\theta + \cos\theta\right)}{2\pi \left(\frac{1}{2}R^2 \sin\theta \cos\theta + \frac{1}{4}\pi R^2 \sin^2\theta\right)} = \dfrac{2R \sin\theta \left(2\sin\theta + \cos\theta\right)}{3 \left(\pi \sin\theta + 2\cos\theta\right)}$

33. (a) The mass contributed by $[x_{i-1}, x_i]$ is approximately $\lambda\left(x_i^*\right) \Delta x_i$ where x_i^* is the midpoint of $[x_{i-1}, x_i]$. The sum of these contributions,
$$\lambda\left(x_1^*\right) \Delta x_1 + \cdots + \lambda\left(x_n^*\right) \Delta x_n,$$
is a Riemann sum, which as $\|P\| \to 0$, tends to the given integral.

(b) Take M_i as the mass contributed by $[x_{i-1}, x_i]$. Then $x_{M_i} M_i \cong x_i^* \lambda\left(x_i^*\right) \Delta x_i$ where x_i^* is the midpoint of $[x_{i-1}, x_i]$. Therefore
$$x_M M = x_{M_1} M_1 + \cdots + x_{M_n} M_n \cong x_1^* \lambda\left(x_1^*\right) \Delta x_1 + \cdots + x_n^* \lambda\left(x_n^*\right) \Delta x_n.$$

As $\|P\| \to 0$, the sum on the right converges to the given integral.

35.

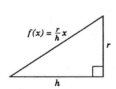

$$\bar{x}\left(\frac{1}{3}\pi r^2 h\right) = \int_0^h \pi x \left(\frac{r}{h}x\right)^2 dx = \frac{1}{4}\pi r^2 h^2$$

$$\bar{x} = \left(\frac{1}{4}\pi r^2 h^2\right) / \left(\frac{1}{3}\pi r^2 h\right) = \frac{3}{4}h.$$

The centroid of the cone lies on the axis of the cone at a distance $\frac{3}{4}h$ from the vertex.

37. (a) $V_x = \displaystyle\int_0^1 \pi \left(\sqrt{x}\right)^2 dx = \frac{1}{2}\pi, \quad \bar{x}V_x = \int_0^1 \pi x \left(\sqrt{x}\right)^2 dx = \frac{1}{3}\pi$

$\bar{x} = \left(\frac{1}{3}\pi\right) / \left(\frac{1}{2}\pi\right) = \frac{2}{3};$ centroid $\left(\frac{2}{3}, 0\right)$

(b) $V_y = \displaystyle\int_0^1 2\pi x \sqrt{x}\, dx = \frac{4}{5}\pi, \quad \bar{y}V_y = \int_0^1 \pi x \left(\sqrt{x}\right)^2 dx = \frac{1}{3}\pi$

$\bar{y} = \left(\frac{1}{3}\pi\right) / \left(\frac{4}{5}\pi\right) = \frac{5}{12};$ centroid $\left(0, \frac{5}{12}\right)$

39. $V_x = \int_0^a \pi \dfrac{b^2}{a^2} \left(a^2 - x^2\right) dx = \dfrac{2}{3}\pi ab^2, \quad \bar{x}V_x = \int_0^a \pi x \dfrac{b^2}{a^2} \left(a^2 - x^2\right) dx = \dfrac{1}{4}\pi a^2 b^2$

$\bar{x} = \left(\tfrac{1}{4}\pi a^2 b^2\right) / \left(\tfrac{2}{3}\pi ab^2\right) = \tfrac{3}{8}a; \quad$ centroid $\left(\tfrac{3}{8}a, 0\right)$

SECTION 6.5

1.   To counteract the restoring force of the spring we must apply a force $F(x) = kx$.
Since $F(4) = 200$, we see that $k = 50$ and therefore $F(x) = 50x$.

(a) $W = \int_0^1 50x\, dx = 25$ ft-lb   (b) $W = \int_0^{3/2} 50x\, dx = \dfrac{225}{4}$ ft-lb

3.   Let L be the natural length of the spring.

$$\int_{2-L}^{2.1-L} kx\, dx = \dfrac{1}{2}\int_{2.1-L}^{2.2-L} kx\, dx$$

$$\left[\tfrac{1}{2}kx^2\right]_{2-L}^{2.1-L} = \tfrac{1}{2}\left[\tfrac{1}{2}kx^2\right]_{2.1-L}^{2.2-L}$$

$$(2.1 - L)^2 - (2 - L)^2 = \tfrac{1}{2}\left[(2.2 - L)^2 - (2.1 - L)^2\right].$$

Solve this equation for L and you will find that $L = 1.95$.

Answer: 1.95 ft

5.

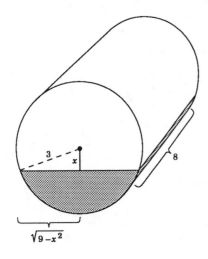

(a) $W = \int_0^3 (x + 3)(60)(8)\left(2\sqrt{9 - x^2}\right) dx$

$= 960 \int_0^3 x\left(9 - x^2\right)^{1/2} dx$

$+ 2880 \underbrace{\int_0^3 \sqrt{9 - x^2}\, dx}_{\substack{\text{area of quarter} \\ \text{circle of radius 3}}}$

$= 960 \left[-\tfrac{1}{3}\left(9 - x^2\right)^{3/2}\right]_0^3 + 2880 \left[\tfrac{9}{4}\pi\right]$

$= (8640 + 6480\pi)$ ft-lb

(b) $W = \int_0^3 (x + 7)(60)(8)^2 \left(2\sqrt{9 - x^2}\right) dx$

$= 960 \int_0^3 x\left(9 - x^2\right)^{1/2} dx + 6720 \int_0^3 \sqrt{9 - x^2}\, dx$

$= (8640 + 15120\pi)$ ft-lb

7.

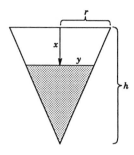

By similar triangles

$$\frac{h}{r} = \frac{h-x}{y} \quad \text{so that} \quad y = \frac{r}{h}(h-x).$$

Thus, the area of a cross section of the fluid at a depth of x feet is

$$\pi y^2 = \pi \frac{r^2}{h^2}(h-x)^2.$$

(a) $W = \displaystyle\int_0^{h/2} x\sigma \left[\pi \frac{r^2}{h^2}(h-x)^2\right] dx = \frac{\sigma \pi r^2}{h^2} \int_0^{h/2}(h^2 x - 2hx^2 + x^3)\, dx = \frac{11}{192}\sigma \pi r^2 h^2$ ft-lb

(b) $W = \displaystyle\int_0^{h/2}(x+k)\sigma \left[\pi \frac{r^2}{h^2}(h-x)^2\right] dx = \frac{11}{192}\pi r^2 h^2 \sigma + \frac{7}{24}\pi r^2 hk\sigma$ ft-lb

9. $W = \displaystyle\int_{r_1}^{r_2} F\, dr = \int_{r_1}^{r_2} -\frac{GmM}{r^2}\, dr = \left[\frac{GmM}{r}\right]_{r_1}^{r_2} = GmM\left[\frac{1}{r_2} - \frac{1}{r_1}\right]$

11. The bag is raised 8 feet and loses a total of 3 pounds at a constant rate. Thus, the bag loses sand at the rate of 3/8 lb/ft. After the bag has been raised x feet it weighs $100 - \dfrac{3x}{8}$ pounds.

$$W = \int_0^8 \left(100 - \frac{3x}{8}\right) dx = \left[100 - \frac{3x^2}{16}\right]_0^8 = 788 \text{ ft-lb.}$$

13. (a) $W = \displaystyle\int_0^l x\sigma\, dx = \frac{1}{2}\sigma l^2$ ft-lb   (b) $W = \displaystyle\int_0^l (x+l)\sigma\, dx = \frac{3}{2}\sigma l^2$ ft-lb

15. Thirty feet of cable and the steel beam weighing a total of

$$800 + 30(6) = 980 \text{ lb}$$

are raised 20 feet. The work requires $(20)(980)$ ft-lb.

Next, the remaining 20 feet of cable is raised a varying distance and wound onto the steel drum. Thus the total work is given by

$$W = (20)(980) + \int_0^{20} 6x\, dx = 19600 + 1200 = 20800 \text{ ft-lb.}$$

SECTION * 6.6

1.

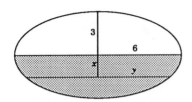

ellipse: $\dfrac{x^2}{3^2} + \dfrac{y^2}{6^2} = 1$

$$F = \int_0^3 (60)\, x \left[12\sqrt{1 - \dfrac{x^2}{9}} \right] dx$$

$$= 240 \int_0^3 x\,(9 - x^2)^{1/2}\, dx$$

$$= 240 \left[-\tfrac{1}{3}(9 - x^2)^{3/2} \right]_0^3 = 2160 \text{ lb}$$

3.

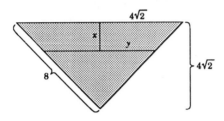

By similar triangles

$$\dfrac{4\sqrt{2}}{4\sqrt{2}} = \dfrac{y}{4\sqrt{2} - x} \quad \text{so} \quad y = 4\sqrt{2} - x.$$

$$F = \int_0^{4\sqrt{2}} (62.5)\, x \left[2 \left(4\sqrt{2} - x \right) \right] dx$$

$$= 125 \int_0^{4\sqrt{2}} \left(4\sqrt{2}\, x - x^2 \right) dx = \dfrac{8000}{3}\sqrt{2} \text{ lb}$$

5.

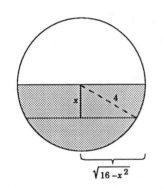

$$F = \int_0^4 60x \left(2\sqrt{16 - x^2} \right) dx$$

$$= 120 \int_0^4 x\,(16 - x^2)^{1/2}\, dx$$

$$= 120 \left[-\tfrac{1}{3}(16 - x^2)^{3/2} \right]_0^4 = 2560 \text{ lb}$$

7. (a) The width of the plate is 10 feet and the depth of the plate ranges from 8 feet to 14 feet. Thus

$$F = \int_8^{14} 62.5x\,(10)\, dx = 41,250 \text{ lb}.$$

(b) The width of the plate is 6 feet and the depth of the plate ranges from 6 feet to 16 feet. Thus

$$F = \int_6^{16} 62.5x\,(6)\, dx = 41,250 \text{ lb}.$$

9. $F = \displaystyle\int_a^b \sigma x w(x)\, dx = \sigma \int_a^b x w(x)\, dx = \sigma \bar{x} A$

where A is the area of the submerged surface and $\bar{x}$ is the depth of the centroid.

SECTION * 6.7

1.

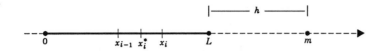

Let $P = \{x_0, x_1, \ldots, x_n\}$ be a partition of $[0, L]$. The mass between x_{i-1} and x_i is the product $\lambda \, \Delta x_i$ where λ is the mass density of the rod M/L. This mass attracts m with a force that is approximately

$$\frac{Gm\lambda \, \Delta x_i}{(L + h - x_i^*)^2}$$

where x_i^* is the midpoint of $[x_{i-1}, x_i]$. The force exerted by the entire rod is thus approximately

$$Gm\lambda \left[\frac{\Delta x_1}{(L + h - x_1^*)^2} + \cdots + \frac{\Delta x_n}{(L + h - x_n^*)^2} \right].$$

As the norm of the partition tends to zero, this expression tends to

$$Gm\lambda \int_0^L \frac{dx}{(L + h - x)^2} = \frac{Gm\lambda L}{h(L + h)} \overset{\lambda L = M}{\underset{}{=}} \frac{GmM}{h(L + h)}.$$

3. A partition $P = \{\theta_0, \theta_1, \ldots, \theta_n\}$ of $[0, 2\pi]$ induces a decomposition of the circular wire into n little arcs of length $r \, \Delta\theta_i$ and mass $\lambda r \, \Delta\theta_i$ where λ is the mass density $M/2\pi R$. This ith little piece attracts m with a force of magnitude

$$\frac{Gm (\lambda R \, \Delta\theta_i)}{h^2 + R^2}.$$

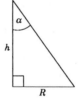

But part of this force is to the side. The magnitude of the downward part is

$$\frac{Gm (\lambda R \, \Delta\theta_i)}{h^2 + R^2} (\cos\alpha) = \frac{Gm (\lambda r \, \Delta\theta_i)}{h^2 + R^2} \frac{h}{\sqrt{h^2 + R^2}} = \frac{Gm\lambda Rh}{(h^2 + R^2)^{3/2}} \, \Delta\theta_i.$$

The sum of all these downward forces is

$$F = \frac{Gm\lambda Rh}{(h^2 + R^2)^{3/2}} \underbrace{(\Delta\theta_1 + \cdots + \Delta\theta_n)}_{2\pi}$$

$$= \frac{Gm (2\pi R\lambda) h}{(h^2 + R^2)^{3/2}} = \frac{GmMh}{(h^2 + R^2)^{3/2}}.$$

$$2\pi r \lambda = M$$

5. For h small compared to R $$\frac{GM}{(R + h)^2} \cong \frac{GM}{R^2} = g.$$

CHAPTER 7

SECTION 7.2

1. $\ln 20 = \ln 2 + \ln 10 \cong 2.99$

3. $\ln 1.6 = \ln \frac{16}{10} = 2\ln 4 - \ln 10 \cong 0.48$

5. $\ln 0.1 = \ln \frac{1}{10} = \ln 1 - \ln 10 \cong -2.30$

7. $\ln 7.2 = \ln \frac{72}{10} = \ln 8 + \ln 9 - \ln 10 \cong 1.98$

9. $\ln \sqrt{2} = \frac{1}{2}\ln 2 \cong 0.35$

11. $\ln 2^5 = 5\ln 2 \cong 3.45$

13.

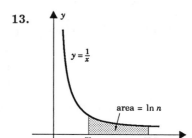

15. $\frac{1}{2}[L_f(P) + U_f(P)] = \frac{1}{2}\left[\frac{763}{1980} + \frac{1691}{3960}\right] \cong 0.406$

17. (a) $\ln 5.2 \cong \ln 5 + \frac{1}{5}(0.2) \cong 1.61 + 0.04 = 1.65$

   (b) $\ln 4.8 \cong \ln 5 - \frac{1}{5}(0.2) \cong 1.61 - 0.04 = 1.57$

   (c) $\ln 5.5 \cong \ln 5 + \frac{1}{5}(0.5) \cong 1.61 + 0.1 = 1.71$

19. $x = e^2$

21. $2 - \ln x = 0$ or $\ln x = 0$. Thus $x = e^2$ or $x = 1$.

23.
$$\ln[(2x+1)(x+2)] = 2\ln(x+2)$$
$$\ln[(2x+1)(x+2)] = \ln\left[(x+2)^2\right]$$
$$(2x+1)(x+2) = (x+2)^2$$
$$x^2 + x - 2 = 0$$
$$(x+2)(x-1) = 0$$
$$x = -2 \text{ or } x = 1.$$

We disregard the solution $x = -2$ since it does not satisfy' the initial equation. Thus, the only solution is $x = 1$.

25. $l \cong 1$

27. $l \cong 0$

SECTION 7.3

1. $\operatorname{dom}(f) = (0, \infty)$, $f'(x) = \frac{1}{4x}(4) = \frac{1}{x}$

3. $\operatorname{dom}(f) = (-1, \infty)$, $f'(x) = \frac{1}{x^3 + 1}\frac{d}{dx}(x^3 + 1) = \frac{3x^2}{x^3 + 1}$

5. $\operatorname{dom}(f) = (-\infty, \infty)$, $f(x) = \frac{1}{2}\ln(1 + x^2)$ so $f'(x) = \frac{1}{2}\left[\frac{1}{1 + x^2}(2x)\right] = \frac{x}{1 + x^2}$

7. $\operatorname{dom}(f) = \{x \mid x \neq \pm 1\}$, $\quad f'(x) = \dfrac{1}{x^4 - 1}\dfrac{d}{dx}(x^4 - 1) = \dfrac{4x^3}{x^4 - 1}$

9. $\operatorname{dom}(f) = (0, \infty)$, $\quad f'(x) = x\dfrac{d}{dx}(\ln x) + 1(\ln x) = 1 + \ln x$

11. $\operatorname{dom}(f) = (0,1) \cup (1, \infty)$, $\quad f(x) = (\ln x)^{-1}$ $\quad$ so $\quad f'(x) = -(\ln x)^{-2}\dfrac{d}{dx}(\ln x) = -\dfrac{1}{x(\ln x)^2}$

13. $\operatorname{dom}(f) = (-1, \infty)$, $\quad f'(x) = \dfrac{(x+1)\dfrac{d}{dx}[\ln(x+1)] - 1[\ln(x+1)]}{(x+1)^2} = \dfrac{1 - \ln(x+1)}{(x+1)^2}$

15. $\displaystyle\int \dfrac{dx}{x+1} = \ln|x+1| + C$

17. $\left\{ \begin{array}{c} u = 3 - x^2 \\ du = -2x\,dx \end{array} \right\}$; $\quad \displaystyle\int \dfrac{x}{3 - x^2}\,dx = -\dfrac{1}{2}\int \dfrac{du}{u} = -\dfrac{1}{2}\ln|u| + C = -\dfrac{1}{2}\ln|3 - x^2| + C$

19. $\left\{ \begin{array}{c} u = 3 - x^2 \\ du = -2x\,dx \end{array} \right\}$; $\quad \displaystyle\int \dfrac{x}{(3 - x^2)^2}\,dx = -\dfrac{1}{2}\int \dfrac{du}{u^2} = \dfrac{1}{2u} + C = \dfrac{1}{2(3 - x^2)} + C$

21. $\displaystyle\int \left(\dfrac{1}{x+2} - \dfrac{1}{x-2} \right) dx = \ln|x+2| - \ln|x-2| + C = \ln\left| \dfrac{x+2}{x-2} \right| + C$

23. $\left\{ u = \ln x, \ du = \dfrac{dx}{x} \right\}$; $\quad \displaystyle\int \dfrac{dx}{x(\ln x)^2} = \int \dfrac{du}{u^2} = -\dfrac{1}{u} + C = -\dfrac{1}{\ln x} + C$

25. $\left\{ u = 1 + x\sqrt{x}, \ du = \dfrac{3}{2}x^{1/2}\,dx \right\}$; $\quad \displaystyle\int \dfrac{\sqrt{x}}{1 + x\sqrt{x}}\,dx = \dfrac{2}{3}\int \dfrac{du}{u} = \dfrac{2}{3}\ln|u| + C = \dfrac{2}{3}\ln|1 + x\sqrt{x}| + C$

27. $\displaystyle\int_1^e \dfrac{dx}{x} = [\ln x]_1^e = \ln e - \ln 1 = 1 - 0 = 1$

29. $\displaystyle\int_e^{e^2} \dfrac{dx}{x} = [\ln x]_e^{e^2} = \ln e^2 - \ln e = 2 - 1 = 1$

31. $\displaystyle\int_4^5 \dfrac{x}{x^2 - 1}\,dx = \left[\dfrac{1}{2}\ln|x^2 - 1| \right]_4^5 = \dfrac{1}{2}(\ln 24 - \ln 15) = \dfrac{1}{2}\ln\dfrac{8}{5}$

33. $\left\{ \begin{array}{c} u = \ln(x+1) \\ du = \dfrac{dx}{x+1} \end{array} \ \left| \begin{array}{ccc} x = 0 & \Longrightarrow & u = 0 \\ x = 1 & \Longrightarrow & u = \ln 2 \end{array} \right. \right\}$;

$\displaystyle\int_0^1 \dfrac{\ln(x+1)}{x+1}\,dx = \int_0^{\ln 2} u\,du = \dfrac{1}{2}[u^2]_0^{\ln 2} = \dfrac{1}{2}(\ln 2)^2$

35.
$$\ln |g(x)| = 2\ln (x^2 + 1) + 5\ln |x - 1| + 3\ln x$$

$$\frac{g'(x)}{g(x)} = 2\left(\frac{2x}{x^2 + 1}\right) + \frac{5}{x - 1} + \frac{3}{x}$$

$$g'(x) = (x^2 + 1)^2 (x - 1)^5 x^3 \left(\frac{4x}{x^2 + 1} + \frac{5}{x - 1} + \frac{3}{x}\right)$$

37.
$$\ln |g(x)| = 4\ln |x| + \ln |x - 1| - \ln |x + 2| - \ln (x^2 + 1)$$

$$\frac{g'(x)}{g(x)} = \frac{4}{x} + \frac{1}{x - 1} - \frac{1}{x + 2} - \frac{2x}{x^2 + 1}$$

$$g'(x) = \frac{x^4 (x - 1)}{(x + 2)(x^2 + 1)} \left(\frac{4}{x} + \frac{1}{x - 1} - \frac{1}{x + 2} - \frac{2x}{x^2 + 1}\right)$$

39.
$$\ln |g(x)| = \tfrac{1}{2} \left(\ln |x - 1| + \ln |x - 2| - \ln |x - 3| - \ln |x - 4|\right)$$

$$\frac{g'(x)}{g(x)} = \frac{1}{2}\left(\frac{1}{x - 1} + \frac{1}{x - 2} - \frac{1}{x - 3} - \frac{1}{x - 4}\right)$$

$$g'(x) = \frac{1}{2}\sqrt{\frac{(x - 1)(x - 2)}{(x - 3)(x - 4)}} \left(\frac{1}{x - 1} + \frac{1}{x - 2} - \frac{1}{x - 3} - \frac{1}{x - 4}\right)$$

41. $A = \displaystyle\int_1^4 \left[\frac{5 - x}{4} - \frac{1}{x}\right] dx = \left[\frac{5}{4}x - \frac{1}{8}x^2 - \ln x\right]_1^4 = \frac{15}{8} - \ln 4$

43. $v(t) = \displaystyle\int a(t)\, dt = \int -(t + 1)^{-2}\, dt = \frac{1}{t + 1} + C.$

Since $v(0) = 1$,   we get $1 = 1 + C$ so that $C = 0$. Then

$$s = \int_0^4 |v(t)|\, dt = \int_0^4 \frac{dt}{t + 1} = [\ln (t + 1)]_0^4 = \ln 5.$$

The particle traveled $\ln 5$ ft.

45. $\dfrac{d}{dx}(\ln x) = \dfrac{1}{x}$

$\dfrac{d^2}{dx^2}(\ln x) = -\dfrac{1}{x^2}$

$\dfrac{d^3}{dx^3}(\ln x) = \dfrac{2}{x^3}$

$\dfrac{d^4}{dx^4}(\ln x) = -\dfrac{2\cdot 3}{x^4}$

$\dfrac{d^5}{dx^5}(\ln x) = \dfrac{2\cdot 3\cdot 4}{x^5}$

$\vdots$

$\dfrac{d^n}{dx^n}(\ln x) = (-1)^{n-1}\dfrac{(n-1)!}{x^n}$

47. $\dfrac{d^n}{dx^n}(\ln 2x) = \dfrac{d^n}{dx^n}[\ln 2 + \ln x]$

$= 0 + \dfrac{d^n}{dx^n}(\ln x)$

$= (-1)^{n-1}\dfrac{(n-1)!}{x^n}$

⌞— See Exercise 45

49. (a) for $t \in (1, x)$ (b) for $x > 1$

$\dfrac{1}{t} < \dfrac{1}{\sqrt{t}}$ $0 < \ln x = \displaystyle\int_1^x \dfrac{dt}{t} < \int_1^x \dfrac{dt}{\sqrt{t}} = \left[2\sqrt{t}\right]_1^x = 2\left(\sqrt{x} - 1\right)$

(c) for $0 < x < 1$

$$0 < \ln \dfrac{1}{x} < 2\left(\sqrt{\dfrac{1}{x}} - 1\right) \qquad \text{by (b)}$$

$$0 < -\ln x < 2\left(\dfrac{1}{\sqrt{x}} - 1\right)$$

$$2\left(1 - \dfrac{1}{\sqrt{x}}\right) < \ln x < 0$$

$$2x\left(1 - \dfrac{1}{\sqrt{x}}\right) < x\ln x < 0$$

(d) Use part (c) and the pinching theorem for one-sided limits.

51. $f(x) = \ln 2x, \; x > 0$

$f'(x) = \dfrac{1}{x}$

$f''(x) = \dfrac{-1}{x^2}$

(i)   domain $(0, \infty)$
(ii) increases on $(0, \infty)$
(iii) no extreme values
(iv) concave down on $(0, \infty)$;
      no pts of inflection

$f':$

$f'':$

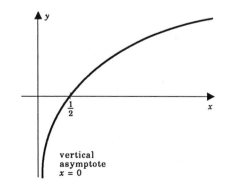

vertical
asymptote
$x = 0$

53. $f(x) = \ln(4 - x), \quad x < 4$

f' :

$f'(x) = \dfrac{1}{x - 4}$

f'' :
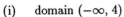

$f''(x) = \dfrac{-1}{(x - 4)^2}$

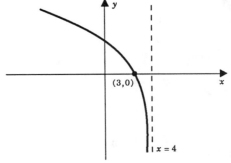

(i)   domain $(-\infty, 4)$
(ii)   decreases throughout
(iii)   no extreme values
(iv)   concave down throughout:
        no pts of inflection

55. $f(x) = x \ln x, \quad x > 0$

f' :
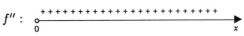

$f'(x) = 1 + \ln x$

f'' :
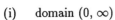

$f''(x) = \dfrac{1}{x}$

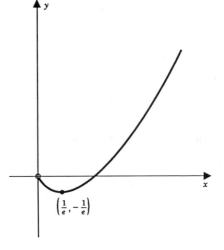

(i)   domain $(0, \infty)$
(ii)   decreases on $(0, 1/e]$,   increases on $[1/e, \infty)$
(iii) $f(1/e) = -1/e$ local and absolute min
(iv)   concave up throughout;   no pts of inflection

57. $f(x) = \ln\left(\dfrac{x}{1 + x^2}\right), \quad x > 0$

f' :
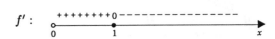

$f'(x) = \dfrac{1 - x^2}{x(1 + x^2)}$

f'' :
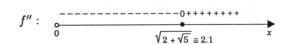

$f''(x) = \dfrac{x^4 - 4x^2 - 1}{x^2(1 + x^2)^2}$

(i)   domain $(0, \infty)$
(ii)   increases on $(0, 1]$,   decreases on $[1, \infty)$
(iii) $f(1) = -\ln 2$ local and absolute max
(iv)   concave down on $\left(0, \sqrt{2 + \sqrt{5}}\,\right)$,

concave up on $\left(\sqrt{2 + \sqrt{5}}, \infty\right)$;

pt of inflection at $x = \sqrt{2 + \sqrt{5}}$

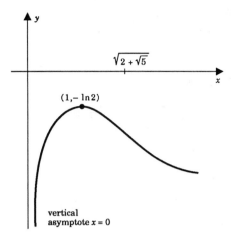

SECTION 7.4

1. $\dfrac{dy}{dx} = e^{-2x}\dfrac{d}{dx}(-2x) = -2e^{-2x}$

3. $\dfrac{dy}{dx} = e^{x^2-1}\dfrac{d}{dx}\left(x^2 - 1\right) = 2xe^{x^2-1}$

5. $\dfrac{dy}{dx} = e^x\dfrac{d}{dx}(\ln x) + \ln x\dfrac{d}{dx}(e^x) = e^x\left(\dfrac{1}{x} + \ln x\right)$

7. $\dfrac{dy}{dx} = x^{-1}\dfrac{d}{dx}\left(e^{-x}\right) + e^{-x}\dfrac{d}{dx}\left(x^{-1}\right) = -x^{-1}e^{-x} - e^{-x}x^{-2} = -\left(x^{-1} + x^{-2}\right)e^{-x}$

9. $\dfrac{dy}{dx} = \dfrac{1}{2}\left(e^x - e^{-x}\right)$

11. $\dfrac{dy}{dx} = e^{\sqrt{x}}\dfrac{d}{dx}\left(\ln\sqrt{x}\,\right) + \ln\sqrt{x}\,\dfrac{d}{dx}\left(e^{\sqrt{x}}\right)$

$\qquad = e^{\sqrt{x}}\left(\dfrac{1}{\sqrt{x}}\cdot\dfrac{1}{2\sqrt{x}}\right) + \ln\sqrt{x}\,\dfrac{e^{\sqrt{x}}}{2\sqrt{x}} = \dfrac{1}{2}e^{\sqrt{x}}\left(\dfrac{1}{x} + \dfrac{\ln\sqrt{x}}{\sqrt{x}}\right)$

13. $\dfrac{dy}{dx} = 2\left(e^x + e^{-x}\right)\dfrac{d}{dx}\left(e^x + e^{-x}\right) = 2\left(e^x + e^{-x}\right)\left(e^x - e^{-x}\right) = 2\left(e^{2x} - e^{-2x}\right)$

15. $\dfrac{dy}{dx} = 2\left(e^{x^2} + 1\right)\dfrac{d}{dx}\left(e^{x^2} + 1\right) = 2\left(e^{x^2} + 1\right)e^{x^2}\dfrac{d}{dx}\left(x^2\right) = 4xe^{x^2}\left(e^{x^2} + 1\right)$

17. $\dfrac{dy}{dx} = \left(x^2 - 2x + 2\right)\dfrac{d}{dx}(e^x) + e^x(2x - 2) = x^2e^x$

19. $\dfrac{dy}{dx} = \dfrac{(e^x + 1)e^x - (e^x - 1)e^x}{(e^x + 1)^2} = \dfrac{2e^x}{(e^x + 1)^2}$

21. $\dfrac{dy}{dx} = \dfrac{\left(e^{ax} + e^{bx}\right)\left(ae^{ax} - be^{bx}\right) - \left(e^{ax} - e^{bx}\right)\left(ae^{ax} + be^{bx}\right)}{\left(e^{ax} + e^{bx}\right)^2} = 2(a - b)\dfrac{e^{(a+b)x}}{\left(e^{ax} + e^{bx}\right)^2}$

23. $y = e^{4\ln x} = \left(e^{\ln x}\right)^4 = x^4$ so $\dfrac{dy}{dx} = 4x^3$.

25. $\displaystyle\int e^{2x}\,dx = \frac{1}{2}e^{2x} + C$
  27. $\displaystyle\int e^{kx}\,dx = \frac{1}{k}e^{kx} + C$

29. $\{u = x^2, \quad du = 2x\,dx\};\quad \displaystyle\int xe^{x^2}\,dx = \frac{1}{2}\int e^u\,du = \frac{1}{2}e^u + C = \frac{1}{2}e^{x^2} + C$

31. $\left\{u = \dfrac{1}{x}, \quad du = -\dfrac{1}{x^2}\,dx\right\};\quad \displaystyle\int \frac{e^{1/x}}{x^2}\,dx = -\int e^u\,du = -e^u + C = -e^{1/x} + C$

33. $\displaystyle\int \left(e^x + e^{-x}\right)^2\,dx = \int \left(e^{2x} + 2 + e^{-2x}\right)\,dx = \frac{1}{2}e^{2x} + 2x - \frac{1}{2}e^{-2x} + C$

35. $\displaystyle\int \ln e^x\,dx = \int x\,dx = \frac{1}{2}x^2 + C$
  37. $\displaystyle\int \frac{4}{\sqrt{e^x}}\,dx = \int 4e^{-x/2}\,dx = -8e^{-x/2} + C$

39. $\left\{\begin{array}{c} u = e^x + 1 \\ du = e^x\,dx \end{array}\right\};\quad \displaystyle\int \frac{e^x}{\sqrt{e^x + 1}}\,dx = \int \frac{du}{\sqrt{u}} = \int u^{-1/2}\,du = 2u^{1/2} + C = 2\sqrt{e^x + 1} + C$

41. $\left\{\begin{array}{c} u = 2e^{2x} + 3 \\ du = 4e^{2x}\,dx \end{array}\right\};\quad \displaystyle\int \frac{e^{2x}}{2e^{2x} + 3}\,dx = \frac{1}{4}\int \frac{du}{u} = \frac{1}{4}\ln u + C = \frac{1}{4}\ln\left(2e^{2x} + 3\right) + C$

43. $\displaystyle\int_0^1 e^x\,dx = [e^x]_0^1 = e - 1$

45. $\displaystyle\int_0^{\ln \pi} e^{-6x}\,dx = \left[-\frac{1}{6}e^{-6x}\right]_0^{\ln \pi} = -\frac{1}{6}e^{-6\ln \pi} + \frac{1}{6}e^0 = \frac{1}{6}\left(1 - \pi^{-6}\right)$

47. $\displaystyle\int_0^1 \frac{e^x + 1}{e^x}\,dx = \int_0^1 \left(1 + e^{-x}\right)\,dx = \left[x - e^{-x}\right]_0^1 = \left(1 - e^{-1}\right) - (0 - 1) = 2 - \frac{1}{e}$

49. $\displaystyle\int_0^{\ln 2} \frac{e^x}{e^x + 1}\,dx = [\ln\left(e^x + 1\right)]_0^{\ln 2} = \ln\left(e^{\ln 2} + 1\right) - \ln\left(e^0 + 1\right) = \ln 3 - \ln 2 = \ln \frac{3}{2}$

51. $\displaystyle\int_{\ln 2}^{\ln 3} \left(e^x - e^{-x}\right)^2\,dx = \int_{\ln 2}^{\ln 3} \left(e^{2x} - 2 + e^{-2x}\right)\,dx$

$$= \left[\tfrac{1}{2}e^{2x} - 2x - \tfrac{1}{2}e^{-2x}\right]_{\ln 2}^{\ln 3}$$

$$= \left[\tfrac{1}{2}(9) - 2\ln 3 - \tfrac{1}{2}\left(\tfrac{1}{9}\right)\right] - \left[\tfrac{1}{2}(4) - 2\ln 2 - \tfrac{1}{2}\left(\tfrac{1}{4}\right)\right]$$

$$= \tfrac{185}{72} + \ln \tfrac{4}{9}$$

53. $\displaystyle\int_0^1 x\left(e^{x^2}+2\right)\,dx = \int_0^1 \left(xe^{x^2}+2x\right)\,dx = \left[\frac{1}{2}e^{x^2}+x^2\right]_0^1 = \left(\frac{1}{2}e+1\right)-\left(\frac{1}{2}+0\right)=\frac{1}{2}\left(e+1\right)$

55. $e^{-0.4}=\dfrac{1}{e^{0.4}}\cong\dfrac{1}{1.49}\cong 0.67$ 　　　**57.** $e^{2.8}=\left(e^2\right)\left(e^{0.8}\right)\cong(7.39)\,(2.23)\cong 16.48$

59. $e^{2.03}\cong e^2+e^2\,(0.03)$ 　　　　　　　**61.** $e^{3.15}\cong e^3+e^3\,(0.15)$

　　　　$\cong 7.39+(7.39)\,(0.03)$ 　　　　　　　　　　$\cong 20.09+(20.09)\,(0.15)$

　　　　$\cong 7.61$ 　　　　　　　　　　　　　　　　　　$\cong 23.10$

63. 　　　　$x(t)=Ae^{ct}+Be^{-ct}$

　　　　　　$x'(t)=Ace^{ct}-Bce^{-ct}$

　　　　　　$x''(t)=Ac^2e^{ct}+Bc^2e^{-ct}$

　　　　　　　　　$=c^2\left(Ae^{ct}+Be^{-ct}\right)$

　　　　　　　　　$=c^2x\,(t)$

65.

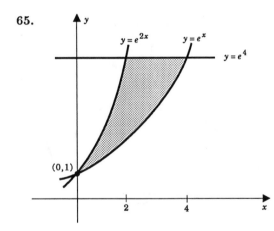

$\displaystyle A=\int_0^2\left(e^{2x}-e^x\right)\,dx+\int_2^4\left(e^4-e^x\right)\,dx$

　　$=\left[\frac{1}{2}e^{2x}-e^x\right]_0^2+\left[e^4x-e^x\right]_2^4$

　　$=\left(\frac{1}{2}e^4-e^2-\frac{1}{2}+1\right)+\left(4e^4-e^4-2e^4+e^2\right)$

　　$=\frac{1}{2}\left(3e^4+1\right)$

67.

$\displaystyle A=\int_1^2\left(e^y-2\right)\,dy$

　　$=\left[e^y-2y\right]_1^2=e^2-e-2$

69. $f(x) = \frac{1}{2}(e^x + e^{-x})$

$f' :$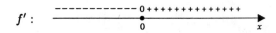

$f'(x) = \frac{1}{2}(e^x - e^{-x})$

$f'' :$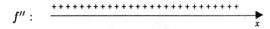

$f''(x) = \frac{1}{2}(e^x + e^{-x})$

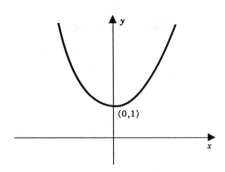

(i)   domain $(-\infty, \infty)$

(ii)   decreases on $(-\infty, 0]$,   increases on $[0, \infty)$

(iii) $f(0) = 1$ local and absolute min

(iv)   concave up everywhere

71. $f(x) = xe^x$

$f' :$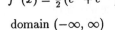

$f'(x) = (x + 1)e^x$

$f'' :$

$f''(x) = (x + 2)e^x$

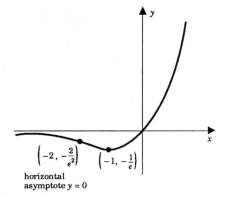

(i)   domain $(-\infty, \infty)$

(ii)   decreases on $(-\infty, -1]$,   increases on $[-1, \infty)$

(iii) $f(-1) = -\dfrac{1}{e}$ local and absolute min

(iv)   concave down on $(-\infty, -2)$,   concave up on $(-2, \infty)$;

       pt of inflection $\left(-2, -\dfrac{2}{e^2}\right)$

73. $f(x) = e^{(1/x)^2}$

$f' :$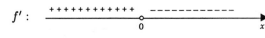

$f'(x) = \dfrac{-2}{x^3}e^{(1/x)^2}$

$f'' :$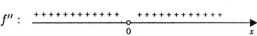

$f''(x) = \dfrac{6x^2 + 4}{x^6}e^{(1/x)^2}$

(i)   domain $(-\infty, 0) \cup (0, \infty)$

(ii) increases on $(-\infty, 0)$,   decreases on $(0, \infty)$

(iii) no extreme values

(iv) concave up on $(-\infty, 0)$ and on $(0, \infty)$

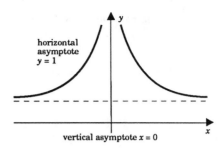

horizontal asymptote $y = 1$

vertical asymptote $x = 0$

75. (a) For $y = e^{ax}$ we have $dy/dx = ae^{ax}$. Therefore the line tangent to the curve $y = e^{ax}$ at an arbitrary point (x_0, e^{ax_0}) has equation

$$y - e^{ax_0} = ae^{ax_0} (x - x_0).$$

The line passes through the origin iff $e^{ax_0} = (ae^{ax_0}) x_0$ iff $x_0 = 1/a$. The point of tangency is $(1/a, e)$. This is point B. By symmetry, point A is $(-1/a, e)$.

(b) The tangent line at B has equation $y = aex$.   By symmetry

$$A_{\mathrm{I}} = 2 \int_0^{1/a} (e^{ax} - aex) \, dx = 2 \left[\frac{1}{a} e^{ax} - \frac{1}{2} aex^2 \right]_0^{1/a} = \frac{1}{a} (e - 2).$$

(c) The normal at B has equation

$$y - e = -\frac{1}{ae} \left(x - \frac{1}{a} \right).$$

This can be written

$$y = -\frac{1}{ae} x + \frac{a^2 e^2 + 1}{a^2 e}.$$

Therefore

$$A_{\mathrm{II}} = 2 \int_0^{1/a} \left(-\frac{1}{ae} x + \frac{a^2 e^2 + 1}{a^2 e} - e^{ax} \right) dx = \frac{1 + 2a^2 e}{a^3 e}.$$

77. For $x > (n+1)!$

$$e^x > 1 + x + \cdots + \frac{x^{n+1}}{(n+1)!} > \frac{x^{n+1}}{(n+1)!} = x^n \left[\frac{x}{(n+1)!} \right] > x^n.$$

79. Numerically, $8.15 \le l \le 8.16$. The limit is the derivitive of $f(x) = e^{x^3}$ at $x = 1$. Note that
$$f'(x) = 3x^2 e^{x^3} \quad \text{and} \quad f'(1) = 3e \cong 8.15485.$$

SECTION 7.5

1. $\log_2 64 = \log_2 (2^6) = 6$

3. $\log_{64} (1/2) = \dfrac{\ln (1/2)}{\ln 64} = \dfrac{-\ln 2}{6 \ln 2} = -\dfrac{1}{6}$

5. $\log_5 1 = \log_5 (5^0) = 0$

7. $\log_5 (125) = \log_5 (5^3) = 3$

9. $\log_{32} 8 = \dfrac{\ln 8}{\ln 32} = \dfrac{3 \ln 2}{5 \ln 2} = \dfrac{3}{5}$

11. $\log_{10} 100^{-4/5} = \log_{10} 10^{-8/5} = -\dfrac{8}{5}$

13. $\log_p xy = \dfrac{\ln xy}{\ln p} = \dfrac{\ln x + \ln y}{\ln p} = \dfrac{\ln x}{\ln p} + \dfrac{\ln y}{\ln p} = \log_p x + \log_p y$

15. $\log_p x^y = \dfrac{\ln x^y}{\ln p} = y\dfrac{\ln x}{\ln p} = y \log_p x$

17.

$$10^x = e^x$$
$$\left(e^{\ln 10}\right)^x = e^x$$
$$e^{x \ln 10} = e^x$$
$$x \ln 10 = x$$
$$x \left(\ln 10 - 1\right) = 0$$

Thus, $x = 0$.

19.

$$\log_x 10 = \log_4 100$$
$$\dfrac{\ln 10}{\ln x} = \dfrac{\ln 100}{\ln 4}$$
$$\dfrac{\ln 10}{\ln x} = \dfrac{2\ln 10}{2\ln 2}$$
$$\ln x = \ln 2$$

Thus, $x = 2$.

21. $\log_2 x = \displaystyle\int_2^x \dfrac{dt}{t}$

$\dfrac{\ln x}{\ln 2} = [\ln t]_2^x = \ln x - \ln 2 \quad$ so $\quad \ln x \left(\ln 2 - 1\right) = (\ln 2)^2$

Thus, $x = e^c$ where $c = \dfrac{(\ln 2)^2}{\ln 2 - 1}$.

23. The logarithm function is increasing. Thus,

$$e^{t_1} < a < e^{t_2} \quad \Longrightarrow \quad t_1 = \ln e^{t_1} < \ln a < \ln e^{t_2} = t_2.$$

25. $\displaystyle\int 3^x \, dx = \dfrac{3^x}{\ln 3} + C$

27. $\displaystyle\int \left(x^3 + 3^{-x}\right) \, dx = \dfrac{1}{4}x^4 - \dfrac{3^{-x}}{\ln 3} + C$

29. $\displaystyle\int \dfrac{dx}{x \ln 5} = \dfrac{1}{\ln 5} \int \dfrac{dx}{x} = \dfrac{\ln|x|}{\ln 5} + C = \log_5 |x| + C$

31.

$$\int \dfrac{\log_2 x^3}{x} \, dx = \dfrac{1}{\ln 2} \int \dfrac{\ln x^3}{x} \, dx = \dfrac{3}{\ln 2} \int \dfrac{\ln x}{x} \, dx$$

$$= \dfrac{3}{\ln 2} \left[\dfrac{1}{2} (\ln x)^2\right] + C = \dfrac{3}{\ln 4} (\ln x)^2 + C$$

33. $\left\{ \begin{array}{l} u = x^2 \\ du = 2x \, dx \end{array} \right\}; \quad \displaystyle\int x 10^{x^2} \, dx = \dfrac{1}{2} \int 10^u \, du = \dfrac{1}{2\ln 10} 10^u + C = \dfrac{10^{x^2}}{2\ln 10} + C$

35. $f'(x) = \dfrac{1}{x \ln 3} \quad$ so $\quad f'(e) = \dfrac{1}{e \ln 3}$

37. $f'(x) = \dfrac{1}{x \ln x} \quad$ so $\quad f'(e) = \dfrac{1}{e \ln e} = \dfrac{1}{e}$

39.
$$f(x) = p^x$$

$$\ln f(x) = x \ln p$$

$$\frac{f'(x)}{f(x)} = \ln p$$

$$f'(x) = f(x) \ln p$$

$$f'(x) = p^x \ln p$$

41.
$$y = (x + 1)^x$$

$$\ln y = x \ln (x + 1)$$

$$\frac{1}{y}\frac{dy}{dx} = \frac{x}{x+1} + \ln (x + 1)$$

$$\frac{dy}{dx} = (x + 1)^x \left[\frac{x}{x+1} + \ln (x + 1)\right]$$

43.
$$y = (\ln x)^{\ln x}$$

$$\ln y = \ln x \left[\ln (\ln x)\right]$$

$$\frac{1}{y}\frac{dy}{dx} = \ln x \left[\frac{1}{x \ln x}\right] + \frac{1}{x}\left[\ln (\ln x)\right]$$

$$\frac{dy}{dx} = (\ln x)^{\ln x}\left[\frac{1 + \ln (\ln x)}{x}\right]$$

45.
$$y = (x^2 + 2)^{\ln x}$$

$$\ln y = (\ln x) \ln (x^2 + 2)$$

$$\frac{1}{y}\frac{dy}{dx} = \frac{1}{x}\ln (x^2 + 2) + \frac{2x}{x^2 + 2}\ln x$$

$$\frac{dy}{dx} = (x^2 + 2)^{\ln x}\left[\frac{2x \ln x}{x^2 + 2} + \frac{\ln (x^2 + 2)}{x}\right]$$

47.
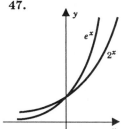

49.
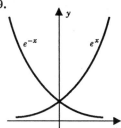

51.
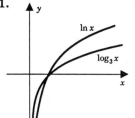

53.
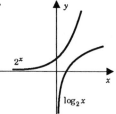

55. $f'(x) = -2x(\ln 10)10^{1-x^2}$;

f' :

domain $(-\infty, \infty)$;   increases on $(-\infty, 0]$, decreases on $[0, \infty)$; $f(0) = 10$ local and absolute max

57. $f'(x) = \dfrac{-x \ln 10}{\sqrt{1 - x^2}}10^{\sqrt{1-x^2}}$;

f' : ++++++++++0----------
       •―――――•――――――•
      -1   0   1

domain $[-1, 1]$;   increases on $[-1, 0]$, decreases on $[0, 1]$; $f(0) = 10$ local and absolute max, $f(-1) = f(1) = 1$ endpt and absolute min

59. $\displaystyle\int_1^2 2^{-x}\,dx = \left[-\frac{2^{-x}}{\ln 2}\right]_1^2 = \frac{1}{4\ln 2}$

61. $\displaystyle\int_1^4 \frac{dx}{x \ln 2} = [\log_2 x]_1^4 = \log_2 4 - 0 = 2$

63. $\displaystyle\int_0^1 x10^{1+x^2}\,dx = \left[\frac{1}{2\ln 10}10^{1+x^2}\right]_0^1 = \frac{1}{2\ln 10}(100 - 10) = \frac{45}{\ln 10}$

65. $\left\{\begin{array}{l} u = \log_{10} x \\ du = \dfrac{dx}{x \ln 10} \end{array}\right.\left|\begin{array}{l} x = 10 \implies u = 1 \\ x = 100 \implies u = 2 \end{array}\right\}$; $\displaystyle\int_{10}^{100} \frac{dx}{x \log_{10} x} = \ln 10 \int_{1}^{2} \frac{du}{u} = \ln 10 \left[\ln u\right]_{1}^{2}$

$$= (\ln 2)(\ln 10)$$

67. $\displaystyle\int_{0}^{1} (2^x + x^2)\, dx = \left[\frac{2^x}{\ln 2} + \frac{x^3}{3}\right]_{0}^{1} = \frac{1}{3} + \frac{1}{\ln 2}$ **69.** $\log_{10} 7 = \dfrac{\ln 7}{\ln 10} \cong \dfrac{1.95}{2.30} \cong 0.85$

71. $\log_{10} 45 = \dfrac{\ln 45}{\ln 10} = \dfrac{\ln 9 + \ln 5}{\ln 10} \cong \dfrac{2.20 + 1.61}{2.30} \cong 1.66$

73. approx 16.99999; $5^{\ln 17 / \ln 5} = \left(e^{\ln 5}\right)^{\ln 17 / \ln 5} = e^{\ln 17} = 17$

75. approx 54.59815; $16^{1/\ln 2} = \left(e^{\ln 16}\right)^{1/\ln 2} = e^{\ln 16 / \ln 2} = e^{4 \ln 2 / \ln 2} = e^4 \cong 54.59815$

SECTION 7.6

1. We begin with

$$A(t) = A_0 e^{rt}$$

and take $A_0 = \$500$ and $t = 10$. The interest earned is given by

$$A(10) - A_0 = 500\left(e^{10r} - 1\right).$$

Thus, (a) $500\left(e^{0.6} - 1\right) \cong \$411.06$ (b) $500\left(e^{0.8} - 1\right) \cong \$612.77$

(c) $500\left(e - 1\right) \cong \$859.14$.

3. We begin with

$$A(t) = A_0 e^{rt}$$

and take $t = 5$ and $A(5) = \$1000$. We solve for A_0 to obtain

$$A_0 = 1000 e^{-5r}.$$

Thus, (a) $1000 e^{-0.3} \cong \$740.82$ (b) $1000 e^{-0.4} \cong \$670.32$

(c) $1000 e^{-0.5} \cong \$606.53$

5. In general

$$A(t) = A_0 e^{rt}.$$

We set

$$3 A_0 = A_0 e^{20r}$$

and solve for r:

$$3 = e^{20r}, \quad \ln 3 = 20r, \quad r = \frac{\ln 3}{20} \cong 5\tfrac{1}{2}\%.$$

7. Let $s(t)$ be the number of pounds of salt present after t minutes. Since

$$s'(t) = \text{rate in} - \text{rate out} = 3\,(0.2) - 3\left(\frac{s(t)}{100}\right),$$

we have

$$s'(t) + 0.03 s(t) = 0.6.$$

Using the integrating factor $e^{\int 0.03 dt} = e^{0.03t}$, we obtain

$$e^{0.03t} s'(t) + 0.03 e^{0.03t} s(t) = 0.6 e^{0.03t}$$

$$\frac{d}{dt}\left[e^{0.03t} s(t)\right] = 0.6 e^{0.03t}$$

$$e^{0.03t} s(t) = 20 e^{0.03t} + C$$

$$s(t) = 20 + C e^{-0.03t}.$$

We use the initial condition $s(0) = 100(0.25) = 25$ to determine C:

$$25 = 20 + C e^0 \quad \text{so} \quad C = 5.$$

Thus,

$$s(t) = 20 + 5 e^{-0.03t} \text{ lb}.$$

9.

$$V'(t) = ktV(t)$$

$$V'(t) - ktV(t) = 0$$

$$e^{-kt^2/2} V'(t) - kt e^{-kt^2/2} V(t) = 0$$

$$\frac{d}{dt}\left[e^{-kt^2/2} V(t)\right] = 0$$

$$e^{-kt^2/2} V(t) = C$$

$$V(t) = C e^{kt^2/2}.$$

Since $V(0) = C = 200$,

$$V(t) = 200 e^{kt^2/2}.$$

Since $V(5) = 160$,

$$200 e^{k(25/2)} = 160, \quad e^{k(25/2)} = \tfrac{4}{5}, \quad e^k = \left(\tfrac{4}{5}\right)^{2/25}$$

and therefore

$$V(t) = 200\left(\tfrac{4}{5}\right)^{t^2/25} \text{ liters}.$$

11. Let $A(t)$ be the amount of radioactive substance present after t years.

In general

$$A(t) = A_0 e^{kt}.$$

Since one quarter of the substance decays in 4 years,

$$A(4) = \tfrac{3}{4}A_0$$

and thus

$$\tfrac{3}{4}A_0 = A_0 e^{4k} \quad \text{so that} \quad e^k = \left(\tfrac{3}{4}\right)^{1/4}.$$

To determine the half-life of the substance we must find the value of t for which

$$\tfrac{1}{2}A_0 = A_0 e^{kt} = A_0(\tfrac{3}{4})^{t/4}.$$

Solving for t, we have

$$\frac{1}{2} = \left(\frac{3}{4}\right)^{t/4}, \quad \ln\frac{1}{2} = \frac{t}{4}\ln\frac{3}{4}, \quad t = \frac{4\ln(1/2)}{\ln(3/4)} \cong 9.64.$$

The half-life is a little more than $9\tfrac{1}{2}$ years.

13.   Take two years ago as time $t = 0$. In general

(*) $A(t) = A_0 e^{kt}.$

We are given that

$$A_0 = 5 \quad \text{and} \quad A(2) = 4.$$

Thus,

$$4 = 5e^{2k} \quad \text{so that} \quad \tfrac{4}{5} = e^{2k} \quad \text{or} \quad e^k = \left(\tfrac{4}{5}\right)^{1/2}.$$

We can write

$$A(t) = 5\left(\tfrac{4}{5}\right)^{t/2}$$

and compute $A(5)$ as follows:

$$A(5) = 5\left(\tfrac{4}{5}\right)^{5/2} = 5e^{\frac{5}{2}\ln(4/5)} \cong 5e^{-0.56} \cong 2.86.$$

About 2.86 gm will remain 3 years from now.

15.   A fundamental property of radioactive decay is that the percentage of substance that decays during any year is constant:

$$100\left[\frac{A(t) - A(t+1)}{A(t)}\right] = 100\left[\frac{A_0 e^{kt} - A_0 e^{k(t+1)}}{A_0 e^{kt}}\right] = 100(1 - e^k)$$

If the half-life is n years, then

$$\tfrac{1}{2}A_0 = A_0 e^{kn} \quad \text{so that} \quad e^k = \left(\tfrac{1}{2}\right)^{1/n}.$$

Thus, $100\left[1 - \left(\tfrac{1}{2}\right)^{1/n}\right]\%$ of the material decays during any one year.

17.   (a) $x_1(t) = 10^6 t, \quad x_2(t) = e^t - 1$

      (b) $\dfrac{d}{dt}[x_1(t) - x_2(t)] = \dfrac{d}{dt}[10^6 t - (e^t - 1)] = 10^6 - e^t$

      This derivative is zero at $t = 6\ln 10 \cong 13.8$. After that the derivative is negative.

(c) $x_2(15) < e^{15} = (e^3)^5 \cong 20^5 = 2^5(10^5) = 3.2(10^6) < 15(10^6) = x_1(15)$

   $x_2(18) = e^{18} - 1 = (e^3)^6 - 1 \cong 20^6 - 1 = 64(10^6) - 1 > 18(10^6) = x_1(18)$

   $x_2(18) - x_1(18) \cong 64(10^6) - 1 - 18(10^6) \cong 46(10^6)$

(d)   If by time t_1 EXP has passed LIN, then $t_1 > 6 \ln 10$. For all $t \geq t_1$ the speed of EXP is greater than the speed of LIN:

$$\text{for}\quad t \geq t_1 > 6 \ln 10, \quad v_2(t) = e^t > 10^6 = v_1(t).$$

19.   Let $p(h)$ denote the pressure at altitude h. The equation $\dfrac{dp}{dh} = kp$ gives

(∗) $p(h) = p_0 e^{kh}$

where p_0 is the pressure at altitude zero (sea level).

Since $p_0 = 15$ and $p(10000) = 10$,

$$10 = 15e^{10000k}, \quad \tfrac{2}{3} = e^{10000k}, \quad \tfrac{1}{10000}\ln\tfrac{2}{3} = k.$$

Thus, (∗) can be written

$$p(h) = 15\left(\tfrac{2}{3}\right)^{h/10000}$$

(a) $p(5000) = 15\left(\tfrac{2}{3}\right)^{1/2} \cong 12.25$ lb/in.^{2}.

(b) $p(15000) = 15\left(\tfrac{2}{3}\right)^{3/2} \cong 8.16$ lb/in.^{2}.

21.   With inflation at $r\%$ the \$6000 tuition will rise in 3 years to $6000e^{3r}$.

Thus, (a) $6000\,e^{0.15} \cong \$6971$ (b) $6000\,e^{0.24} \cong \$7627$

   (c) $6000\,e^{0.36} \cong \$8600$.

23.   By Exercise 22

(∗) $v(t) = Ce^{-kt}, \quad t$ in seconds.

We use the initial conditions

$$v(0) = C = 4 \text{ mph} = \tfrac{1}{900} \text{ mi/sec} \quad\text{and}\quad v(60) = 2 = \tfrac{1}{1800} \text{ mi/sec}$$

to determine e^{-k}:

$$\tfrac{1}{1800} = \tfrac{1}{900}e^{-60k}, \quad e^{60k} = 2, \quad e^k = 2^{1/60}.$$

Thus,   (∗) can be written

$$v(t) = \tfrac{1}{900}\,2^{-t/60}.$$

The distance traveled by the boat is

$$s = \int_0^{60} \frac{1}{900}2^{-t/60}\,dt = \frac{1}{900}\left[\frac{-60}{\ln 2}2^{-t/60}\right]_0^{60} = \frac{1}{30\ln 2} \text{ mi} = \frac{176}{\ln 2} \text{ ft} \quad (\text{about 254 ft}).$$

25.   about 284 million; $203\,(227/203)^3 \cong 283.85$

27.   Future dollars are discounted by a factor of $e^{-0.05t}$. We want to maximize the function

$$f(t) = V(t)e^{-0.05t}$$

$$= V_0 \left(\tfrac{3}{2}\right)^{\sqrt{t}} e^{-0.05t}$$

$$= V_0 e^{\sqrt{t}\left(\ln \tfrac{3}{2}\right) - 0.05t}, \qquad t \geq 0.$$

Differentiation gives

$$f'(t) = V_0 e^{\sqrt{t}\left(\ln \tfrac{3}{2}\right) - 0.05t} \left[\frac{1}{2\sqrt{t}} \ln \left(\frac{3}{2}\right) - 0.05\right].$$

Setting $f'(t) = 0$, we find that

$$\frac{\ln(3/2)}{2\sqrt{t}} = \frac{1}{20}, \quad \sqrt{t} = 10 \ln \frac{3}{2}, \quad t = 100 \left(\ln \frac{3}{2}\right)^2 \cong 16.44.$$

Clearly this value of t maximizes f. The company should wait about $16\tfrac{1}{2}$ years before cutting the timber.

29.   Since the amount $A(t)$ of raw sugar present after t hours decreases at a rate proportional to A, we have

$$A(t) = A_0 e^{kt}.$$

We are given $A_0 = 1000$ and $A(10) = 800$.   Thus,

$$800 = 1000 e^{10k}, \quad \tfrac{4}{5} = e^{10k}, \quad e^k = \left(\tfrac{4}{5}\right)^{1/10}$$

so that

$$A(t) = 1000 \left(\tfrac{4}{5}\right)^{t/10}.$$

Then, $A(20) = 1000 \left(\tfrac{4}{5}\right)^{20/10} = 640$. After 10 more hours of inversion there will remain 640 pounds.

31.   Proceeding from the hint, we know from Theorem 7.6.1 that

$$f(t) + k_2/k_1 = A e^{k_1(t + k_2/k_1)} = A e^{k_1 t + k_2} = \left(A e^{k_2}\right) e^{k_1 t} = C e^{k_1 t}.$$
$$\underset{\text{where } A \text{ is an arbitrary constant}}{\big\uparrow} \qquad\qquad \underset{\text{set } A e^{k_2} = C}{\big\uparrow}$$

Therefore $f(t) = C e^{k_1 t} - k_2/k_1$, C an arbitrary constant.

33.   From Exercise 31 you can determine that

$$v(t) = \frac{32}{K} \left(1 - e^{-Kt}\right).$$

At each time t, $1 - e^{-Kt} < 1$.   With $K > 0$,

$$v(t) = \frac{32}{K} \left(1 - e^{-Kt}\right) < \frac{32}{K}.$$

35.   From Exercise 34

$$\frac{dT}{dt} = -k\,(T - \tau),$$

which can be written

$$\frac{d}{dt}\,(T - \tau) = -k\,(T - \tau).$$

Integration gives

$$T - \tau = Ce^{-kt} \quad \text{and therefore} \quad T(t) = \tau + Ce^{-kt}.$$

Since $\tau = 65$ and $T(0) = 185$, we have

$$185 = 65 + C \quad \text{and therefore} \quad C = 120.$$

Therefore

$$T(t) = 65 + 120e^{-kt}.$$

Since $T(2) = 155$,

$$155 = 65 + 120e^{-2k}, \quad \tfrac{90}{120} = e^{-2k}, \quad e^k = \left(\tfrac{4}{3}\right)^{1/2}.$$

Thus,

$$T(t) = 65 - 120\left(\tfrac{4}{3}\right)^{-t/2}.$$

We want to find t so that $T = 105$:

$$105 = 65 + 120\left(\frac{4}{3}\right)^{-t/2}, \quad \frac{1}{3} = \left(\frac{4}{3}\right)^{-t/2}, \quad \ln\frac{1}{3} = -\frac{t}{2}\ln\frac{4}{3}.$$

Thus, $t = \dfrac{2\ln 3}{\ln(4/3)}$. You would expect to wait $\dfrac{2\ln 3}{\ln(4/3)} - 2 \cong 5.64$ minutes more for the

coffee to cool to $105°\,F$.

SECTION 7.7

1.
$$\boxed{\begin{array}{ll} u = x, & dv = e^{-x}\,dx \\ du = dx, & v = -e^{-x} \end{array}}$$
$$\int xe^{-x}\,dx = -xe^{-x} - \int -e^{-x}\,dx = -xe^{-x} - e^{-x} + C$$

3.
$$\int x^2 \ln x\,dx = \frac{1}{3}x^3 \ln x - \int \frac{1}{3}x^3\left(\frac{1}{x}\right)dx$$
$$\boxed{\begin{array}{ll} u = \ln x, & dv = x^2\,dx \\ du = \dfrac{dx}{x}, & v = \dfrac{1}{3}x^3 \end{array}}$$
$$= \frac{1}{3}x^3 \ln x - \frac{1}{3}\int x^2\,dx$$
$$= \tfrac{1}{3}x^3 \ln x - \tfrac{1}{9}x^3 + C$$

5. $\left\{ \begin{array}{l} t = -x^3 \\ dt = -3x^2 \, dx \end{array} \right\}$; $\displaystyle\int x^2 e^{-x^3} \, dx = -\frac{1}{3} \int e^t \, dt = -\frac{1}{3} e^t + C = -\frac{1}{3} e^{-x^3} + C$

7. $\displaystyle\int x^2 e^{-x} \, dx = -x^2 e^{-x} - \int -2x e^{-x} \, dx = -x^2 e^{-x} + 2 \int x e^{-x} \, dx$

$$\begin{array}{ll} u = x^2, & dv = e^{-x} \, dx \\ du = 2x \, dx, & v = -e^{-x} \end{array}$$

$$\begin{array}{ll} u = x, & dv = e^{-x} \, dx \\ du = dx, & v = -e^{-x} \end{array}$$

$$= -x^2 e^{-x} + 2 \left[-x e^{-x} - \int -e^{-x} \, dx \right]$$

$$= -x^2 e^{-x} + 2 \left(-x e^{-x} - e^{-x} \right) + C$$

$$= -e^{-x} \left(x^2 + 2x + 2 \right) + C$$

9. $\displaystyle\int x^2 (1-x)^{-1/2} \, dx = -2x^2 (1-x)^{1/2} + 4 \int x (1-x)^{1/2} \, dx$

$$\begin{array}{ll} u = x^2, & dv = (1-x)^{-1/2} \, dx \\ du = 2x \, dx, & v = -2(1-x)^{1/2} \end{array}$$

$$\begin{array}{ll} u = x, & dv = (1-x)^{1/2} \, dx \\ du = dx, & v = -\frac{2}{3}(1-x)^{3/2} \end{array}$$

$$= -2x^2 (1-x)^{1/2} + 4 \left[-\frac{2x}{3} (1-x)^{3/2} + \int \frac{2}{3} (1-x)^{3/2} \, dx \right]$$

$$= -2x^2 (1-x)^{1/2} - \frac{8x}{3} (1-x)^{3/2} - \frac{16}{15} (1-x)^{5/2} + C$$

Or, use the substitution $t = 1 - x$ (no integration by parts needed) to obtain:

$$-2(1-x)^{1/2} + \tfrac{4}{3} (1-x)^{3/2} - \tfrac{2}{5} (1-x)^{5/2} + C.$$

11. $\displaystyle\int x \ln \sqrt{x} \, dx = \frac{1}{2} \int x \ln x \, dx = \frac{1}{2} \left[\frac{1}{2} x^2 \ln x - \frac{1}{2} \int x \, dx \right]$

$$\begin{array}{ll} u = \ln x, & dv = x \, dx \\ du = \dfrac{dx}{x}, & v = \dfrac{1}{2} x^2 \end{array}$$

$$= \tfrac{1}{4} x^2 \ln x - \tfrac{1}{8} x^2 + C$$

13. $\displaystyle\int \frac{\ln (x+1)}{\sqrt{x+1}} \, dx = 2\sqrt{x+1} \ln (x+1) - \int \frac{2 \, dx}{\sqrt{x+1}}$

$$\begin{array}{ll} u = \ln (x+1), & dv = \dfrac{dx}{\sqrt{x+1}} \\ du = \dfrac{dx}{x+1}, & v = 2\sqrt{x+1} \end{array}$$

$$= 2\sqrt{x+1} \ln (x+1) - 4\sqrt{x+1} + C$$

15.

$$\int (\ln x)^2 \, dx = x \, (\ln x)^2 - 2 \int \ln x \, dx$$

$$\boxed{\begin{aligned} u &= (\ln x)^2 \quad, \quad dv = dx \\ du &= \frac{2\ln x}{x} \, dx \, , \quad v = x \end{aligned}}$$

$$= x \, (\ln x)^2 - 2 \left[x \ln x - \int \, dx \right]$$

$$\boxed{\begin{aligned} u &= \ln x \, , \quad dv = dx \\ du &= \frac{dx}{x} \, , \quad v = x \end{aligned}}$$

$$= x \, (\ln x)^2 - 2x \ln x + 2x + C$$

17.

$$\int x^3 \, 3^x \, dx = \frac{x^3 \, 3^x}{\ln 3} - \frac{3}{\ln 3} \int x^2 \, 3^x \, dx$$

$$\boxed{\begin{aligned} u &= x^3 \quad, \quad dv = 3^x \, dx \\ du &= 3x^2 \, dx \, , \quad v = \frac{3^x}{\ln 3} \end{aligned}}$$

$$= \frac{x^3 \, 3^x}{\ln 3} - \frac{3}{\ln 3} \left[\frac{x^2 \, 3^x}{\ln 3} - \frac{2}{\ln 3} \int x \, 3^x \, dx \right]$$

$$\boxed{\begin{aligned} u &= x^2 \quad, \quad dv = 3^x \, dx \\ du &= 2x \, dx \, , \quad v = \frac{3^x}{\ln 3} \end{aligned}}$$

$$= \frac{x^3 \, 3^x}{\ln 3} - \frac{3x^2 \, 3^x}{(\ln 3)^2} + \frac{6}{(\ln 3)^2} \int x \, 3^x \, dx$$

$$= \frac{x^3 \, 3^x}{\ln 3} - \frac{3x^2 \, 3^x}{(\ln 3)^2} + \frac{6}{(\ln 3)^2} \left[\frac{x3^x}{\ln 3} - \frac{1}{\ln 3} \int 3^x \, dx \right]$$

$$\boxed{\begin{aligned} u &= x \, , \quad dv = 3^x \, dx \\ du &= dx \, , \quad v = \frac{3^x}{\ln 3} \end{aligned}}$$

$$= 3^x \left[\frac{x^3}{\ln 3} - \frac{3x^2}{(\ln 3)^2} + \frac{6x}{(\ln 3)^3} - \frac{6}{(\ln 3)^4} \right] + C$$

19.

$$\int x \, (x+5)^{14} \, dx = \frac{x}{15} \, (x+5)^{15} - \frac{1}{15} \int (x+5)^{15} \, dx$$

$$\boxed{\begin{aligned} u &= x \, , \quad dv = (x+5)^{14} \, dx \\ d &= dx \, , \quad v = \tfrac{1}{15} \, (x+5)^{15} \end{aligned}}$$

$$= \tfrac{1}{15} x \, (x+5)^{15} - \tfrac{1}{240} \, (x+5)^{16} + C$$

Or, use the substitution $t = x + 5$ (integration by parts not needed) to obtain:

$$\tfrac{1}{16} \, (x+5)^{16} - \tfrac{1}{3} \, (x+5)^{15} + C.$$

21.

$$\int x \cos x \, dx = x \sin x - \int \sin x \, dx$$

$$\boxed{\begin{array}{ll} u = x \,, & dv = \cos x \, dx \\ du = dx \,, & v = \sin x \end{array}}$$

$$= x \sin x + \cos x + C$$

23.

$$\int x^2 (x+1)^9 \, dx = \frac{x^2}{10} (x+1)^{10} - \frac{1}{5} \int x (x+1)^{10} \, dx$$

$$\boxed{\begin{array}{ll} u = x^2 \,, & dv = (x+1)^9 \, dx \\ du = 2x \, dx \,, & v = \frac{1}{10}(x+1)^{10} \end{array}}$$

$$= \frac{x^2}{10}(x+1)^{10} - \frac{1}{5}\left[\frac{x}{11}(x+1)^{11} - \frac{1}{11}\int (x+1)^{11} \, dx\right]$$

$$\boxed{\begin{array}{ll} u = x \,, & dv = (x+1)^{10} \, dx \\ du = dx \,, & v = \frac{1}{11}(x+1)^{11} \end{array}}$$

$$= \frac{x^2}{10}(x+1)^{10} - \frac{x}{55}(x+1)^{11} + \frac{1}{660}(x+1)^{12} + C$$

25.

$$\int e^x \sin x \, dx = -e^x \cos x + \int e^x \cos x \, dx$$

$$\boxed{\begin{array}{ll} u = e^x \,, & dv = \sin x \, dx \\ du = e^x \, dx \,, & v = -\cos x \end{array}}$$

$$= -e^x \cos x + e^x \sin x - \int e^x \sin x \, dx$$

$$\boxed{\begin{array}{ll} u = e^x \,, & dv = \cos x \, dx \\ du = e^x \, dx \,, & v = \sin x \end{array}}$$

Adding $\int e^x \sin x \, dx$ to both sides, we get

$$2 \int e^x \sin x \, dx = -e^x \cos x + e^x \sin x$$

so that

$$\int e^x \sin x \, dx = \frac{1}{2} e^x (\sin x - \cos x) + C.$$

27

$$\int \ln (1+x^2) \, dx = x \ln (1+x^2) - 2 \int \frac{x^2}{1+x^2} \, dx$$

$$\boxed{\begin{array}{ll} u = \ln (1+x^2) \,, & dv = dx \\ du = \dfrac{2x}{1+x^2} \, dx \,, & v = x \end{array}}$$

$$= x \ln (1+x^2) - 2 \int \frac{x^2 + 1 - 1}{1+x^2} \, dx$$

$$= x \ln (1+x^2) - 2 \int \left(1 - \frac{1}{1+x^2}\right) dx$$

$$= x \ln (1+x^2) - 2x + 2 \tan^{-1} x + C$$

29.

$$\int x^n \ln x \, dx = \frac{x^{n+1} \ln x}{n+1} - \frac{1}{n+1} \int x^n \, dx$$

$$\boxed{\begin{array}{ll} u = \ln x\,, & dv = x^n \, dx \\[2mm] du = \dfrac{dx}{x}\,, & v = \dfrac{x^{n+1}}{n+1} \end{array}}$$

$$= \frac{x^{n+1} \ln x}{n+1} - \frac{x^{n+1}}{(n+1)^2} + C$$

31.

$$\{t = x^2, \quad dt = 2x \, dx\} \,; \qquad \int x^3 \sin x^2 \, dx = \frac{1}{2} \int t \sin t \, dt$$

$$\boxed{\begin{array}{ll} u = t\,, & dv = \sin t \, dt \\[2mm] du = dt\,, & v = -\cos t \end{array}}$$

$$= \frac{1}{2} \left[-t \cos t + \int \cos t \, dt \right]$$

$$= \frac{1}{2}(-t \cos t + \sin t) + C$$

$$= -\tfrac{1}{2} x^2 \cos x^2 + \tfrac{1}{2} \sin x^2 + C$$

33. To calculate $\int x^4 e^x \, dx$ we could integrate by parts four times, each time selecting $dv = e^x \, dx$. Instead we shall guess the antiderivative. We have seen that

$$\int x e^x \, dx = (x - 1) e^x + C \quad \text{and} \quad \int x^2 e^x \, dx = (x^2 - 2x + 2) e^x + C.$$

Thus we guess that

$$\int x^4 e^x \, dx = P(x) e^x + C$$

with P a polynomial of degree 4:

$$P(x) = Ax^4 + Bx^3 + Dx^2 + Ex + F.$$

From $\dfrac{d}{dx}[P(x)e^x] = x^4 e^x$,

$$\left(Ax^4 + Bx^3 + Dx^2 + Ex + F \right) e^x + \left(4Ax^3 + 3Bx^2 + 2Dx + E \right) e^x = x^4 e^x$$

or

$$Ax^4 + (B + 4A)\, x^3 + (D + 3B)\, x^2 + (E + 2D)\, x + (F + E) + x^4.$$

Comparing coefficients, we get

$$A = 1, \quad B + 4A = 0 \implies B = -4, \quad D + 3B = 0 \implies D = 12,$$
$$E + 2D = 0 \implies E = -24, \quad F + E = 0 \implies F = 24.$$

Thus,

$$\int x^4 e^x \, dx = (x^4 - 4x^3 + 12x^2 - 24x + 24) e^x + C.$$

35. $\bar{x} = \dfrac{1}{e-1}, \qquad \bar{y} = \dfrac{1}{4}(e+1)$ **37.** $\bar{x} = \tfrac{1}{2}\pi, \qquad \bar{y} = \tfrac{1}{8}\pi$

39. (a) $M = \displaystyle\int_0^1 e^{kx} \, dx = \dfrac{1}{k}(e^k - 1)$

(b) $x_M M = \displaystyle\int_0^1 x e^{kx} \, dx = \dfrac{(k-1)e^k + 1}{k^2}, \qquad x_M = \dfrac{(k-1)e^k + 1}{k(e^k - 1)}$

41. $V_y = \int_0^1 2\pi x e^{\alpha x}\, dx = 2\pi \left[\frac{1}{\alpha^2}\left(\alpha x e^{\alpha x} - e^{\alpha x}\right)\right]_0^1 = \frac{2\pi}{\alpha^2}\left(\alpha e^\alpha - e^\alpha + 1\right)$

43. $V_y = \int_0^1 2\pi x \cos \frac{1}{2}\pi x\, dx = \left[4x \sin \frac{1}{2}\pi x + \frac{8}{\pi}\cos \frac{1}{2}\pi x\right]_0^1 = 4 - \frac{8}{\pi}$

45. $V_y = \int_0^1 2\pi x^2 e^x\, dx = 2\pi\left(e - 2\right)$ (Problem 5)

47.

$$V_x = \int_0^1 \pi e^{2x}\, dx = \pi\left[\frac{1}{2}e^{2x}\right]_0^1 = \frac{1}{2}\pi\left(e^2 - 1\right)$$

$$\bar{x}V_x = \int_0^1 \pi x e^{2x}\, dx = \pi\left[\frac{1}{2}x e^{2x} - \frac{1}{4}e^{2x}\right]_0^1 = \frac{1}{4}\pi\left(e^2 + 1\right)$$

$$\bar{x} = \frac{e^2 + 1}{2\left(e^2 - 1\right)}$$

SECTION 7.8

1. The equation of motion is of the form

$$x\left(t\right) = A \sin\left(\omega t + \phi_0\right).$$

The period is $T = 2\pi/\omega = \pi/4$. Therefore $\omega = 8$. Thus

$$x\left(t\right) = A \sin\left(8t + \phi_0\right) \text{ and } v\left(t\right) = 8A \cos\left(8t + \phi_0\right).$$

Since $x\left(0\right) = 1$ and $v\left(0\right) = 0$, we have

$$1 = A \sin \phi_0 \quad \text{and} \quad 0 = 8A \cos \phi_0.$$

These equations are satisfied by taking $A = 1$ and $\phi_0 = \pi/2$.

Therefore the equation of motion reads

$$x\left(t\right) = \sin\left(8t + \tfrac{1}{2}\pi\right).$$

The amplitude is 1 and the frequency is $8/2\pi = 4/\pi$.

3. We can write the equation of motion as

$$x\left(t\right) = A \sin\left(\frac{2\pi}{T}t\right).$$

Differentiation gives

$$v\left(t\right) = \frac{2\pi A}{T}\cos\left(\frac{2\pi}{T}t\right).$$

The object passes through the origin whenever $\sin\left[\left(2\pi/T\right)\right] = 0$.

Then $\cos\left[\left(2\pi/T\right)t\right] = \pm 1$ and $v = \pm 2\pi A/T$.

5. In this case $\phi_0 = 0$ and, measuring t_2 in seconds, $T = 6$.

Therefore $\omega = 2\pi/6 = \pi/3$ and we have

$$x(t) = A\sin\left(\frac{\pi}{3}t\right), \quad v(t) = \frac{\pi A}{3}\cos\left(\frac{\pi}{3}t\right).$$

Since $v(0) = 5$, we have $\pi A/3 = 5$ and therefore $A = 15/\pi$.

The equation of motion can be written

$$x(t) = (15/\pi)\sin\left(\tfrac{1}{3}\pi t\right)$$

7. $x(t) = x_0\sin\left(\sqrt{k/m}\,t + \tfrac{1}{2}\pi\right)$

9. The equation of motion for the bob of Problem 3 reads

$$x(t) = x_0\sin\left(\sqrt{k/m}\,t + \tfrac{1}{2}\pi\right). \qquad \text{(Exercise 7)}$$

Since $v(t) = \sqrt{k/m}\,x_0\cos\left(\sqrt{k/m}\,t + \tfrac{1}{2}\pi\right)$, the maximum speed is $\sqrt{k/m}\,x_0$.

The bob takes on half of that speed where $\left|\cos\left(\sqrt{k/m}\,t + \tfrac{1}{2}\pi\right)\right| = \tfrac{1}{2}$. Therefore

$$\left|\sin\left(\sqrt{k/m}\,t + \tfrac{1}{2}\pi\right)\right| = \sqrt{1 - \tfrac{1}{4}} = \tfrac{1}{2}\sqrt{3} \quad \text{and} \quad x(t) = \pm\tfrac{1}{2}\sqrt{3}\,x_0.$$

11. $\quad \text{KE} = \tfrac{1}{2}m[v(t)]^2 = \tfrac{1}{2}m(k/m)x_0{}^2\cos^2\left(\sqrt{k/m}\,t + \tfrac{1}{2}\pi\right)$

$$= \tfrac{1}{4}kx_0{}^2\left[1 + \cos\left(2\sqrt{k/m}\,t + \tfrac{1}{2}\pi\right)\right].$$

$$\text{Average KE} = \frac{1}{2\pi\sqrt{m/k}}\int_0^{2\pi\sqrt{m/k}} \tfrac{1}{4}kx_0{}^2\left[1 + \cos\left(2\sqrt{k/m}\,t + \frac{1}{2}\pi\right)\right]\,dt$$

$$= \tfrac{1}{4}kx_0{}^2.$$

13. Setting $\quad y(t) = x(t) - 2, \quad$ we can write $\quad x''(t) = 8 - 4x(t) \quad$ as $y''(t) + 4y(t) = 0$.

This is simple harmonic motion about the point $y = 0$; that is, about the point $x = 2$. The equation of motion is of the form

$$y(t) = A\sin(2t + \phi_0).$$

Since $y(0) = x(0) - 2 = -2$, the amplitude A is 2. Since $\omega = 2$, the period T is $2\pi/2 = \pi$.

15. **(a)**   Take the downward direction as positive. We begin by analyzing the forces on the buoy at a general position x cm beyond equilibrium. First there is the weight of the buoy: $F_1 = mg$. This is a downward force. Next there is the buoyancy force equal to the weight of the fluid displaced; this force is in the opposite direction: $F_2 = -\pi r^2 (L + x)\rho$. We are neglecting friction so the total force is

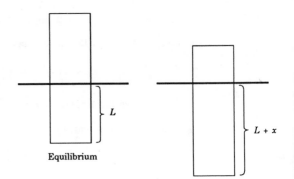

Equilibrium

$$F = F_1 + F_2 = mg - \pi r^2 (L + x)\rho = \left(mg - \pi r^2 L\rho\right) - \pi r^2 x\rho.$$

We are assuming at the equilibrium point that the forces (weight of buoy and buoyant force of fluid) are in balance:

$$mg - \pi r^2 L\rho = 0.$$

Thus,

$$F = -\pi r^2 x\rho.$$

By Newton's

$$F = ma \qquad\qquad \text{(force = mass × acceleration)}$$

we have

$$ma = -\pi r^2 x\rho \qquad \text{and thus} \qquad a + \frac{\pi r^2 \rho}{m}x = 0.$$

Thus, at each time t,

$$x''(t) + \frac{\pi r^2 \rho}{m}x(t) = 0.$$

(b)   The usual procedure shows that

$$x(t) = x_0 \sin\left(r\sqrt{\pi\rho/m}\, t + \tfrac{1}{2}\pi\right).$$

The amplitude A is x_0 and the period T is $(2/r)\sqrt{m\pi/\rho}$.

SECTION 7.9

1. $\left\{ \begin{matrix} u = 3x \\ du = 3dx \end{matrix} \right\}$; $\displaystyle\int \tan 3x\, dx = \frac{1}{3}\int \tan u\, du = \frac{1}{3}\ln|\sec u| + C = \frac{1}{3}\ln|\sec 3x| + C$

3. $\left\{ \begin{matrix} u = \pi x \\ du = \pi\, dx \end{matrix} \right\}$; $\displaystyle\int \csc \pi x\, dx = \frac{1}{\pi}\int \csc u\, du = \frac{1}{\pi}\ln|\csc u - \cot u| + C$

$$= \frac{1}{\pi}\ln|\csc \pi x - \cot \pi x| + C$$

5. $\left\{ \begin{array}{l} u = e^x \\ du = e^x \, dx \end{array} \right\}$; $\displaystyle\int e^x \cot e^x \, dx = \int \cot u \, du = \ln|\sin u| + C = \ln|\sin e^x| + C$

7. $\left\{ \begin{array}{l} u = 3 - 2\cos 2x \\ du = 4\sin 2x \, dx \end{array} \right\}$; $\displaystyle\int \frac{\sin 2x}{3 - 2\cos 2x} \, dx = \frac{1}{4}\int \frac{du}{u} = \frac{1}{4}\ln|u| + C$

$$= \tfrac{1}{4}\ln|3 - 2\cos 2x| + C$$

9. $\left\{ \begin{array}{l} u = \tan 3x \\ du = 3\sec^2 3x \, dx \end{array} \right\}$; $\displaystyle\int e^{\tan 3x} \sec^2 3x \, dx = \frac{1}{3}\int e^u \, du = \frac{1}{3}e^u + C = \frac{1}{3}e^{\tan 3x} + C$

11. $\left\{ \begin{array}{l} u = x^2 \\ du = 2x \, dx \end{array} \right\}$; $\displaystyle\int x \sec x^2 \, dx = \frac{1}{2}\int \sec u \, du = \frac{1}{2}\ln|\sec u + \tan u| + C$

$$= \tfrac{1}{2}\ln|\sec x^2 + \tan x^2| + C$$

13. $\left\{ \begin{array}{l} u = \ln|\sin x| \\ du = \cot x \, dx \end{array} \right\}$; $\displaystyle\int \cot x \ln|\sin x| \, dx = \int u \, du = \frac{1}{2}u^2 + C$

$$= \tfrac{1}{2}\left(\ln|\sin x|\right)^2 + C$$

15. $\displaystyle\int (1 + \sec x)^2 \, dx = \int \left(1 + 2\sec x + \sec^2 x\right) \, dx = x + 2\ln|\sec x + \tan x| + \tan x + C$

17. $\left\{ \begin{array}{l} u = 1 + \cot x \\ du = -\csc^2 x \, dx \end{array} \right\}$; $\displaystyle\int \left(\frac{\csc x}{1 + \cot x}\right)^2 \, dx = -\int \frac{du}{u^2} = \frac{1}{u} + C = \frac{1}{1 + \cot x} + C$

19. $\left\{ \begin{array}{l} u = 1 + \sin x \\ du = \cos x \, dx \end{array} \right. \left| \begin{array}{l} x = \pi/6 \implies u = 3/2 \\ x = \pi/2 \implies u = 2 \end{array} \right\}$; $\displaystyle\int_{\pi/6}^{\pi/2} \frac{\cos x}{1 + \sin x} \, dx = \int_{3/2}^{2} \frac{du}{u} = [\ln u]_{3/2}^{2} = \ln \frac{4}{3}$

21. $\displaystyle\int_{\pi/4}^{\pi/2} \cot x \, dx = [\ln|\sin x|]_{\pi/4}^{\pi/2} = \ln 1 - \ln \frac{\sqrt{2}}{2} = \ln \sqrt{2} = \frac{1}{2}\ln 2$

23. $\left\{ \begin{array}{l} u = e^x \\ du = e^x \, dx \end{array} \right. \left| \begin{array}{l} x = 0 \implies u = 1 \\ x = \ln\dfrac{\pi}{4} \implies u = \dfrac{\pi}{4} \end{array} \right\}$; $\displaystyle\int_{0}^{\ln \pi/4} e^x \sec e^x \, dx$

$$= \int_{1}^{\pi/4} \sec u \, du = [\ln|\sec u + \tan u|]_{1}^{\pi/4} = \ln|\sqrt{2} + 1| - \ln|\sec 1 + \tan 1|$$

$$= \ln \left| \frac{1 + \sqrt{2}}{\sec 1 + \tan 1} \right| \cong -0.345$$

25.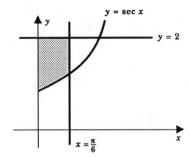

$$A = \int_0^{\pi/6} (2 - \sec x)\, dx$$

$$= \left[2x - \ln|\sec x + \tan x|\right]_0^{\pi/6}$$

$$= \frac{\pi}{3} - \ln\left|\frac{2}{\sqrt{3}} + \frac{1}{\sqrt{3}}\right| = \frac{\pi}{3} - \frac{1}{2}\ln 3$$

27.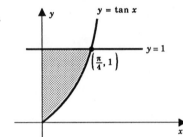

$$A = \int_0^{\pi/4} (1 - \tan x)\, dx$$

$$= \left[x - \ln|\sec x|\right]_0^{\pi/4}$$

$$= \frac{\pi}{4} - \ln\sqrt{2} = \frac{\pi}{4} - \frac{1}{2}\ln 2$$

SECTION 7.10

1. 0   **3.** $\pi/3$ **5.** $-\pi/3$ **7.** 0   **9.** $-\pi/6$

11. 1/2   **13.** $-\pi/4$ **15.** $\frac{1}{2}\sqrt{3}$ **17.** $\pi/4$ **19.** $\frac{1}{2}\sqrt{3}$

21. 1.16   **23.** -0.46 **25.** 0.28   **27.** $1/\sqrt{1+x^2}$ **29.** x

31. $\sqrt{1+x^2}$ **33.** $\sqrt{1-x^2}$ **35.** $\frac{1}{x}\sqrt{1+x^2}$ **37.** $\dfrac{x}{\sqrt{1-x^2}}$

39. $\dfrac{dy}{dx} = \dfrac{1}{1+(x+1)^2} = \dfrac{1}{x^2+2x+2}$ **41.** $\dfrac{dy}{dx} = \dfrac{1}{\sqrt{1-(x^2)^2}}\dfrac{d}{dx}(x^2) = \dfrac{2x}{\sqrt{1-x^4}}$

43. $f'(x) = \sin^{-1} 2x + x\dfrac{1}{\sqrt{1-(2x)^2}}\dfrac{d}{dx}(2x) = \sin^{-1} 2x + \dfrac{2x}{\sqrt{1-4x^2}}$

45. $\dfrac{du}{dx} = 2\left(\sin^{-1} x\right)\dfrac{d}{dx}\left(\sin^{-1} x\right) = \dfrac{2\sin^{-1} x}{\sqrt{1-x^2}}$

47. $\dfrac{dy}{dx} = \dfrac{x\left(\dfrac{1}{1+x^2}\right) - (1)\tan^{-1} x}{x^2} = \dfrac{x - (1+x^2)\tan^{-1} x}{x^2(1+x^2)}$

49. $f'(x) = \dfrac{1}{2}\left(\tan^{-1} 2x\right)^{-1/2}\dfrac{d}{dx}\left(\tan^{-1} 2x\right) = \dfrac{1}{2}\left(\tan^{-1} 2x\right)^{-1/2}\dfrac{2}{1+(2x)^2} = \dfrac{1}{(1+4x^2)\sqrt{\tan^{-1} 2x}}$

51. $\dfrac{dy}{dx} = \dfrac{1}{1+(\ln x)^2}\dfrac{d}{dx}(\ln x) = \dfrac{1}{x\left[1+(\ln x)^2\right]}$

53. $\dfrac{d\theta}{dr} = \dfrac{1}{\sqrt{1-\left(\sqrt{1-r^2}\,\right)^2}}\dfrac{d}{dr}\left(\sqrt{1-r^2}\,\right) = \dfrac{1}{\sqrt{r^2}}\cdot\dfrac{-r}{\sqrt{1-r^2}} = -\dfrac{r}{|r|\sqrt{1-r^2}}$

55. $\dfrac{d\theta}{dr} = \dfrac{1}{1+\left(\dfrac{c+r}{1-cr}\right)^2}\left[\dfrac{(1-cr)(1)-(c+r)(-c)}{(1-cr)^2}\right] = \dfrac{1+c^2}{(1-rc)^2+(c+r)^2} = \dfrac{1}{1+r^2}$

57. $f'(x) = \dfrac{-x}{\sqrt{c^2-x^2}} + \dfrac{c}{\sqrt{1-(x/c)^2}}\cdot\left(\dfrac{1}{c}\right) = \dfrac{c-x}{\sqrt{c^2-x^2}} = \sqrt{\dfrac{c-x}{c+x}}$

59. $\dfrac{dy}{dx} = \dfrac{\sqrt{c^2-x^2}\,(1)-x\left(\dfrac{-x}{\sqrt{c^2-x^2}}\right)}{\left(\sqrt{c^2-x^2}\,\right)^2} - \dfrac{1}{\sqrt{1-(x/c)^2}}\left(\dfrac{1}{c}\right)$

$\qquad = \dfrac{c^2}{(c^2-x^2)^{3/2}} - \dfrac{1}{(c^2-x^2)^{1/2}} = \dfrac{x^2}{(c^2-x^2)^{3/2}}$

61. $\left\{\begin{array}{l} au = x+b \\ a\,du = dx \end{array}\right\};\quad \displaystyle\int \dfrac{dx}{\sqrt{a^2-(x+b)^2}} = \int \dfrac{a\,du}{\sqrt{a^2-a^2u^2}} = \int \dfrac{du}{\sqrt{1-u^2}}$

$\qquad\qquad\qquad = \sin^{-1}u + C = \sin^{-1}\left(\dfrac{x+b}{a}\right) + C$

63. $\displaystyle\int_0^1 \dfrac{dx}{1+x^2} = \left[\tan^{-1}x\right]_0^1 = \dfrac{\pi}{4}$ **65.** $\displaystyle\int_0^{1/\sqrt{2}} \dfrac{dx}{\sqrt{1-x^2}} = \left[\sin^{-1}x\right]_0^{1/\sqrt{2}} = \dfrac{\pi}{4}$

67. $\displaystyle\int_0^5 \dfrac{dx}{25+x^2} = \left[\dfrac{1}{5}\tan^{-1}\dfrac{x}{5}\right]_0^5 = \dfrac{\pi}{20}$

$\qquad\qquad\qquad\quad\longleftarrow (7.10.9)$

69. $\left\{\begin{array}{l|l} 3u = 2x & x=0 \\ 3\,du = 2\,dx & x=3/2 \end{array}\begin{array}{l} \Longrightarrow\ u=0 \\ \Longrightarrow\ u=1 \end{array}\right\};\ \displaystyle\int_0^{3/2}\dfrac{dx}{9+4x^2} = \dfrac{1}{6}\int_0^1\dfrac{du}{1+u^2} = \dfrac{1}{6}\left[\tan^{-1}u\right]_0^1 = \dfrac{\pi}{24}$

71. $\displaystyle\int_{-3}^{-2}\dfrac{dx}{\sqrt{4-(x+3)^2}} = \left[\sin^{-1}\left(\dfrac{x+3}{2}\right)\right]_{-3}^{-2} = \dfrac{\pi}{6}$

$\qquad\qquad\quad\longleftarrow (7.10.11)$

73. $\left\{\begin{array}{l|l} u = e^x & x=0 \\ du = e^x\,dx & x=\ln 2 \end{array}\begin{array}{l} \Longrightarrow\ u=1 \\ \Longrightarrow\ u=2 \end{array}\right\};\ \displaystyle\int_0^{\ln 2}\dfrac{e^x}{1+e^{2x}}\,dx = \int_1^2\dfrac{du}{1+u^2}$

$\qquad\qquad\qquad = \left[\tan^{-1}u\right]_1^2 = \tan^{-1}2 - \dfrac{\pi}{4} \cong 0.322$

75. $\left\{ \begin{array}{ll} u = \sin^{-1} x, & dv = dx \\ du = \dfrac{dx}{\sqrt{1-x^2}}, & v = x \end{array} \right\}$; $\displaystyle \int \sin^{-1} x\, dx = uv - \int v\, du$

$$= x \sin^{-1} x - \int \frac{x}{\sqrt{1-x^2}}\, dx = x \sin^{-1} x + \sqrt{1-x^2} + C$$

77. $\left\{ \begin{array}{l} u = x^2 \\ du = 2x\, dx \end{array} \right\}$; $\displaystyle \int \frac{x}{\sqrt{1-x^4}}\, dx = \frac{1}{2} \int \frac{du}{\sqrt{1-u^2}} = \frac{1}{2} \sin^{-1} u + C = \frac{1}{2} \sin^{-1} x^2 + C$

79. $\left\{ \begin{array}{l} u = x^2 \\ du = 2x\, dx \end{array} \right\}$; $\displaystyle \int \frac{x}{1+x^4}\, dx = \frac{1}{2} \int \frac{du}{1+u^2} = \frac{1}{2} \tan^{-1} u + C = \frac{1}{2} \tan^{-1} x^2 + C$

81. $\left\{ \begin{array}{l} u = \tan x \\ du = \sec^2 x\, dx \end{array} \right\}$; $\displaystyle \int \frac{\sec^2 x}{9 + \tan^2 x}\, dx = \int \frac{du}{9 + u^2} = \tan^{-1}\left(\frac{u}{3}\right) + C = \tan^{-1}\left(\frac{\tan x}{3}\right) + C$

83.   Set $\theta = \sin^{-1} x$. Then θ is the only number in $\left[-\frac{1}{2}\pi, \frac{1}{2}\pi\right]$ for which $\sin \theta = x$.
Since $\cos\left(\frac{1}{2}\pi - \theta\right) = \sin \theta = x$ and $\left(\frac{1}{2}\pi - \theta\right) \in [0, \pi]$, $\cos^{-1} x = \frac{1}{2}\pi - \theta$. Therefore

$$\sin^{-1} x + \cos^{-1} x = \theta + \left(\frac{1}{2}\pi - \theta\right) = \frac{1}{2}\pi.$$

85.   By Exercise 83

$$\frac{d}{dx}\left(\cos^{-1} x\right) = \frac{d}{dx}\left(\frac{1}{2}\pi - \sin^{-1} x\right) = -\frac{d}{dx}\left(\sin^{-1} x\right) = -\frac{1}{\sqrt{1-x^2}}.$$

87. $y = \sec^{-1} x$

$\sec y = x$

$\sec y \tan y \dfrac{dy}{dx} = 1$

$y = \sec^{-1} x;$ $x < -1,\ x > 1$

$$\frac{dy}{dx} = \cos y \cot y = \left(\frac{1}{x}\right)\left(\frac{1}{x\sqrt{1-1/x^2}}\right) = \frac{1}{|x|\sqrt{x^2-1}}$$

89.   estimate $\cong 0.523$, $\sin 0.523 \cong 0.499$

Explanation: the integral $= \sin^{-1} 0.5;$ therefore $\sin(\text{integral}) = 0.5$.

SECTION 7.11

1. $\dfrac{dy}{dx} = \cosh x^2 \dfrac{d}{dx}\left(x^2\right) = 2x \cosh x^2$

3. $\dfrac{dy}{dx} = \dfrac{1}{2}\left(\cosh ax\right)^{-1/2}\left(a \sinh ax\right) = \dfrac{a \sinh ax}{2\sqrt{\cosh ax}}$

5. $\dfrac{dy}{dx} = \dfrac{\left(\cosh x - 1\right)\left(\cosh x\right) - \sinh x\left(\sinh x\right)}{\left(\cosh x - 1\right)^2} = \dfrac{1}{1 - \cosh x}$

7. $\dfrac{dy}{dx} = ab \cosh bx - ab \sinh ax = ab\left(\cosh bx - \sinh ax\right)$

9. $\dfrac{dy}{dx} = \dfrac{1}{\sinh ax}\left(a \cosh ax\right) = \dfrac{a \cosh ax}{\sinh ax}$

11. $\cosh^2 t - \sinh^2 t = \left(\dfrac{e^t + e^{-t}}{2}\right)^2 - \left(\dfrac{e^t + e^{-t}}{2}\right)^2 = \dfrac{1}{4}\left\{\left(e^{2t} + 2 + e^{-2t}\right) - \left(e^{2t} - 2 + e^{-2t}\right)\right\} = \dfrac{4}{4} = 1$

13. $\cosh t \cosh s + \sinh t \sinh s$

$$= \left(\dfrac{e^t + e^{-t}}{2}\right)\left(\dfrac{e^s + e^{-s}}{2}\right) + \left(\dfrac{e^t - e^{-t}}{2}\right)\left(\dfrac{e^s - e^{-s}}{2}\right)$$

$$= \tfrac{1}{4}\left\{\left(e^{t+s} + e^{t-s} + e^{s-t} + e^{-t-s}\right) + \left(e^{t+s} - e^{t-s} - e^{s-t} + e^{-s-t}\right)\right\}$$

$$= \dfrac{1}{4}\left\{2e^{t+s} + 2e^{-(t+s)}\right\} = \dfrac{e^{t+s} + e^{-(t+s)}}{2} = \cosh\left(t + s\right)$$

15. Set $s = t$ in $\cosh\left(t + s\right) = \cosh t \cosh s + \sinh t \sinh s$.

17. $$y = -5 \cosh x + 4 \sinh x = -\tfrac{5}{2}\left(e^x + e^{-x}\right) + \tfrac{4}{2}\left(e^x - e^{-x}\right) = -\tfrac{1}{2}e^x - \tfrac{9}{2}e^{-x}$$

$$\dfrac{dy}{dx} = -\dfrac{1}{2}e^x + \dfrac{9}{2}e^{-x} = \dfrac{e^{-x}}{2}\left(9 - e^{2x}\right); \quad \dfrac{dy}{dx} = 0 \quad \Longrightarrow \quad e^x = 3 \text{ or } x = \ln 3$$

$$\dfrac{d^2y}{dx^2} = -\dfrac{1}{2}e^x - \dfrac{9}{2}e^{-x} < 0 \quad \text{all} \quad x \quad \text{so abs max occurs when} \quad x = \ln 3.$$

The abs max is $y = -\tfrac{1}{2}e^{\ln 3} - \tfrac{9}{2}e^{-\ln 3} = -\tfrac{1}{2}\left(3\right) - \tfrac{9}{2}\left(\tfrac{1}{3}\right) = -3.$

19. $$\left[\cosh x + \sinh x\right]^n = \left[\dfrac{e^x + e^{-x}}{2} + \dfrac{e^x - e^{-x}}{2}\right]^n$$

$$= \left[e^x\right]^n = e^{nx} = \dfrac{e^{nx} + e^{-nx}}{2} + \dfrac{e^{nx} - e^{-nx}}{2} = \cosh nx + \sinh nx$$

21.
$$y = A \cosh cx + B \sinh cx; \qquad\qquad y(0) = 2 \quad \Longrightarrow \quad 2 = A.$$

$$y' = Ac \sinh cx + Bc \cosh cx; \qquad\qquad y'(0) = 1 \quad \Longrightarrow \quad 1 = Bc.$$

$$y'' = Ac^2 \cosh cx + Bc^2 \sinh cx = c^2 y; \qquad y'' - 9y = 0 \quad \Longrightarrow \quad (c^2 - 9)\, y = 0.$$

Thus, $c = 3$, $B = \frac{1}{3}$, and $A = 2$.

23. $\dfrac{1}{a} \sinh ax + C$ 　　　　　　　　　　**25.** $\dfrac{1}{3a} \sinh^3 ax + C$

27. $\dfrac{1}{a} \ln(\cosh ax) + C$ 　　　　　　　**29.** $-\dfrac{1}{a \cosh ax} + C$

31.
$$\left\{ \begin{array}{ll} u = x, & dv = \sinh x\, dx \\[2mm] du = dx, & v = \cosh x \end{array} \right\}; \quad \int x \sinh x\, dx = uv - \int v\, du$$

$$= x \cosh x - \int \cosh x\, dx = x \cosh x - \sinh x + C$$

33.
$$\left\{ \begin{array}{ll} u = \cosh x, & dv = \cosh x\, dx \\[2mm] du = \sinh x\, dx; & v = \sinh x \end{array} \right\}; \quad \int \cosh^2 x\, dx = uv - \int v\, du$$

$$= \sinh x \cosh x - \int \sinh^2 x\, dx = \sinh x \cosh x - \int (\cosh^2 x - 1)\, dx$$

$$= \sinh x \cosh x - \int \cosh^2 x\, dx + x + C_1$$

$$2 \int \cosh^2 x\, dx = \sinh x \cosh x + x + C_1$$

$$\int \cosh^2 x\, dx = \frac{1}{2} (\sinh x \cosh x + x) + C$$

35.
$$A = \int_0^1 \cosh x\, dx = [\sinh x]_0^1 = \sinh 1 = \frac{e - e^{-1}}{2} = \frac{e^2 - 1}{2e}$$

$$\bar{x} A = \int_0^1 x \cosh x\, dx = [x \sinh x - \cosh x]_0^1 = \sinh 1 - \cosh 1 + 1 = \frac{2(e - 1)}{2e}$$

$$\bar{y} A = \int_0^1 \frac{1}{2} \cosh^2 x\, dx = \frac{1}{4} [\sinh x \cosh x + x]_0^1 = \frac{1}{4} (\sinh 1 \cosh 1 + 1) = \frac{e^4 + 4e^2 - 1}{16e^2}$$

Therefore $\bar{x} = \dfrac{2}{e + 1}$ and $\bar{y} = \dfrac{e^4 + 4e^2 - 1}{8e(e^2 - 1)}$.

37. (a) $V_x = \displaystyle\int_0^1 \pi (\cosh^2 x - \sinh^2 x)\, dx = \int_0^1 \pi\, dx = \pi$

(b) $V_y = \displaystyle\int_0^1 2\pi x (\cosh x - \sinh x)\, dx$

$$\int_0^1 2\pi x \cosh x \, dx = 2\pi \left[x \sinh x - \cosh x\right]_0^1 = 2\pi \left(\sinh 1 - \cosh 1 + 1\right)$$

$$\int_0^1 2\pi x \sinh x \, dx = 2\pi \left[x \cosh x - \sinh x\right]_0^1 = 2\pi \left(\cosh 1 - \sinh 1\right)$$

$$V_y = 2\pi \left(2\sinh 1 - 2\cosh 1 + 1\right) = \frac{2\pi}{e}\left(e - 2\right)$$

SECTION * 7.12

1. $\dfrac{dy}{dx} = 2\tanh x \operatorname{sech}^2 x$

3. $\dfrac{dy}{dx} = \dfrac{1}{\tanh x}\operatorname{sech}^2 x = \operatorname{sech} x + \operatorname{csch} x$

5. $\dfrac{dy}{dx} = \cosh\left(\tan^{-1} e^{2x}\right)\dfrac{d}{dx}\left(\tan^{-1} e^{2x}\right) = \dfrac{2e^{2x}\cosh\left(\tan^{-1} e^{2x}\right)}{1 + e^{4x}}$

7. $\dfrac{dy}{dx} = -\operatorname{csch}^2\left(\sqrt{x^2+1}\right)\dfrac{d}{dx}\left(\sqrt{x^2+1}\right) = -\dfrac{x}{\sqrt{x^2+1}}\operatorname{csch}^2\left(\sqrt{x^2+1}\right)$

9. $\dfrac{dy}{dx} = \dfrac{\left(1 + \cosh x\right)\left(-\operatorname{sech} x \tanh x\right) - \operatorname{sech} x \left(\sinh x\right)}{\left(1 + \cosh x\right)^2}$

$$= \dfrac{-\operatorname{sech} x \left(\tanh x + \cosh x \tanh x + \sinh x\right)}{\left(1 + \cosh x\right)^2} = \dfrac{-\operatorname{sech} x \left(\tanh x + 2\sinh x\right)}{\left(1 + \cosh x\right)^2}$$

11.
$$\frac{d}{dx}\left(\coth x\right) = \frac{d}{dx}\left[\frac{\cosh x}{\sinh x}\right] = \frac{\sinh x \left(\sinh x\right) - \cosh x \left(\cosh x\right)}{\sinh^2 x}$$

$$= -\frac{\cosh^2 x - \sinh^2 x}{\sinh^2 x} = \frac{-1}{\sinh^2 x} = -\operatorname{csch}^2 x$$

13. $\dfrac{d}{dx}\left(\operatorname{csch} x\right) = \dfrac{d}{dx}\left[\dfrac{1}{\sinh x}\right] = -\dfrac{\cosh x}{\sinh^2 x} = -\operatorname{csch} x \coth s$

15. (a) By the hint $\operatorname{sech}^2 x_0 = \dfrac{9}{25}$. Take $\operatorname{sech} x_0 = \dfrac{3}{5}$ since $\operatorname{sech} x = \dfrac{1}{\cosh x} > 0$ for all x.

(b) $\cosh x_0 = \dfrac{1}{\operatorname{sech} x_0} = \dfrac{5}{3}$

(c) $\sinh x_0 = \cosh x_0 \tanh x_0 = \left(\dfrac{5}{3}\right)\left(\dfrac{4}{5}\right) = \dfrac{4}{3}$

(d) $\coth x_0 = \dfrac{\cosh x_0}{\sinh x_0} = \dfrac{5/3}{4/3} = \dfrac{5}{4}$

(e) $\operatorname{csch} x_0 = \dfrac{1}{\sinh x_0} = \dfrac{3}{4}$

17.   If $x \le 0$, the result is obvious. Suppose then that $x > 0$. Since $x^2 \ge 1$, we have $x \ge 1$. Consequently

$$x - 1 = \sqrt{x-1}\,\sqrt{x-1} \le \sqrt{x-1}\,\sqrt{x+1} = \sqrt{x^2-1}$$

and therefore

$$x - \sqrt{x^2-1} \le 1.$$

19.   By Theorem 7.12.2,

$$\frac{d}{dx}\left(\sinh^{-1} x\right) = \frac{d}{dx}\left[\ln\left(x + \sqrt{x^2+1}\,\right)\right] = \frac{1}{x + \sqrt{x^2+1}}\left(1 + \frac{x}{\sqrt{x^2+1}}\right) = \frac{1}{\sqrt{x^2+1}}.$$

21.   By Theorem 7.12.2

$$\frac{d}{dx}\left(\tan^{-1} x\right) = \frac{d}{dx}\left[\frac{1}{2}\ln\left(\frac{1+x}{1-x}\right)\right] = \frac{1}{2}\frac{1}{\left(\dfrac{1+x}{1-x}\right)}\left(\frac{(1-x)(1) - (1+x)(-1)}{(1-x)^2}\right)$$

$$= \frac{1}{\left(\dfrac{1+x}{1-x}\right)(1-x)^2} = \frac{1}{1-x^2}.$$

23.   (a) $\tan\phi = \sinh x$ (b) $\sinh x = \tan\phi$

$$\phi = \tan^{-1}(\sinh x)$$ $$x = \sinh^{-1}(\tan\phi)$$

$$\frac{d\phi}{dx} = \frac{\cosh x}{1 + \sinh^2 x}$$ $$= \ln\left(\tan\phi + \sqrt{\tan^2\phi + 1}\,\right)$$

$$= \frac{\cosh x}{\cosh^2 x} = \frac{1}{\cosh x} = \mathrm{sech}\, x$$ $$= \ln\left(\tan\phi + \sec\phi\right)$$

$$= \ln\left(\sec\phi + \tan\phi\right)$$

   (c) $x = \ln(\sec\phi + \tan\phi)$

$$\frac{dx}{d\phi} = \frac{\sec\phi\tan\phi + \sec^2\phi}{\tan\phi + \sec\phi} = \sec\phi$$

SECTION 7.13

1. $y = ae^{kx} + be^{-kx}, \quad a > 0, \quad b > 0$

$$\frac{dy}{dx} = ake^{kx} - bke^{-kx} = ke^{-kx}\left[ae^{2kx} - b\right]; \qquad \frac{dy}{dx} = 0 \implies e^{kx} = \sqrt{b/a}.$$

Since $\dfrac{d^2y}{dx^2} = ak^2 e^{kx} + bk^2 e^{-kx} > 0$ all real x, y is minimal when $e^{kx} = \sqrt{b/a}$;

the minimum is $y = a\sqrt{b/a} + b\sqrt{a/b} = 2\sqrt{ab}$.

3. $y = e^{-x^2}, \qquad \dfrac{dy}{dx} = -2xe^{-x^2}, \qquad \dfrac{d^2y}{dx^2} = (4x^2 - 2)\,e^{-x^2}.$

The points of inflection are $\left(\pm \dfrac{1}{\sqrt{2}}, \dfrac{1}{\sqrt{e}}\right).$

5. Set $y = x^2 \ln(1/x) = -x^2 \ln x, \quad x > 0.$

Then, $\dfrac{dy}{dx} = -2x \ln x - x^2 \left(\dfrac{1}{x}\right) = -x\,(1 + 2\ln x); \qquad \dfrac{dy}{dx} = 0 \implies x = \dfrac{1}{\sqrt{e}}.$

Since y increases on $(0, 1/\sqrt{e}]$ and decreases on $[1/\sqrt{e}, \infty)$, the abs. max occurs when $x = 1/\sqrt{e}.$

7.
$$\text{For } f(x) = e^x,$$
$$f'(x) = \lim_{h \to 0} \frac{e^{x+h} - e^x}{h} = e^x.$$
Note that
$$f'(0) = \lim_{h \to 0} \frac{e^{0+h} - e^0}{h} = e^0.$$
This gives
$$\lim_{h \to 0} \frac{1}{h}\left(e^h - 1\right) = 1.$$

9. $A = \displaystyle\int_0^1 \frac{x}{x^2 + 1}\,dx$

$= \left[\tfrac{1}{2} \ln\left(x^2 + 1\right)\right]_0^1 = \tfrac{1}{2} \ln 2$

11. $A = \displaystyle\int_0^{1/2} \frac{dx}{\sqrt{1 - x^2}}$

$= \left[\sin^{-1} x\right]_0^{1/2} = \dfrac{\pi}{6}$

13. $x(t) = \displaystyle\int v(t)\,dt = \int \frac{dt}{1 + t^2} = \tan^{-1} t + C$

$x(0) = 0 \implies C = 0.$ Then, $x(1) = \tan^{-1} 1 = \pi/4.$

15. $x(t) = \tan^{-1}(1 + t), \qquad v(t) = \dfrac{1}{1 + (t + 1)^2} = \dfrac{1}{t^2 + 2t + 2},$

$a(t) = \dfrac{-(2t + 2)}{(t^2 + 2t + 2)^2}.$ Thus, $a(0) = -1/2.$

17. Set $f(x) = \ln x.$ Average slope of f on $[a, b]$:

$$\frac{1}{b - a} \int_a^b f'(x)\,dx = \frac{1}{b - a} \int_a^b \frac{dx}{x} = \frac{1}{b - a}\,[\ln x]_a^b = \frac{1}{b - a} \ln \frac{b}{a}$$

19. $A = \displaystyle\int_a^{2a} \frac{a^2}{x}\,dx = a^2\,[\ln x]_a^{2a} = a^2\,(\ln 2a - \ln a) = a^2 \ln 2$

21. (a) $\dfrac{dy}{dx} = 1 - \dfrac{1}{50}x; \qquad$ at $x = 0, \quad \dfrac{dy}{dx} = 1, \quad \tan\theta = 1, \quad \theta = \dfrac{1}{4}\pi.$

Answer: $45°.$

(b)   At $x = 75$, $\dfrac{dy}{dx} = 1 - \dfrac{75}{50} = -\dfrac{1}{2}$, $\tan\theta = -\dfrac{1}{2}$,

$153° \le \theta \le 154°$, $63° \le \phi \le 64°$.

Answer: between 63° and 64°.

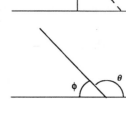

(c) $16 = x - \frac{1}{100}x^2$

$x^2 - 100x - 1600 = 0$

$x = 20$, $x = 80$.

At $x = 20$, $\dfrac{dy}{dx} > 0$ and the ball is still rising.

At $x = 80$, $\dfrac{dy}{dx} = -\dfrac{3}{5}$, $\tan\theta = -\dfrac{3}{5}$, $\theta \cong 149°$, $\phi \cong 31°$.

Answer: about 31°.

23.   (a) $V_x = \displaystyle\int_0^{\sqrt{3}} \dfrac{\pi}{1+x^2}\,dx = \pi\left[\tan^{-1}x\right]_0^{\sqrt{3}} = \dfrac{1}{3}\pi^2$

(b) $V_y = \displaystyle\int_0^{\sqrt{3}} \dfrac{2\pi x}{\sqrt{1+x^2}}\,dx = 2\pi\left[\sqrt{1+x^2}\right]_0^{\sqrt{3}} = 2\pi$

25. $A = \displaystyle\int_{-1/2}^{1/2} \sec\frac{1}{2}\pi x\,dx = 2\int_0^{1/2}\sec\frac{1}{2}\pi x\,dx = \frac{4}{\pi}\left[\ln\left|\sec\frac{1}{2}\pi x + \tan\frac{1}{2}\pi x\right|\right]_0^{1/2} = \frac{2\ln 2}{\pi}$

$\overline{x} = 0$ by symmetry

$\overline{y}A = \displaystyle\int_{-1/2}^{1/2}\frac{1}{2}\sec^2\frac{1}{2}\pi x = \int_0^{1/2}\sec^2\frac{1}{2}\pi x\,dx = \frac{2}{\pi}\left[\tan\frac{1}{2}\pi x\right]_0^{1/2} = \frac{2}{\pi}$

$\overline{y} = \left(\dfrac{2}{\pi}\right)\Big/\left(\dfrac{2\ln 2}{\pi}\right) = \dfrac{1}{\ln 2}$

27. $\left\{\begin{array}{l} u = ax \\ du = a\,dx \end{array}\right.$ $\left|\begin{array}{l} x = 0 \implies u = 0 \\ x = 1 \implies u = a \end{array}\right\}$; $\displaystyle\int_0^1 \frac{a}{1+a^2x^2}\,dx = \int_0^a \frac{du}{1+u^2} = \int_0^a \frac{dx}{1+x^2}$.

(Both integrals represent $\tan^{-1}a$.)

29.   By the hint $\ln(m+1) - \ln m < \dfrac{1}{m}$

$\ln(m+2) - \ln(m+1) < \dfrac{1}{m+1}$

$\vdots$

$\ln(n+1) - \ln n < \dfrac{1}{n}$.

By addition

$$\underbrace{\ln{(n+1)} - \ln m}_{\ln \frac{n+1}{m}} < \frac{1}{m} + \frac{1}{m+1} + \cdots + \frac{1}{n}.$$

Again by the hint

$$\frac{1}{m} < \ln m - \ln{(m-1)}$$

$$\frac{1}{m+1} < \ln{(m+1)} - \ln m$$

$$\vdots$$

$$\frac{1}{n} < \ln n - \ln{(n-1)}.$$

This time addition gives

$$\frac{1}{m} + \frac{1}{m+1} + \cdots + \frac{1}{n} < \underbrace{\ln n - \ln{(m-1)}}_{\ln \frac{n}{m-1}}.$$

31. $\dfrac{d}{dx}\left[e^{-x}P(x)\right] = e^{-x}\left[p'(x) + \cdots + p^{(n)}(x)\right] - e^{-x}\left[p(x) + \cdots + p^{(n)}(x)\right] = e^{-x}p(x)$

  $\overset{\displaystyle\longleftarrow}{} \quad p^{(n+1)}(x) = 0$

33.
$$A = \int_0^1 \sin^{-1}x\,dx = \left[x\sin^{-1}x + \sqrt{1-x^2}\,\right]_0^1 = \frac{1}{2}(\pi - 2)$$

$$\bar{x}A = \int_0^1 x\sin^{-1}x\,dx = \frac{1}{8}\pi \qquad \text{[using Exercise 32 (a)]}$$

$$\bar{y}A = \int_0^1 \frac{1}{2}\left(\sin^{-1}x\right)^2 dx = \frac{1}{8}\left(\pi^2 - 8\right) \qquad \text{[using Exercise 32 (b)]}$$

$$\bar{x} = \frac{\pi}{4(\pi-2)}, \quad \bar{y} = \frac{\pi^2 - 8}{4(\pi-2)}$$

35. (a) $A = \displaystyle\int_1^e \ln x\,dx = [x\ln x - x]_1^e = 1$

(b) $\bar{x}A = \displaystyle\int_1^e x\ln x\,dx = \left[\frac{1}{2}x^2\ln x - \frac{1}{4}x^2\right]_1^e = \frac{1}{4}\left(e^2 + 1\right), \quad \bar{x} = \frac{1}{4}\left(e^2 + 1\right)$

$\bar{y}A = \displaystyle\int_1^e \frac{1}{2}\left(\ln x\right)^2 dx = \frac{1}{2}\left[x\left(\ln x\right)^2 - 2x\ln x + 2x\right]_1^e = \frac{1}{2}e - 1, \quad \bar{y} = \frac{1}{2}e - 1$

(c) $V_x = 2\pi\bar{y}A = \pi(e - 2), \quad V_y = 2\pi\bar{x}A = \frac{1}{2}\pi\left(e^2 + 1\right)$

(d) The distance from (a, b) to the line $y = x$ is the minimum value of $f(x) = \sqrt{(a-x)^2 + (b-x)^2}$. As you can check by differentiation, this is $\left(\sqrt{2}/2\right)|a - b|$. Thus

$$\bar{R} = \frac{1}{2}\sqrt{2}\,\left|\frac{1}{4}\left(e^2 + 1\right) - \left(\frac{1}{2}e - 1\right)\right| = \frac{1}{8}\sqrt{2}\,\left(e^2 - 2 + 5\right).$$

(e) $V_l = 2\pi\bar{R}A = \frac{1}{4}\sqrt{2}\,\pi\left(e^2 - 2e + 5\right)$

37. (a) P.V. $= \int_0^4 1000e^{-0.04t}\,dt = \left[-\dfrac{1000}{0.04}e^{-0.04t}\right]_0^4 = 25000\left(1 - e^{-0.16}\right) \cong \$3696$

(b) P.V. $= \int_0^8 1000e^{-0.08t}\,dt = \left[-\dfrac{1000}{0.08}e^{-0.08t}\right]_0^4 = 12500\left(1 - e^{-0.32}\right) \cong \$3423$

39.

(a) P.V. $= \int_0^2 (1000 + 60t)\,e^{-0.05t}\,dt$

$$= 1000\int_0^2 e^{-t/20}\,dt + 60\int_0^2 te^{-t/20}\,dt$$

$$= 1000\left[-20e^{-t/20}\right]_0^2 + 60\left[-20te^{-t/20} - 400e^{-t/20}\right]_0^2$$
$$\underset{\text{(by parts)}}{\uparrow\!\!\_\_\_}$$

$$= \left[-(44000 + 1200t)\,e^{-t/20}\right]_0^2 = 44000 - 46400\,e^{-0.10} \cong \$2016$$

(b) P.V. $= \int_0^2 (1000 + 60t)\,e^{-0.1t}\,dt$

$$= 1000\left[-10e^{-t/10}\right]_0^2 + 60\left[-10te^{-t/10} - 100e^{-t/10}\right]_0^2$$
$$\underset{\text{(by parts)}}{\uparrow\!\!\_\_\_}$$

$$= \left[-(16000 + 600t)\,e^{-t/10}\right]_0^2 = 16000 - 17200e^{-0.20} \cong \$1918$$

41. (a) P.V. $= \int_3^4 (1000 + 80t)\,e^{-0.06t}\,dt$

$$= 1000\int_3^4 e^{-3t/50}\,dt + 80\int_3^4 te^{-3t/50}\,dt$$

$$= 1000\left[-\dfrac{50}{3}e^{-3t/50}\right]_3^4 + 80\left[-\dfrac{50}{3}te^{-3t/50} - \dfrac{2500}{9}e^{-3t/50}\right]_3^4$$

$$= \left[-\left(\dfrac{350000}{9} + \dfrac{4000}{3}t\right)e^{-3t/50}\right]_3^4 = \left[-\dfrac{1000}{9}(350 + 12t)\,e^{-3t/50}\right]_3^4$$

$$= \tfrac{1000}{9}\left(386e^{-0.18} - 398\,e^{-0.24}\right) \cong \$1037$$

(b) P.V. $= \int_3^4 (1000 + 80t)\,e^{-0.08t}\,dt$

$$= 1000\int_3^4 e^{-2t/25}\,dt + 80\int_3^4 te^{-2t/25}\,dt$$

$$= 1000\left[-\dfrac{25}{2}e^{-2t/25}\right]_3^4 + 80\left[-\dfrac{25}{2}te^{-2t/25} - \dfrac{625}{4}e^{-2t/25}\right]_3^4$$
$$\underset{\text{(by parts)}}{\uparrow\!\!\_\_\_}$$

$$= \left[-1000(25 + t)\,e^{-2t/25}\right]_3^4 = 1000\left(28\,e^{-0.24} - 29\,e^{-0.32}\right) \cong \$967$$

43.   (a) $V_x = \displaystyle\int_0^1 \pi \sinh^2 x\,dx = \frac{\pi}{2}\left[\sinh x \cosh x - x\right]_0^1 = \frac{\pi}{8e^2}\left(e^4 - 4e^2 - 1\right)$

$V_y = \displaystyle\int_0^1 2\pi x \sinh x\,dx = 2\pi\left[x \cosh x - \sinh x\right]_0^1 = \frac{2\pi}{e}$

45.   (a) $n_1 \sin\theta_1 = n \sin\theta = n_2 \sin\theta_2$

(b)   Think of n and θ as functions of altitude y. Then
$$n \sin\theta = C.$$
Differentiation with respect to y gives

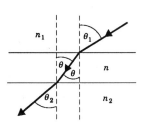

$$n \cos\theta \frac{d\theta}{dy} + \frac{dn}{dy}\sin\theta = 0, \quad \cot\theta\frac{d\theta}{dy} + \frac{1}{n}\frac{dn}{dy} = 0$$

and thus

$$\underbrace{\frac{1}{n}\frac{dn}{dy} - \cot\theta\frac{d\theta}{dy}}_{\alpha = \frac{1}{2}\pi - \theta} = \left(-\frac{dy}{dx}\right)\left(-\frac{d\alpha}{dy}\right) = \frac{d\alpha}{dy}\frac{dy}{dx} = \underbrace{\frac{d\alpha}{dx}}_{\tan\alpha = dy/dx,\ \alpha = \tan^{-1}(dy/dx)} = \frac{d^2y/dx^2}{1+(dy/dx)^2}.$$

(c) $1 + (dy/dx)^2 = 1 + \tan^2\alpha = 1 + \cot^2\theta = \csc^2\theta = 1/\sin^2\theta \underbrace{= (1/C)\,n^2}_{n \sin\theta = C}$

(d) $n(y) = k/|y + b|$ where b and k are constants, $k > 0$

SECTION 8.1

1. $\int e^{2-x}\,dx = -e^{2-x} + C$

3. $\int_0^1 \sin \pi x\,dx = \left[-\dfrac{1}{\pi}\cos \pi x\right]_0^1 = \dfrac{2}{\pi}$

5. $\int \sec^2 (1-x)\,dx = -\tan (1-x) + C$

7. $\int_{\pi/6}^{\pi/3} \cot x\,dx = \left[\ln (\sin x)\right]_{\pi/6}^{\pi/3} = \ln \dfrac{\sqrt{3}}{2} - \ln \dfrac{1}{2} = \dfrac{1}{2}\ln 3$

9. $\left\{\begin{array}{l} u = 1-x^2 \\ du = -2x\,dx \end{array}\right\};\quad \int \dfrac{x\,dx}{\sqrt{1-x^2}} = -\dfrac{1}{2}\int u^{-1/2}\,du = -u^{1/2} + C = -\sqrt{1-x^2} + C$

11. $\int_{-\pi/4}^{\pi/4} \dfrac{\sin x}{\cos^2 x}\,dx = \int_{-\pi/4}^{\pi/4} \sec x \tan x\,dx = \left[\sec x\right]_{-\pi/4}^{\pi/4} = 0$

13. $\left\{\begin{array}{ll} u = 1/x\,|\,x=1 &\Longrightarrow\quad u=1 \\ du = -\dfrac{dx}{x^2}\,|\,x=2 &\Longrightarrow\quad u=1/2 \end{array}\right\};$

$\int_1^2 \dfrac{e^{1/x}}{x^2}\,dx = \int_1^{1/2} -e^u\,du = \left[-e^u\right]_1^{1/2} = e - \sqrt{e}$

15. $\int \dfrac{x\,dx}{x^2+1} = \dfrac{1}{2}\ln (x^2+1) + C$

17. $\int_0^c \dfrac{dx}{x^2+c^2} = \left[\dfrac{1}{c}\tan^{-1}\left(\dfrac{x}{c}\right)\right]_0^c = \dfrac{\pi}{4c}$

19. $\left\{\begin{array}{l} u = 3\tan\theta + 1 \\ du = 3\sec^2\theta\,d\theta \end{array}\right\};$

$\int \dfrac{\sec^2\theta}{\sqrt{3\tan\theta+1}}\,d\theta = \dfrac{1}{3}\int u^{-1/2}\,du = \dfrac{2}{3}u^{1/2} + C = \dfrac{2}{3}\sqrt{1+3+\tan\theta} + C$

21. $\int \dfrac{e^x}{ae^x - b}\,dx = \dfrac{1}{a}\ln |ae^x - b| + C$

23. $\int \dfrac{1+\cos 2x}{\sin^2 2x}\,dx = \int (\csc^2 2x + \csc 2x \cot 2x)\,dx = -\dfrac{1}{2}\cot 2x - \dfrac{1}{2}\csc 2x + C$

25. $\left\{\begin{array}{l} u = x+1 \\ du = dx \end{array}\right\};$

$\int \dfrac{x}{(x+1)^2 + 4}\,dx = \int \dfrac{u-1}{u^2+4}\,du = \int \dfrac{u}{u^2+4}\,du - \int \dfrac{du}{u^2+4}$

$\qquad = \dfrac{1}{2}\ln |u^2+4| - \dfrac{1}{2}\tan^{-1}\dfrac{u}{2} + C$

$\qquad = \dfrac{1}{2}\ln |(x+1)^2+4| - \dfrac{1}{2}\tan^{-1}\left(\dfrac{x+1}{2}\right) + C$

27. $\left\{ \begin{array}{l} u = x^2 \\ du = 2x\,dx \end{array} \right\};$ $\displaystyle\int \frac{x}{\sqrt{1-x^4}}\,dx = \frac{1}{2}\int \frac{du}{\sqrt{1-u^2}} = \frac{1}{2}\sin^{-1} u + C$

$$= \tfrac{1}{2}\sin^{-1}(x^2) + C$$

29. $\left\{ \begin{array}{l} u = x + 3 \\ du = dx \end{array} \right\};$ $\displaystyle\int \frac{dx}{x^2 + 6x + 10} = \int \frac{dx}{(x+3)^2 + 1} = \int \frac{du}{u^2 + 1}$

$$= \tan^{-1} u + C = \tan^{-1}(x+3) + C$$

31. $\displaystyle\int x \sin x^2\,dx = -\frac{1}{2}\cos x^2 + C$

33. $\displaystyle\int \tan^2 x\,dx = \int (\sec^2 x - 1)\,dx = \tan x - x + C$

35. $\left\{ \begin{array}{ll} u = -\frac{1}{2} x^2, & dv = -2xe^{-x^2}\,dx \\ du = -x\,dx, & v = e^{-x^2} \end{array} \right\};$

$$\int x^3\,e^{-x^2}\,dx = uv - \int v\,du = -\frac{1}{2}x^2 e^{-x^2} + \int x\,e^{-x^2}\,dx = -\frac{1}{2}x^2 e^{-x^2} - \frac{1}{2}e^{-x^2} + C$$

SECTION 8.2

1. $\displaystyle\frac{7}{(x-2)(x+5)} = \frac{A}{x-2} + \frac{B}{x+5}$

$$7 = A(x+5) + B(x-2)$$

$x = -5$: $7 = -7B$ $\implies$ $B = -1$

$x = 2$: $7 = 7A$ $\implies$ $A = 1$

$$\int \frac{7}{(x-2)(x+5)}\,dx = \int \left(\frac{1}{x-2} - \frac{1}{x+5} \right) = \ln|x-2| - \ln|x+5| + C = \ln\left| \frac{x-2}{x+5} \right| + C$$

3. $\displaystyle\frac{2x^2 + 3}{x^2(x-1)} = \frac{A}{x} + \frac{B}{x^2} + \frac{C}{x-1}$

$$2x^2 + 3 = Ax(x-1) + B(x-1) + Cx^2$$

$x = 0$: $3 = -B$ $\implies$ $B = -3$

$x = 1$: $5 = C$ $\implies$ $C = 5$

$x = -1$: $5 = 2A - 2B + C$ $\implies$ $A = -3$

$$\int \frac{2x^2 + 3}{x^2(x-1)}\,dx = \int \left(-\frac{3}{x} - \frac{3}{x^2} + \frac{5}{x-1} \right)\,dx = -3\ln|x| + \frac{3}{x} + 5\ln|x-1| + C$$

5.   We carry out the division until the numerator has degree smaller than the denominator:

$$\frac{x^5}{(x-2)^2} = \frac{x^5}{x^2 - 4x + 4} = x^3 + 4x^2 + 12x + 32 + \frac{80x - 128}{(x-2)^2}.$$

Then,

$$\frac{80x - 128}{(x-2)^2} = \frac{80x - 160 + 32}{(x-2)^2} = \frac{80}{x-2} + \frac{32}{(x-2)^2}$$

$$\int \frac{x^5}{(x-2)^2}\,dx = \int \left(x^3 + 4x^2 + 12x + 32 + \frac{80}{x-2} + \frac{32}{(x-2)^2} \right)\,dx$$

$$= \frac{1}{4}x^4 + \frac{4}{3}x^3 + 6x^2 + 32x + 80\ln|x-2| - \frac{32}{x-2} + C.$$

7. $$\frac{x+3}{x^2 - 3x + 2} = \frac{A}{x-1} + \frac{B}{x-2}$$

$$x + 3 = A(x-2) + B(x-1)$$

$x = 1$: $4 = -A \implies A = -4$
$x = 2$: $5 = B \implies B = 5$

$$\int \frac{x+3}{x^2 - 3x + 2}\,dx = \int \left(\frac{-4}{x-1} + \frac{5}{x-2} \right)\,dx = -4\ln|x-1| + 5\ln|x-2| + C$$

9. $\displaystyle \int \frac{dx}{(x-1)^3} = \int (x-1)^{-3}\,dx = -\frac{1}{2}(x-1)^{-2} + C = -\frac{1}{2(x-1)^2} + C$

11. $$\frac{x^2}{(x-1)^2(x+1)} = \frac{A}{x-1} + \frac{B}{(x-1)^2} + \frac{C}{x+1}$$

$$x^2 = A(x-1)(x+1) + B(x+1) + C(x-1)^2$$

$x = 1$: $1 = 2B \implies B = 1/2$
$x = -1$: $1 = 4C \implies C = 1/4$
$x = 0$: $0 = -A + B + C \implies A = 3/4$

$$\int \frac{x^2}{(x-1)^2(x+1)}\,dx = \int \left(\frac{3/4}{x-1} + \frac{1/2}{(x-1)^2} + \frac{1/4}{x+1} \right)\,dx$$

$$= \frac{3}{4}\ln|x-1| - \frac{1}{2(x-1)} + \frac{1}{4}\ln|x+1| + C$$

13.
$$x^4 - 16 = (x^2 - 4)(x^2 + 4) = (x - 2)(x + 2)(x^2 + 4)$$

$$\frac{1}{x^4 - 16} = \frac{A}{x - 2} + \frac{B}{x + 2} + \frac{Cx + D}{x^2 + 4}$$

$$1 = A(x + 2)(x^2 + 4) + B(x - 2)(x^2 + 4) + (Cx + D)(x^2 - 4)$$

$x = 2$: $\quad 1 = 32A \quad \Longrightarrow \quad A = 1/32$

$x = -2$: $\quad 1 = -32B \quad \Longrightarrow \quad B = -1/32$

$x = 0$: $\quad 1 = 8A - 8B - 4D \quad \Longrightarrow \quad D = -1/8$

$x = 1$: $\quad 1 = 15A - 5B - 3C - 3D \quad \Longrightarrow \quad C = 0$

$$\int \frac{dx}{x^4 - 16} = \int \left(\frac{1/32}{x - 2} - \frac{1/32}{x + 2} - \frac{1/8}{x^2 + 4} \right) dx$$

$$= \frac{1}{32} \ln|x - 2| - \frac{1}{32} \ln|x + 2| - \frac{1}{8} \left(\frac{1}{2} \tan^{-1} \frac{x}{2} \right) + C$$

$$= \frac{1}{32} \ln \left| \frac{x - 2}{x + 2} \right| - \frac{1}{16} \tan^{-1} \frac{x}{2} + C$$

15.
$$\frac{x^3 + 4x^2 - 4x - 1}{(x^2 + 1)^2} = \frac{Ax + B}{x^2 + 1} + \frac{Cx + D}{(x^2 + 1)^2}$$

$$x^3 + 4x^2 - 4x - 1 = (Ax + B)(x^2 + 1) + (Cx + D)$$

$x = 0$: $\quad -1 = B + D \quad \Longrightarrow \quad D = -B - 1 \quad \Big\}$

$x = 1$: $\quad 0 = 2A + 2B + C + D \quad \Big\} \Longrightarrow 6 = 4B + 2D \quad \Big\} \Longrightarrow B = 4, \ D = -5$

$x = -1$: $\quad 6 = -2A + 2B - C + D \quad \Big\}$

$\qquad\qquad\qquad\qquad\qquad\qquad\qquad 6 = -2A + 8 - C - 5 \quad \Big\}$

$x = 2$: $\quad 15 = 10A + 5B + 2C + D \quad \Big\}\qquad 15 = 10A + 20 + 2C - 5 \quad \Big\} \Longrightarrow \begin{matrix} A = 1, \\ C = -5 \end{matrix}$

$$\int \frac{x^3 + 4x^2 - 4x - 1}{(x^2 + 1)^2} \, dx = \int \left(\frac{x}{x^2 + 1} + \frac{4}{x^2 + 1} - \frac{5x}{(x^2 + 1)^2} - \frac{5}{(x^2 + 1)^2} \right) dx$$

(*)
$$= \frac{1}{2} \ln(x^2 + 1) + 4 \tan^{-1} x + \frac{5}{2(x^2 + 1)} - 5 \int \frac{dx}{(x^2 + 1)^2}$$

For this last integral we set

$$\left\{\begin{array}{l} x = \tan u \\ dx = \sec^2 u\, du \end{array}\right\}; \quad \int \frac{dx}{(x^2+1)^2} = \int \frac{\sec^2 u\, du}{(1+\tan^2 u)^2} = \int \cos^2 u\, du$$

$$= \frac{1}{2}\int (1+\cos 2u)\, du$$

$$= \frac{1}{2}\left(u + \tfrac{1}{2}\sin 2u\right) + C = \tfrac{1}{2}(u + \sin u \cos u) + C$$

$$= \frac{1}{2}\left(\tan^{-1} x + \frac{x}{1+x^2}\right) + C.$$

Substituting this result in (∗) and rearranging the terms, we get

$$\int \frac{x^3 + 4x^2 - 4x + 1}{(x^2+1)^2}\, dx = \frac{1}{2}\ln\left(x^2+1\right) + \frac{3}{2}\tan^{-1} x + \frac{5(1-x)}{2(1+x^2)} + C.$$

17.
$$\frac{1}{x^4+4} = \frac{Ax+B}{x^2+2x+2} + \frac{Cx+D}{x^2-2x+2} \qquad \text{(using the hint)}$$

$$1 = (Ax+B)(x^2 - 2x + 2) + (Cx+D)(x^2 + 2x + 2)$$

$$
\left.\begin{array}{ll}
x = 0: & 1 = 2B + 2D \\
x = 1: & 1 = A + B + 5C + 5D \\
x = -1: & 1 = -5A + 5B - C + D \\
x = 2: & 1 = 4A + 2B + 20C + 10D
\end{array}\right\}
\implies
\begin{array}{l}
A = 1/8 \\
B = 1/4 \\
C = -1/8 \\
D = 1/4
\end{array}
$$

$$\int \frac{dx}{x^4+4} = \frac{1}{8}\int \frac{x+2}{x^2+2x+2}\, dx - \frac{1}{8}\int \frac{x-2}{x^2-2x+2}\, dx$$

$$= \frac{1}{8}\int \frac{x+1}{x^2+2x+2}\, dx + \frac{1}{8}\int \frac{dx}{(x+1)^2+1} - \frac{1}{8}\int \frac{x-1}{x^2-2x+2}\, dx + \frac{1}{8}\int \frac{dx}{(x-1)^2+1}$$

$$= \tfrac{1}{16}\ln\left(x^2+2x+2\right) + \tfrac{1}{8}\tan^{-1}(x+1) - \tfrac{1}{16}\ln\left(x^2-2x+2\right) + \tfrac{1}{8}\tan^{-1}(x-1) + C$$

$$= \frac{1}{16}\ln\left(\frac{x^2+2x+2}{x^2-2x+2}\right) + \frac{1}{8}\tan^{-1}(x+1) + \frac{1}{8}\tan^{-1}(x-1) + C$$

19. Note that

$$y = \frac{1}{x^2-1} = \frac{1}{2}\left[\frac{1}{x-1} - \frac{1}{x+1}\right]$$

and thus

$$\frac{d^0 y}{dx^0} = \left(\frac{1}{2}\right)(-1)^0\, 0!\left[\frac{1}{(x-1)^{0+1}} - \frac{1}{(x+1)^{0+1}}\right].$$

The rest is a routine induction.

21.

$$A = \int_0^1 \frac{dx}{x^2 + 1} = \left[\tan^{-1} x\right]_0^1 = \frac{1}{4}\pi$$

$$\bar{x}A = \int_0^1 \frac{x}{x^2 + 1}\, dx = \frac{1}{2}\left[\ln\left(x^2 + 1\right)\right]_0^1 = \frac{1}{2}\ln 2$$

$$\bar{y}A = \int_0^1 \frac{dx}{2(x^2 + 1)^2} = \frac{1}{2}\left[\tan^{-1} x + \frac{x}{x^2 + 1}\right]_0^1 = \frac{1}{8}(\pi + 2)$$

$$\bar{x} = \frac{2\ln 2}{\pi}; \quad \bar{y} = \frac{\pi + 2}{2\pi}$$

23. **(a)** By the hint

$$\int \frac{dC}{\left(A_0 - \frac{1}{2}C\right)^2} = \int k\, dt$$

$$\frac{2}{A_0 - \frac{1}{2}C} = kt + K.$$

First, $C(0) = 0 \implies K = 2/A_0$. Then, $C(1) = A_0 \implies k = 2/A_0$. Thus,

$$\frac{2}{A_0 - \frac{1}{2}C} = \frac{2}{A_0}(t + 1), \quad \text{which gives} \quad C(t) = 2A_0\left(\frac{t}{t + 1}\right).$$

(b) By the hint

$$\int \frac{dC}{\left(A_0 - \frac{1}{2}C\right)\left(2A_0 - \frac{1}{2}C\right)} = \int k\, dt$$

$$\frac{1}{A_0}\int \left[\frac{1}{A_0 - \frac{1}{2}C} - \frac{1}{2A_0 - \frac{1}{2}C}\right] dC = \int k\, dt$$

$$\frac{1}{A_0}\left[-2\ln\left|A_0 - \frac{1}{2}C\right| + 2\ln\left|2A_0 - \frac{1}{2}C\right|\right] = kt + K$$

$$\frac{2}{A_0}\ln\left|\frac{2A_0 - \frac{1}{2}C}{A_0 - \frac{1}{2}C}\right| = kt + K.$$

First, $C(0) = 0 \implies K = \dfrac{2}{A_0}\ln 2$. Then,

$$C(1) = A_0 \implies \frac{2}{A_0}\ln 3 = k + \frac{2}{A_0}\ln 2 \implies k = \frac{2}{A_0}\ln\frac{3}{2}.$$

Thus,

$$\frac{2}{A_0}\ln\left|\frac{2A_0 - \frac{1}{2}C}{A_0 - \frac{1}{2}C}\right| = \frac{2}{A_0}t\ln\frac{3}{2} + \frac{2}{A_0}\ln 2 = \frac{2}{A_0}\ln\left[2\left(\frac{3}{2}\right)^t\right]$$

so that

$$\frac{2A_0 - \frac{1}{2}C}{A_0 - \frac{1}{2}C} = 2\left(\frac{3}{2}\right)^t \quad \text{and therefore} \quad C(t) = 4A_0\frac{3^t - 2^t}{2(3^t) - 2^t}.$$

(c) By the hint

$$\int \frac{dC}{\left(A_0 - \dfrac{m}{m+n}C\right)\left(A_0 - \dfrac{n}{m+n}C\right)} = \int k\, dt$$

$$\int \frac{1}{A_0(m-n)}\left[\frac{m}{A_0 - \dfrac{m}{m+n}C} - \frac{n}{A_0 - \dfrac{n}{m+n}C}\right] dC = \int k\, dt$$

$$\frac{1}{A_0(m-n)}\left[-(m+n)\ln\left|A_0 - \frac{m}{m+n}C\right| + (m+n)\ln\left|A_0 - \frac{n}{m+n}C\right|\right] = kt + K$$

$$\frac{m+n}{A_0(m-n)}\ln\left|\frac{A_0 - \dfrac{n}{m+n}C}{A_0 - \dfrac{m}{m+n}C}\right| = kt + K.$$

First, $C(0) = 0 \implies K = \dfrac{m+n}{A_0(m-n)}\ln\left|\dfrac{A_0}{A_0}\right| = 0.$ Then,

$$C(1) = A_0 \implies k = \frac{m+n}{A_0(m-n)}\ln\left|\frac{A_0 - \dfrac{n}{m+n}A_0}{A_0 - \dfrac{m}{m+n}A_0}\right| = \frac{m+n}{A_0(m-n)}\ln\frac{m}{n}.$$

Thus,

$$\frac{m+n}{A_0(m-n)}\ln\left|\frac{A_0 - \dfrac{n}{m+n}C}{A_0 - \dfrac{m}{m+n}C}\right| = \frac{m+n}{A_0(m-n)}\ln\left(\frac{m}{n}\right)(t) + 0$$

so that

$$\frac{A_0 - \dfrac{n}{m+n}C}{A_0 - \dfrac{m}{m+n}C} = \left(\frac{m}{n}\right)^t \quad \text{and therefore} \quad C(t) = A_0(m+n)\left[\frac{m^t - n^t}{m^{t+1} - n^{t+1}}\right].$$

25.   The density function $\lambda(x) = kx$ leads to

$$F = Gmk\int_0^L \frac{x}{(L+h-x)^2}\, dx = Gmk\left[\frac{L}{h} + \ln\left(\frac{h}{L+h}\right)\right].$$

Since

$$M = \int_0^L kx\, dx = \frac{1}{2}kL^2,$$

we have $k = 2M/L^2$ and thus

$$F = \frac{2GmM}{L^2}\left[\frac{L}{h} + \ln\left(\frac{h}{L+h}\right)\right].$$

SECTION 8.3

1. $\displaystyle\int \sin^3 x \, dx = \int (1 - \cos^2 x) \sin x \, dx = \frac{1}{3} \cos^3 x - \cos x + C$

3. $\displaystyle\int \sin^2 3x \, dx = \int \frac{1 - \cos 6x}{2} \, dx = \frac{1}{2} x - \frac{1}{12} \sin 6x + C$

5.
$$\int \cos^4 x \sin^3 x \, dx = \int \cos^4 x \, (1 - \cos^2 x) \sin x \, dx$$
$$= \int (\cos^4 x - \cos^6 x) \sin x \, dx$$
$$= -\tfrac{1}{5} \cos^5 x + \tfrac{1}{7} \cos^7 x + C$$

7.
$$\int \sin^3 x \cos^3 x \, dx = \int \sin^3 x \, (1 - \sin^2 x) \cos x \, dx$$
$$= \int (\sin^3 x - \sin^5 x) \cos x \, dx$$
$$= \tfrac{1}{4} \sin^4 x - \tfrac{1}{6} \sin^6 x + C$$

9
$$\int \sin^2 x \cos^3 x \, dx = \int \sin^2 x \, (1 - \sin^2 x) \cos x \, dx$$
$$= \int (\sin^2 x - \sin^4 x) \cos x \, dx$$
$$= \tfrac{1}{3} \sin^3 x - \tfrac{1}{5} \sin^5 x + C$$

11.
$$\int \sin^4 x \, dx = \int \left(\frac{1 - \cos 2x}{2} \right)^2 dx$$
$$= \frac{1}{4} \int \left(1 - 2 \cos 2x + \cos^2 2x \right) dx$$
$$= \frac{1}{4} \int \left(1 - 2 \cos 2x + \frac{1 + \cos 4x}{2} \right) dx$$
$$= \int \left(\frac{3}{8} - \frac{1}{2} \cos 2x + \frac{1}{8} \cos 4x \right) dx$$
$$= \tfrac{3}{8} x - \tfrac{1}{4} \sin 2x + \tfrac{1}{32} \sin 4x + C$$

13.
$$\int \sin 2x \cos 3x \, dx = \int \frac{1}{2} \left[\sin(-x) + \sin 5x \right] dx$$
$$= \int \frac{1}{2} (- \sin x + \sin 5x) \, dx$$
$$= \tfrac{1}{2} \cos x - \tfrac{1}{10} \cos 5x + C$$

184   **SECTION 8.3**

15.
$$\int \sin^2 x \sin 2x \, dx = \int \sin^2 x \, (2 \sin x \cos x) \, dx$$
$$= 2 \int \sin^3 x \cos x \, dx$$
$$= \tfrac{1}{2} \sin^4 x + C$$

17.
$$\int \sin^4 x \cos^4 x \, dx = \int \left(\frac{\sin 2x}{2}\right)^4 dx = \frac{1}{16} \int \left(\frac{1 - \cos 4x}{2}\right)^2 dx$$
$$= \frac{1}{64} \int (1 - 2\cos 4x + \cos^2 4x) \, dx$$
$$= \frac{1}{64} \int \left(1 - 2\cos 4x + \frac{1 + \cos 8x}{2}\right) dx$$
$$= \frac{1}{64} \left(\frac{3x}{2} - \frac{\sin 4x}{2} + \frac{\sin 8x}{16}\right) + C$$

19.
$$\int \sin^6 x \, dx = \int \left(\frac{1 - \cos 2x}{2}\right)^3 dx$$
$$= \frac{1}{8} \int (1 - 3\cos 2x + 3\cos^2 2x - \cos^3 2x) \, dx$$
$$= \frac{1}{8} \int \left[1 - 3\cos 2x + 3\left(\frac{1 + \cos 4x}{2}\right) - \cos 2x \, (1 - \sin^2 2x)\right] dx$$
$$= \frac{1}{8} \int \left(\frac{5}{2} - 4\cos 2x + \frac{3}{2}\cos 4x + \sin^2 2x \cos 2x\right) dx$$
$$= \tfrac{5}{16}x - \tfrac{1}{4}\sin 2x + \tfrac{3}{64}\sin 4x + \tfrac{1}{48}\sin^3 2x + C$$

21.
$$\int \cos^7 x \, dx = \int (1 - \sin^2 x)^3 \cos x \, dx$$
$$= \int (1 - 3\sin^2 x + 3\sin^4 x - \sin^6 x) \cos x \, dx$$
$$= \int (\cos x - 3\sin^2 x \cos x + 3\sin^4 x \cos x - \sin^6 x \cos x) \, dx$$
$$= \sin x - \sin^3 x + \tfrac{3}{5}\sin^5 x - \tfrac{1}{7}\sin^7 x + C$$

23.
$$\int \cos 3x \cos 2x \, dx = \int \frac{1}{2}(\cos x + \cos 5x) \, dx$$
$$= \tfrac{1}{2}\sin x + \tfrac{1}{10}\sin 5x + C$$

25. $\left\{\begin{array}{l} x = \tan u \\ dx = \sec^2 u\, du \end{array}\right\};$

$$\int \frac{dx}{(x^2+1)^3} = \int \frac{\sec^2 u\, du}{(1+\tan^2 u)^3} = \int \cos^4 u\, du$$

$$= \int \left(\frac{1+\cos 2u}{2}\right)^2 du$$

$$= \frac{1}{4}\int (1 + 2\cos 2u + \cos^2 2u)\, du$$

$$= \frac{1}{4}\int \left(1 + 2\cos 2u + \frac{1+\cos 4u}{2}\right) du$$

$$= \frac{1}{4}\left(u + \sin 2u + \frac{u}{2} + \frac{\sin 4u}{8}\right) + C$$

$$= \frac{3}{8}u + \frac{\sin u \cos u}{2} + \frac{\sin 2u \cos 2u}{16} + C$$

$$\left[\begin{array}{l} \sin u = \dfrac{x}{\sqrt{1+x^2}} \\ \cos u = \dfrac{x}{\sqrt{1+x^2}} \end{array}\right]$$

$$\left\{\begin{array}{l} \sin 2u = 2\sin u \cos u \\ \cos 2u = \cos^2 u = \sin^2 u \end{array}\right\}$$

$$= \frac{3}{8}\tan^{-1} x + \frac{x}{2(x^2+1)} + \frac{x(1-x^2)}{8(x^2+1)^2} + C$$

27. $\left\{\begin{array}{l} x+1 = \tan u \\ dx = \sec^2 u\, du \end{array}\right\};$

$$\int \frac{dx}{[(x+1)^2+1]^2} = \int \frac{\sec^2 u\, du}{(1+\tan^2 u)^2}$$

$$= \int \cos^2 u\, du$$

$$= \int \frac{1+\cos 2u}{2}\, du$$

$$= \frac{u}{2} + \frac{\sin 2u}{4} + C$$

$$= \tfrac{1}{2}(u + \sin u \cos u) + C$$

$$= \frac{1}{2}\left[\tan^{-1}(x+1) + \frac{x+1}{x^2+2x+2}\right] + C$$

29. $\dfrac{1}{\pi}\displaystyle\int_0^{2\pi} \cos^2 nx\, dx = \dfrac{1}{2\pi}\int_0^{2\pi} (1+\cos 2nx)\, dx = \dfrac{1}{2\pi}\left[x + \dfrac{\sin 2nx}{2n}\right]_0^{2\pi} = 1$

31. $\displaystyle\int_0^{2\pi} \sin mx \cos nx\, dx = \left[-\dfrac{\cos[(m+n)x]}{2(m+n)} - \dfrac{\cos[(m-n)x]}{2(m-n)}\right]_0^{2\pi} = 0$

33. $\displaystyle\int_0^{2\pi} \cos mx \cos nx\, dx = \left[\dfrac{\sin[(m+n)x]}{2(m+n)} + \dfrac{\sin[(m-n)x]}{2(m-n)}\right]_0^{2\pi} = 0$

SECTION 8.4

1. $\displaystyle\int \tan^2 3x\, dx = \int (\sec^2 3x - 1)\, dx = \frac{1}{3}\tan 3x - x + C$

3. $\displaystyle\int \sec^2 \pi x\, dx = \frac{1}{\pi}\tan \pi x + C$

5. $\displaystyle\int \tan^3 x\, dx = \int (\sec^2 x - 1)\tan x\, dx$

$$= \int \tan x \sec^2 x\, dx - \int \tan x\, dx$$

$$= \tfrac{1}{2}\tan^2 x + \ln|\cos x| + C$$

7. $\displaystyle\int \tan^2 x \sec^2 x\, dx = \frac{1}{3}\tan^3 x + C$

9. $$\int \csc^3 x\, dx = -\csc x \cot x - \int \csc x \cot^2 x\, dx$$

$$\boxed{\begin{array}{ll} u = \csc x, & dv = \csc^2 x\, dx \\ du = -\csc x \cot x\, dx, & v = -\cot x \end{array}}$$

$$= -\csc x \cot x - \int \csc x\,(\csc^2 x - 1)\, dx$$

Thus

$$2\int \csc^3 x\, dx = -\csc x \cot x + \int \csc x\, dx$$

so that

$$\int \csc^3 x\, dx = -\frac{1}{2}\csc x \cot x + \frac{1}{2}\ln|\csc x - \cot x| + C.$$

11. $$\int \cot^4 x\, dx = \int \cot^2 x\,(\csc^2 x - 1)\, dx$$

$$= \int (\cot^2 x \csc^2 x - \cot^2 x)\, dx$$

$$= \int (\cot^2 x \csc^2 x - \csc^2 x + 1)\, dx$$

$$= -\tfrac{1}{3}\cot^3 x + \cot x + x + C$$

13. $$\int \cot^3 x \csc^3 x\, dx = \int (\csc^2 x - 1)\csc^3 x \cot x\, dx$$

$$= \int (\csc^4 x - \csc^2 x)\csc x \cot x\, dx$$

$$= -\tfrac{1}{5}\csc^5 x + \tfrac{1}{3}\csc^3 x + C$$

15. $\displaystyle\int \csc^4 2x\, dx = \int (1 + \cot^2 2x)\csc^2 2x\, dx = -\frac{1}{2}\cot 2x - \frac{1}{6}\cot^3 2x + C$

17.

$$\int \cot^2 x \csc x \, dx = -\cot x \csc x - \int \csc^3 x \, dx$$

$$\boxed{\begin{aligned} u &= \cot x, & dv &= \cot x \csc x \, dx \\ du &= -\csc x^2 \, dx, & v &= -\csc x \end{aligned}}$$

$$= -\cot x \csc x - \int \csc x \left(1 + \cot^2 x\right) dx.$$

Thus,

$$2 \int \cot^2 x \csc x \, dx = -\cot x \csc x - \int \csc x \, dx$$

so that

$$\int \cot^2 x \csc x \, dx = -\frac{1}{2} \cot x \csc x - \frac{1}{2} \ln |\csc x - \cot x| + C.$$

19.

$$\int \tan^5 3x \, dx = \int \tan^3 3x \left(\sec^2 3x - 1\right) dx$$

$$= \int \tan^3 3x \sec^2 3x \, dx - \int \tan^3 3x \, dx$$

$$= \int \tan^3 3x \sec^2 3x \, dx - \int \left(\tan 3x \sec^2 3x - \tan 3x\right) dx$$

$$= \tfrac{1}{12} \tan^4 3x - \tfrac{1}{6} \tan^2 3x + \tfrac{1}{3} \ln |\sec 3x| + C$$

21.

$$\int \sec^5 x \, dx = \tan x \sec^3 x - 3 \int \sec^3 x \tan^2 x \, dx$$

$$\boxed{\begin{aligned} u &= \sec^3 x, & dv &= \sec^2 x \, dx \\ du &= 3 \sec^3 x \tan x \, dx, & v &= \tan x \end{aligned}}$$

$$= \tan x \sec^3 x - 3 \int \sec^5 x \, dx + 3 \int \sec^3 x \, dx.$$

Rearranging terms, we get

$$4 \int \sec^5 x \, dx = \tan x \sec^3 x + 3 \int \sec^3 x \, dx.$$

We have already seen that

$$\int \sec^3 x \, dx = \frac{1}{2} \sec x \tan x + \frac{1}{2} \ln |\sec x + \tan x| + C.$$

Therefore

$$\int \sec^5 x \, dx = \frac{1}{4} \tan x \sec^3 x + \frac{3}{8} \sec x \tan x + \frac{3}{8} \ln |\sec x + \tan x| + C.$$

23.
$$\int \tan^4 x \sec^4 x \, dx = \int \tan^4 x \, (\tan^2 x + 1) \sec^2 x \, dx$$
$$= \int (\tan^6 x + \tan^4 x) \sec^2 x \, dx$$
$$= \tfrac{1}{7} \tan^7 x + \tfrac{1}{5} \tan^5 x + C$$

SECTION 8.5

1. $\left\{ \begin{array}{l} x = a \sin u \\ dx = a \cos u \, du \end{array} \right\};$

$$\int \frac{dx}{\sqrt{a^2 - x^2}} = \int \frac{a \cos u \, du}{a \cos u}$$
$$= \int du = u + C = \sin^{-1}\left(\frac{x}{a}\right) + C$$

3. $\left\{ \begin{array}{l} x = \sqrt{5} \sin u \\ dx = \sqrt{5} \cos u \, du \end{array} \right\};$

$$\int \frac{dx}{(5 - x^2)^{3/2}} = \int \frac{\sqrt{5} \cos u \, du}{(5 \cos^2 u)^{3/2}}$$
$$= \frac{1}{5} \int \sec^2 u \, du$$
$$= \frac{1}{5} \tan u + C = \frac{x}{5\sqrt{5 - x^2}} + C$$

5. $\left\{ \begin{array}{l} x = \sec u \\ dx = \sec u \tan u \, du \end{array} \right\};$

$$\int \sqrt{x^2 - 1} \, dx = \int \tan^2 u \sec u \, du$$
$$= \int (\sec^3 u - \sec u) \, du$$
$$= \tfrac{1}{2} \sec u \tan u - \tfrac{1}{2} \ln |\sec u + \tan u| + C$$
$$\uparrow$$
$$\llcorner \text{Section 8.4}$$
$$= \tfrac{1}{2} x \sqrt{x^2 - 1} - \tfrac{1}{2} \ln |x + \sqrt{x^2 - 1}| + C$$

7. $\left\{ \begin{array}{l} x = 2 \sin u \\ dx = 2 \cos u \, du \end{array} \right\};$

$$\int \frac{x^2}{\sqrt{4 - x^2}} \, dx = \int \frac{4 \sin^2 u}{2 \cos u} 2 \cos u \, du$$
$$= 2 \int (1 - \cos 2u) \, du$$
$$= 2u - \sin 2u + C$$
$$= 2u - 2 \sin u \cos u + C$$
$$= 2 \sin^{-1}\left(\frac{x}{2}\right) - \frac{1}{2} x \sqrt{4 - x^2} + C$$

9. $\left\{ \begin{array}{l} u = 1 - x^2 \\ du = -2x \, dx \end{array} \right\};$ $\int \frac{x}{(1 - x^2)^{3/2}} \, dx = -\frac{1}{2} \int \frac{du}{u^{3/2}} = u^{-1/2} + C = \frac{1}{\sqrt{1 - x^2}} + C$

11. $\left\{\begin{array}{l} x = \sin u \\ dx = \cos u\, du \end{array}\right\}$; $\displaystyle\int \frac{x^2}{(1-x^2)^{3/2}}\, dx = \int \frac{\sin^2 u}{\cos^3 u} \cos u\, du = \int \tan^2 u\, du$

$\qquad\qquad\qquad\qquad\qquad\quad = \displaystyle\int (\sec^2 u - 1)\, du = \tan u - u + C$

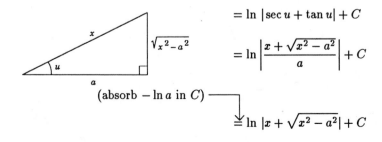

$\qquad\qquad\qquad\qquad\qquad\quad = \dfrac{x}{\sqrt{1-x^2}} - \sin^{-1} x + C$

13. $\left\{\begin{array}{l} u = 4 - x^2 \\ du = -2x\, dx \end{array}\right\}$; $\displaystyle\int x\sqrt{4-x^2}\, dx = -\frac{1}{2}\int u^{1/2}\, du = -\frac{1}{3}u^{3/2} + C$

$\qquad\qquad\qquad\qquad\qquad\qquad\quad = -\tfrac{1}{3}(4-x^2)^{3/2} + C$

15. $\left\{\begin{array}{l} x = \sqrt{8}\tan u \\ dx = \sqrt{8}\sec^2 u\, du \end{array}\right\}$; $\displaystyle\int \frac{x^2}{(x^2+8)^{3/2}}\, dx = \int \frac{8\tan^2 u}{(8\sec^2 u)^{3/2}}\sqrt{8}\sec^2 u\, du$

$\qquad\qquad\qquad\qquad\qquad\qquad\qquad = \displaystyle\int \frac{\tan^2 u}{\sec u}\, du$

$\qquad\qquad\qquad\qquad\qquad\qquad\qquad = \displaystyle\int \frac{\sec^2 u - 1}{\sec u}\, du$

$\qquad\qquad\qquad\qquad\qquad\qquad\qquad = \displaystyle\int (\sec u - \cos u)\, du$

$\qquad\qquad\qquad\qquad\qquad\qquad\qquad = \ln|\sec u + \tan u| - \sin u + C$

$\qquad\qquad\qquad\qquad\qquad\qquad\qquad = \ln\left(\dfrac{\sqrt{x^2+8}+x}{\sqrt{8}}\right) - \dfrac{x}{\sqrt{x^2+8}} + C$

(absorb $-\ln\sqrt{8}$ in C) ——┐

$\qquad\qquad\qquad\qquad\qquad\qquad\overset{\cdot}{=} \ln\left(\sqrt{x^2+8}+x\right) - \dfrac{x}{\sqrt{x^2+8}} + C$

17. $\left\{\begin{array}{l} x = a\sin u \\ dx = a\cos u\, du \end{array}\right\}$; $\displaystyle\int \frac{dx}{x\sqrt{a^2-x^2}} = \int \frac{a\cos u\, du}{a\sin u\,(a\cos u)} = \frac{1}{a}\int \csc u\, du$

$\qquad\qquad\qquad\qquad\qquad\qquad\quad = \dfrac{1}{a}\ln|\csc u - \cot u| + C$

$\qquad\qquad\qquad\qquad\qquad\qquad\quad = \dfrac{1}{a}\ln\left|\dfrac{a-\sqrt{a^2-x^2}}{x}\right| + C$

19. $\left\{\begin{array}{l} x = a\sec u \\ dx = a\sec u\tan u\, du \end{array}\right\}$; $\displaystyle\int \frac{dx}{\sqrt{x^2-a^2}} = \int \frac{a\sec u\tan u\, du}{a\tan u} = \int \sec u\, du$

$\qquad\qquad\qquad\qquad\qquad\qquad\quad = \ln|\sec u + \tan u| + C$

$\qquad\qquad\qquad\qquad\qquad\qquad\quad = \ln\left|\dfrac{x+\sqrt{x^2-a^2}}{a}\right| + C$

(absorb $-\ln a$ in C) ——┐

$\qquad\qquad\qquad\qquad\qquad\qquad\overset{\cdot}{=} \ln\left|x+\sqrt{x^2-a^2}\right| + C$

21. $\left\{ \begin{array}{l} e^x = \sec u \\ e^x\,dx = \sec u \tan u\,du \end{array} \right\}$; $\displaystyle\int e^x \sqrt{e^{2x}-1}\,dx = \int \tan^2 u \sec u\,du$

$$= \int (\sec^3 u - \sec u)\,du$$

$$= \tfrac{1}{2}\sec u \tan u - \tfrac{1}{2}\ln|\sec u + \tan u| + C$$

$$= \tfrac{1}{2}e^x\sqrt{e^{2x}-1} - \tfrac{1}{2}\ln\left(e^x + \sqrt{e^{2x}-1}\right) + C$$

23. $\left\{ \begin{array}{l} x = a\tan u \\ dx = a\sec^2 u\,du \end{array} \right\}$; $\displaystyle\int \frac{dx}{x^2\sqrt{a^2+x^2}} = \int \frac{a\sec^2 u\,du}{a^2\tan^2 u\,(a\sec u)}$

$$= \frac{1}{a^2}\int \frac{\sec u}{\tan^2 u}\,du$$

$$= \frac{1}{a^2}\int \cot u \csc u\,du$$

$$= -\frac{1}{a^2}\cos u + C$$

$$= -\frac{1}{a^2 x}\sqrt{a^2+x^2} + C$$

25. $\left\{ \begin{array}{l} x = a\sec u \\ dx = a\sec u \tan u\,du \end{array} \right\}$; $\displaystyle\int \frac{dx}{x^2\sqrt{x^2-a^2}} = \int \frac{a\sec u \tan u\,du}{a^2\sec^2 u\,(a\tan u)}$

$$= \frac{1}{a^2}\int \cos u\,du$$

$$= \frac{1}{a^2}\sin u + C$$

$$= \frac{1}{a^2 x}\sqrt{x^2-a^2} + C$$

27. $\left\{ \begin{array}{l} e^x = 3\sec u \\ e^x\,dx = 3\sec u \tan u\,du \end{array} \right\}$; $\displaystyle\int \frac{dx}{e^x\sqrt{e^{2x}-9}} = \int \frac{\tan u\,du}{3\sec u\,(3\tan u)}$

$$= \frac{1}{9}\int \cos u\,du$$

$$= \tfrac{1}{9}\sin u + C$$

$$= \tfrac{1}{9}e^{-x}\sqrt{e^{2x}-9} + C$$

29. $\displaystyle\int \frac{dx}{(x^2-4x+4)^{3/2}} = \int \frac{dx}{(x-2)^3} = -\frac{1}{2(x-2)^2} + C$

31. $\left\{\begin{array}{l} x - 3 = \sin u \\ dx = \cos u\, du \end{array}\right\}; \quad \displaystyle\int x\sqrt{6x - x^2 - 8}\, dx = \int x\sqrt{1 - (x-3)^2}\, dx$

$$= \int (3 + \sin u)(\cos u)\cos u\, du$$

$$= \int (3\cos^2 u + \cos^2 u \sin u)\, du$$

$$= \int \left[3\left(\frac{1 + \cos 2u}{2}\right) + \cos^2 u \sin u \right] du$$

$$= \frac{3u}{2} + \frac{3}{4}\sin 2u - \frac{1}{3}\cos^3 u + C$$

$$= \tfrac{3}{2}\sin^{-1}(x-3) + \tfrac{3}{2}(x-3)\sqrt{6x - x^2 - 8} - \tfrac{1}{3}(6x - x^2 - 8)^{3/2} + C$$

33. $\left\{\begin{array}{l} x + 1 = 2\tan u \\ dx = 2\sec^2 u\, du \end{array}\right\}; \quad \displaystyle\int \frac{x}{(x^2 + 2x + 5)^2}\, dx = \int \frac{x}{[(x+1)^2 + 4]^2}\, dx$

$$= \int \frac{2\tan u - 1}{(4\sec^2 u)^2}\, 2\sec^2 u\, du$$

$$= \frac{1}{8}\int \frac{2\tan u - 1}{\sec^2 u}\, du$$

$$= \frac{1}{8}\int (2\sin u \cos u - \cos^2 u)\, du$$

$$= \frac{1}{8}\int \left(2\sin u \cos u - \frac{1 + \cos 2u}{2} \right) du$$

$$= \frac{1}{8}\left(\sin^2 u - \frac{u}{2} - \frac{\sin 2u}{4} \right) + C$$

$$= \frac{1}{8}\left[\left(\frac{x+1}{\sqrt{x^2 + 2x + 5}}\right)^2 - \frac{1}{2}\tan^{-1}\left(\frac{x+1}{2}\right) - \frac{1}{4}\overbrace{(2)\left(\frac{x+1}{\sqrt{x^2 + 2x + 5}}\right)\left(\frac{2}{\sqrt{x^2 + 2x + 5}}\right)}^{\sin 2u = 2\sin u \cos u} \right] + C$$

$$= \frac{x^2 + x}{8(x^2 + 2x + 5)} - \frac{1}{16}\tan^{-1}\left(\frac{x+1}{2}\right) + C$$

35. $\left\{\begin{array}{l} x + 2 = 3\tan u \\ dx = 3\sec^2 u\, du \end{array}\right\}; \quad \displaystyle\int \frac{x + 3}{\sqrt{x^2 + 4x + 13}}\, dx = \int \frac{x + 3}{\sqrt{(x+2)^2 + 9}}\, dx$

$$= \int \frac{3\tan u + 1}{3\sec u}\, 3\sec^2 u\, du$$

$$= \int (3\sec u \tan u + \sec u)\, du$$

$$= 3\sec u + \ln|\sec u + \tan u| + C$$

$$= \sqrt{x^2 + 4x + 13} + \ln\left|\frac{\sqrt{x^2 + 4x + 13}}{3} + \frac{x + 2}{3}\right| + C$$

(absorb $- \ln 3$ in C)

$$= \sqrt{x^2 + 4x + 13} + \ln|x + 2 + \sqrt{x^2 + 4x + 13}| + C$$

37. $\left\{\begin{array}{l} x - 3 = \sin u \\ dx = \cos u\, du \end{array}\right\};\quad \displaystyle\int \sqrt{6x - x^2 - 8}\, dx = \int \sqrt{1 - (x - 3)^2}\, dx$

$$= \int \cos^2 u\, du$$

$$= \int \frac{1 + \cos 2u}{2}\, du$$

$$= \frac{u}{2} + \frac{\sin 2u}{4} + C$$

$$= \tfrac{1}{2}\left[\sin^{-1}(x - 3) + (x - 3)\sqrt{6x - x^2 - 8}\right] + C$$

39. $\left\{\begin{array}{l} x + 1 = 3\sin u \\ dx = 3\cos u\, du \end{array}\right\};\quad \displaystyle\int x(8 - 2x - x^2)^{3/2}\, dx = \int x[9 - (x + 1)^2]^{3/2}\, dx$

$$= \int (3\sin u - 1)[9\cos^2 u]^{3/2}\, 3\cos u\, du$$

$$= 81\int (3\cos^4 u \sin u - \cos^4 u)\, du$$

$$\left[\cos^4 u = \left(\frac{1 + \cos 2u}{2}\right)^2 = \frac{1}{4}(1 + 2\cos 2u + \cos^2 2u)\right.$$
$$\left. = \frac{1}{4}\left(1 + 2\cos 2u + \frac{1 + \cos 4u}{2}\right)\right]$$

$$= 81\int\left(3\cos^4 u \sin u - \tfrac{3}{8} - \tfrac{1}{2}\cos 2u - \tfrac{1}{8}\cos 4u\right)\, du$$

$$= 81\left(-\tfrac{3}{5}\cos^5 u - \tfrac{3}{8}u - \tfrac{1}{4}\sin 2u - \tfrac{1}{32}\sin 4u\right) + C$$

$$= 81\left(-\tfrac{3}{5}\cos^5 u - \tfrac{3}{8}u - \tfrac{1}{4}\sin 2u - \tfrac{1}{16}\sin 2u \cos 2u\right) + C$$

$$= 81\left(-\tfrac{3}{5}\cos^5 u - \tfrac{3}{8}u - \tfrac{1}{16}\sin 2u\,(4 + \cos 2u)\right) + C$$

$$= 81\left(-\tfrac{3}{5}\cos^5 u - \tfrac{3}{8}u - \tfrac{1}{8}\sin u \cos u\,(5 - 2\sin^2 u)\right) + C$$

$$= -\frac{243}{5}\left[\frac{\sqrt{8 - 2x - x^2}}{3}\right]^5 - \frac{243}{8}\sin^{-1}\left(\frac{x + 1}{3}\right)$$

$$- \frac{81}{8}\left[\frac{(x + 1)\sqrt{8 - 2x - x^2}}{9}\right]\left[5 - 2\left(\frac{x + 1}{3}\right)^2\right] + C$$

$$= -\frac{1}{5}(8 - 2x - x^2)^{5/2} - \frac{243}{8}\sin^{-1}\left(\frac{x + 1}{3}\right) - \frac{1}{8}(x + 1)(43 - 4x - 2x^2)(8 - 2x - x^2)^{1/2} + C$$

41. $M = \int_0^a \frac{dx}{\sqrt{x^2 + a^2}} = \left[\ln\left(x + \sqrt{x^2 + a^2}\right)\right]_0^a = \ln\left(1 + \sqrt{2}\right)$

$x_M M = \int_0^a \frac{x}{\sqrt{x^2 + a^2}}\, dx = \left[\sqrt{x^2 + a^2}\right]_0^a = \left(\sqrt{2} - 1\right)a \qquad x_M = \dfrac{\left(\sqrt{2} - 1\right)a}{\ln\left(1 + \sqrt{2}\right)}$

43.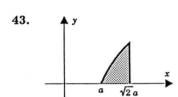

$A = \int_a^{\sqrt{2}a} \sqrt{x^2 - a^2}\, dx = \left[\frac{1}{2}x\sqrt{x^2 - a^2} - \frac{1}{2}a^2 \ln\left|x + \sqrt{x^2 - a^2}\right|\right]_a^{\sqrt{2}a}$

$\qquad\qquad = \frac{1}{2}a^2\left[\sqrt{2} - \ln\left(\sqrt{2} + 1\right)\right]$

$\overline{x}A = \int_a^{\sqrt{2}a} x\sqrt{x^2 - a^2}\, dx = \frac{1}{3}a^3, \qquad \overline{y}A = \int_a^{\sqrt{2}a} \left[\frac{1}{2}(x^2 - a^2)\right]\, dx = \frac{1}{6}a^3(2 - \sqrt{2})$

$\overline{x} = \dfrac{2a}{3\left[\sqrt{2} - \ln\left(\sqrt{2} + 1\right)\right]}, \qquad \overline{y} = \dfrac{\left(2 - \sqrt{2}\right)a}{3\left[\sqrt{2} - \ln\left(\sqrt{2} + 1\right)\right]}$

45. $V_y = 2\pi \overline{R}A = \frac{2}{3}\pi a^3 \qquad \overline{y}V_y = \int_a^{\sqrt{2}a} \pi x(x^2 - a^2)\, dx = \frac{1}{4}\pi a^4, \quad \overline{y} = \frac{3}{8}a$

$\qquad\qquad\qquad\qquad \uparrow\\ \qquad\qquad\qquad\quad \rule{0pt}{0pt}\llcorner (6.4.6)$

SECTION 8.6

1. $\left\{\begin{array}{l} x = u^2 \\ dx = 2u\, du \end{array}\right\}; \quad \int \dfrac{dx}{1 - \sqrt{x}} = \int \dfrac{2u\, du}{1 - u} = 2\int\left(\dfrac{1}{1 - u} - 1\right)\, du$

$\qquad\qquad\qquad\qquad\quad = -2(\ln|1 - u| + u) + C = -2\left(\sqrt{x} + \ln|1 - \sqrt{x}|\right) + C$

3. $\left\{\begin{array}{l} u^2 = 1 + e^x \\ 2u\, du = e^x\, dx \end{array}\right\}; \quad \int \sqrt{1 + e^x}\, dx = \int u \cdot \dfrac{2u\, du}{u^2 - 1} = 2\int\left(1 + \dfrac{1}{u^2 - 1}\right)\, du$

$\qquad\qquad\qquad\qquad\qquad = 2\int\left(1 + \dfrac{1}{2}\left[\dfrac{1}{u - 1} - \dfrac{1}{u + 1}\right]\right)\, du$

$\qquad\qquad\qquad\qquad\qquad = 2u + \ln|u - 1| - \ln|u + 1| + C$

$\qquad\qquad\qquad\qquad\qquad = 2\sqrt{1 + e^x} + \ln\left[\dfrac{\sqrt{1 + e^x} - 1}{\sqrt{1 + e^x} + 1}\right] + C$

$\qquad\qquad\qquad\qquad\qquad = 2\sqrt{1 + e^x} + \ln\left[\dfrac{\left(\sqrt{1 + e^x} - 1\right)^2}{e^x}\right] + C$

$\qquad\qquad\qquad\qquad\qquad = 2\sqrt{1 + e^x} + 2\ln\left(\sqrt{1 + e^x} - 1\right) - x + C$

5 (a) $\left\{\begin{array}{l} u^2 = 1 + x \\ 2u\,du = dx \end{array}\right\};$ $\displaystyle\int x\sqrt{1+x}\,dx = \int (u^2-1)(u)2u\,du$

$$= \int (2u^4 - 2u^2)\,du$$

$$= \tfrac{2}{5}u^5 - \tfrac{2}{3}u^3 + C$$

$$= \tfrac{2}{5}(x+1)^{5/2} - \tfrac{2}{3}(x+1)^{3/2} + C$$

(b) $\left\{\begin{array}{l} u = 1 + x \\ du = dx \end{array}\right\};$ $\displaystyle\int x\sqrt{1+x}\,dx = \int (u-1)\sqrt{u}\,du = \int \left(u^{3/2} - u^{1/2}\right)du$

$$= \tfrac{2}{5}u^{5/2} - \tfrac{2}{3}u^{3/2} + C$$

$$= \tfrac{2}{5}(1+x)^{5/2} - \tfrac{2}{3}(1+x)^{3/2} + C$$

7. $\left\{\begin{array}{l} u^2 = x - 1 \\ 2u\,du = dx \end{array}\right\};$ $\displaystyle\int (x+2)\sqrt{x-1}\,dx = \int (u+3)(u)2u\,du$

$$= \int (2u^4 + 6u^2)\,du$$

$$= \tfrac{2}{5}u^5 + 2u^3 + C$$

$$= \tfrac{2}{5}(x-1)^{5/2} + 2(x-1)^{3/2} + C$$

9. $\left\{\begin{array}{l} u^2 = 1 + x^2 \\ 2u\,du = 2x\,dx \end{array}\right\};$ $\displaystyle\int \frac{x^3}{(1+x^2)^3}\,dx = \int \frac{x^2}{(1+x^2)^3}x\,dx = \int \frac{u^2-1}{u^6}u\,du$

$$= \int (u^{-3} - u^{-5})\,du = \frac{1}{2}u^{-2} + \frac{1}{4}u^{-4} + C$$

$$= \frac{1}{4(1+x^2)^2} - \frac{1}{2(1+x^2)} + C$$

$$= -\frac{1+2x^2}{4(1+x^2)^2} + C$$

11. $\left\{\begin{array}{l} u^2 = x \\ 2u\,du = dx \end{array}\right\};$ $\displaystyle\int \frac{\sqrt{x}}{\sqrt{x}-1}\,dx = \int \left(\frac{u}{u-1}\right)2u\,du = 2\int \left(u+1+\frac{1}{u-1}\right)du$

$$= u^2 + 2u + 2\ln|u-1| + C$$

$$= x + 2\sqrt{x} + 2\ln|\sqrt{x}-1| + C$$

13. $\left\{ \begin{array}{l} u^2 = x - 1 \\ 2u\,du = dx \end{array} \right\}$; $\displaystyle \int \frac{\sqrt{x-1}+1}{\sqrt{x-1}-1}\,dx = \int \frac{u+1}{u-1} 2u\,du = \int \left(2u + 4 + \frac{4}{u-1} \right) du$

$$= u^2 + 4u + 4\ln|u-1| + C$$

$$= x - 1 + 4\sqrt{x-1} + 4\ln|\sqrt{x-1}-1| + C$$

(absorb -1 in C)

$$= x + 4\sqrt{x-1} + 4\ln|\sqrt{x-1}-1| + C$$

15. $\left\{ \begin{array}{l} u^2 = 1 + e^x \\ 2u\,du = e^x\,dx \end{array} \right\}$; $\displaystyle \int \frac{dx}{\sqrt{1+e^x}} = \int \left(\frac{1}{u} \right) \frac{2u\,du}{u^2-1} = \int \left[\frac{1}{u-1} - \frac{1}{u+1} \right] du$

$$= \ln|u-1| - \ln|u+1| + C$$

$$= \ln \left[\frac{\sqrt{1+e^x}-1}{\sqrt{1+e^x}+1} \right] + C$$

$$= \ln \left[\frac{(\sqrt{1+e^x}-1)^2}{e^x} \right] + C$$

$$= 2\ln(\sqrt{1+e^x}-1) - x + C$$

17. $\left\{ \begin{array}{l} u^2 = x + 4 \\ 2u\,du = dx \end{array} \right\}$; $\displaystyle \int \frac{x}{\sqrt{x+4}}\,dx = \int \frac{u^2-4}{u} 2u\,du = \int (2u^2 - 8)\,du$

$$= \tfrac{2}{3}u^3 - 8u + C$$

$$= \tfrac{2}{3}(x+4)^{3/2} - 8(x+4)^{1/2} + C$$

$$= \tfrac{2}{3}(x-8)\sqrt{x+4} + C$$

19. $\left\{ \begin{array}{l} u^2 = 4x + 1 \\ 2u\,du = 4\,dx \end{array} \right\}$; $\displaystyle \int 2x^2(4x+1)^{-5/2}\,dx = \int 2\left(\frac{u^2-1}{4} \right)^2 (u^{-5}) \frac{u}{2}\,du$

$$= \frac{1}{16} \int (1 - 2u^{-2} + u^{-4})\,du = \frac{1}{16}u + \frac{1}{8}u^{-1} - \frac{1}{48}u^{-3} + C$$

$$= \tfrac{1}{16}(4x+1)^{1/2} + \tfrac{1}{8}(4x+1)^{-1/2} - \tfrac{1}{48}(4x+1)^{-3/2} + C$$

21. $\left\{ \begin{array}{l} u^2 = ax + b \\ 2u\,du = a\,dx \end{array} \right\}$; $\displaystyle \int \frac{x}{(ax+b)^{3/2}}\,dx = \int \frac{\dfrac{u^2-b}{a}}{u^3} \frac{2u}{a}\,du$

$$= \frac{2}{a^2} \int (1 - bu^{-2})\,du$$

$$= \frac{2}{a^2}(u + bu^{-1}) + C = \frac{2u^2 + 2b}{a^2 u} + C$$

$$= \frac{4b + 2ax}{a^2\sqrt{ax+b}} + C$$

SECTION 8.7

1. (a) $L_{12} = \frac{12}{12}[0 + 1 + 4 + 9 + 16 + 25 + 36 + 49 + 64 + 81 + 100 + 121] = 506$

    (b) $R_{12} = \frac{12}{12}[1 + 4 + 9 + 16 + 25 + 36 + 49 + 64 + 81 + 100 + 121 + 144] = 650$

    (c) $M_6 = \frac{12}{6}[1 + 9 + 25 + 49 + 81 + 121] = 572$

    (d) $T_{12} = \frac{12}{24}[0 + 2(1 + 4 + 9 + 16 + 25 + 36 + 49 + 64 + 81 + 100 + 121) + 144] = 578$

    (e) $S_6 = \frac{12}{36}[0 + 144 + 2(4 + 16 + 36 + 64 + 100) + 4(1 + 9 + 25 + 49 + 81 + 121)] = 576$

    $$\int_0^{12} x^2\, dx = \left[\frac{1}{3}x^3\right]_0^{12} = 576$$

3. (a) $L_6 = \frac{3}{6}\left[\frac{1}{1+0} + \frac{1}{1+1/8} + \frac{1}{1+1} + \frac{1}{1+27/8} + \frac{1}{1+8} + \frac{1}{1+125/8}\right]$

    $= \frac{1}{2}\left[1 + \frac{8}{9} + \frac{1}{2} + \frac{8}{35} + \frac{1}{9} + \frac{8}{133}\right] \cong 1.39$

    (b) $R_6 = \frac{3}{6}\left[\frac{1}{1+1/8} + \frac{1}{1+1} + \frac{1}{1+27/8} + \frac{1}{1+8} + \frac{1}{1+125/8} + \frac{1}{1+27}\right]$

    $= \frac{1}{2}\left[\frac{8}{9} + \frac{1}{2} + \frac{8}{35} + \frac{1}{9} + \frac{8}{133} + \frac{1}{28}\right] \cong 0.91$

    (c) $M_3 = \frac{3}{3}\left[\frac{1}{1+1/8} + \frac{1}{1+27/8} + \frac{1}{1+125/8}\right] = \frac{8}{9} + \frac{8}{35} + \frac{8}{133} \cong 1.18$

    (d) $T_6 = \frac{3}{12}\left[1 + 2\left(\frac{8}{9} + \frac{1}{2} + \frac{8}{35} + \frac{1}{9} + \frac{8}{133}\right) + \frac{1}{28}\right] \cong 1.15$

    (e) $S_3 = \frac{3}{18}\left\{1 + \frac{1}{28} + 2\left[\frac{1}{2} + \frac{1}{9}\right] + 4\left[\frac{8}{9} + \frac{8}{35} + \frac{8}{133}\right]\right\} \cong 1.16$

5. (a) $\frac{1}{4}\pi \cong T_4 = \frac{1}{8}\left[1 + 2\left(\frac{1}{1+1/16} + \frac{1}{1+1/4} + \frac{1}{1+9/16}\right) + \frac{1}{1+1}\right]$

    $= \frac{1}{8}\left[1 + 2\left(\frac{16}{17} + \frac{4}{5} + \frac{16}{25}\right) + \frac{1}{2}\right] \cong 0.78$

    (b) $\frac{1}{4}\pi \cong S_4 = \frac{1}{24}\left[1 + \frac{1}{2} + 2\left(\frac{16}{17} + \frac{4}{5} + \frac{16}{25}\right) + 4\left(\frac{64}{65} + \frac{64}{73} + \frac{64}{89} + \frac{64}{113}\right)\right] \cong 0.79$

7. Such a curve passes through the three points

    $$(a_1, b_1),\quad (a_2, b_2),\quad (a_3, b_3)$$

    iff

    $$b_1 = a_1{}^2 A + a_1 B + C,\quad b_2 = a_2{}^2 A + a_2 B + C,\quad b_3 = a_3{}^2 A + a_3 B + C,$$

which happens iff

$$A = \frac{b_1(a_2 - a_3) - b_2(a_1 - a_3) + b_3(a_1 - a_2)}{(a_1 - a_3)(a_1 - a_2)(a_2 - a_3)},$$

$$B = -\frac{b_1(a_2{}^2 - a_3{}^2) - b_2(a_1{}^2 - a_3{}^2) + b_3(a_1{}^2 - a_2{}^2)}{(a_1 - a_3)(a_1 - a_2)(a_2 - a_3)},$$

$$C = \frac{a_1{}^2(a_2 b_3 - a_3 b_2) - a_2{}^2(a_1 b_3 - a_3 b_1) + a_3{}^2(a_1 b_2 - a_2 b_1)}{(a_1 - a_3)(a_1 - a_2)(a_2 - a_3)}.$$

9. (a) $\left|\dfrac{(b-a)^3}{12n^2} f''(c)\right| = \dfrac{27}{12n^2} \dfrac{1}{4c^{3/2}} \leq \dfrac{9}{16n^2} < 0.01 \implies n^2 > \left(\dfrac{15}{2}\right)^2 \implies n \geq 8$

   (b) $\left|\dfrac{(b-a)^5}{180n^4} f^{(4)}(c)\right| = \dfrac{243}{180n^4} \dfrac{15}{16c^{7/2}} \leq \dfrac{81}{64n^4} < 0.01 \implies n > \dfrac{3}{2}\sqrt{5} \implies n \geq 4$

11. (a) $\left|\dfrac{(b-a)^3}{12n^2} f''(c)\right| = \dfrac{27}{12n^2} \dfrac{1}{4c^{3/2}} \leq \dfrac{9}{16n^2} < 0.00001 \implies n > 75\sqrt{10} \implies n \geq 238$

   (b) $\left|\dfrac{(b-a)^5}{180n^4} f^{(4)}(c)\right| = \dfrac{243}{180n^4} \dfrac{15}{16c^{7/2}} \leq \dfrac{81}{64n^4} < 0.00001 \implies n > 15\left(\dfrac{5}{2}\right)^{1/4} \implies n \geq 19$

13. (a) $\left|\dfrac{(b-a)^3}{12n^2} f''(c)\right| = \dfrac{\pi^3}{12n^2} \sin c \leq \dfrac{\pi^3}{12n^2} < 0.001 \implies n > 5\pi\sqrt{\dfrac{10\pi}{3}} \implies n \geq 51$

   (b) $\left|\dfrac{(b-a)^5}{180n^4} f^{(4)}(c)\right| = \dfrac{\pi^5}{180n^4} \sin c \leq \dfrac{\pi^5}{180n^4} < 0.001 \implies n > \pi\left(\dfrac{50\pi}{9}\right)^{1/4} \implies n \geq 7$

15. (a) $\left|\dfrac{(b-a)^3}{12n^2} f''(c)\right| = \dfrac{8}{12n^2} e^c \leq \dfrac{8}{12n^2} e^3 < 0.01 \implies n > 10e\sqrt{\dfrac{2e}{3}} \implies n \geq 37$

   (b) $\left|\dfrac{(b-a)^5}{180n^4} f^{(4)}(c)\right| = \dfrac{32}{180n^4} e^c \leq \dfrac{8}{45n^4} e^3 < 0.01 \implies n > 2\left(\dfrac{10e^3}{9}\right)^{1/4} \implies n \geq 5$

17. $f^{(4)}(x) = 0$ for all x; therefore by (8.7.2) the theoretical error is zero

SECTION 8.8

1. $\displaystyle\int 10^{nx}\, dx = \dfrac{10^{nx}}{n \ln 10} + C$

3. $\displaystyle\int \sqrt{2x+1}\, dx = \dfrac{1}{3}(2x+1)^{3/2} + C$

5.
$$\left\{ \begin{array}{l} u = e^x \\ du = e^x\,dx \end{array} \right\};\qquad \int e^x \tan e^x\,dx = \int \tan u\,du = \ln|\sec u| + C$$

$$= \ln|\sec e^x| + C$$

7.
$$\left\{ \begin{array}{l} ax = bu \\ a\,dx = b\,du \end{array} \right\};\qquad \int \frac{dx}{a^2x^2 + b^2} = \frac{1}{ab}\int \frac{du}{u^2 + 1} = \frac{1}{ab}\tan^{-1}u + C$$

$$= \frac{1}{ab}\tan^{-1}\left(\frac{ax}{b}\right) + C$$

9.
$$\int \ln\left(x\sqrt{x}\right)dx = \frac{3}{2}\int \ln x\,dx = \frac{3}{2}\left(x\ln x - \int dx\right) = x\ln\left(x\sqrt{x}\right) - \frac{3}{2}x + C$$

$$\boxed{\begin{array}{l} u = \ln x, \quad dv = dx \\[2mm] du = \dfrac{dx}{x}, \quad v = x \end{array}}$$

11.
$$\left\{ \begin{array}{l} x = \tan u \\ dx = \sec^2 u\,du \end{array} \right\};\qquad \int \frac{x^3}{\sqrt{1+x^2}}\,dx = \int \frac{\tan^3 u}{\sec u}\sec^2 u\,du$$

$$= \int (\sec^2 u - 1)\sec u \tan u\,du$$

$$= \tfrac{1}{3}\sec^3 u - \sec u + C$$

$$= \tfrac{1}{3}(1+x^2)^{3/2} - (1+x^2)^{1/2} + C$$

$$= \tfrac{1}{3}(x^2 - 2)\sqrt{1+x^2} + C$$

13.
$$\int \sin^2\left(\frac{\pi}{n}x\right)dx = \frac{1}{2}\int \left(1 - \cos\frac{2\pi x}{n}\right)dx = \frac{1}{2}x - \frac{n}{4\pi}\sin\left(\frac{2\pi}{n}x\right) + C$$

15.
$$\left\{ \begin{array}{l} u^2 = x \\ 2u\,du = dx \end{array} \right\};\qquad \int \frac{dx}{a\sqrt{x}+b} = \int \frac{2u\,du}{au+b} = \frac{2}{a}\int \left(1 - \frac{b}{au+b}\right)du$$

$$= \frac{2}{a}\left(u - \frac{b}{a}\ln|au+b|\right) + C$$

$$= \frac{2\sqrt{x}}{a} - \frac{2b}{a^2}\ln|a\sqrt{x}+b| + C$$

17.
$$\int \frac{\sin 3x}{2 + \cos 3x}\,dx = -\frac{1}{3}\ln\left(2 + \cos 3x\right) + C$$

19. $\int \dfrac{\sin x}{\cos^3 x}\,dx = \int \tan x \sec^2 x\,dx = \dfrac{1}{2}\tan^2 x + C$

21. $\int \dfrac{dx}{2x^2 - 2x + 1} = \int \dfrac{2\,dx}{4x^2 - 4x + 2} = \int \dfrac{2\,dx}{1 + (2x-1)^2} = \tan^{-1}(2x-1) + C$

23.
$$\int \dfrac{\sqrt{a+x}}{\sqrt{a-x}}\,dx = \int \dfrac{a+x}{\sqrt{a^2 - x^2}}\,dx = \int \left(\dfrac{a}{\sqrt{a^2 - x^2}} + x(a^2 - x^2)^{-1/2} \right) dx$$
$$= a\sin^{-1}\left(\dfrac{x}{a}\right) - \sqrt{a^2 - x^2} + C$$

25. $\left\{ \begin{array}{rcl} x &=& a\sin u \\ dx &=& a\cos u\,du \end{array} \right\};$
$\qquad \displaystyle\int \dfrac{\sqrt{a^2 - x^2}}{x^2}\,dx = \int \dfrac{a\cos u}{a^2 \sin^2 u} a\cos u\,du$

$$= \int \cot^2 u\,du = \int (\csc^2 u - 1)\,du$$
$$= -\cot u - u + C$$
$$= -\left(\dfrac{1}{x}\right)\sqrt{a^2 - x^2} - \sin^{-1}\left(\dfrac{x}{a}\right) + C$$

27.
$$\int \dfrac{x}{(x+1)^2}\,dx = \int \dfrac{x+1-1}{(x+1)^2}\,dx = \int \left(\dfrac{1}{x+1} - \dfrac{1}{(x+1)^2} \right) dx$$
$$= \ln|x+1| + \dfrac{1}{x+1} + C$$

29.
$$\int \dfrac{1-\sin 2x}{1+\sin 2x}\,dx = \int \dfrac{(1-\sin 2x)^2}{1 - \sin^2 2x}\,dx = \int \left(\dfrac{1-\sin 2x}{\cos 2x} \right)^2 dx$$
$$= \int (\sec 2x - \tan 2x)^2\,dx$$
$$= \int (2\sec^2 2x - 2\sec 2x \tan 2x - 1)\,dx$$
$$= \tan 2x - \sec 2x - x + C$$

31.
$$\int x\ln(ax+b)\,dx = \dfrac{1}{2}x^2 \ln(ax+b) - \dfrac{1}{2}\int \dfrac{ax^2}{ax+b}\,dx$$

$$\boxed{\begin{array}{l} u = \ln(ax+b), \quad dv = x\,dx \\[2mm] du = \dfrac{a}{ax+b}\,dx, \quad v = \dfrac{1}{2}x^2 \end{array}}$$

$$= \dfrac{1}{2}x^2 \ln(ax+b) - \dfrac{1}{2}\int \left[x - \dfrac{b}{a} + \dfrac{b^2}{a}\left(\dfrac{1}{ax+b} \right) \right] dx$$
$$= \dfrac{x^2}{2}\ln(ax+b) - \dfrac{x^2}{4} + \dfrac{b}{2a}x - \dfrac{b^2}{2a^2}\ln(ax+b) + C$$

33.
$$\int \frac{dx}{\sqrt{x+1}-\sqrt{x}} = \int \frac{1}{\sqrt{x+1}-\sqrt{x}} \left(\frac{\sqrt{x+1}+\sqrt{x}}{\sqrt{x+1}+\sqrt{x}} \right) dx$$

$$= \int (\sqrt{x+1}+\sqrt{x}) \, dx = \tfrac{2}{3}(x+1)^{3/2} + \tfrac{2}{3}x^{3/2} + C$$

35.
$$\int \frac{x^2}{1+x^2} \, dx = \int \left(1 - \frac{1}{1+x^2} \right) dx = x - \tan^{-1} x + C$$

37.
$$\left\{ \begin{array}{l} x = \sin u \\ dx = \cos u \, du \end{array} \right\} ; \qquad \int \frac{-x^2}{\sqrt{1-x^2}} \, dx = \int \frac{-\sin^2 u}{\cos u} \cos u \, du = \int \frac{\cos 2u - 1}{2} \, du$$

$$= \frac{\sin 2u}{4} - \frac{u}{2} + C = \frac{\sin u \cos u}{2} - \frac{u}{2} + C$$

$$= \tfrac{1}{2}x\sqrt{1-x^2} - \tfrac{1}{2}\sin^{-1} x + C$$

39.
$$\int \sinh^2 x \, dx = \frac{1}{2} \int (\cosh 2x - 1) \, dx = \frac{1}{4} \sinh 2x - \frac{x}{2} + C$$

41.
$$\int e^{-x} \cosh x \, dx = \int e^{-x} \left(\frac{e^x + e^{-x}}{2} \right) dx = \frac{1}{2} \int (1 + e^{-2x}) \, dx = \frac{x}{2} - \frac{1}{4}e^{-2x} + C$$

43.

$$\boxed{ \begin{array}{ll} u = \tan^{-1}(x-3), & dv = x \, dx \\[2mm] du = \dfrac{dx}{x^2 - 6x + 10}, & v = \dfrac{1}{2}x^2 \end{array} } \qquad \int x \tan^{-1}(x-3) \, dx$$

$$= \frac{1}{2}x^2 \tan^{-1}(x-3) - \frac{1}{2} \int \frac{x^2}{x^2 - 6x + 10} \, dx$$

$$= \frac{1}{2}x^2 \tan^{-1}(x-3) - \frac{1}{2} \int \left(1 + \frac{3(2x-6)}{x^2 - 6x + 10} + \frac{8}{x^2 - 6x + 10} \right) dx$$

$$= \frac{1}{2}x^2 \tan^{-1}(x-3) - \frac{x}{2} - \frac{3}{2} \ln |x^2 - 6x + 10| - 4\tan^{-1}(x-3) + C$$

45.
$$\frac{2}{x(1+x^2)} = \frac{A}{x} + \frac{Bx+C}{1+x^2}$$

$$2 = A(1+x^2) + (Bx+C)x$$

$$\left. \begin{array}{lll} x = & 0: & 2 = A \\ x = & 1: & 2 = 2A + B + C \\ x = & -1: & 2 = 2A + B - C \end{array} \right\} \implies \begin{array}{l} A = 2 \\ B = -2 \\ C = 0 \end{array}$$

$$\int \frac{2}{x(1+x^2)} \, dx = \int \left(\frac{2}{x} - \frac{2x}{1+x^2} \right) dx = 2\ln|x| - \ln(1+x^2) + C = \ln \left(\frac{x^2}{1+x^2} \right) + C$$

47.
$$\int \frac{\cos^4 x}{\sin^2 x}\,dx = \int \frac{\cos^2 x\,(1 - \sin^2 x)}{\sin^2 x}\,dx$$

$$= \int (\cot^2 x - \cos^2 x)\,dx = \int \left(\csc^2 x - 1 - \frac{1 + \cos 2x}{2} \right)\,dx$$

$$= -\cot x - \tfrac{3}{2}x - \tfrac{1}{4}\sin 2x + C$$

49. $\left\{ \begin{aligned} x &= 2\tan u \\ dx &= 2\sec^2 u\,du \end{aligned} \right\};$

$$\int \frac{\sqrt{x^2 + 4}}{x}\,dx = \int \frac{2\sec u}{2\tan u}\,2\sec^2 u\,du$$

$$= 2\int \frac{\sec u}{\tan u}\,(1 + \tan^2 u)\,du$$

$$= 2\int (\csc u + \sec u \tan u)\,du$$

$$= 2\ln|\csc u - \cot u| + 2\sec u + C$$

$$= 2\ln\left| \frac{\sqrt{x^2 + 4} - 2}{x} \right| + \sqrt{x^2 + 4} + C$$

$$= 2\ln(\sqrt{x^2 + 4} - 2) - 2\ln x + \sqrt{x^2 + 4} + C$$

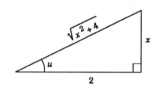

51. $\left\{ \begin{aligned} x &= 3\sin u \\ dx &= 3\cos u\,du \end{aligned} \right\};$

$$\int \frac{dx}{x\sqrt{9 - x^2}} = \int \frac{3\cos u\,du}{9\sin u \cos u} = \frac{1}{3}\int \csc u\,du$$

$$= \tfrac{1}{3}\ln|\csc u - \cot u| + C$$

$$= \frac{1}{3}\ln\left| \frac{3 - \sqrt{9 - x^2}}{x} \right| + C$$

$$= \tfrac{1}{3}\ln(3 - \sqrt{9 - x^2}) - \tfrac{1}{3}\ln|x| + C$$

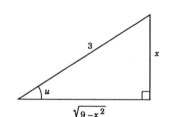

53. $\left\{ \begin{aligned} u &= \ln(1 - \sqrt{x}), & dv &= dx \\ du &= -\frac{dx}{2\sqrt{x}\,\ln(1 - \sqrt{x})}, & v &= x \end{aligned} \right\};$ $\quad \int \ln(1 - \sqrt{x})\,dx = uv - \int v\,du$

$$= x\ln(1 - \sqrt{x}) + \frac{1}{2}\int \frac{\sqrt{x}}{1 - \sqrt{x}}\,dx$$

$$\frac{1}{2}\int \frac{\sqrt{x}}{1 - \sqrt{x}}\,dx = \int \frac{u^2}{1 - u}\,du = \int \left(-u - 1 + \frac{1}{1 - u} \right)\,du$$

$\boxed{u^2 = x,\ 2u\,du = dx}$

$$= -\tfrac{1}{2}u^2 - u - \ln(1 - u) + C = -\tfrac{1}{2}x - \sqrt{x} - \ln(1 - \sqrt{x}) + C$$

$$\int \ln(1 - \sqrt{x})\,dx = (x - 1)\ln(1 - \sqrt{x}) - \frac{1}{2}x - \sqrt{x} + C$$

55. $\left\{ \begin{array}{l} e^x = \sin u \\ e^x\, dx = \cos u\, du \end{array} \right\}$;

$$\int \frac{dx}{\sqrt{1-e^{2x}}} = \int \frac{1}{\cos u}\, \frac{\cos u\, du}{\sin u}$$

$$= \int \csc u\, du$$

$$= \ln |\csc u - \cot u| + C$$

$$= \ln \left| \frac{1 - \sqrt{1-e^{2x}}}{e^x} \right| + C$$

$$= \ln \left| 1 - \sqrt{1-e^{2x}} \right| - x + C$$

57.
$$\int \frac{\sec^3 x}{\tan x}\, dx = \int \frac{\sec x}{\tan x}(1 + \tan^2 x)\, dx = \int (\csc x + \sec x \tan x)\, dx$$

$$= \ln |\csc x - \cot x| + \sec + C$$

59.
$$\int \frac{\sin 4x}{\sin x}\, dx = \int \frac{2 \sin 2x \cos 2x}{\sin x}\, dx = \int \frac{2(2 \sin x \cos x)(1 - 2 \sin^2 x)}{\sin x}\, dx$$

$$= \int 4(1 - 2 \sin^2 x) \cos x\, dx = 4 \sin x - \frac{8}{3} \sin^3 x + C$$

61.
$$\int \sin^5 \left(\frac{x}{2}\right) dx = \int \left[1 - \cos^2 \left(\frac{x}{2}\right)\right]^2 \sin \left(\frac{x}{2}\right) dx$$

$$= \int \left[1 - 2 \cos^2 \left(\frac{x}{2}\right) + \cos^4 \left(\frac{x}{2}\right)\right] \sin \left(\frac{x}{2}\right) dx$$

$$= -2 \cos \left(\frac{x}{2}\right) + \frac{4}{3} \cos^3 \left(\frac{x}{2}\right) - \frac{2}{5} \cos^5 \left(\frac{x}{2}\right) + C$$

63. $\int \cot^2 x \sec x\, dx = \int \cot x \csc x\, dx = -\csc x + C$

65. $\int \frac{\sin x}{\sin 2x}\, dx = \int \frac{\sin x}{2 \sin x \cos x}\, dx = \frac{1}{2} \int \sec x\, dx = \frac{1}{2} \ln |\sec x + \tan x| + C$

67. $\int x^2 \sin^{-1} x\, dx = \frac{1}{3} x^3 \sin^{-1} x - \frac{1}{3} \int \frac{x^3}{\sqrt{1-x^2}}\, dx$

$$\boxed{\begin{array}{ll} u = \sin^{-1} x, & dv = x^2\, dx \\[2mm] du = \dfrac{dx}{\sqrt{1-x^2}}, & v = \dfrac{1}{3} x^3 \end{array}}$$

$$= \tfrac{1}{3}x^3\sin^{-1}x + \tfrac{1}{3}(1-x^2)^{1/2} - \tfrac{1}{9}(1-x^2)^{3/2} + C$$

$$\left\{\begin{array}{l} x = \sin t \\ dx = \cos t\,dt \end{array}\right\};\qquad \int \frac{x^3}{\sqrt{1-x^2}}\,dx = \int \sin^3 t\,dt$$

$$= \int (1 - \cos^2 t)\sin t\,dt$$

$$= -\cos t + \tfrac{1}{3}\cos^3 t + C$$

$$= -(1-x^2)^{1/2} + \tfrac{1}{3}(1-x^2)^{3/2} + C$$

69. $\left\{\begin{array}{l} x+1 = 3\sec u \\ dx = 3\sec u\tan u\,du \end{array}\right\};\quad \int x\sqrt{x^2+2x-8}\,dx = \int x\sqrt{(x+1)^2-9}\,dx$

$$= \int (3\sec u - 1)(3\tan u)3\sec u\tan u\,du$$

$$= \int (27\tan^2 u\sec^2 u - 9\tan^2 u\sec u)\,du$$

$$= 9\tan^3 u - 9\int \tan^2 u\sec u\,du$$

⌐ (by parts)

$$\overset{\downarrow}{=} 9\tan^3 u - 9\left(\tfrac{1}{2}\sec u\tan u - \tfrac{1}{2}\ln|\sec u + \tan u|\right) + C$$

$$= \tfrac{1}{3}(x^2+2x-8)^{3/2} - \tfrac{1}{2}(x+1)\sqrt{x^2+2x-8} + \tfrac{9}{2}\ln\left|\frac{x+1+\sqrt{x^2+2x-8}}{3}\right| + C$$

$$\overset{\uparrow}{=} \tfrac{1}{3}(x^2+2x-8)^{3/2} - \tfrac{1}{2}(x+1)\sqrt{x^2+2x-8} + \tfrac{9}{2}\ln\left|x+1+\sqrt{x^2+2x-8}\right| + C$$

⌐ (absorb $-\tfrac{9}{2}\ln 3$ in C)

71. $\displaystyle\int (\sin^2 x\cos x)^2\,dx = \int (\sin^4 x - 2\sin^2 x\cos x + \cos^2 x)\,dx$

$$= \int\left[\left(\frac{1-\cos 2x}{2}\right)^2 - 2\sin^2 x\cos x + \frac{1+\cos 2x}{2}\right]dx$$

$$= \int\left[\frac{1}{4}\left(1 - 2\cos 2x + \frac{1+\cos 4x}{2}\right) - 2\sin^2 x\cos x + \frac{1}{2} + \frac{1}{2}\cos 2x\right]dx$$

$$= \int\left(\frac{7}{8} + \frac{1}{8}\cos 4x - 2\sin^2 x\cos x\right)dx$$

$$= \tfrac{7}{8}x + \tfrac{1}{32}\sin 4x - \tfrac{2}{3}\sin^3 x + C$$

73. $\left\{\begin{array}{l} 2x + \frac{3}{4} = \left(\frac{1}{4}\sqrt{41}\right)u \\ 2\,dx = \frac{1}{4}\sqrt{41}\,du \end{array}\right\}$;

$$\int \frac{3}{\sqrt{2 - 3x - 4x^2}}\,dx = \int \frac{3}{\sqrt{\frac{41}{16} - \left(2x + \frac{3}{4}\right)^2}}\,dx$$

$$= \frac{3}{2}\int \frac{du}{\sqrt{1 - u^2}} = \frac{3}{2}\sin^{-1}u + C$$

$$= \frac{3}{2}\sin^{-1}\left(\frac{8x + 3}{\sqrt{41}}\right) + C$$

CHAPTER 9

SECTION 9.2

1. use Theorem 9.2.2   (a) $\frac{2}{13}$ (b) $\frac{29}{13}$

3. $d((0,1),l) = 1/\sqrt{113}$, $d((1,0),l) = 2/\sqrt{113}$, $d((-1,1),l) = 7/\sqrt{113}$

    The closest point is $(0,1)$;   the point $(-1,1)$ is farthest away.

5. We select the side joining $A(1,-2)$ and $B(-1,3)$ as the base of the triangle.

    length of side AB : $\sqrt{29}$

    equation of line through A and B : $5x + 2y - 1 = 0$

    length of altitude from vertex $C(2,4)$ to side AB : $\dfrac{|5(2) + 2(4) - 1|}{\sqrt{29}} = \dfrac{17}{\sqrt{29}}$

    area of triangle: $\dfrac{1}{2}\left(\sqrt{29}\right)\left(\dfrac{17}{\sqrt{29}}\right) = \dfrac{17}{2}$

7. Adjust the sign of A and B so that the equation reads

    $$Ax + By - |C| = 0.$$

    Then we have

    $$Ax + By = |C| \quad \text{and thus} \quad x\frac{A}{\sqrt{A^2 + B^2}} + y\frac{B}{\sqrt{A^2 + B^2}} = \frac{|C|}{\sqrt{A^2 + B^2}}.$$

    Now set

    $$\frac{|C|}{\sqrt{A^2 + B^2}} = p, \qquad \frac{A}{\sqrt{A^2 + B^2}} = \cos\alpha, \qquad \frac{B}{\sqrt{A^2 + B^2}} = \sin\alpha.$$

    p is the length of $\overline{OQ}$, the distance between the line and the origin; α is the angle from the positive x-axis to the line segment $\overline{OQ}$.

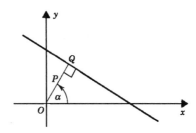

9. (a) $4(x + 1) + 5(y - 2) + 3 = 0$ and thus $4x + 5y - 3 = 0.$

    (b) $4(x - 1) + 5(y - 2) + 3 = 0$ and thus $4x + 5y - 11 = 0.$

    (c) $4(x + 1) + 5(y + 2) + 3 = 0$ and thus $4x + 5y + 17 = 0.$

    (d) $4(x - 1) + 5(y + 2) + 3 = 0$ and thus $4x + 5y + 9 = 0$.

11. (a) $(x - 3)^2 = (y - 4)^3$ (b) $(x + 3)^2 = (y - 4)^3$

     (c) $(x - 4)^2 = (y + 3)^3$ (d) $(x + 4)^2 = (y + 3)^3$

13. At and near the time under consideration: $0 < \theta < \dfrac{\pi}{2}$,

$l: y = x\tan\theta + b^2\tan\theta.$

$$D = \frac{b^2\tan\theta}{\sqrt{\tan^2\theta + 1}} = \frac{b^2\tan\theta}{\sec\theta} = b^2\sin\theta,$$

$$\frac{dD}{dt} = b^2\cos\theta\,\frac{d\theta}{dt}.$$

When $\quad m = \dfrac{3}{4},\quad \dfrac{dD}{dt} = b^2\left(\dfrac{4}{5}\right)(-2\pi) = \dfrac{-8b^2\pi}{5}.$

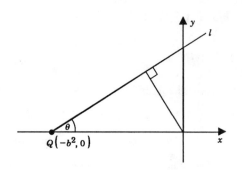

The distance is decreasing at the rate of $\dfrac{8b^2\pi}{5}$ units/min.

SECTION 9.3

1. $y^2 = 8x$

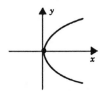

3. $(x+1)^2 = -12(y-3)$

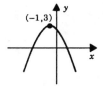

5. $(x-1)^2 = 4y$

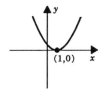

7. $(y-1)^2 = -2\left(x - \frac{3}{2}\right)$

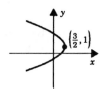

9. $y^2 = 2x$

vertex $(0,0)$

focus $\left(\frac{1}{2}, 0\right)$

axis $y = 0$

directrix $x = -\frac{1}{2}$

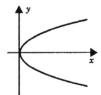

11. $x^2 = \frac{1}{2}\left(y + \frac{1}{2}\right)$

vertex $\left(0, -\frac{1}{2}\right)$

focus $\left(0, -\frac{3}{8}\right)$

axis $x = 0$

directrix $y = -\frac{5}{8}$

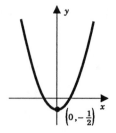

13. $(x+2)^2 = -8(y - \frac{3}{2})$

vertex $(-2, \frac{3}{2})$

focus $(-2, -\frac{1}{2})$

axis $x = -2$

directrix $y = \frac{7}{2}$

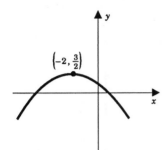

15. $(y + \frac{1}{2})^2 = x - \frac{3}{4}$

vertex $(\frac{3}{4}, -\frac{1}{2})$

focus $(1, -\frac{1}{2})$

axis $y = -\frac{1}{2}$

directrix $x = \frac{1}{2}$

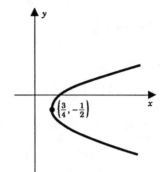

17. $\sqrt{(x-1)^2 + (y-2)^2} = \dfrac{|x+y+1|}{\sqrt{2}}$ simplifies to $(x-y)^2 = 6x + 10y - 9$

19.   Directrix has equation $x - y - 6 = 0$ since it has slope 1 and passes through the point $(4, -2)$.

$$\sqrt{x^2 + (y-2)^2} = \dfrac{|x - y - 6|}{\sqrt{2}}$$

This simplifies to $(x+y)^2 = -12x + 20y + 28$.

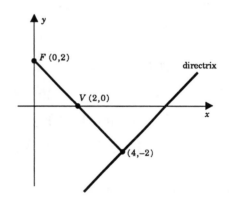

21. $P(x, y)$ is on the parabola with directrix l: $Ax + By + C = 0$ and focus $F(a, b)$ iff $d(P, l) = d(P, F)$

which happens iff $\dfrac{|Ax + By + C|}{\sqrt{A^2 + B^2}} = \sqrt{(x-a)^2 + (y-b)^2}.$

Squaring both sides of this equation and simplifying, we obtain

$$(Ay - Bx)^2 = (2aS + 2AC)x + (2bS + 2BC)y + c^2 - (a^2 + b^2)S$$

with $S = A^2 + B^2 \neq 0$.

23. We can choose the coordinate system so that the
     parabola has an equation of the form $y = \alpha x^2, \alpha >$
     0. One of the points of intersection is then the origin
     and the other is of the form $(c, \alpha c^2)$. We will assume
     that $c > 0$.

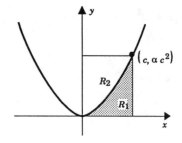

$$\text{area of } R_1 = \int_0^c \alpha x^2 dx = \frac{1}{3}\alpha c^3 = \frac{1}{3}A,$$
$$\text{area of } R_2 = A - \frac{1}{3}A = \frac{2}{3}A.$$

25. There are two possible positions for the focus:

$$(2, 2) \text{ and } (2, 10).$$

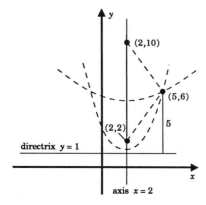

[The point $(5, 6)$ is equidistant from the focus and
the directrix. This distance is 5. The points on the
line $x = 2$ which are 5 units from $(5, 6)$ are $(2, 2)$ and
$(2, 10)$.] These in turn give rise to two parabolas.

Focus $(2, 2)$, vertex $(2, 3/2)$:

$$(x - 2)^2 = 4(\tfrac{1}{2})(y - \tfrac{3}{2}), \quad \text{which simplifies to} \quad x^2 - 4x + 7 = 2y.$$

Focus $(2, 10)$, vertex $(2, 11/2)$:

$$(x - 2)^2 = 4(\tfrac{9}{2})(y - \tfrac{11}{2}), \quad \text{which simplifies to} \quad x^2 - 4x + 103 = 18y.$$

27. In this case the length of the latus rectum is the width of the parabola at height $y = c$. With
     $y = c$, $4c^2 = x^2$, and $x = \pm 2c$. The length of the latus rectum is thus $4c$.

29.
$$A = \int_{-2c}^{2c} \left(c - \frac{x^2}{4c}\right) dx = 2\int_0^{2c} \left(c - \frac{x^2}{4c}\right) dx = 2\left[cx - \frac{x^3}{12c}\right]_0^{2c} = \frac{8}{3}c^2$$

$\bar{x} = 0$ by symmetry

$$\bar{y}A = \int_{-2c}^{2c} \frac{1}{2}\left(c^2 - \frac{x^4}{16c^2}\right) dx = \int_0^{2c} \left(c^2 - \frac{x^4}{16c^2}\right) dx = \left[c^2 x - \frac{x^5}{80c^2}\right]_0^{2c} = \frac{8}{5}c^3$$

$$\bar{y} = (\tfrac{8}{5}c^3)/(\tfrac{8}{3}c^2) = \frac{3}{5}c$$

31. Equation (9.3.4) with $x_0 = 0$, $y_0 = 0$, $g = 32$: $\quad y = -\dfrac{16}{v_0^2}(\sec^2 \theta)x^2 + (\tan \theta)x.$

33. $y = 0$ (and $x \neq 0$) $\implies$ $\frac{1}{16}v_0^2 \cos\theta \sin\theta$

35. The range $\frac{1}{16}v_0^2 \sin\theta \cos\theta = \frac{1}{32}v_0^2 \sin 2\theta$ is clearly maximal when $\theta = \frac{1}{4}\pi$ for then $\sin 2\theta = 1$.

37. $\dfrac{kx}{p(0)} = \tan \theta = \dfrac{dy}{dx}, \quad y = \dfrac{k}{2\,p(0)}x^2 + C$

In our figure $C = y(0) = 0$. Thus the equation of the cable is $y = kx^2/2p(0)$, the equation of a parabola.

39. Start with any two parabolas γ_1, γ_2. By moving them we can see to it that they have equations of the following form:

$$\gamma_1: x^2 = 4c_1 y, \quad c_1 > 0; \qquad \gamma_2: x^2 = 4c_2 y, \quad c_2 > 0.$$

Now we change the scale for γ_2 so that the equation for γ_2 will look exactly like the equation for γ_1. Set $X = (c_1/c_2)\,x, \quad Y = (c_1/c_2)\,y.$ Then

$$x^2 = 4c_2 y \quad \Longrightarrow \quad (c_2/c_1)^2 X^2 = 4c_2\,(c_2/c_1)\,Y \quad \Longrightarrow \quad X^2 = 4c_1 Y.$$

Now γ_2 has exactly the same equation as γ_1; only the scale, the units by which we measure distance, has changed.

SECTION 9.4

1. $\dfrac{x^2}{9} + \dfrac{y^2}{4} = 1$

center $(0,0)$

foci $(\pm\sqrt{5},\,0)$

length of major axis 6

length of minor axis 4

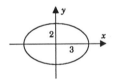

3. $\dfrac{x^2}{4} + \dfrac{y^2}{6} = 1$

center $(0,0)$

foci $(0,\,\pm\sqrt{2})$

length of major axis $2\sqrt{6}$

length of minor axis 4

5. $\dfrac{x^2}{9} + \dfrac{(y-1)^2}{4} = 1$

center $(0,1)$

foci $(\pm\sqrt{5},\,1)$

length of major axis 6

length of minor axis 4

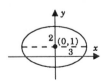

7. $\dfrac{(x-1)^2}{16} + \dfrac{y^2}{64} = 1$

center $(1,0)$

foci $(1,\,\pm4\sqrt{3})$

length of major axis 16

length of minor axis 8

9. Foci $(-1, 0), (1, 0)$ $\implies$ center $(0, 0)$, $c = 1$, and major axis parallel to x-axis.
   Major axis 6 $\implies$ $a = 3$. Thus, $b = \sqrt{8}$.

   Equation: $\dfrac{x^2}{9} + \dfrac{y^2}{8} = 1$.

11. Foci at $(1, 3)$ and $(1, 9)$ $\implies$ center $(1, 6)$, $c = 3$, and major axis parallel to y-axis.
    Minor axis 8 $\implies$ $b = 4$. Thus, $a = 5$.

    Equation: $\dfrac{(x-1)^2}{16} + \dfrac{(y-6)^2}{25} = 1$.

13. Focus $(1, 1)$ and center $(1, 3)$ $\implies$ $c = 2$ and major axis parallel to y-axis.
    Major axis 10 $\implies$ $a = 5$. Thus, $b = \sqrt{21}$.

    Equation: $\dfrac{(x-1)^2}{21} + \dfrac{(y-3)^2}{25} = 1$.

15. Major axis 10 $\implies$ $a = 5$. Vertices at $(3, 2)$ and $(3, -4)$ are then on minor axis parallel to y-axis.
    Then, $b = 3$ and center is $(3, -1)$.

    Equation: $\dfrac{(x-3)^2}{25} + \dfrac{(y+1)^2}{9} = 1$.

17. $d(F_1, F_2) + k = 2(c + a)$ 19. $2\sqrt{\pi^2 a^4 - A^2}/\pi a$

21. The equation of the ellipse is of the form
$$\frac{(x-5)^2}{25} + \frac{y^2}{25 - c^2} = 1.$$
    Substitute $x = 3$ and $y = 4$ in that equation and you find that $c = \pm\frac{5}{21}\sqrt{5}$.
    The foci are at $\left(5 \pm \frac{5}{21}\sqrt{5}, 0\right)$.

23. $e = \frac{3}{5}$ 25. $e = \frac{4}{5}$

27. E_1 is fatter than E_2, more like a circle.

29. The ellipse tends to a line segment of length $2a$.

31. $x^2/9 + y^2 = 1$

SECTION 9.5

1. Foci $(-5, 0)$ and $(5, 0)$ $\implies$ $c = 5$ and center $(0, 0)$.
   Transverse axis 6 $\implies$ $a = 3$. Thus, $b = 4$.

   Equation: $\dfrac{x^2}{9} - \dfrac{y^2}{16} = 1$.

3. Foci $(0, -13)$ and $(0, 13)$ $\implies$ $c = 13$ and center $(0, 0)$.
   Transverse axis 10 $\implies$ $a = 5$. Thus, $b = 12$.

   Equation: $\dfrac{y^2}{25} - \dfrac{x^2}{144} = 1$.

5. Foci $(-5, 1)$ and $(5, 1)$ $\implies$ $c = 5$ and center $(0, 1)$.

Transverse axis 6 $\implies$ $a = 3$. Thus, $b = 4$.

Equation: $\dfrac{x^2}{9} - \dfrac{(y-1)^2}{16} = 1$.

7. Foci $(-1, -1)$ and $(-1, 1)$ $\implies$ $c = 1$ and center $(-1, 0)$.

Transverse axis $\frac{1}{2}$ $\implies$ $a = \frac{1}{4}$. Thus, $b = \frac{1}{4}\sqrt{15}$.

Equation: $\dfrac{y^2}{1/16} - \dfrac{(x+1)^2}{15/16} = 1$.

9. $x^2 - y^2 = 1$

   center $(0, 0)$
   transverse axis 2
   vertices $(\pm 1, 0)$
   foci $(\pm\sqrt{2}, 0)$
   asymptotes $y = \pm x$

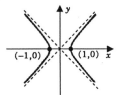

11. $\dfrac{x^2}{9} - \dfrac{y^2}{16} = 1$

   center $(0, 0)$
   transverse axis 6
   vertices $(\pm 3, 0)$
   foci $(\pm 5, 0)$
   asymptotes $y = \pm\frac{4}{3}x$

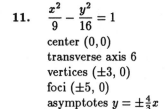

13. $\dfrac{y^2}{16} - \dfrac{x^2}{9} = 1$

   center $(0, 0)$
   transverse axis 8
   vertices $(0, \pm 4)$
   foci $(0, \pm 5)$
   asymptotes $y = \pm\frac{4}{3}x$

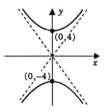

15. $\dfrac{(x-1)^2}{9} - \dfrac{(y-3)^2}{16} = 1$

   center $(1, 3)$
   transverse axis 6
   vertices $(4, 3)$ and $(-2, 3)$
   foci $(6, 3)$ and $(-4, 3)$
   asymptotes $y - 3 = \pm\frac{4}{3}(x - 1)$

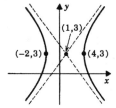

17. $\dfrac{(y-3)^2}{4} - \dfrac{(x-1)^2}{1} = 1$

   center $(1, 3)$
   transverse axis 4
   vertices $(1, 5)$ and $(1, 1)$
   foci $(1, 3 \pm \sqrt{5})$
   asymptotes $y - 3 = \pm 2(x - 1)$

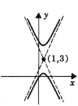

19. By the hint, $xy = X^2 - Y^2 = 1$. In the XY-system $a = 1$, $b = 1$, $c = \sqrt{2}$. We have center $(0,0)$, vertices $(\pm 1, 0)$, foci $(\pm\sqrt{2}, 0)$ and asymptotes $Y = \pm X$. Using

$$x = X + Y \quad \text{and} \quad y = X - Y$$

to convert to the xy-system, we find center $(0, 0)$, vertices $(1, 1)$ and $(-1, -1)$, foci $(\sqrt{2}, \sqrt{2})$ and $(-\sqrt{2}, -\sqrt{2})$, asymptotes $y = 0$ and $x = 0$, tranverse axis $2\sqrt{2}$.

21.
$$A = \frac{2b}{a} \int_a^{2a} \sqrt{x^2 - a^2}\, dx = \frac{2b}{a} \left[\frac{x}{2}\sqrt{x^2 - a^2} - \frac{a^2}{2} \ln\left(x + \sqrt{x^2 - a^2} \right) \right]_a^{2a}$$
$$= [2\sqrt{3} - \ln(2 + \sqrt{3})]ab$$

23. $e = \frac{5}{3}$ **25.** $e = \sqrt{2}$

27. The branches of H_1 open up less quickly than the branches of H_2.

29. The hyperbola tends to a pair of parallel lines separated by the tranverse axis.

31. Measure distances in miles and time in seconds. Place the origin at A and let $P(x, y)$ be the site of the crash. Then

$$d(P, B) - d(P, A) = (4)(0.20) = 0.80.$$

This places P on the right branch of the hyperbola
$$\frac{(x+1)^2}{(0.4)^2} - \frac{y^2}{1 - (0.4)^2} = 1.$$

Also

$$d(P, C) - d(P, A) = 6(0.20) = 1.20.$$

This places P on the left branch of the hyperbola
$$\frac{(x-1)^2}{(0.6)^2} - \frac{y^2}{1 - (0.6)^2} = 1.$$

Solve the two equations simultaneously keeping in mind the conditions of the problem and you will find that $x \cong -0.248$ and $y \cong 1.459$. The impact takes place about a quarter of a mile west of A and one and a half miles north.

SECTION * 9.6

1. $\alpha = \frac{1}{4}\pi$, $\frac{1}{2}(X^2 - Y^2) = 1$ **3.** $\alpha = \frac{1}{6}\pi$, $4X^2 - Y^2 = 1$

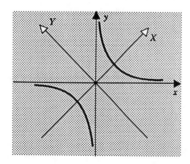   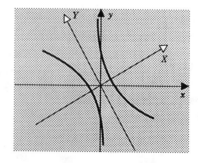

5. $\alpha = \frac{1}{4}\pi,\quad 2Y^2 + \sqrt{2}X = 0$

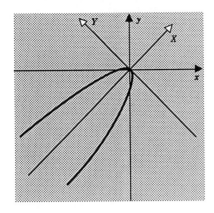

7. $\alpha = -\frac{1}{6}\pi,\quad Y^2 + X = 0$

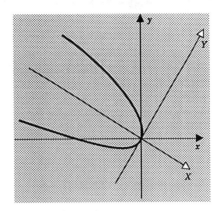

9. $\alpha = \frac{1}{8}\pi,\quad \cos\alpha = \frac{1}{2}\sqrt{2+\sqrt{2}},\ \ \sin\alpha = \frac{1}{2}\sqrt{2-\sqrt{2}}$

SECTION 9.7

1.
$$x^2 - 4y - 4 = 0$$
$$x^2 = 4(y+1)$$

parabola with vertex at $(0,-1)$ and focus at the origin

3.
$$x^2 - 4y^2 - 10x + 41 = 0$$
$$x^2 - 10x + 25 - 4y^2 = -41 + 25$$
$$(x-5)^2 - 4y^2 = -16$$
$$\frac{y^2}{4} - \frac{(x-5)^2}{16} = 1$$

hyperbola with foci at $(5, \pm 2\sqrt{5})$ and tranverse axis 4

5.
$$x^2 + 3y^2 + 6x + 8 = 0$$
$$x^2 + 6x + 9 + 3y^2 = -8 + 9$$
$$(x+3)^2 + \frac{y^2}{1/3} = 1$$

ellipse with foci at $\left(-3 \pm \frac{1}{3}\sqrt{6}, 0\right)$ and major axis 2

7.
$$y^2 + 4y + 2x + 1 = 0$$

$$y^2 + 4y + 4 = -2x - 1 + 4$$

$$(y+2)^2 = -2(x - \tfrac{3}{2})$$

parabola with vertex at $\left(\tfrac{3}{2}, -2\right)$ and focus at $(1, -2)$

9.
$$9x^2 + 25y^2 + 100y + 99 = 0$$

$$9x^2 + 25(y^2 + 4y + 4) = -99 + 100$$

$$\frac{x^2}{1/9} + \frac{(y+2)^2}{1/25} = 1$$

ellipse with foci at $\left(\pm\tfrac{4}{15}, -2\right)$ and major axis $\tfrac{2}{3}$

11.
$$7x^2 - 5y^2 + 14x - 40y = 118$$

$$7(x^2 + 2x + 1) - 5(y^2 + 8y + 16) = 118 + 7 - 80$$

$$\frac{(x+1)^2}{45/7} - \frac{(y+4)^2}{9} = 1$$

hyperbola with foci at $\left(-1 \pm \tfrac{6}{7}\sqrt{21}, -4\right)$ and tranverse axis $\tfrac{6}{7}\sqrt{35}$

13. $(x^2 - 4y)(4x^2 + 9y^2 - 36) = 0$

the union of the parabola $x^2 = 4y$ and the ellipse $\dfrac{x^2}{9} + \dfrac{y^2}{4} = 1$

CHAPTER 10

SECTION 10.1

1–7.

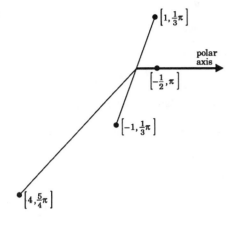

$\left[1, \frac{1}{3}\pi\right]$

polar axis

$\left[-\frac{1}{2}, \pi\right]$

$\left[-1, \frac{1}{3}\pi\right]$

$\left[4, \frac{5}{4}\pi\right]$

9. $x = 3\cos\frac{1}{2}\pi = 0$

$y = 3\sin\frac{1}{2}\pi = 3$

11. $x = -\cos(-\pi) = 1$

$y = -\sin(-\pi) = 0$

13. $x = -3\cos\left(-\frac{1}{3}\pi\right) = -\frac{3}{2}$

$y = -3\sin\left(-\frac{1}{3}\pi\right) = \frac{3}{2}\sqrt{3}$

15. $x = 3\cos\left(-\frac{1}{2}\pi\right) = 0$

$y = 3\sin\left(-\frac{1}{2}\pi\right) = -3$

17. $r^2 = 0^2 + 1^2, \quad r = \pm 1$

$r = 1: \quad \cos\theta = 0 \text{ and } \sin\theta = 1$
$\theta = \frac{1}{2}\pi$ $\Big\}$ $\left[1, \frac{1}{2}\pi + 2n\pi\right], \quad \left[-1, \frac{3}{2}\pi + 2n\pi\right]$

19. $r^2 = (-3)^2 + 0^2 = 9, \quad r = \pm 3$

$r = 3: \quad \cos\theta = -1 \text{ and } \sin\theta = 0$
$\theta = \pi$ $\Big\}$ $[3, \pi + 2n\pi], \quad [3, 2n\pi]$

21. $r^2 = 2^2 + (-2)^2 = 8, \quad r = \pm 2\sqrt{2}$

$r + 2\sqrt{2}: \quad \cos\theta = \frac{1}{2}\sqrt{2}, \quad \sin\theta = -\frac{1}{2}\sqrt{2}$
$\theta = \frac{7}{4}\pi$ $\Big\}$ $\left[2\sqrt{2}, \frac{7}{4}\pi + 2n\pi\right], \quad \left[-2\sqrt{2}, \frac{3}{4}\pi + 2n\pi\right]$

23. $r^2 = \left(4\sqrt{3}\right)^2 + 4^2 = 64, \quad r \pm 8$

$r = 8: \quad \cos\theta = \frac{1}{2}\sqrt{3}, \quad \sin\theta = \frac{1}{2}$
$r = \frac{1}{6}\pi$ $\Big\}$ $\left[8, \frac{1}{6}\pi + 2n\pi\right], \quad \left[-8, \frac{7}{6}\pi + 2n\pi\right]$

25.
$$d^2 = (x_1 - x_2)^2 + (y_1 - y_2)^2 = (r_1\cos\theta_1 - r_2\cos\theta_2)^2 + (r_1\sin\theta_1 - r_2\sin\theta_2)^2$$

$$= r_1{}^2\cos^2\theta_1 - 2r_1r_2\cos\theta_1\cos\theta_2 + r_2{}^2\cos^2\theta_2$$

$$+ r_1{}^2\sin^2\theta_1 - 2r_1r_2\sin\theta_1\sin\theta_2 + r_2{}^2\sin^2\theta_2$$

$$= r_1{}^2 + r_2{}^2 - 2r_1r_2\left(\cos\theta_1\cos\theta_2 + \sin\theta_1\sin\theta_2\right)$$

$$= r_1{}^2 + r_2{}^2 - 2r_1r_2\cos\left(\theta_1 - \theta_2\right)$$

$$d = \sqrt{r_1{}^2 + r_2{}^2 - 2r_1r_2\cos\left(\theta_1 - \theta_2\right)}$$

27. (a) $\left[\frac{1}{2}, \frac{11}{6}\pi\right]$ (b) $\left[\frac{1}{2}, \frac{5}{6}\pi\right]$ (c) $\left[\frac{1}{2}, \frac{7}{6}\pi\right]$

29. (a) $\left[2, \frac{2}{3}\pi\right]$ (b) $\left[2, \frac{5}{3}\pi\right]$ (c) $\left[2, \frac{1}{3}\pi\right]$

31. about the x-axis?: $r = 2 + \cos(-\theta) \implies r = 2 + \cos\theta,$ yes.

about the y-axis?: $r = 2 + \cos(\pi - \theta) \implies r = 2 - \cos\theta,$ no.

about the origin?: $r = 2 + \cos(\pi + \theta) \implies r = 2 - \cos\theta,$ no.

33. about the x-axis?: $r(\sin(-\theta) + \cos(-\theta)) = 1 \implies r(-\sin\theta + \cos\theta) = 1,$ no.

about the y-axis?: $r(\sin(\pi - \theta) + \cos(\pi - \theta)) = 1 \implies r(\sin\theta - \cos\theta) = 1,$ no.

about the origin?: $r(\sin(\pi + \theta) + \cos(\pi + \theta)) = 1 \implies r(-\sin\theta - \cos\theta) = 1,$ no.

35. about the x-axis?: $r^2 \sin(-2\theta) = 1 \implies -r^2 \sin 2\theta = 1,$ no.

about the y-axis?: $r^2 \sin(2(\pi - \theta)) = 1 \implies -r^2 \sin 2\theta = 1,$ no.

about the origin?: $r^2 \sin(2(\pi + \theta)) = 1 \implies r^2 \sin 2\theta = 1,$ yes.

37. $$x = 2$$
$$r\cos\theta = 2$$

39. $$2xy = 1$$
$$2(r\cos\theta)(r\sin\theta) = 1$$
$$r^2 \sin 2\theta = 1$$

41. $$x^2 + (y - 2)^2 = 4$$
$$x^2 + y^2 - 4y = 0$$
$$r^2 - 4r\sin\theta = 0$$
$$r = 4\sin\theta$$

[note: division by r okay since $[0, 0,]$ on curve]

43. $$y = x$$
$$r\sin\theta = r\cos\theta$$
$$\tan\theta = 1$$
$$\theta = \pi/4$$

45. $$x^2 + y^2 + x = \sqrt{x^2 + y^2}$$
$$r^2 + r\cos\theta = r$$
$$r = 1 - \cos\theta$$

47. $$(x^2 + y^2)^2 = 2xy$$
$$r^4 = 2(r\cos\theta)(r\sin\theta)$$
$$r^2 = \sin 2\theta$$

49. the horizontal line $y = 4$

51. the line $y = \sqrt{3}x$

53. $$r = 2(1 - \cos\theta)^{-1}$$
$$r - r\cos\theta = 2$$
$$\sqrt{x^2 + y^2} - x = 2$$
$$x^2 + y^2 = (x + 2)^2$$
$$y^2 = 4(x + 1)$$

a parabola

55. $$r = 3\cos\theta$$
$$r^2 = 3r\cos\theta$$
$$x^2 + y^2 = 3x$$

a circle

57.   the line $y = 2x$ **59.**   the vertical line $x = 0$

SECTION 10.2

1.

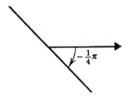

3.

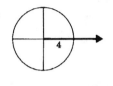

5.

7.

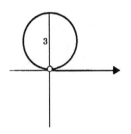

9.

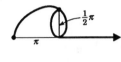

11.

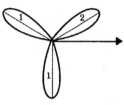

13.

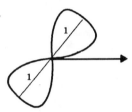

15.

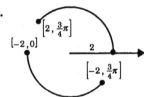

17.

19.

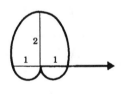

21.

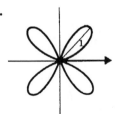

23.

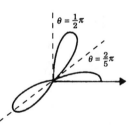

25.

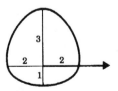

27.

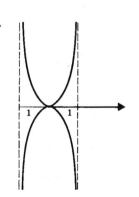

29.

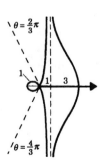

31.

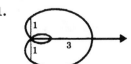

SECTION * 10.3

1. (a) $\frac{1}{2}$ unit to the right of the pole

(b) 2 units

(c)

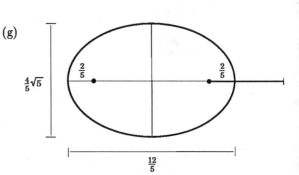

3. the parabola of Exercise 1 rotated by π radians

5. (a) $e = \frac{2}{3}$ (b) 2 units to the left of the pole and $\frac{2}{5}$ units to the right of the pole

(c) $\frac{4}{5}$ units to the left of the pole

(d) $\frac{8}{5}$ units to the left of the pole

(e) $\frac{4}{5}\sqrt{5}$ units (about 1.79 units)

(f) $\frac{4}{3}$ units

(g)

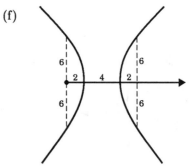

7. the ellipse of Exercise 5 rotated by $\frac{1}{2}\pi$ radians

9. (a) $e = 2$ (b) 2 units to the right of the pole and 6 units to the right of the pole

(c) 4 units to the right of the pole

(d) 8 units to the right of the pole   (e) 12 units

(f)

11. the hyperbola of Exercise 9 rotated by $\frac{3}{2}\pi$ radians

13. ellipse: $\dfrac{x^2}{48} + \dfrac{(y+4)^2}{64} = 1$

15. ellipse: $x^2/9 + (y-4)^2/25 = 1$

SECTION 10.4

1. yes; $[1, \pi] = [-1, 0]$ and the pair $r = -1,\ \theta = 0$ satisfies the equation

3. yes;   the pair $r = \frac{1}{2},\ \theta = \frac{1}{2}\pi$ satisfies the equation

5. $[2, \pi] = [-2, 0]$. The coordinates of $[-2, 0]$ satisfy the equation $r^2 = 4\cos\theta$, and the coordinates of $[2, \pi]$ satisfy the equation $r = 3 + \cos\theta$.

7. $\begin{bmatrix} r = \sin\theta & \Longrightarrow & x^2 + y^2 = y \\ r = -\cos\theta & \Longrightarrow & x^2 + y^2 = -x \end{bmatrix} \Longrightarrow \{x + y = 0, 2x^2 = -x\} \Longrightarrow x = 0, -\tfrac{1}{2}; \quad (0, 0), \left(-\tfrac{1}{2}, \tfrac{1}{2}\right)$

9. $\begin{bmatrix} r = \cos^2\theta & \Longrightarrow & (x^2 + y^2)^{3/2} = x^2 \\ r = -1 & \Longrightarrow & x^2 + y^2 = 1 \end{bmatrix} \Longrightarrow x = \pm 1, \ y = 0; \quad (1, 0), (-1, 0)$

11. $(0, 0), \quad \left(\tfrac{1}{4}, \tfrac{1}{4}\sqrt{3}\right), \quad \left(\tfrac{1}{4}, -\tfrac{1}{4}\sqrt{3}\right)$ 13. $\left(\tfrac{3}{2}, 2\right)$

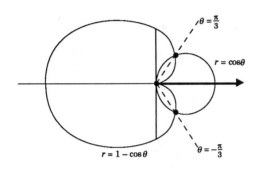
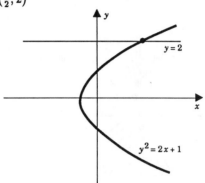

15. $(0, 0), \quad (1, 0)$

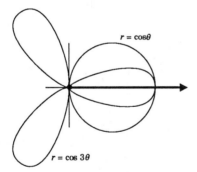

SECTION 10.5

1.

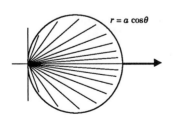

$$A = \int_{-\pi/2}^{\pi/2} \frac{1}{2} [a\cos\theta]^2 \, d\theta$$

$$= a^2 \int_0^{\pi/2} \frac{1 + \cos 2\theta}{2} \, d\theta$$

$$= a^2 \left[\frac{\theta}{2} + \frac{\sin 2\theta}{4}\right]_0^{\pi/2} = \frac{1}{4}\pi a^2$$

3.

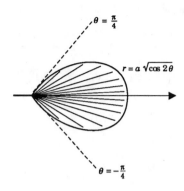

$$A = \int_{-\pi/4}^{\pi/4} \frac{1}{2} \left[a\sqrt{\cos 2\theta} \right]^2 d\theta$$

$$= a^2 \int_0^{\pi/4} \cos 2\theta \, d\theta$$

$$= a^2 \left[\frac{\sin 2\theta}{2} \right]_0^{\pi/4} = \frac{1}{2} a^2$$

5.

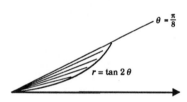

$$A = 2 \int_0^{\pi} \frac{1}{2} \left(a^2 \sin^2 \theta \right) d\theta$$

$$= a^2 \int_0^{\pi} \frac{1 - \cos 2\theta}{2} d\theta$$

$$= a^2 \left[\frac{\theta}{2} - \frac{\sin 2\theta}{4} \right]_0^{\pi} = \frac{1}{2} \pi a^2$$

7.

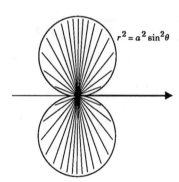

$$A = \int_0^{\pi/8} \frac{1}{2} [\tan 2\theta]^2 d\theta$$

$$= \frac{1}{2} \int_0^{\pi/8} \left(\sec^2 2\theta - 1 \right) d\theta$$

$$= \frac{1}{2} \left[\frac{1}{2} \tan 2\theta - \theta \right]_0^{\pi/8} = \frac{1}{4} - \frac{\pi}{16}$$

9.

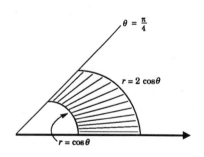

$$A = \int_0^{\pi/4} \frac{1}{2} \left([2\cos\theta]^2 - [\cos\theta]^2 \right) d\theta$$

$$= \frac{3}{2} \int_0^{\pi/4} \frac{1 + \cos 2\theta}{2} d\theta$$

$$= \frac{3}{2} \left[\frac{\theta}{2} + \frac{\sin 2\theta}{4} \right]_0^{\pi/4} = \frac{3}{16} \pi + \frac{3}{8}$$

11.

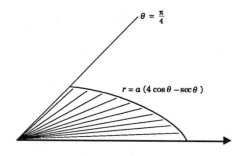

$$A = \int_0^{\pi/4} \frac{1}{2} \left[a \left(4\cos\theta - \sec\theta \right) \right]^2 d\theta$$

$$= \frac{a^2}{2} \int_0^{\pi/4} \left[16\cos^2\theta - 8 + \sec^2\theta \right] d\theta$$

$$= \frac{a^2}{2} \int_0^{\pi/4} \left[8 \left(1 + \cos 2\theta \right) - 8 + \sec^2\theta \right] d\theta$$

$$= \frac{a^2}{2} \left[4\sin 2\theta + \tan\theta \right]_0^{\pi/4} = \frac{5}{2} a^2$$

13.

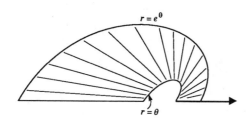

$$A = \int_0^{\pi} \frac{1}{2} \left(\left[e^{\theta} \right]^2 - \left[\theta \right]^2 \right) d\theta$$

$$= \frac{1}{2} \int_0^{\pi} \left(e^{2\theta} - \theta^2 \right) d\theta$$

$$= \frac{1}{2} \left[\frac{1}{2} e^{2\theta} - \frac{1}{3}\theta^3 \right]_0^{\pi} = \frac{1}{12} \left(3e^{2\pi} - 3 - 2\pi^3 \right)$$

15.

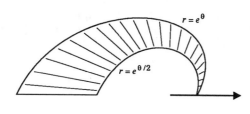

$$A = \int_0^{\pi} \frac{1}{2} \left(\left[e^{\theta} \right]^2 - \left[e^{\theta/2} \right]^2 \right) d\theta$$

$$= \frac{1}{2} \int_0^{\pi} \left(e^{2\theta} - e^{\theta} \right) d\theta$$

$$= \frac{1}{2} \left[\frac{1}{2} e^{2\theta} - e^{\theta} \right]_0^{\pi} = \frac{1}{4} \left(e^{2\pi} + 1 - 2e^{\pi} \right)$$

17.

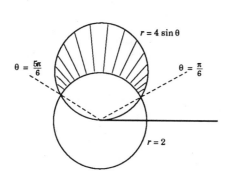

$$A = \int_{\pi/6}^{5\pi/6} \frac{1}{2} \left(\left[4\sin\theta \right]^2 - \left[2 \right]^2 \right) d\theta$$

19.

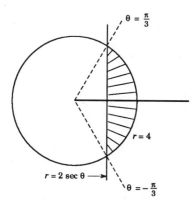

$$A = \int_{-\pi/3}^{\pi/3} \frac{1}{2} \left(\left[4 \right]^2 - \left[2\sec\theta \right]^2 \right) d\theta$$

21.

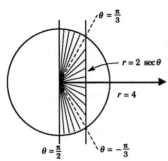

$$A = 2\left\{\int_0^{\pi/3} \frac{1}{2}(2\sec\theta)^2\,d\theta + \int_{\pi/3}^{\pi/2} \frac{1}{2}(4)^2\,d\theta\right\}$$

23.

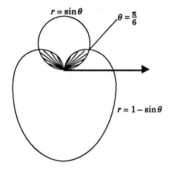

$$A = \int_0^{\pi/3} \frac{1}{2}(2\sin 3\theta)^2\,d\theta$$

25.

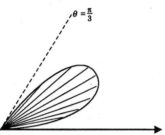

$$A = 2\left\{\int_0^{\pi/6} \frac{1}{2}(\sin\theta)^2\,d\theta + \int_{\pi/6}^{\pi/2} \frac{1}{2}(1-\sin\theta)^2\,d\theta\right\}$$

SECTION 10.6

1. $4x = (y-1)^2$

3. $y = 4x^2 + 1, \quad x \geq 0$

5. $9x^2 + 4y^2 = 36$

7. $1 + x^2 = y^2$

9. $y = 2 - x^2, \quad -1 \leq x \leq 1$

11. $2y - 6 = x, \quad -4 \leq x \leq 4$

13. $y = x - 1$

15. $xy = 1$

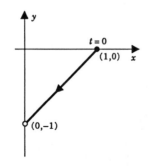

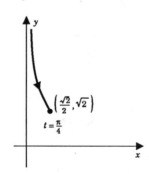

17. $2x + y = 11$

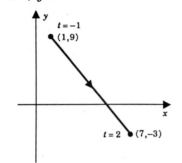

19. $x = \sin \frac{1}{2}\pi y$

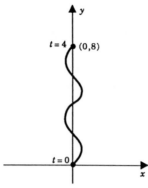

21. $1 + x^2 = y^2$

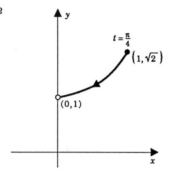

23. (a) $x(t) = -\sin 2\pi t,\quad y(t) = \cos 2\pi t$ (b) $x(t) = \sin 4\pi t,\quad y(t) = \cos 4\pi t$

(c) $x(t) = \cos \frac{1}{2}\pi t,\quad y(t) = \sin \frac{1}{2}\pi t$ (d) $x(t) = \cos \frac{3}{2}\pi t,\quad y(t) = -\sin \frac{3}{2}\pi t$

25. $x(t) = \tan \frac{1}{2}\pi t,\quad y(t) = 2$ **27.** $x(t) = 3 + 5t,\quad y(t) = 7 - 2t$

29. $x(t) = \sin^2 \pi t,\quad y(t) = -\cos \pi t$ **31.** $x(t) = (2 - t)^2,\quad y(t) = (2 - t)^3$

33. $x(t) = t(b - a) + a,\quad y(t) = f(t(b - a) + a)$

35.

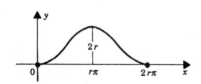

$$A = \int_0^{2\pi} x(t)\, y'(t)\, dt$$

$$= r^2 \int_0^{2\pi} (1 - \cos t)\, dt$$

$$= r^2 \left[t - \sin t\right]_0^{2\pi} = 2\pi r^2$$

37. (a) $V_x = 2\pi \bar{y} A = 2\pi \left(\frac{3}{4}r\right)\left(2\pi r^2\right) = 3\pi^2 r^3$

(b) $V_y = 2\pi \bar{x} A = 2\pi\,(\pi r)\,2\pi r^2 = 4\pi^3 r^3$

39. $x(t) = -a \cos t,\quad y(t) = b \sin t \qquad t \in [0, \pi]$

41. (a) Equation for the ray: $y + 2x = 17, \quad x \ge 6$.

Equation for the circle: $(x - 3)^2 + (y - 1)^2 = 25$.

Simultaneous solution of these equations gives the points of intersection: $(6, 5)$ and $(8, 1)$.

(b)   The particle on the ray is at $(6,5)$ when $t = 0$.   However, when $t = 0$ the particle on the circle is at the point $(-2,1)$. Thus, the intersection point $(6,5)$ is not a collision point. The particle on the ray is at $(8,1)$ when $t = 1$. Since the particle on the circle is also at $(8,1)$ when $t = 1$, the intersection point $(8,1)$ is a collision point.

43.   If $x(r) = x(s)$ and $r \neq s$, then

$$r^2 - 2r = s^2 - 2s$$

$$r^2 - s^2 = 2r - 2s$$

(1) $$r + s = 2.$$

If $y(r) = y(s)$ and $r \neq s$, then

$$r^3 - 3r^2 + 2r = s^3 - 3s^2 + 2s$$

$$\left(r^3 - s^3\right) - 3\left(r^2 - s^2\right) + 2\left(r - s\right) = 0$$

(2) $$\left(r^2 + rs + s^2\right) - 3\left(r + s\right) + 2 = 0.$$

Simultaneous solution of (1) and (2) gives $r = 0$ and $r = 2$. Since $(x(0), y(0)) = (0,0) = (x(2), y(2))$, the curve intersects itself at the origin.

45.   Suppose that $r, s \in [0,4]$ and $r \neq s$.

$$x(r) = x(s) \implies \sin 2\pi r = \sin 2\pi s.$$

$$y(r) = y(s) \implies 2r - r^2 = 2s - s^2 \implies 2(r - s) = r^2 - s^2 \implies 2 = r + s.$$

Now we solve the equations simultaneously:

$$\sin 2\pi r = \sin\left[2\pi\left(2 - r\right)\right] = -\sin 2\pi r$$

$$2\sin 2\pi r = 0$$

$$\sin 2\pi r = 0.$$

Since $r \in [0,4]$, $r = 0, \frac{1}{2}, 1, \frac{3}{2}, 2, \frac{5}{2}, 3, \frac{7}{2}, 4$.

Since $s \in [0,4]$ and $r \neq s$ and $r + s = 2$,   we are left with $r = 0, \frac{1}{2}, \frac{3}{2}, 2$. Note that

$$(x(0), y(0)) = (0,0) = (x(2), y(2)) \quad \text{and} \quad \left(x\left(\tfrac{1}{2}\right), y\left(\tfrac{1}{2}\right)\right) = \left(0, \tfrac{3}{4}\right) = \left(x\left(\tfrac{3}{2}\right), y\left(\tfrac{3}{2}\right)\right).$$

The curve intersects itself at $(0,0)$ and $\left(0, \tfrac{3}{4}\right)$.

SECTION 10.7

1. $x'(1) = 1$, $y'(1) = 3$,   slope 3,   point $(1,0)$;   tangent $y = 3(x - 1)$

3. $x'(0) = 2$, $y'(0) = 0$,   slope 0,   point $(0,1)$;   tangent $y = 1$

5. $x'(1/2) = 1$, $y'(1/2) = -3$,   slope -3,   point $\left(\frac{1}{4}, \frac{9}{4}\right)$;   tangent $y - \frac{9}{4} = -3\left(x - \frac{1}{4}\right)$

7. $x'\left(\frac{\pi}{4}\right) = -\frac{3}{4}\sqrt{2}$, $y'\left(\frac{\pi}{4}\right) = \frac{3}{4}\sqrt{2}$,   slope -1,   point $\left(\frac{1}{4}\sqrt{2}, \frac{1}{4}\sqrt{2}\right)$;

tangent $y - \frac{1}{4}\sqrt{2} = -\left(x - \frac{1}{4}\sqrt{2}\right)$

9. $x(\theta) = \cos\theta\,(4 - 2\sin\theta)$, $y(\theta) = \sin\theta\,(4 - 2\sin\theta)$,   point $(4, 0)$

$x'(\theta) = -4\sin\theta - 2\left(\cos^2\theta - \sin^2\theta\right)$, $y'(\theta) = 4\cos\theta - 4\sin\theta\cos\theta$

$x'(0) = -2$, $y'(0) = 4$,   slope -2,   tangent $y = -2\,(x - 4)$

11. $x(\theta) = \dfrac{4\cos\theta}{5 - \cos\theta}$, $y(\theta) = \dfrac{4\sin\theta}{5 - \cos\theta}$,   point $\left(0, \dfrac{4}{5}\right)$

$x'(\theta) = \dfrac{-20\sin\theta}{(5 - \cos\theta)^2}$, $y'(\theta) = \dfrac{4\,(5\cos\theta - 1)}{(5 - \cos\theta)^2}$

$x'\left(\dfrac{\pi}{2}\right) = -\dfrac{4}{5}$, $y'\left(\dfrac{\pi}{2}\right) = -\dfrac{4}{25}$,   slope $\dfrac{1}{5}$,   tangent $y - \dfrac{4}{5} = \dfrac{1}{5}x$

13. $x(\theta) = \dfrac{\cos\theta\,(\sin\theta - \cos\theta)}{\sin\theta + \cos\theta}$, $y(\theta) = \dfrac{\sin\theta\,(\sin\theta - \cos\theta)}{\sin\theta + \cos\theta}$,   point $(-1, 0)$

$x'(\theta) = \dfrac{\sin\theta\,\cos 2\theta + 2\cos\theta}{(\sin\theta + \cos\theta)^2}$, $y'(\theta) = \dfrac{2\sin\theta - \cos\theta\,\cos 2\theta}{(\sin\theta + \cos\theta)^2}$

$x'(0) = 2$, $y'(0) = -1$,   slope $-\dfrac{1}{2}$,   tangent $y = -\dfrac{1}{2}(x + 1)$

15. $x(t) = t$, $y(t) = t^3$

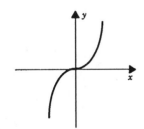

$x'(0) = 1$, $y'(0) = 0$,   slope 0

tangent $y = 0$

17. $x(t) = t^{5/3}$, $y(t) = t$

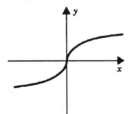

$x'(0) = 0$, $y'(0) = 1$,   slope undefined

tangent $x = 0$

19. $x'(t) = 3 - 3t^2, \quad y'(t) = 1$

$x'(t) = 0 \implies t = \pm 1; \quad y'(t) \neq 0$

(a)   none

(b)   at $(2, 2)$ and $(-2, 0)$

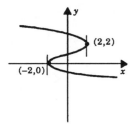

21.   curve traced once completely with $t \in [0, 2\pi)$

$x'(t) = -4 \cos t, \quad y'(t) = -3 \sin t$

$x'(t) = 0 \implies t = \dfrac{\pi}{2}, \dfrac{3\pi}{2};$

$y'(t) = 0 \implies t = 0, \pi$

(a)   at $(3, 7)$ and $(3, 1)$

(b)   at $(-1, 4)$ and $(7, 4)$

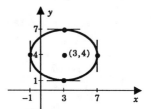

23. $x'(t) = 2t - 2, \quad y'(t) = 3t^2 - 6t + 2$

$x'(t) = 0 \implies t = 1$

$y'(t) = 0 \implies t = 1 \pm \frac{1}{3}\sqrt{3}$

(a)   at $\left(-\frac{2}{3}, \pm\frac{2}{9}\sqrt{3}\right)$

(b)   at $(-1, 0)$

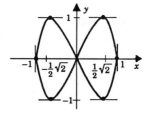

25.   curve traced completely with $t \in [0, 2\pi)$

$x'(t) = -\sin t, \quad y'(t) = 2\cos 2t$

$x'(t) = 0 \implies t = 0, \pi$

$y'(t) = 0 \implies t = \dfrac{\pi}{4}, \dfrac{3\pi}{4}, \dfrac{5\pi}{4}, \dfrac{7\pi}{4}$

(a)   at $\left(\pm\frac{1}{2}\sqrt{2}, \pm 1\right)$

(b)   at $(\pm 1, 0)$

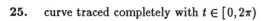

27.   First, we find the values of t when the curve passes through $(2, 0)$.

$$y(t) = 0 \implies t^4 - 4t^2 = 0 \implies t = 0, \pm 2.$$

$$x(-2) = 2, \quad x(0) = 2, \quad x(2) = -2.$$

The curve passes through $(2, 0)$ at $t = -2$ and $t = 0$.

$$x'(t) = -1 - \frac{\pi}{2}\sin\frac{\pi t}{4}, \quad y'(t) = 4t^3 - 8t.$$

At $t = -2$, $\quad x'(-2) = \dfrac{\pi}{2} - 1, \quad y'(t) = -16, \quad$ tangent $y = \dfrac{32}{2 - \pi}(x - 2).$

At $t = 0$, $\quad x'(0) = -1, \quad y'(0) = 0, \quad$ tangent $y = 0$.

29. The slope of $\overline{OP}$ is $\tan\theta_1$. The curve $r = f(\theta)$ can be parametrized by setting

$$x(\theta) = f(\theta)\cos\theta, \qquad y(\theta) = f(\theta)\sin\theta.$$

Differentiation gives

$$x'(\theta) = -f(\theta)\sin\theta + f'(\theta)\cos\theta, \qquad y'(\theta) = f(\theta)\cos\theta + f'(\theta)\sin\theta.$$

If $f'(\theta_1) = 0$, then

$$x'(\theta_1) = -f(\theta_1)\sin\theta_1, \qquad y'(\theta_1) = f(\theta_1)\cos\theta_1.$$

Since $f(\theta_1) \neq 0$, we have

$$m = \frac{y'(\theta_1)}{x'(\theta_1)} = -\cot\theta_1 = -\frac{1}{\text{slope of } \overline{OP}}.$$

31. $x'(t) = 3t^2, \quad y'(t) = 2t$

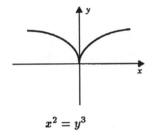

$$x^2 = y^3$$

33. $x'(t) = 5t^4, \quad y'(t) = 3t^2$

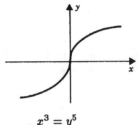

$$x^3 = y^5$$

35. $x'(t) = 2t, \quad y'(t) = 2t$

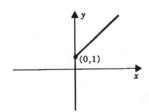

(0,1)

ray: $y = x + 1, \quad x \geq 0$

37. By (10.7.5), $\dfrac{d^2y}{dx^2} = \dfrac{(-\sin t)(-\sin t) - (\cos t)(-\cos t)}{(-\sin t)^3} = \dfrac{-1}{\sin^3 t}$. At $t = \dfrac{\pi}{6}$, $\dfrac{d^2y}{dx^2} = -8$.

39. By (10.7.5), $\dfrac{d^2y}{dx^2} = \dfrac{(e^t)(e^{-t}) - (-e^{-t})(e^t)}{(e^t)^3} = 2e^{-3t}$. At $t = 0$, $\dfrac{d^2y}{dx^2} = 2$.

SECTION 10.8

1. lub = 2; glb = 0

3. no lub; glb = 0

5. lub = 2; glb = -2

7. no lub; glb = 2

9. lub = $2\frac{1}{2}$; glb = 2

11. lub = 1; glb = 0.9

13.   lub $= e$;   glb $= 0$ **15.**   lub $= \frac{1}{2}(-1+\sqrt{5})$;   glb $= \frac{1}{2}(-1-\sqrt{5})$

17.   no lub;   no glb   **19.**   no lub;   no glb

21.   glb $S = 0$, $0 \le \left(\frac{1}{11}\right)^3 < 0 + 0.001$

23.   glb $S = 0$, $0 \le \left(\frac{1}{10^{2n-1}}\right) < 0 + \left(\frac{1}{10^k}\right)$ $\left(n > \frac{k+1}{2}\right)$

25.   Let $\epsilon > 0$. The condition $m \le s$ is satisfied by all numbers s in S. All we have to show therefore is that there is some number s in S such that

$$s < m + \epsilon.$$

Suppose on the contrary that there is no such number in S. We then have

$$m + \epsilon \le x \quad \text{for all} \quad x \in S.$$

This makes $m + \epsilon$ a lower bound for S. But this cannot be, for then $m + \epsilon$ is a lower bound for S that is *greater* than m, and by assumption, m is the *greatest* lower bound.

SECTION 10.9

1. $L = \displaystyle\int_0^1 \sqrt{1 + 2^2}\, dx = \sqrt{5}$

3. $L = \displaystyle\int_1^4 \sqrt{1 + \left[\frac{3}{2}\left(x - \frac{4}{9}\right)^{1/2}\right]^2}\, dx = \int_1^4 \frac{3}{2}\sqrt{x}\, dx = \left[x^{3/2}\right]_1^4 = 7$

5. $L = \displaystyle\int_0^3 \sqrt{1 + \left(\frac{1}{2}\sqrt{x} - \frac{1}{2\sqrt{x}}\right)^2}\, dx = \int_0^3 \left(\frac{1}{2}\sqrt{x} + \frac{1}{2\sqrt{x}}\right) dx = \left[\frac{1}{3}x^{3/2} + x^{1/2}\right]_0^3 = 2\sqrt{3}$

7. $L = \displaystyle\int_0^1 \sqrt{1 + \left[x\left(x^2 + 2\right)^{1/2}\right]^2}\, dx = \int_0^1 \left(x^2 + 1\right) dx = \left[\frac{1}{3}x^3 + x\right]_0^1 = \frac{4}{3}$

9. $L = \displaystyle\int_1^5 \sqrt{1 + \left[\frac{1}{2}\left(x - \frac{1}{x}\right)\right]^2}\, dx = \int_1^5 \frac{1}{2}\left(x + \frac{1}{x}\right) dx = \left[\frac{1}{2}\left(\frac{1}{2}x^2 + \ln x\right)\right]_1^5 = 6 + \frac{1}{2}\ln 5 \cong 6.80$

11. $L = \displaystyle\int_1^8 \sqrt{1 + \left[\frac{1}{2}\left(x^{1/3} - x^{-1/3}\right)\right]^2}\, dx = \int_1^8 \frac{1}{2}\left(x^{1/3} + x^{-1/3}\right) dx = \frac{1}{2}\left[\frac{3}{4}x^{4/3} + \frac{3}{2}x^{2/3}\right]_1^8 = \frac{63}{8}$

13. $L = \displaystyle\int_0^{\pi/4} \sqrt{1 + \tan^2 x}\, dx = \int_0^{\pi/4} \sec x\, dx = [\ln |\sec x + \tan x|]_0^{\pi/4} = \ln\left(1 + \sqrt{2}\right) \cong 0.88$

15. $L = \displaystyle\int_1^2 \sqrt{1 + \left(\sqrt{x^2 - 1}\right)^2}\, dx = \int_1^2 x\, dx = \left[\frac{1}{2}x^2\right]_1^2 = \frac{3}{2}$

17. $L = \int_0^1 \sqrt{1 + \left[\sqrt{3 - x^2}\right]^2}\, dx = \int_0^1 \sqrt{4 - x^2}\, dx = \int_0^{\pi/6} 4\cos^2 u\, du$

$(x = 2\sin u)$

$= 2\int_0^{\pi/6} (1 + \cos 2u)\, du = 2\left[u + \frac{\sin 2u}{2}\right]_0^{\pi/6} = \frac{1}{3}\pi + \frac{1}{2}\sqrt{3} \cong 1.91$

19. $v(t) = \sqrt{(2t)^2 + 2^2} = 2\sqrt{t^2 + 1}$

initial speed $= v(0) = 2,$ terminal speed $= v\left(\sqrt{3}\right) = 4$

$s = \int_0^{\sqrt{3}} 2\sqrt{t^2 + 1}\, dt = 2\int_0^{\pi/3} \sec^3 u\, du = 2\left[\frac{1}{2}\sec u \tan u + \frac{1}{2}\ln |\sec u + \tan u|\right]_0^{\pi/3}$

$(t = \tan u)$ (by parts)

$= 2\sqrt{3} + \ln\left(2 + \sqrt{3}\right) \cong 4.78$

21. $v(t) = \sqrt{(2t)^2 + (3t^2)^2}\, dt = t\left(4 + 9t^2\right)^{1/2}$

initial speed $= v(0) = 0,$ terminal speed $= v(1) = \sqrt{13}$

$s = \int_0^1 t\left(4 + 9t^2\right)^{1/2} dt = \left[\frac{1}{27}\left(4 + 9t^2\right)^{3/2}\right]_0^1 = \frac{1}{27}\left(13\sqrt{13} - 8\right)$

23. $v(t) = \sqrt{[e^t \cos t + e^t \sin t]^2 + [e^t \cos t - e^t \sin t]^2} = \sqrt{2}\, e^t$

initial speed $= v(0) = \sqrt{2},$ terminal speed $= \sqrt{2}\, e^\pi$

$s = \int_0^\pi \sqrt{2}\, e^t\, dt = \left[\sqrt{2}\, e^t\right]_0^\pi = \sqrt{2}\left(e^\pi - 1\right)$

25. $L =$ circumference of circle of radius $1 = 2\pi$

27. $L = \int_0^{4\pi} \sqrt{[e^\theta]^2 + [e^\theta]^2}\, d\theta = \int_0^{4\pi} \sqrt{2}\, e^\theta\, d\theta = \left[\sqrt{2}\, e^\theta\right]_0^{4\pi} = \sqrt{2}\left(e^{4\pi} - 1\right)$

29. $L = \int_0^{2\pi} \sqrt{[e^{2\theta}]^2 + [2e^{2\theta}]^2}\, d\theta = \int_0^{2\pi} \sqrt{5}\, e^{2\theta}\, d\theta = \left[\frac{1}{2}\sqrt{5}\, e^{2\theta}\right]_0^{2\pi} = \frac{1}{2}\sqrt{5}\left(e^{4\pi} - 1\right)$

31. $L = \int_0^{\pi/2} \sqrt{(1 - \cos\theta)^2 + \sin^2\theta}\, d\theta = \int_0^{\pi/2} \sqrt{2 - 2\cos\theta}\, d\theta$

$= \int_0^{\pi/2} \left(2\sin\frac{1}{2}\theta\right) d\theta = \left[-4\cos\frac{1}{2}\theta\right]_0^{\pi/2} = 4 - 2\sqrt{2}$

33. $s = \int_0^1 \sqrt{\left[\dfrac{1}{1+t^2}\right]^2 + \left[\dfrac{-t}{1+t^2}\right]^2}\, dt = \int_0^1 \dfrac{dt}{\sqrt{1+t^2}}$

$$= \int_0^{\pi/4} \sec u\, du = [\ln |\sec u + \tan u|]_0^{\pi/4} = \ln\left(1 + \sqrt{2}\right)$$

$\uparrow\!\llcorner (t = \tan u)$

initial speed $= v(0) = 1,$ terminal speed $= v(1) = \tfrac{1}{2}\sqrt{2}$

35. $c = 1;$ the curve $y = e^x$ is the curve $y = \ln x$ reflected in the line $y = x$

37. $L = \int_a^b \sqrt{1 + \sinh^2 x}\, dx = \int_a^b \sqrt{\cosh^2 x}\, dx = \int_a^b \cosh x\, dx = A$

39. (a)   Express GPE + KE as a function of t and verify that the derivative with respect to t is zero.

(b)   From (a) we learn that throughout the motion

$$32my + \tfrac{1}{2}mv^2 = C.$$

At the time of firing $y = 0$ and $v = |v_0| = v_0$. Therefore

$$32my + \tfrac{1}{2}mv^2 = \tfrac{1}{2}mv_0{}^2.$$

At impact $y = 0,$ $\tfrac{1}{2}mv^2 = \tfrac{1}{2}mv_0{}^2,$ and $v = v_0$.

41. $\sqrt{1 + [f(x)]^2} = \sqrt{1 + \tan^2 [\alpha(x)]} = |\sec [\alpha(x)]|$

SECTION 10.10

1. $L = $ length of the line segment $= 1$

$(\bar{x}, \bar{y}) = \left(\tfrac{1}{2}, 4\right)$ (the midpoint of the line segment)

$A_x = $ lateral surface area of cylinder of radius 4 and side $1 = 16\pi$.

3. $L = \int_0^3 \sqrt{1 + \left(\dfrac{4}{3}\right)^2}\, dx = \left(\dfrac{5}{3}\right)^3 = 5$

$\bar{x}L = \int_0^3 x\sqrt{1 + \left(\dfrac{4}{3}\right)^2}\, dx = \dfrac{5}{3}\left[\dfrac{1}{2}x^2\right]_0^3 = \dfrac{15}{2},\quad \bar{x} = \dfrac{3}{2}$

$\bar{y}L = \int_0^3 \dfrac{4}{3}x\sqrt{1 + \left(\dfrac{4}{3}\right)^2}\, dx = \left(\dfrac{4}{3}\right)\left(\dfrac{15}{2}\right) = 10,\quad \bar{y} = 2$

$A_x = 2\pi\bar{y}L = 2\pi(2)(5) = 20\pi$

5. $L = \int_0^2 \sqrt{(3)^2 + (4)^2}\, dt = (2)(5) = 10$

$\bar{x}L = \int_0^2 3t\sqrt{(3)^2 + (4)^2}\, dt = 15\left[\frac{1}{2}t^2\right]_0^2 = 30, \quad \bar{x} = 3$

$\bar{y}L = \int_0^2 4t\sqrt{(3)^2 + (4)^2}\, dt = 20\left[\frac{1}{2}t^2\right]_0^2 = 40, \quad \bar{y} = 4$

$A_x = 2\pi\bar{y}L = 2\pi(4)(10) = 80\pi$

7. $L = \int_0^{\pi/6} \sqrt{4\sin^2 t + 4\cos^2 t}\, dt = 2\left(\frac{\pi}{6}\right) = \frac{1}{3}\pi$

$\bar{x}L = \int_0^{\pi/6} 2\cos t\sqrt{4\sin^2 t + 4\cos^2 t}\, dt = 4\left[\sin t\right]_0^{\pi/6} = 2, \quad \bar{x} = \frac{6}{\pi}$

$\bar{y}L = \int_0^{\pi/6} 2\sin t\sqrt{4\sin^2 t + 4\cos^2 t}\, dt = 4\left[-\cos t\right]_0^{\pi/6} = 4 - 2\sqrt{3}, \quad \bar{y} = 6\left(2 - \sqrt{3}\right)/\pi$

$A_x = 2\pi\bar{y}L = 2\pi(6(2 - \sqrt{3})/\pi)\frac{1}{3}\pi = 4\pi(2 - \sqrt{3})$

9. $x(t) = a\cos t, \quad y = a\sin t; \quad t \in [\frac{1}{3}\pi, \frac{2}{3}\pi]$

$L = \int_{\pi/3}^{2\pi/3} \sqrt{a^2\sin^2 t + a^2\cos^2 t}\, dt = \frac{1}{3}\pi a$

by symmetry $\bar{x} = 0$

$\bar{y}L = \int_{\pi/3}^{2\pi/3} a\sin t\, \sqrt{a^2\sin^2 t + a^2\cos^2 t}\, dt = a^2\int_{\pi/3}^{2\pi/3} \sin t\, dt$

$= a^2\left[-\cos t\right]_{\pi/3}^{2\pi/3} = a^2, \quad \bar{y} = 3a/\pi$

$A_x = 2\pi\bar{y}L = 2\pi a^2$

11. $A_x = \int_0^2 \frac{2}{3}\pi x^3\sqrt{1 + x^4}\, dx = \frac{1}{9}\pi\left[(1 + x^4)^{3/2}\right]_0^2 = \frac{1}{9}(17\sqrt{17} - 1) \cong 7.68$

13. $A_x = \int_0^1 \frac{1}{2}\pi x^3\sqrt{1 + \frac{9}{16}x^4}\, dx = \frac{4}{27}\pi\left[\left(1 + \frac{9}{16}x^4\right)\right]_0^1 = \frac{61}{432}\pi$

15. $A_x = \displaystyle\int_0^{\pi/2} 2\pi \cos x \sqrt{1 + \sin^2 x}\, dx = \int_0^1 2\pi \sqrt{1 + u^2}\, du$

$u = \sin x$

$= 2\pi \left[\tfrac{1}{2} u \sqrt{1 + u^2} + \tfrac{1}{2} \ln \left(u + \sqrt{1 + u^2} \right) \right]_0^1 = \pi \left[\sqrt{2} + \ln \left(1 + \sqrt{2} \right) \right]$

(8.5.1)

17. $A_x = \displaystyle\int_0^{\pi/2} 2\pi (e^\theta \sin \theta) \sqrt{[e^\theta \cos \theta - e^\theta \sin \theta]^2 + [e^\theta \sin \theta + e^\theta \cos \theta]^2}\, d\theta$

$= 2\pi\sqrt{2} \displaystyle\int_0^{\pi/2} e^{2\theta} \sin \theta\, d\theta$

$= 2\pi\sqrt{2} \left[\tfrac{1}{5} \left(2 e^{2\theta} \sin \theta - e^{2\theta} \cos \theta \right) \right]_0^{\pi/2} = \tfrac{2}{5}\sqrt{2}\,\pi \left(2e^\pi + 1 \right)$

(by parts twice)

19. $A = \tfrac{1}{2}\theta {s_2}^2 - \tfrac{1}{2}\theta {s_1}^2$

$= \tfrac{1}{2}(\theta s_2 + \theta s_1)(s_2 - s_1)$

$= \tfrac{1}{2}(2\pi R + 2\pi r)s = \pi(R + r)s$

21.   (a)   The centroids of the 3, 4, 5 sides are the midpoints $\left(\tfrac{3}{2}, 0\right)$, $(3, 2)$, $\left(\tfrac{3}{2}, 2\right)$.

(b) $\overline{x}(3 + 4 + 5) = \tfrac{3}{2}(3) + 3(4) + \tfrac{3}{2}(5)$, $12\overline{x} = 24$, $\overline{x} = 2$

$\overline{y}(3 + 4 + 5) = 0(3) + 2(4) + 2(5)$, $12\overline{y} = 18$, $\overline{y} = \tfrac{3}{2}$

(c) $A = \tfrac{1}{2}(3)(4) = 6$

$\overline{x}A = \displaystyle\int_0^3 x \left(\tfrac{4}{3} x \right) dx = \int_0^3 \tfrac{4}{3} x^2 dx = \tfrac{4}{9} \left[x^3 \right]_0^3 = 12$, $\overline{x} = 2$

$\overline{y}A = \displaystyle\int_0^3 \tfrac{1}{2} \left(\tfrac{4}{3} x \right)^2 dx = \int_0^3 \tfrac{8}{9} x^2 dx = \tfrac{8}{27} \left[x^3 \right]_0^3 = 8$, $\overline{x} = \tfrac{4}{3}$

(d) $\overline{x}(4 + 5) = 3(4) + \tfrac{3}{2}(5)$, $9\overline{x} = \tfrac{39}{2}$, $\overline{x} = \tfrac{13}{6}$

$\overline{y}(4 + 5) = 2(4) + 2(5)$, $9\overline{y} = 18$, $\overline{y} = 2$

(e) $A_x = 2\pi(2)(5) = 20\pi$ (f) $A_x = 2\pi(2)(4 + 5) = 36\pi$

23. $A_x = 2\pi \bar{y} L = 2\pi(b)(2\pi a) = 4\pi^2 ab$

25.   The band can be obtained by revolving about the x-axis the graph of the function

$$f(x) = \sqrt{r^2 - x^2}, \qquad x \in [a,b].$$

A straightforward calculation shows that the surface area of the band is $2\pi r(b-a)$.

27.   (a)   Parametrize the upper half of the ellipse by

$$x(t) = a\cos t, \quad y(t) = b\sin t; \qquad t \in [0, \pi].$$

Here

$$\sqrt{[x'(t)]^2 + [y'(t)]^2} = \sqrt{a^2 \sin^2 t + b^2 \cos^2 t} = \sqrt{a^2 - (a^2 - b^2)\cos^2 t},$$

which, with $c = \sqrt{a^2 - b^2}$, can be written $\sqrt{a^2 - c^2 \cos^2 t}$.   Therefore,

$$A = \int_0^\pi 2\pi b \sin t \sqrt{a^2 - c^2 \cos^2 t}\, dt = 4\pi b \int_0^{\pi/2} \sin t \sqrt{a^2 - c^2 \cos^2 t}\, dt.$$

Setting $u = c\cos t$, we have $du = -c\sin t$ and

$$A = -\frac{4\pi b}{c} \int_c^0 \sqrt{a^2 - u^2}\, du = \frac{4\pi b}{c} \left[\frac{u}{2} \sqrt{a^2 - u^2} + \frac{a^2}{2} \sin^{-1}\left(\frac{u}{a}\right) \right]_0^c$$

$$= 2\pi b^2 + \frac{2\pi a^2 b}{c} \sin^{-1}\left(\frac{c}{a}\right) = 2\pi b^2 + \frac{2\pi ab}{e} \sin^{-1} e$$

where e is the eccentricity of ellipse: $e = c/a$.

(b)   Parametrize the right half of the ellipse by

$$x(t) = a\cos t, \quad y(t) = b\sin t; \qquad t \in [-\tfrac{1}{2}\pi, \tfrac{1}{2}\pi].$$

Again $\sqrt{[x'(t)]^2 + [y'(t)]^2} = \sqrt{a^2 - c^2 \cos^2 t}$ where $c = \sqrt{a^2 - b^2}$.

Therefore

$$A = \int_{-\pi/2}^{\pi/2} 2\pi a \cos t \sqrt{a^2 - c^2 \cos^2 t}\, dt.$$

Set $u = c\sin t$.   Then $du = c\cos t\, dt$ and

$$A = \frac{2\pi a}{c} \int_{-c}^c \sqrt{b^2 + u^2}\, du = \frac{2\pi a}{c} \left[\frac{u}{2} \sqrt{b^2 + u^2} + \frac{b^2}{2} \ln \left| u + \sqrt{b^2 + u^2} \right| \right]_{-c}^c$$

Routine calculation gives

$$A = 2\pi a^2 + \frac{\pi b^2}{e} \ln \left| \frac{1+e}{1-e} \right|.$$

29. Such a hemisphere can be obtained by revolving about the x-axis the curve

$$x(t) = r \cos t, \quad y(t) = r \sin t; \quad t \in [0, \tfrac{1}{2}\pi].$$

Therefore

$$\bar{x}A = \int_0^{\pi/2} 2\pi(r \cos t)(r \sin t)\sqrt{r^2 \sin^2 t + r^2 \cos^2 t}\, dt$$

$$= \int_0^{\pi/2} 2\pi r^3 \sin t \cos t \, dt = \pi r^3 \left[\sin^2 t \right]_0^{\pi/2} = \pi r^3.$$

$$A = 2\pi r^2; \quad \bar{x} = \bar{x}A/A = \tfrac{1}{2}r.$$

The centroid lies on the midpoint of the axis of the hemisphere.

31. Such a surface can be obtained by revolving about the x-axis the graph of the function

$$f(x) = \left(\frac{R-r}{h} \right)x + r, \quad x \in [0, h].$$

Formula (10.10.8) gives

$$\bar{x}A = \int_0^h 2\pi x f(x)\sqrt{1 + [f'(x)]^2}\, dx$$

$$= \frac{2\pi}{h}\sqrt{h^2 + (R-h)^2} \int_0^h \left[\left(\frac{R-r}{h} \right)x^2 + rx \right] dx$$

$$= \frac{\pi}{3}\sqrt{h^2 + (R-r)^2}\,(2R+r)h$$

$$A = \pi(R+r)s = \pi(R+r)\sqrt{h^2 + (R-r)^2}$$

$$\bar{x} = \frac{\bar{x}A}{A} = \left(\frac{2R+r}{R+r} \right)\frac{h}{3}.$$

The centroid of the surface lies on the axis of the cone $\left(\dfrac{2R+r}{R+r} \right)\dfrac{h}{3}$ units from the base of radius r.

SECTION * 10.11

1. Referring to Figure 10.11.1 we have

$$x(\theta) = \overline{OB} - \overline{AB} = R\theta - R\sin\theta = R(\theta - \sin\theta)$$

$$y(\theta) = \overline{BQ} - \overline{QC} = R - R\cos\theta = R(1 - \cos\theta).$$

3. (a) The slope at P is

$$m = \frac{y'(\theta)}{x'(\theta)} = \frac{\sin\theta}{1 - \cos\theta}.$$

The line tangent to the cycloid at P has equation

$$y - R(1 - \cos \theta) = \frac{\sin \theta}{1 - \cos \theta}[x - R(\theta - \sin \theta)].$$

The top of the circle is the point $(R\theta, 2R)$. Its coordinates satisfy the equation for the tangent:

$$2R - R(1 - \cos \theta) \stackrel{?}{=} \frac{\sin \theta}{1 - \cos \theta}[R\theta - R(\theta - \sin \theta)]$$

$$R(1 + \cos \theta) \stackrel{?}{=} \frac{R \sin^2 \theta}{1 - \cos \theta}$$

$$1 - \cos^2 \theta \stackrel{\checkmark}{=} \sin^2 \theta.$$

(b) In view of the symmetry and repetitiveness of the curve we can assume that $\theta \in (0, \pi)$. Then

$$\tan \alpha = \frac{\sin \theta}{1 - \cos \theta} = \frac{\sin \theta}{2 \sin^2 \frac{1}{2}\theta} = \frac{\sin \frac{1}{2}\theta \cos \frac{1}{2}\theta}{\sin^2 \frac{1}{2}\theta} = \cot \frac{1}{2}\theta$$

and

$$\alpha = \tfrac{1}{2}\pi - \tfrac{1}{2}\theta = \tfrac{1}{2}(\pi - \theta).$$

5. $\displaystyle L = \int_0^{2\pi} \sqrt{[R(1 - \cos \theta)]^2 + [R \sin \theta]^2}\, d\theta$

$\displaystyle = R \int_0^{2\pi} \sqrt{2 - 2 \cos \theta}\, d\theta$

$\displaystyle = R \int_0^{2\pi} \sqrt{4 \sin^2 \left(\frac{\theta}{2}\right)}\, d\theta = 2R \int_0^{2\pi} \sin \frac{\theta}{2}\, d\theta = 4R \left[-\cos \frac{\theta}{2}\right]_0^{2\pi} = 8R$

7. $\bar{x} = \pi R$ by symmetry

$\displaystyle \bar{y}A = \int_0^{2\pi} \frac{1}{2}[y(\theta)]^2 x'(\theta)\, d\theta$

$\displaystyle = \int_0^{2\pi} \frac{1}{2}R^2(1 - \cos \theta)^2 [R(1 - \cos \theta)]\, d\theta$

$\displaystyle = \frac{1}{2}R^3 \int_0^{2\pi} (1 - 3 \cos \theta + 3 \cos^2 \theta - \cos^3 \theta)\, d\theta = \frac{5}{2}\pi R^3$

$A = 3\pi R^2$ (by Exercise 6) $\bar{y} = \left(\frac{5}{2}\pi R^3\right) / (3\pi R^2) = \frac{5}{6}R$

9. $V_y = 2\pi \bar{x} A = 2\pi(\pi R)(3\pi R^2) = 6\pi^3 R^3$

11. $A_x = 2\pi \bar{y} L = 2\pi \left(\frac{4}{3}R\right)(8R) = \frac{64}{3}\pi R^2$ $\left(\bar{y} = \frac{4}{3}R \text{ by Exercise 10}\right)$

13. $\dfrac{dy}{dx} = \dfrac{y'(\phi)}{x'(\phi)} = \dfrac{R\sin\phi}{R(1+\cos\phi)} = \dfrac{2\sin\frac{1}{2}\phi\cos\frac{1}{2}\phi}{2\cos^2\frac{1}{2}\phi} = \tan\dfrac{1}{2}\phi; \quad \alpha = \dfrac{1}{2}\phi$

15. **(a)** Already shown more generally in Problem 7 of Section 10.9.

(b) Combining $d^2s/dt^2 = -g\sin\alpha$ with $s = 4R\sin\alpha$, we have

$$\frac{d^2s}{dt^2} = -\frac{g}{4R}s.$$

This is simple harmonic motion (Section 7.8) with angular frequency $\omega = \frac{1}{2}\sqrt{g/R}$ and period $T = 2\pi/\omega = 4\pi\sqrt{R/g}$.

SECTION 10.12

1.

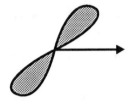

$r^2 = 4\sin 2\theta$

$A = 2\displaystyle\int_0^{\pi/2} \frac{1}{2}[4\sin 2\theta]\,d\theta$

$\quad = [-2\cos 2\theta]_0^{\pi/2} = 4$

3.

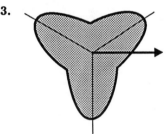

$r = 2 + \sin 3\theta$

$A = \displaystyle\int_0^{2\pi} \frac{1}{2}[2 + \sin 3\theta]^2\,d\theta$

$\quad = \dfrac{1}{2}\displaystyle\int_0^{2\pi}\left[4 + 4\sin 3\theta + \dfrac{1-\cos 6\theta}{2}\right]d\theta$

$\quad = \dfrac{1}{2}\left[\dfrac{9}{2}\theta - \dfrac{4}{3}\cos 3\theta - \dfrac{\sin 6\theta}{12}\right]_0^{2\pi} = \dfrac{9}{2}\pi$

5. $L = \displaystyle\int_0^{5/9}\sqrt{1+\left[\dfrac{3}{2}x^{1/2}\right]^2}\,dx = \displaystyle\int_0^{5/9}\left(1+\dfrac{9}{4}x\right)^{1/2}dx$

$\quad = \left[\dfrac{8}{27}\left(1+\dfrac{9}{4}x\right)^{3/2}\right]_0^{5/9} = \dfrac{19}{27}$

7. $L = \displaystyle\int_0^{5a}\sqrt{1+\left[\dfrac{3}{2\sqrt{a}}x^{1/2}\right]^2}\,dx = \displaystyle\int_0^{5a}\left(1+\dfrac{9x}{4a}\right)^{1/2}dx$

$\quad = \left[\dfrac{8a}{27}\left(1+\dfrac{9x}{4a}\right)^{3/2}\right]_0^{5a} = \dfrac{335}{27}a \quad (a>0)$

9.

$$L = \int_0^{\pi/2} \sqrt{\left[2\left(1+\cos\theta\right)^{-1}\right]^2 + \left[2\sin\theta\left(1+\cos\theta\right)^{-2}\right]^2}\, d\theta$$

$$= \int_0^{\pi/2} \sqrt{4(1+\cos\theta)^{-4}\left[(1+\cos\theta)^2 + \sin^2\theta\right]}\, d\theta$$

$$= \int_0^{\pi/2} \sqrt{8\left(1+\cos\theta\right)^{-3}}\, d\theta = \int_0^{\pi/2} \sqrt{8\left(2\cos^2\frac{\theta}{2}\right)^{-3}}\, d\theta$$

$$= \int_0^{\pi/2} \sec^3\frac{\theta}{2}\, d\theta = \left[\sec\frac{\theta}{2}\tan\frac{\theta}{2} + \ln\left|\sec\frac{\theta}{2} + \tan\frac{\theta}{2}\right|\right]_0^{\pi/2} = \sqrt{2} + \ln\left(1+\sqrt{2}\right)$$

— (by parts)

11.

$$L = \int_0^{2\pi} \sqrt{\left[r\left(1-\cos t\right)\right]^2 + \left[r\sin t\right]^2}\, dt = r\int_0^{2\pi} \sqrt{2 - 2\cos t}\, dt$$

$$= r\int_0^{2\pi} \sqrt{4\sin^2\left(\frac{t}{2}\right)}\, dt = 2r\int_0^{2\pi} \sin\frac{t}{2}\, dt = 4r\left[-\cos\frac{t}{2}\right]_0^{2\pi} = 8r$$

13.

$$A_x = \int_0^{4p} 2\pi\sqrt{px}\sqrt{1 + \left[\frac{p}{\sqrt{2px}}\right]^2}\, dx = 2\pi\int_0^{4p} \sqrt{2px + p^2}\, dx$$

$$= \left[\frac{2\pi}{3p}\left(2px + p^2\right)^{3/2}\right]_0^{4p} = \frac{52}{3}\pi p^2$$

15.

$$A_x = \int_a^{2a} 2\pi\left(\frac{x^3}{6a^2} + \frac{a^2}{2x}\right)\sqrt{1 + \left[\frac{x^2}{2a^2} - \frac{a^2}{2x^2}\right]^2}\, dx$$

$$= \int_a^{2a} 2\pi\left(\frac{x^3}{6a^2} + \frac{a^2}{2x}\right)\left(\frac{x^2}{2a^2} + \frac{a^2}{2x^2}\right)\, dx = \frac{\pi}{6a^4}\int_a^{2a}\left(x^5 + 4a^4 x + 3a^8 x^{-3}\right)\, dx$$

$$= \frac{\pi}{6a^4}\left[\frac{1}{6}x^6 + 2a^4 x^2 - \frac{3}{2}a^8 x^{-2}\right]_a^{2a} = \frac{47}{16}\pi a^2$$

17.

$$A_x = \int_3^8 2\pi y(t)\sqrt{[x'(t)]^2 + [y'(t)]^2}\, dt$$

$$= 2\pi\int_3^8 t\sqrt{1+t}\, dt \qquad (\text{set } u^2 = 1+t,\quad 2u\, du = dt)$$

$$= 4\pi\int_2^3 \left(u^4 - u^2\right) du = 4\pi\left[\frac{1}{5}u^5 - \frac{1}{3}u^3\right]_2^3 = \frac{2152}{15}\pi$$

19. $L = \int_{-a}^{a} \sqrt{1 + \sinh^2\left(\frac{x}{a}\right)}\, dx = 2\int_0^a \cosh\left(\frac{x}{a}\right) dx = \frac{a(e^2 - 1)}{e}$

$\bar{x} = 0$ (by symmetry)

$\bar{y}L = \int_{-a}^{a} a\cosh\left(\frac{x}{a}\right)\sqrt{1 + \sinh^2\left(\frac{x}{a}\right)}\, dx = 2a\int_0^a \cosh^2\left(\frac{x}{a}\right) dx = \frac{a^2(e^4 + 4e^2 - 1)}{4e^2}$

$\bar{y} = \dfrac{\bar{y}L}{L} = \dfrac{a(e^4 + 4e^2 - 1)}{4e(e^2 - 1)}$

21. $L = 4\int_0^{\pi/2} \sqrt{9r^2\cos^4\theta\sin^2\theta + 9r^2\sin^4\theta\cos^2\theta}\, d\theta$

$= 4\int_0^{\pi/2} 3r\sin\theta\cos\theta\, d\theta = 12r\left[\frac{1}{2}\sin^2\theta\right]_0^{\pi/2} = 6r$

23.   upper half: $x(\theta) = -r\cos^3\theta, \quad y(\theta) = r\sin^3\theta$ with $\theta \in [0, \pi]$

$A = 2\int_0^{\pi} y(\theta)x'(\theta)\, d\theta = 2\int_0^{\pi} \left(r\sin^3\theta\right)\left(3r\cos^2\theta\sin\theta\right) d\theta$

$= 6r^2\int_0^{\pi} \sin^4\theta\cos^2\theta\, d\theta = 6r^2\left[\frac{1}{16}\theta - \frac{1}{64}\sin 4\theta - \frac{1}{48}\sin^3 2\theta\right]_0^{\pi} = \frac{3}{8}\pi r^2$

⎣ Example 5, Section 8.3

25. $L = 8a$ (Example 5, Section 10.9)

$x(\theta) = a(1 - \cos\theta)\cos\theta, \quad y(\theta) = a(1 - \cos\theta)\sin\theta$

$\bar{x}L = \int_0^{2\pi} x(\theta)\sqrt{[x'(\theta)]^2 + [y'(\theta)]^2}\, d\theta$

$= \int_0^{2\pi} a(1 - \cos\theta)\cos\theta\left(\sqrt{2}\,a\sqrt{1 - \cos\theta}\right) d\theta$

$= 4a^2\int_0^{2\pi} \left(\frac{1 - \cos\theta}{2}\right)^{3/2}\cos\theta\, d\theta = -\frac{32}{5}a^2$

⎣ expressing integrand in terms of $\frac{1}{2}\theta$

$\bar{x} = \left(-\frac{32}{5}a^2\right)/(8a) = -\frac{4}{5}a, \quad \bar{y} = 0$ by symmetry

*27.   (a) $x(\theta) = R\theta - b\sin\theta, \quad y(\theta) = R - b\cos\theta$

(b)

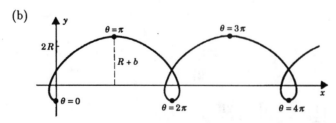

CHAPTER 11

SECTION 11.1

1. decreasing;   bounded below by 0 and above by 2

3. increasing;   bounded below by 1 but not bounded above

5. $\dfrac{n+(-1)^n}{n}=1+(-1)^n\dfrac{1}{n}$: not monotonic;   bounded below by 0 and above by $\dfrac{3}{2}$

7. decreasing; bounded below by 0 and above by 0.9

9. $\dfrac{n^2}{n+1}=n-1+\dfrac{1}{n+1}$: increasing;   bounded below by $\dfrac{1}{2}$ but not bounded above

11. $\dfrac{4n}{\sqrt{4n^2+1}}=\dfrac{2}{\sqrt{1+1/4n^2}}$ and $\dfrac{1}{4n^2}$ decreases to 0: increasing;

 bounded below by $\frac{4}{5}\sqrt{5}$ and above by 2

13. increasing;   bounded below by $\frac{2}{51}$ but not bounded above

15. decreasing;   bounded below by 0 and above by $\frac{1}{2}(10^{10})$

17. $\dfrac{2n}{n+1}=2-\dfrac{2}{n+1}$ increases toward 2: increasing;   bounded below by 0 and above by ln 2

19. decreasing;   bounded below by 1 and above by 4

21. increasing;   bounded below by $\sqrt{3}$ and above by 2

23. $(-1)^{2n+1}\sqrt{n}=-\sqrt{n}$: decreasing;   bounded above by -1 but not bounded below

25. $\dfrac{2^n-1}{2^n}=1-\dfrac{1}{2^n}$: increasing;   bounded below by $\dfrac{1}{2}$ and above by 1

27. consider $\sin x$ as $x\to 0^+$: decreasing;   bounded below by 0 and above by 1

29. decreasing;   bounded below by 0 and above by $\frac{5}{6}$

31. $\dfrac{1}{n}-\dfrac{1}{n+1}=\dfrac{1}{n(n+1)}$: decreasing;   bounded below by 0 and above by $\dfrac{1}{2}$

33. Set $f(x)=\dfrac{\ln x}{x}$.   Then, $f'(x)=\dfrac{1-\ln x}{x^2}<0$ for $x>e$: decreasing;

 bounded below by 0 and above by $\frac{1}{3}\ln 3$.

35. Set $a_n=\dfrac{3^n}{(n+1)^2}$.   Then, $\dfrac{a_{n+1}}{a_n}=3\left(\dfrac{n+1}{n+2}\right)^2>1$: increasing;

 bounded below by $\frac{3}{4}$ but not bounded above.

37. For $n \geq 5$

$$\frac{a_{n+1}}{a_n} = \frac{5^{n+1}}{(n+1)!} \cdot \frac{n!}{5^n} = \frac{5}{n+1} < 1 \quad \text{and thus} \quad a_{n+1} < a_n.$$

Sequence is not nonincreasing: $a_1 = 5 < \frac{25}{2} = a_2$.

39. boundedness: $\quad 0 < (c^n + d^n)^{1/n} < (2d^n)^{1/n} = 2^{1/n} d \leq 2d$

monotonicity : $\quad a_{n+1}^{n+1} = c^{n+1} + d^{n+1} = cc^n + dd^n$

$$< (c^n + d^n)^{1/n} c^n + (c^n + d^n)^{1/n} d^n$$

$$= (c^n + d^n)^{1+1/n}$$

$$= (c^n + d^n)^{(n+1)/n}$$

$$= a_n^{n+1}$$

Taking the $(n+1)^{\text{th}}$ root of each side we have $\quad a_{n+1} < a_n.$ The sequence is monotonic decreasing.

41. $a_1 = 1, \quad a_2 = \frac{1}{2}, \quad a_3 = \frac{1}{6}, \quad a_4 = \frac{1}{24}, \quad a_5 = \frac{1}{120}, \quad a_6 = \frac{1}{720}; \quad a_n = 1/n!$

43. $a_1 = a_2 = a_3 = a_4 = a_5 = a_6 = 1; \quad a_n = 1$

45. $a_1 = 1, \quad a_2 = 3, \quad a_3 = 5, \quad a_4 = 7, \quad a_5 = 9, \quad a_6 = 11; \quad a_n = 2n - 1$

47. $a_1 = 1, \quad a_2 = 4, \quad a_3 = 13, \quad a_4 = 40, \quad a_5 = 121, \quad a_6 = 364; \quad a_n = \frac{1}{2}(3^n - 1)$

49. $a_1 = 1, \quad a_2 = 4, \quad a_3 = 9, \quad a_4 = 16, \quad a_5 = 25, \quad a_6 = 36; \quad a_n = n^2$

51. $a_1 = 1, \quad a_2 = 3, \quad a_3 = 4, \quad a_4 = 8, \quad a_5 = 16, \quad a_6 = 32; \quad a_n = 2^{n-1} \quad (n \geq 3)$

53. $a_1 = 2, \quad a_2 = 1, \quad a_3 = 2, \quad a_4 = 1, \quad a_5 = 2, \quad a_6 = 1; \quad a_n = \frac{1}{2}[3 - (-1)^n]$

55. $a_1 = 1, \quad a_2 = 3, \quad a_3 = 5, \quad a_4 = 7, \quad a_5 = 9, \quad a_6 = 11; \quad a_n = 2n - 1$

57. First $a_1 = 2^1 - 1 = 1.$ Next suppose $a_k = 2^k - 1$ for some $k \geq 1.$ Then

$$a_{k+1} = 2a_k + 1 = 2\left(2^k - 1\right) + 1 = 2^{k+1} - 1.$$

59. First $a_1 = \frac{1}{2^0} = 1.$ Next suppose $a_k = \frac{k}{2^{k-1}}$ for some $k \geq 1.$ Then

$$a_{k+1} = \frac{k+1}{2k} a_k = \frac{k+1}{2k} \frac{k}{2^{k-1}} = \frac{k+1}{2^k}.$$

SECTION 11.2

1. diverges

3. converges to 0

5. converges to 1: $\quad \dfrac{n-1}{n} = 1 - \dfrac{1}{n} \to 1$

7. converges to 0: $\quad \dfrac{n+1}{n^2} = \dfrac{1}{n} + \dfrac{1}{n^2} \to 0$

9. converges to 0: $\quad 0 < \dfrac{2^n}{4^n + 1} < \dfrac{2^n}{4^n} = \dfrac{1}{2^n} \to 0$

11. diverges

13. converges to 0

15. converges to 1: $\dfrac{n\pi}{4n+1} \to \dfrac{\pi}{4}$ so $\tan \dfrac{n\pi}{4n+1} \to \tan \dfrac{\pi}{4} = 1$

17. converges to $\dfrac{4}{9}$: $\dfrac{(2n+1)^2}{(3n-1)^2} = \dfrac{4+4/n+1/n^2}{9-6/n+1/n^2} \to \dfrac{4}{9}$

19. converges to $\dfrac{1}{2}\sqrt{2}$: $\dfrac{n^2}{\sqrt{2n^4+1}} = \dfrac{1}{\sqrt{2+1/n^4}} \to \dfrac{1}{\sqrt{2}}$

21. diverges: $\cos n\pi = (-1)^n$ **23.** converges to 1: $\dfrac{1}{\sqrt{n}} \to 0$ so $e^{1/\sqrt{n}} \to e^0 = 1$

25. diverges

27. converges to 0 : $\ln n - \ln(n+1) = \ln\left(\dfrac{n}{n+1}\right) \to \ln 1 = 0$

29. converges to $\dfrac{1}{2}$: $\dfrac{\sqrt{n+1}}{2\sqrt{n}} = \dfrac{1}{2}\sqrt{1+\dfrac{1}{n}} \to \dfrac{1}{2}$

31. converges to e^2: $\left(1+\dfrac{1}{n}\right)^{2n} = \left[\left(1+\dfrac{1}{n}\right)^n\right]^2 \to e^2$

33. diverges;   since $2^n > n^3$ for $n \geq 10$, $\dfrac{2^n}{n^2} > \dfrac{n^3}{n^2} = n$

35. converges to 0: $\left|\dfrac{\sqrt{n}\sin(e^n\pi)}{n+1}\right| = \dfrac{|\sin(e^n\pi)|}{\sqrt{n}+\dfrac{1}{\sqrt{n}}} \leq \dfrac{1}{\sqrt{n}+\dfrac{1}{\sqrt{n}}} \to 0$

37. Set $\epsilon > 0$. Since $a_n \to l$, there exists N_1 such that

$$\text{if}\quad n \leq N_1, \quad \text{then}\quad |a_n - l| < \epsilon/2.$$

Since $b_n \to m$, there exists N_2 such that

$$\text{if}\quad n \geq N_2, \quad \text{then}\quad |b_n - l| < \epsilon/2.$$

Now set $N = \max\{N_1, N_2\}$. Then, for $n \geq N$,

$$|(a_n + b_n) - (l + m)| \leq |a_n - l| + |b_n - m| < \frac{\epsilon}{2} + \frac{\epsilon}{2} = \epsilon.$$

39. Since $\left(1+\dfrac{1}{n}\right) \to 1$ and $\left(1+\dfrac{1}{n}\right)^n \to e$,

$$\left(1+\frac{1}{n}\right)^{n+1} = \left(1+\frac{1}{n}\right)^n \left(1+\frac{1}{n}\right) \to (e)(1) = e.$$

41. Suppose that $\{a_n\}$ is bounded and non-increasing. If l is the greatest lower bound of the range of this sequence, then $a_n \geq l$ for all n. Set $\epsilon > 0$. By Theorem 10.8.4 there exists a_k such that $a_k < l + \epsilon$. Since the sequence is non-increasing, $a_n \leq a_k$ for all $n \geq k$. Thus,

$$l \leq a_n < l + \epsilon \quad \text{or} \quad |a_n - l| < \epsilon \quad \text{for all} \quad n \geq k$$

and $a_n \to l$.

43. Let $\epsilon > 0$.   Choose k so that, for $n \geq k$,

$$l - \epsilon < a_n < l + \epsilon, \quad l - \epsilon < c_n < l + \epsilon \quad \text{and} \quad a_n \leq b_n \leq c_n.$$

For such n,

$$l - \epsilon < b_n < l + \epsilon.$$

45. Set $f(x) = x^{1/p}$. Since $\dfrac{1}{n} \to 0$ and f is continuous at 0, it follows by Theorem 11.2.12 that

$$\left(\frac{1}{n}\right)^{1/p} \to 0.$$

47. $a_n = e^{1-n} \to 0$

49. $a_n = \dfrac{1}{n!} \to 0$

51. $a_n = \dfrac{1}{2}[1 - (-1)^n] \quad$ diverges

53. $a_n = \dfrac{2^n - 1}{2^{n-1}} \to 2$

55. $l = 0, \quad n = 32$

57. $l = 0, \quad n = 4$

59. $l = 0, \quad n = 7$

61. $l = 0, \quad n = 65$

SECTION 11.3

1.   converges to 1: $2^{2/n} = (2^{1/n})^2 \to 1^2 = 1$

3.   converges to 0:   for $n > 3$, $0 < \left(\dfrac{2}{n}\right)^n < \left(\dfrac{2}{3}\right)^n \to 0$

5.   converges to 0: $\dfrac{\ln(n+1)}{n} = \left[\dfrac{\ln(n+1)}{n+1}\right]\left(\dfrac{n+1}{n}\right) \to (0)(1) = 0$

7.   converges to 0: $\dfrac{x^{100n}}{n!} = \dfrac{(x^{100})^n}{n!} \to 0$

9.   converges to 1: $n^{\alpha/n} = (n^{1/n})^\alpha \to 1^\alpha = 1$

11.   converges to 0: $\dfrac{3^{n+1}}{4^{n-1}} = 12\left(\dfrac{3^n}{4^n}\right) = 12\left(\dfrac{3}{4}\right)^n \to 12(0) = 0$

13.   converges to 1: $(n+2)^{1/n} = e^{\frac{1}{n}\ln(n+2)}$ and, since

$$\frac{1}{n}\ln(n+2) = \left[\frac{\ln(n+2)}{n+2}\right]\left(\frac{n+2}{n}\right) \to (0)(1) = 0,$$

it follows that

$$(n + 2)^{1/n} \to e^0 = 1.$$

15. converges to 1: $\displaystyle\int_0^n e^{-x}\,dx = 1 - \frac{1}{e^n} \to 1$

17. converges to π: $\text{integral} = 2\displaystyle\int_0^n \frac{dx}{1 + x^2} = 2\tan^{-1} n \to 2\left(\frac{\pi}{2}\right) = \pi$

19. converges to 1: recall (11.3.6)

21. converges to 0: $\dfrac{\ln\left(n^2\right)}{n} = 2\dfrac{\ln n}{n} \to 2(0) = 0$

23. diverges: since $\displaystyle\lim_{x \to 0} \frac{\sin x}{x} = 1,$

$$\frac{n}{\pi}\sin\frac{\pi}{n} = \frac{\sin\left(\pi/n\right)}{\pi/n} \to 1$$

and, for n sufficiently large,

$$n^2 \sin\frac{\pi}{n} = n\pi\left(\frac{n}{\pi}\sin\frac{\pi}{n}\right) > n\pi\left(\frac{1}{2}\right) = \frac{n\pi}{2}$$

25. converges to 0: $\dfrac{5^{n+1}}{4^{2n-1}} = 20\left(\dfrac{5}{16}\right)^n \to 0$

27. converges to e^{-1}: $\left(\dfrac{n+1}{n+2}\right)^n = \left(1 - \dfrac{1}{n+2}\right)^n = \dfrac{\left(1 + \dfrac{(-1)}{n+2}\right)^{n+2}}{\left(1 + \dfrac{(-1)}{n+2}\right)^2} \to \dfrac{e^{-1}}{1} = e^{-1}$

29. converges to 0: $0 < \displaystyle\int_n^{n+1} e^{-x^2}\,dx \le e^{-n^2}[(n+1) - n] = e^{-n^2} \to 0$

31. converges to 0: $\dfrac{n^n}{2^{n^2}} = \left(\dfrac{n}{2^n}\right)^n \to 0$ since $\dfrac{n}{2^n} \to 0$

33. converges to e^x: recall (11.3.7)

35. converges to 0: $\left|\displaystyle\int_{-1/n}^{1/n} \sin x^2\,dx\right| \le \displaystyle\int_{-1/n}^{1/n} |\sin x^2|\,dx \le \displaystyle\int_{-1/n}^{1/n} 1\,dx = \dfrac{2}{n} \to 0$

37. $\sqrt{n+1} - \sqrt{n} = \dfrac{\sqrt{n+1} - \sqrt{n}}{\sqrt{n+1} + \sqrt{n}}\left(\sqrt{n+1} + \sqrt{n}\right) = \dfrac{1}{\sqrt{n+1} + \sqrt{n}} \to 0$

39. $\displaystyle\lim_{n \to \infty} 2n\sin\left(\pi/n\right) = 2\pi$; the number $2n\sin\left(\pi/n\right)$ is the perimeter of a regular polygon of n sides inscribed in the unit circle. As n tends to ∞, the perimeter of the polygon tends to the circumference of the circle.

41. By the hint, $\displaystyle\lim_{n\to\infty}\frac{1+2+\cdots+n}{n^2}=\lim_{n\to\infty}\frac{n(n+1)}{2n^2}=\lim_{n\to\infty}\frac{1+1/n}{2}=\frac{1}{2}.$

43. By the hint, $\displaystyle\lim_{n\to\infty}\frac{1^3+2^3+\cdots+n^3}{2n^4+n-1}=\lim_{n\to\infty}\frac{n^2(n+1)^2}{4(2n^4+n-1)}=\lim_{n\to\infty}\frac{1+2/n+1/n^2}{8+4/n^3-4/n^4}=\frac{1}{8}.$

45. (a)
$$m_{n+1}-m_n=\frac{1}{n+1}(a_1+\cdots+a_n+a_{n+1})-\frac{1}{n}(a_1+\cdots+a_n)$$

$$=\frac{1}{n(n+1)}\big[\,na_{n+1}-(\overbrace{a_1+\cdots+a_n}^{n})\,\big]$$

$$>0\quad\text{since }\{a_n\}\text{ is increasing.}$$

(b)   We begin with the hint
$$m_n<\frac{|a_1+\cdots+a_j|}{n}+\frac{\epsilon}{2}\left(\frac{n-j}{n}\right).$$

Since j is fixed,
$$\frac{|a_1+\cdots+a_j|}{n}\to 0$$

and therefore for n sufficiently large
$$\frac{|a_1+\cdots+a_j|}{n}<\frac{\epsilon}{2}.$$

Since
$$\frac{\epsilon}{2}\left(\frac{n-j}{n}\right)<\frac{\epsilon}{2},$$

we see that, for n sufficiently large, $|m_n|<\epsilon$. This shows that $m_n\to 0$.

47. The numerical work suggests $l\cong 1$. Justification: Set $f(x)=\sin x-x^2$. Note that $f(0)=0$ and $f'(x)=\cos x-2x>0$ for x close to 0. Therefore $\sin x-x^2>0$ for x close to 0 and $\sin 1/n-1/n^2>0$ for n large. Thus, for n large,

$$\frac{1}{n^2}<\sin\frac{1}{n}<\frac{1}{n}$$

$$\llcorner\quad |\sin x|\le|x|\quad\text{for all }x$$

$$\left(\frac{1}{n^2}\right)^{1/n}<\left(\sin\frac{1}{n}\right)^{1/n}<\left(\frac{1}{n}\right)^{1/n}$$

$$\left(\frac{1}{n^{1/n}}\right)^2<\left(\sin\frac{1}{n}\right)^{1/n}<\frac{1}{n^{1/n}}.$$

As $n\to\infty$ both bounds tend to 1 and therefore the middle term also tends to 1.

SECTION 11.4

(We'll use $\star$ to indicate differentiation of numerator and denominator.)

1. $\displaystyle\lim_{x\to0+}\frac{\sin x}{\sqrt{x}}\overset{\star}{=}\lim_{x\to0+}2\sqrt{x}\cos x=0$ **3.** $\displaystyle\lim_{x\to0}\frac{e^x-1}{\ln(1+x)}\overset{\star}{=}\lim_{x\to0}(1+x)e^x=1$

5. $\displaystyle\lim_{x\to\pi/2}\frac{\cos x}{\sin 2x}\overset{\star}{=}\lim_{x\to\pi/2}\frac{-\sin x}{2\cos 2x}=\frac{1}{2}$

7. $\displaystyle\lim_{x\to 0}\frac{2^x-1}{x}\overset{\star}{=}\lim_{x\to 0}2^x\ln 2=\ln 2$

9. $\displaystyle\lim_{x\to 1}\frac{x^{1/2}-x^{1/4}}{x-1}\overset{\star}{=}\lim_{x\to 1}\left(\frac{1}{2}x^{-1/2}-\frac{1}{4}x^{-3/4}\right)=\frac{1}{4}$

11. $\displaystyle\lim_{x\to 0}\frac{e^x-e^{-x}}{\sin x}\overset{\star}{=}\lim_{x\to 0}\frac{e^x+e^{-x}}{\cos x}=2$

13. $\displaystyle\lim_{x\to 0}\frac{x+\sin\pi x}{x-\sin\pi x}\overset{\star}{=}\lim_{x\to 0}\frac{1+\pi\cos\pi x}{1-\pi\cos\pi x}=\frac{1+\pi}{1-\pi}$

15. $\displaystyle\lim_{x\to 0}(e^x+x)^{1/x}=e^2$ since $\displaystyle\lim_{x\to 0}\ln[(e^x+x)^{1/x}]=\lim_{x\to 0}\frac{\ln(e^x+x)}{x}\overset{\star}{=}\lim_{x\to 0}\frac{e^x+1}{e^x+x}=2$

17. $\displaystyle\lim_{x\to 0}\frac{\tan\pi x}{e^x-1}\overset{\star}{=}\lim_{x\to 0}\frac{\pi\sec^2\pi x}{e^x}=\pi$

19. $\displaystyle\lim_{x\to 0}\frac{1+x-e^x}{x(e^x-1)}\overset{\star}{=}\lim_{x\to 0}\frac{1-e^x}{xe^x+e^x-1}\overset{\star}{=}\lim_{x\to 0}\frac{-e^x}{xe^x+2e^x}=-\frac{1}{2}$

21. $\displaystyle\lim_{x\to 0}\frac{x-\tan x}{x-\sin x}\overset{\star}{=}\lim_{x\to 0}\frac{1-\sec^2 x}{1-\cos x}\overset{\star}{=}\lim_{x\to 0}\frac{-2\sec^2 x\tan x}{\sin x}=\lim_{x\to 0}\frac{-2\sec^2 x}{\cos x}=-2$

23. $\displaystyle\lim_{x\to 1^-}\frac{\sqrt{1-x^2}}{\sqrt{1-x^3}}=\lim_{x\to 1^-}\sqrt{\frac{1-x^2}{1-x^3}}=\sqrt{\frac{2}{3}}=\frac{1}{3}\sqrt{6}$ since $\displaystyle\lim_{x\to 1^-}\frac{1-x^2}{1-x^3}\overset{\star}{=}\lim_{x\to 1^-}\frac{2x}{3x^2}=\frac{2}{3}$

25. $\displaystyle\lim_{x\to\pi/2}\frac{\ln(\sin x)}{(\pi-2x)^2}\overset{\star}{=}\lim_{x\to\pi/2}\frac{-\cot x}{4(\pi-2x)}\overset{\star}{=}\lim_{x\to\pi/2}\frac{\csc^2 x}{-8}=-\frac{1}{8}$

27. $\displaystyle\lim_{x\to 1}\left(\frac{1}{\ln x}-\frac{x}{x-1}\right)=\lim_{x\to 1}\frac{x-1-x\ln x}{(x-1)\ln x}\overset{\star}{=}\lim_{x\to 1}\frac{-x\ln x}{x-1+x\ln x}$

$$\overset{\star}{=}\lim_{x\to 1}\frac{-1-\ln x}{1+1+\ln x}=-\frac{1}{2}$$

29. $\displaystyle\lim_{x\to\pi/4}\frac{\sec^2 x-2\tan x}{1+\cos 4x}\overset{\star}{=}\lim_{x\to\pi/4}\frac{2\sec^2 x\,\tan x-2\sec^2 x}{-4\sin 4x}$

$$\overset{\star}{=}\lim_{x\to\pi/4}\frac{2\sec^4 x+4\sec^2 x\tan^2 x-4\sec^2 x\tan x}{-16\cos 4x}=\frac{1}{2}$$

31. $1;\quad\displaystyle\lim_{x\to\infty}x(\pi/2-\tan^{-1}x)=\lim_{x\to\infty}\frac{\pi/2-\tan^{-1}x}{1/x}\overset{\star}{=}\lim_{x\to\infty}\frac{x^2}{1+x^2}=1$

33. $1;\quad\displaystyle\lim_{x\to\infty}\frac{1}{x[\ln(x+1)-\ln x]}=\lim_{x\to\infty}\frac{1/x}{\ln(1+1/x)}=\lim_{t\to 0^+}\frac{t}{\ln(1+t)}\overset{\star}{=}\lim_{t\to 0^+}(1+t)=1$

35. $\displaystyle\lim_{x\to 0}(2+x+\sin x)\neq 0,\quad\lim_{x\to 0}(x^3+x-\cos x)\neq 0$

37. $\displaystyle\lim_{x\to 0}\frac{1}{x}\int_0^x f(t)\,dt \overset{\star}{=} \lim_{x\to 0}\frac{f(x)}{1} = f(0)$

39. The numerical work suggests $l \cong -\frac{1}{6}$.   Justification:

$$\lim_{x\to\infty}\left[x^3\left(\sin\frac{1}{x}-\frac{1}{x}\right)\right] = \lim_{x\to 0+}\frac{\sin x - x}{x^3} \overset{\star}{=} \lim_{x\to 0+}\frac{\cos x - 1}{3x^2} \overset{\star}{=} \lim_{x\to 0+}\frac{-\sin x}{6x} = -\frac{1}{6}.$$

SECTION 11.5

(We'll use $\star$ to indicate differentiation of numerator and denominator.)

1. $\displaystyle\lim_{x\to-\infty}\frac{x^2+1}{1-x} \overset{\star}{=} \lim_{x\to-\infty}\frac{2x}{-1} = \infty$

3. $\displaystyle\lim_{x\to\infty}\frac{x^3}{1-x^3} = \lim_{x\to\infty}\frac{1}{1/x^3-1} = -1$

5. $\displaystyle\lim_{x\to\infty}x^2\sin\frac{1}{x} = \lim_{h\to 0+}\left[\left(\frac{1}{h}\right)\left(\frac{\sin h}{h}\right)\right] = \infty$

7. $\displaystyle\lim_{x\to\frac{\pi}{2}-}\frac{\tan 5x}{\tan x} = \lim_{x\to\frac{\pi}{2}-}\left[\left(\frac{\sin 5x}{\sin x}\right)\left(\frac{\cos x}{\cos 5x}\right)\right] = \frac{1}{5}$ since

$$\lim_{x\to\frac{\pi}{2}-}\frac{\sin 5x}{\sin x} = 1 \quad\text{and}\quad \lim_{x\to\frac{\pi}{2}-}\frac{\cos x}{\cos 5x} \overset{\star}{=} \lim_{x\to\frac{\pi}{2}-}\frac{\sin x}{5\sin 5x} = \frac{1}{5}$$

9. $\displaystyle\lim_{x\to 0+}x^{2x} = \lim_{x\to 0+}(x^x)^2 = 1^2 = 1$ [see (11.5.4)]

11. $\displaystyle\lim_{x\to 0}x(\ln|x|)^2 = \lim_{x\to 0}\frac{(\ln|x|)^2}{1/x} \overset{\star}{=} \lim_{x\to 0}\frac{2\ln|x|}{-1/x} \overset{\star}{=} \lim_{x\to 0}\frac{2}{1/x} = 0$

13. $\displaystyle\lim_{x\to\infty}\frac{1}{x}\int_0^x e^{t^2}\,dt \overset{\star}{=} \lim_{x\to\infty}\frac{e^{x^2}}{1} = \infty$

15. $$\lim_{x\to 0}\left[\frac{1}{\sin^2 x}-\frac{1}{x^2}\right] = \lim_{x\to 0}\frac{x^2-\sin^2 x}{x^2\sin^2 x} \overset{\star}{=} \lim_{x\to 0}\frac{2x-2\sin x\cos x}{2x^2\sin x\cos x + 2x\sin^2 x}$$

$$= \lim_{x\to 0}\frac{2x-\sin 2x}{x^2\sin 2x + 2x\sin^2 x}$$

$$\overset{\star}{=} \lim_{x\to 0}\frac{2-2\cos 2x}{2x^2\cos 2x + 4x\sin 2x + 2\sin^2 x}$$

$$\overset{\star}{=} \lim_{x\to 0}\frac{4\sin 2x}{-4x^2\sin 2x + 12x\cos 2x + 6\sin 2x}$$

$$\overset{\star}{=} \lim_{x\to 0}\frac{8\cos 2x}{-8x^2\cos 2x - 32x\sin 2x + 24\cos 2x} = \frac{1}{3}$$

17. $\lim\limits_{x\to 1} x^{1/(x-1)} = e$ since $\lim\limits_{x\to 1} \ln\left[x^{1/(x-1)}\right] = \lim\limits_{x\to 1} \dfrac{\ln x}{x-1} \stackrel{*}{=} \lim\limits_{x\to 1} \dfrac{1}{x} = 1$

19. $\lim\limits_{x\to\infty}\left(\cos\dfrac{1}{x}\right)^x = 1$ since $\lim\limits_{x\to\infty} \ln\left[\left(\cos\dfrac{1}{x}\right)^x\right] = \lim\limits_{x\to\infty} \dfrac{\ln\left(\cos\dfrac{1}{x}\right)}{(1/x)}$

$$\stackrel{*}{=} \lim\limits_{x\to\infty} \left(-\dfrac{\sin(1/x)}{\cos(1/x)}\right) = 0$$

21 $\lim\limits_{x\to 0}\left[\dfrac{1}{\ln(1+x)} - \dfrac{1}{x}\right] = \lim\limits_{x\to 0} \dfrac{x - \ln(1+x)}{x\ln(1+x)} \stackrel{*}{=} \lim\limits_{x\to 0} \dfrac{x}{x+(1+x)\ln(1+x)}$

$$\stackrel{*}{=} \lim\limits_{x\to 0} \dfrac{1}{1+1+\ln(1+x)} = \dfrac{1}{2}$$

23. $\lim\limits_{x\to 1}\left[\dfrac{x}{x-1} - \dfrac{1}{\ln x}\right] = \lim\limits_{x\to 1} \dfrac{x\ln x - x + 1}{(x-1)\ln x} \stackrel{*}{=} \lim\limits_{x\to 1} \dfrac{\ln x}{1 + \ln x - 1/x}$

$$\stackrel{*}{=} \lim\limits_{x\to 1} \dfrac{1/x}{1/x + 1/x^2} = \dfrac{1}{2}$$

25. $\lim\limits_{x\to\infty}\left(\sqrt{x^2+2x} - x\right) = \lim\limits_{x\to\infty}\left[\left(\sqrt{x^2+2x} - x\right)\left(\dfrac{\sqrt{x^2+2x}+x}{\sqrt{x^2+2x}+x}\right)\right]$

$$= \lim\limits_{x\to\infty} \dfrac{2x}{\sqrt{x^2+2x}+x} = \lim\limits_{x\to\infty} \dfrac{2}{\sqrt{1+2/x}+1} = 1$$

27. $\lim\limits_{x\to\infty}\left(x^3+1\right)^{1/\ln x} = e^3$ since

$$\lim\limits_{x\to\infty} \ln\left[\left(x^3+1\right)^{1/\ln x}\right] = \lim\limits_{x\to\infty} \dfrac{\ln\left(x^3+1\right)}{\ln x} \stackrel{*}{=} \lim\limits_{x\to\infty} \dfrac{\left(\dfrac{3x^2}{x^3+1}\right)}{1/x} = \lim\limits_{x\to\infty} \dfrac{3}{1+1/x^3} = 3.$$

29. $\lim\limits_{x\to\infty}(\cosh x)^{1/x} = e$ since

$$\lim\limits_{x\to\infty} \ln\left[(\cosh x)^{1/x}\right] = \lim\limits_{x\to\infty} \dfrac{\ln(\cosh x)}{x} \stackrel{*}{=} \lim\limits_{x\to\infty} \dfrac{\sinh x}{\cosh x} = 1.$$

31. $0;$ $\dfrac{1}{n}\ln\dfrac{1}{n} = -\dfrac{\ln n}{n} \to 0$ **33.** $1;$ $\ln\left[(\ln n)^{1/n}\right] = \dfrac{1}{n}\ln(\ln n) \to 0$

35. $1;$ $\ln\left[\left(n^2+n\right)^{1/n}\right] = \dfrac{1}{n}\ln[n(n+1)] = \dfrac{\ln n}{n} + \dfrac{\ln(n+1)}{n} \to 0$

37. $0;$ $0 \le \dfrac{n^2\ln n}{e^n} < \dfrac{n^3}{e^n},$ $\lim\limits_{x\to\infty} \dfrac{x^3}{e^x} = 0$

39.

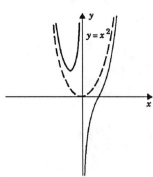

$y = x^2$

vertical asymptote y-axis

41.

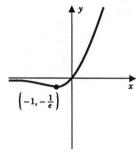

$\left(-1, -\frac{1}{e}\right)$

horizontal asymptote x-axis

43.

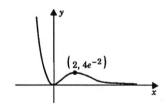

$\left(2, 4e^{-2}\right)$

horizontal asymptote x-axis

45. $\dfrac{b}{a}\sqrt{x^2 - a^2} - \dfrac{b}{a}x = \dfrac{\sqrt{x^2 - a^2} + x}{\sqrt{x^2 - a^2} + x}\left(\dfrac{b}{a}\right)\left(\sqrt{x^2 - a^2} - x\right) = \dfrac{-ab}{\sqrt{x^2 - a^2} + x} \to 0$ as $x \to \infty$

47. for instance, $f(x) = x^3 + \dfrac{1}{x}$ **49.** for instance, $f(x) = x + \dfrac{\sin x}{x}$

51. $\displaystyle\lim_{x \to 0^-} -\dfrac{2x}{\cos x} \neq \lim_{x \to 0^-} \dfrac{2}{-\sin x}$. L'Hospital's rule does not apply here since $\displaystyle\lim_{x \to 0^-} \cos x = 1$.

SECTION 11.6

1. 1; $\displaystyle\int_1^\infty \dfrac{dx}{x^2} = \lim_{b \to \infty} \int_1^b \dfrac{dx}{x^2} = \lim_{b \to \infty}\left[-\dfrac{1}{x}\right]_1^b = \lim_{b \to \infty}\left[1 - \dfrac{1}{b}\right] = 1$

3. $\dfrac{\pi}{4}$; $\displaystyle\int_0^\infty \dfrac{dx}{4 + x^2} = \lim_{b \to \infty} \int_0^b \dfrac{dx}{4 + x^2} = \lim_{b \to \infty}\left[\dfrac{1}{2}\tan^{-1}\dfrac{x}{2}\right]_0^b = \lim_{b \to \infty}\dfrac{1}{2}\tan^{-1}\left(\dfrac{b}{2}\right) = \dfrac{\pi}{4}$

5. diverges; $\displaystyle\int_0^\infty e^{px}\,dx = \lim_{b \to \infty} \int_0^b e^{px}\,dx = \lim_{b \to \infty}\left[\dfrac{1}{p}e^{px}\right]_0^b = \lim_{b \to \infty}\dfrac{1}{p}\left(e^{pb} - 1\right) = \infty$

7. 6; $\displaystyle\int_0^8 \dfrac{dx}{x^{2/3}} = \lim_{a \to 0^+} \int_a^8 x^{-2/3}\,dx = \lim_{a \to 0^+}\left[3x^{1/3}\right]_a^8 = \lim_{a \to 0^+}\left[6 - 3a^{1/3}\right] = 6$

9. $\frac{\pi}{2}$; $\displaystyle\int_0^1 \frac{dx}{\sqrt{1-x^2}} = \lim_{b\to 1^-}\int_0^b \frac{dx}{\sqrt{1-x^2}} = \lim_{b\to 1^-}\sin^{-1}b = \frac{\pi}{2}$

11. 2; $\displaystyle\int_0^2 \frac{x}{\sqrt{4-x^2}}\,dx = \lim_{b\to 2^-}\int_0^b x\left(4-x^2\right)^{-1/2}dx = \lim_{b\to 2^-}\left[-\left(4-x^2\right)^{1/2}\right]_0^b$

$$= \lim_{b\to 2^-}\left(2-\sqrt{4-b^2}\right) = 2$$

13. diverges; $\displaystyle\int_e^\infty \frac{\ln x}{x}\,dx = \lim_{b\to\infty}\int_e^b \frac{\ln x}{x}\,dx = \lim_{b\to\infty}\left[\frac{1}{2}\left(\ln x\right)^2\right]_e^b$

$$= \lim_{b\to\infty}\left[\frac{1}{2}\left(\ln b\right)^2 - \frac{1}{2}\right] = \infty$$

15. $-\frac{1}{4}$; $\displaystyle\int_0^1 x\ln x\,dx = \lim_{a\to 0^+}\int_a^1 x\ln x\,dx = \lim_{a\to 0^+}\left[\frac{1}{2}x^2\ln x - \frac{1}{4}x^2\right]_a^1$

(by parts)

$$= \lim_{a\to 0^+}\left[\frac{1}{4}a^2 - \frac{1}{2}a^2\ln a - \frac{1}{4}\right] = -\frac{1}{4}$$

Note: $\displaystyle\lim_{t\to 0^+}t^2\ln t = \lim_{t\to 0^+}\frac{\ln t}{1/t^2} \overset{*}{=} \lim_{t\to 0^+}\frac{1/t}{-2/t^3} = -\frac{1}{2}\lim_{t\to 0^+}t^2 = 0.$

17. π; $\displaystyle\int_{-\infty}^\infty \frac{dx}{1+x^2} = \lim_{a\to -\infty}\int_a^0 \frac{dx}{1+x^2} + \lim_{b\to\infty}\int_0^b \frac{dx}{1+x^2}$

$$= \lim_{a\to -\infty}\left[\tan^{-1}x\right]_a^0 + \lim_{b\to\infty}\left[\tan^{-1}x\right]_0^b = -\left(-\frac{\pi}{2}\right) + \frac{\pi}{2} = \pi$$

19. diverges; $\displaystyle\int_{-\infty}^\infty \frac{dx}{x^2} = \lim_{a\to -\infty}\int_a^{-1}\frac{dx}{x^2} + \lim_{b\to 0^-}\int_{-1}^b \frac{dx}{x^2} + \lim_{c\to 0^+}\int_c^1 \frac{dx}{x^2} + \lim_{d\to\infty}\int_1^d \frac{dx}{x^2}$;

and, $\displaystyle\lim_{c\to 0^+}\int_c^1 \frac{dx}{x^2} = \lim_{c\to 0^+}\left[-\frac{1}{x}\right]_c^1 = \lim_{c\to 0^+}\left[\frac{1}{c} - 1\right] = \infty$

21. $\ln 2$;

$$\int_1^\infty \frac{dx}{x(x+1)} = \lim_{b\to\infty} \int_1^b \left[\frac{1}{x} - \frac{1}{x+1}\right] dx$$

$$= \lim_{b\to\infty} \left[\ln\left(\frac{x}{x+1}\right)\right]_1^b = \lim_{b\to\infty} \left[\ln\left(\frac{b}{b+1}\right) - \ln\left(\frac{1}{2}\right)\right]$$

$$= 0 - \ln\tfrac{1}{2} = \ln 2$$

23. 4;

$$\int_3^5 \frac{x}{\sqrt{x^2-9}}\, dx = \lim_{a\to3-} \int_a^5 x\left(x^2-9\right)^{-1/2} dx$$

$$= \lim_{a\to3-} \left[(x^2-9)^{1/2}\right]_a^5 = \lim_{a\to3-} \left[4 - (a^2-9)^{1/2}\right] = 4$$

25. $\int_{-3}^3 \frac{dx}{x(x+1)}$ diverges since $\int_0^3 \frac{dx}{x(x+1)}$ diverges:

$$\int_0^3 \frac{dx}{x(x+1)} = \int_0^3 \left(\frac{1}{x} - \frac{1}{x+1}\right) dx = \lim_{a\to0+} \left[\ln|x| - \ln|x+1|\right]_a^3$$

$$= \lim_{a\to0+} \left[\ln 3 - \ln 4 - \ln a + \ln(a+1)\right] = \infty.$$

27. $\int_{-3}^1 \frac{dx}{x^2-4}$ diverges since $\int_{-2}^1 \frac{dx}{x^2-4}$ diverges:

$$\int_{-2}^1 \frac{dx}{x^2-4} = \int_{-2}^1 \frac{1}{4}\left[\frac{1}{x-2} - \frac{1}{x+2}\right] dx$$

$$= \lim_{a\to-2+} \left[\frac{1}{4}\left(\ln|x-2| - \ln|x+2|\right)\right]_a^1$$

$$= \lim_{a\to-2+} \frac{1}{4}\left[-\ln 3 - \ln|a-2| + \ln|a+2|\right] = -\infty.$$

29.   diverges: $\int_0^\infty \cosh x\, dx = \lim_{b\to\infty} \int_0^b \cosh x\, dx = \lim_{b\to\infty} [\sinh x]_0^b = \infty$

31.   (a)

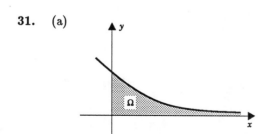

(b) $A = \int_0^\infty e^{-x}\, dx = 1$

(c) $V_x = \int_0^\infty \pi e^{-2x}\, dx = \pi/2$

(d)
$$V_y = \int_0^\infty 2\pi x e^{-x}\,dx = \lim_{b\to\infty}\int_0^b 2\pi x e^{-x}\,dx = \lim_{b\to\infty}\left[2\pi(-x-1)e^{-x}\right]_0^b$$
$$\text{(by parts)}$$
$$= 2\pi\left(1 - \lim_{b\to\infty}\frac{b+1}{e^b}\right) = 2\pi(1-0) = 2\pi$$

(e) $A = \int_0^\infty 2\pi e^{-x}\sqrt{1+e^{-2x}}\,dx = \lim_{b\to\infty}\int_0^b 2\pi e^{-x}\sqrt{1+e^{-2x}}\,dx$

$$\int_0^b 2\pi e^{-x}\sqrt{1+e^{-2x}}\,dx = -2\pi\int_1^{e^{-b}}\sqrt{1+u^2}\,du$$
$$u = e^{-x}$$

$$= -\pi\left[u\sqrt{1+u^2} + \ln\left(1+\sqrt{1+u^2}\right)\right]_1^{e^{-b}}$$
$$= \pi\left[\sqrt{2} + \ln\left(1+\sqrt{2}\right) - e^{-b}\sqrt{1+e^{-2b}} - \ln\left(1+\sqrt{1+e^{-2b}}\right)\right]$$

Taking the limit of this last expression as $b\to\infty$, we have

$$A = \pi\left[\sqrt{2} + \ln\left(1+\sqrt{2}\right)\right].$$

33. (a) The interval $[0,1]$ causes no problem. For $x \ge 1$, $e^{-x^2} \le e^{-x}$ and $\int_1^\infty e^{-x}\,dx$ is finite.

(b) $V_y = \int_0^\infty 2\pi x e^{-x^2}\,dx = \lim_{b\to\infty}\int_0^b 2\pi x e^{-x^2}\,dx = \lim_{b\to\infty}\pi\left[-e^{-x^2}\right]_0^b = \lim_{b\to\infty}\pi\left(1 - e^{-b^2}\right) = \pi$

35. (a)

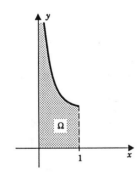

(b) $A = \lim_{a\to 0+}\int_a^1 x^{-1/4}\,dx = \lim_{a\to 0+}\left[\frac{4}{3}x^{3/4}\right]_a^1 = \frac{4}{3}$

(c) $V_x = \lim_{a\to 0+}\int_a^1 \pi x^{-1/2}\,dx = \lim_{a\to 0+}\left[2\pi x^{1/2}\right]_a^1 = 2\pi$

(d) $V_y = \lim_{a\to 0+}\int_a^1 2\pi x^{3/4}\,dx = \lim_{a\to 0+}\left[\frac{8\pi}{7}x^{7/4}\right]_a^1 = \frac{8}{7}\pi$

37. converges by comparison with $\displaystyle\int_1^\infty \frac{dx}{x^{3/2}}$

39. diverges since for x large the integrand is greater than $\dfrac{1}{x}$ and $\displaystyle\int_1^\infty \frac{dx}{x}$ diverges

41. converges by comparison with $\displaystyle\int_1^\infty \frac{dx}{x^{3/2}}$

43. $\rho(\theta) = ae^{c\theta}, \quad \rho'(\theta) = ace^{c\theta}$

$$L = \int_{-\infty}^{\theta_1} \sqrt{a^2 e^{2c\theta} + a^2 c^2 e^{2c\theta}}\, d\theta$$

$$\underset{(10.9.4)}{\Big\uparrow}$$

$$= \left(a\sqrt{1+c^2}\right)\left(\lim_{b\to -\infty}\int_b^{\theta_1} e^{c\theta}\, d\theta\right)$$

$$= \left(a\sqrt{1+c^2}\right)\left(\lim_{b\to -\infty}\left[\frac{e^{c\theta}}{c}\right]_b^{\theta_1}\right)$$

$$= \left(\frac{a\sqrt{1+c^2}}{c}\right)\left(\lim_{b\to\infty}\left[e^{c\theta_1} - e^{cb}\right]\right) = \left(\frac{a\sqrt{1+c^2}}{c}\right)e^{c\theta_1}$$

SECTION 11.7

1. converges to 1: recall (11.3.1)

3. converges to 0: $\cos n\pi \, \sin n\pi = (-1)^n(0) = 0$

5. converges to 1: $n^{1/n} \to 1$ and $(1+n)^{1/n} \to 1$, [recall (11.3.6)]

7. converges to 0: $\cos\dfrac{\pi}{n}\,\sin\dfrac{\pi}{n} = \dfrac{1}{2}\sin\dfrac{2\pi}{n} \to 0$

9. diverges: $\left(2+\dfrac{1}{n}\right)^n > 2^n$

11. converges to 0: $\dfrac{\ln\left[n\,(n+1)\right]}{n} = \dfrac{\ln(n)}{n} + \dfrac{\ln(n+1)}{n+1}\left(\dfrac{n+1}{n}\right) \to 0 + 0(1) = 0$

13. converges to 0: recall (11.3.5)   **15.** converges to 0: $\dfrac{n}{\pi}\sin n\pi = \dfrac{n}{\pi}(0) = 0$

17. converges to 0: $\displaystyle\int_n^{n+1} e^{-x}\, dx = e^{-n} - e^{-(n+1)} \to 0$

19. If $|a_n - l| < \epsilon$ for all $n \ge k$, then $|b_n - l| < \epsilon$ for all $n \ge k-1$.

21. $l \cong 0.7390851, \quad \cos(0.7390851) \cong 0.7390851$

23. $l \cong 0.7681691, \quad \cos\sin 0.7681691 \cong 0.7681691$

25.   Observe that

$$F(t) = \int_l^t f(x)\, dx$$

is continuous and increasing, that

$$a_n = \int_l^n f(x)\, dx$$

is increasing, and that

(*) $a_n \leq \int_l^t f(x)\, dx \leq a_{n+1} \quad \text{for} \quad t \in [n, n+1]$.

If

$$\int_l^\infty f(x)\, dx$$

converges, then F, being continuous, is bounded and, by (*), $\{a_n\}$ is bounded and therefore convergent. If $\{a_n\}$ converges, then $\{a_n\}$ is bounded and, by (*), F is bounded. Being increasing, F is also convergent; i.e., $\int_1^\infty f(x)\, dx$ converges.

27.   (a)   If $\displaystyle\int_{-\infty}^\infty f(x)\, dx = L,$ then both

$$\int_0^\infty f(x)\, dx = \lim_{c \to \infty} \int_0^c f(x)\, dx \quad \text{and} \quad \int_{-\infty}^0 f(x)\, dx = \lim_{c \to \infty} \int_{-c}^0 f(x)\, dx$$

exist and their sum is L. Thus

$$\lim_{c \to \infty} \int_{-c}^c f(x)\, dx = \lim_{c \to \infty} \left\{ \int_{-c}^0 f(x)\, dx + \int_0^c f(x)\, dx \right\}$$

$$= \lim_{c \to \infty} \int_{-c}^0 f(x)\, dx + \lim_{c \to \infty} \int_0^c f(x)\, dx = L.$$

(b) $\displaystyle\lim_{c \to \infty} \int_{-c}^c x\, dx = \lim_{c \to \infty} \left[\frac{x^2}{2}\right]_{-c}^c = \lim_{c \to \infty} 0 = 0,$

but $\displaystyle\int_{-\infty}^\infty x\, dx$ diverges

since

$$\int_0^\infty x\, dx \quad \text{and} \quad \int_{-\infty}^0 x\, dx \quad \text{both diverge.}$$

29.   P.V. $= \displaystyle\lim_{b \to \infty} \int_0^b 100 e^{-0.05t}\, dt = \lim_{b \to \infty} \left[-2000 e^{-0.05t}\right]_0^b = \$2000$

31. The numerical work suggests $l \cong -12.5$:

$$l = \lim_{x \to 0+} \frac{\ln(1+5x) - 5x}{x^2} \overset{\star}{=} \lim_{x \to 0+} \frac{5/(1+5x) - 5}{2x}$$

$$= \lim_{x \to 0+} \frac{-25}{2(1+5x)} = -12.5.$$

33. Note that

$$\sin \theta_i^* = \frac{a}{(x_i^*)^2 + a^2} \quad \text{and} \quad \frac{1}{(r_i^*)^2} = \frac{1}{[(x_i^*)^2 + a^2]^2}.$$

Take $\{x_0, x_1, \cdots, x_n\}$ as a partition of $[-c, c]$.   The contribution of $[-c, c]$ is approximately

$$k_M I \sum_{i=1}^{n} \frac{\sin \theta_i^*}{(r_i^*)^2} \Delta x_i = k_M I a \sum_{i=1}^{n} \frac{1}{[(x_i^*)^2 + a^2]^{3/2}} \Delta x_i.$$

As $\max \Delta x_i \to 0$,   this sum tends to

$$k_M I a \int_{-c}^{c} \frac{dx}{(x^2 + a^2)^{3/2}} = 2 k_M I a \int_{0}^{c} \frac{dx}{(x^2 + a^2)^{3/2}} = \frac{2 k_M I}{a} \left[\frac{x}{(a^2 + x^2)^{1/2}} \right]_0^c = \frac{2 k_M I}{a} \left[\frac{c}{(a^2 + c^2)^{1/2}} \right].$$

Thus

$$B = \lim_{c \to \infty} \frac{2 k_M I}{a} \left[\frac{c}{(a^2 + c^2)^{1/2}} \right] = \frac{2 k_M I}{a}.$$

CHAPTER 12

SECTION 12.1

1. $1 + 4 + 7 = 12$ **3.** $1 + 2 + 4 + 8 = 15$

5. $2 - 4 + 8 - 16 = -10$ **7.** $\frac{1}{2} + \frac{1}{4} + \frac{1}{8} + \frac{1}{16} = \frac{15}{16}$

9. $-\frac{1}{6} + \frac{1}{24} - \frac{1}{120} = -\frac{2}{15}$ **11.** $1 + \frac{1}{4} + \frac{1}{16} + \frac{1}{64} = \frac{85}{64}$

13. $\displaystyle\sum_{k=1}^{n} m_k \, \Delta x_k$ **15.** $\displaystyle\sum_{k=1}^{n} f(x_k^*) \, \Delta x_k$ **17.** $\displaystyle\sum_{k=0}^{5} (-1)^k a^{5-k} b^k$

19. $\displaystyle\sum_{k=0}^{4} a_k x^{4-k}$ **21.** $\displaystyle\sum_{k=0}^{4} (-1)^k (k+1) x^k$ **23.** $\displaystyle\sum_{k=3}^{10} \frac{1}{2^k}, \quad \sum_{i=0}^{7} \frac{1}{2^{i+3}}$

25. $\displaystyle\sum_{k=3}^{10} (-1)^{k+1} \frac{k}{k+1}, \quad \sum_{i=0}^{7} (-1)^i \frac{i+3}{i+4}$

27.   (a) $(1 - x) \displaystyle\sum_{k=0}^{n} x^k = \sum_{k=0}^{n} (x^k - x^{k+1}) = (1 - x) + (x - x^2) + \cdots + (x^n - x^{n+1}) = 1 - x^{n+1}$

      (b) $a_n = \dfrac{1 - (\frac{1}{3})^{n+1}}{1 - \frac{1}{3}} = \dfrac{3}{2} - \dfrac{3}{2}\left(\dfrac{1}{3}\right)^{n+1} \to \dfrac{3}{2}$

29. $|a_n - l| < \epsilon$ for $n \geq k$ iff $|a_{n-p} - l| < \epsilon$ for $n \geq k + p$

31.   True for $n = 1$:

$$\sum_{k=1}^{1} k = 1 = \frac{1}{2}(1)(2).$$

Suppose true for $n = p$. Then

$$\sum_{k=1}^{p+1} k = \sum_{k=1}^{p} k + (p+1)$$

$$= \tfrac{1}{2}(p)(p+1) + (p+1)$$

$$= \tfrac{1}{2}[p(p+1) + 2(p+1)]$$

$$= \tfrac{1}{2}(p+1)(p+2)$$

$$= \tfrac{1}{2}(p+1)[(p+1)+1]$$

and thus true for $n = p + 1$.

33.   True for $n = 1$:

$$\sum_{k=1}^{1} k^2 = 1 = \frac{1}{6}(1)(2)(3).$$

Suppose true for $n = p$. Then

$$\sum_{k=1}^{p+1} k^2 = \sum_{k=1}^{p} k^2 + (p+1)^2$$

$$= \tfrac{1}{6}(p)(p+1)(2p+1) + (p+1)^2$$

$$= \tfrac{1}{6}(p+1)[p(2p+1) + 6(p+1)]$$

$$= \tfrac{1}{6}(p+1)[2p^2 + 7p + 6]$$

$$= \tfrac{1}{6}(p+1)[(p+2)(2p+3)]$$

$$= \tfrac{1}{6}(p+1)[(p+1)+1][2(p+1)+1]$$

and thus true for $n = p + 1$.

SECTION 12.2

1. $\frac{1}{4}$; $\quad s_n = \dfrac{1}{4 \cdot 5} + \dfrac{1}{5 \cdot 6} + \cdots + \dfrac{1}{(n+1)(n+2)}$

$$= \left(\frac{1}{4} - \frac{1}{5}\right) + \left(\frac{1}{5} - \frac{1}{6}\right) + \cdots + \left(\frac{1}{n+1} - \frac{1}{n+2}\right) = \frac{1}{4} - \frac{1}{n+2} \to \frac{1}{4}$$

3. $\frac{1}{2}$; $\quad s_n = \dfrac{1}{2}\left[\dfrac{1}{1 \cdot 2} + \dfrac{1}{2 \cdot 3} + \cdots + \dfrac{1}{(n)(n+1)}\right]$

$$= \frac{1}{2}\left[\left(1 - \frac{1}{2}\right) + \left(\frac{1}{2} - \frac{1}{3}\right) + \cdots + \left(\frac{1}{n} - \frac{1}{n+1}\right)\right] = \frac{1}{2}\left[1 - \frac{1}{n+1}\right] \to \frac{1}{2}$$

5. $\frac{11}{18}$; $\quad s_n = \dfrac{1}{1 \cdot 4} + \dfrac{1}{2 \cdot 5} + \cdots + \dfrac{1}{n(n+3)}$

$$= \frac{1}{3}\left[\left(1 - \frac{1}{4}\right) + \left(\frac{1}{2} - \frac{1}{5}\right) + \cdots + \left(\frac{1}{n} - \frac{1}{n+3}\right)\right]$$

$$= \frac{1}{3}\left[1 + \frac{1}{2} + \frac{1}{3} - \frac{1}{n+1} - \frac{1}{n+2} - \frac{1}{n+3}\right] \to \frac{1}{3}\left(1 + \frac{1}{2} + \frac{1}{3}\right) = \frac{11}{18}$$

7. $\frac{10}{3}$; $\quad \displaystyle\sum_{k=0}^{\infty} \frac{3}{10^k} = 3\sum_{k=0}^{\infty}\left(\frac{1}{10}\right)^k = 3\left(\frac{1}{1 - 1/10}\right) = \frac{30}{9} = \frac{10}{3}$

9. $\frac{67000}{999}$; $\quad \displaystyle\sum_{k=0}^{\infty} \frac{67}{1000^k} = 67\sum_{k=0}^{\infty}\left(\frac{1}{1000}\right)^k = 67\left(\frac{1}{1 - 1/1000}\right) = \frac{67000}{999}$

11. 4; $\quad \displaystyle\sum_{k=0}^{\infty}\left(\frac{3}{4}\right)^k = \frac{1}{1 - 3/4} = 4$

13. $-\frac{3}{2}$; $\quad \displaystyle\sum_{k=0}^{\infty} \frac{1 - 2^k}{3^k} = \sum_{k=0}^{\infty}\left(\frac{1}{3}\right)^k - \sum_{k=0}^{\infty}\left(\frac{2}{3}\right)^k = \frac{1}{1 - 1/3} - \frac{1}{1 - 2/3} = \frac{3}{2} - 3 = -\frac{3}{2}$

15. $\frac{1}{2}$; geometric series with $a = \dfrac{1}{4}$ and $r = \dfrac{1}{2}$, sum $= \dfrac{a}{1-r} = \dfrac{1}{2}$

17. 24; geometric series with $a = 8$ and $r = \dfrac{2}{3}$, sum $= \dfrac{a}{1-r} = 24$

19. $\displaystyle\sum_{k=1}^{\infty} \frac{7}{10^k} = \frac{7/10}{1 - 1/10} = \frac{7}{9}$

21. $\displaystyle\sum_{k=1}^{\infty} \frac{24}{100^k} = \frac{24/100}{1 - 1/100} = \frac{8}{33}$

23. $\displaystyle\sum_{k=1}^{\infty} \frac{112}{1000^k} = \frac{112/1000}{1 - 1/1000} = \frac{112}{999}$

25. $\dfrac{62}{100} + \dfrac{1}{100} \displaystyle\sum_{k=1}^{\infty} \dfrac{45}{100^k} = \dfrac{62}{100} + \dfrac{1}{100}\left(\dfrac{45/100}{1 - 1/100}\right) = \dfrac{687}{1100}$

27. Let $x = 0.\overbrace{a_1 a_2 \cdots a_n}\,\overbrace{a_1 a_2 \cdots a_n} \cdots$.   Then

$$x = \sum_{k=1}^{\infty} \frac{a_1 a_2 \cdots a_n}{(10^n)^k} = a_1 a_2 \cdots a_n \sum_{k=1}^{\infty} \left(\frac{1}{10^n}\right)^k$$

$$= a_1 a_2 \cdots a_n \left[\frac{1}{1 - 1/10^n} - 1\right] = \frac{a_1 a_2 \cdots a_n}{10^n - 1}.$$

29. $\dfrac{1}{1+x} = \dfrac{1}{1-(-x)} = \displaystyle\sum_{k=0}^{\infty}(-x)^k = \sum_{k=0}^{\infty}(-1)^k x^k$

31. $\dfrac{x}{1-x} = x\left(\dfrac{1}{1-x}\right) = x\displaystyle\sum_{k=0}^{\infty}(x^k) = \sum_{k=0}^{\infty} x^{k+1}$

33. $\dfrac{x}{1+x^2} = x\left[\dfrac{1}{1-(-x^2)}\right] = x\displaystyle\sum_{k=0}^{\infty}(-x^2)^k = \sum_{k=0}^{\infty}(-1)^k x^{2k+1}$

35. $\dfrac{1}{1+4x^2} = \dfrac{1}{1-(-4x^2)} = \displaystyle\sum_{k=0}^{\infty}(-4x^2)^k = \sum_{k=0}^{\infty}(-1)^k (2x)^{2k}$

37. $4 + \dfrac{1}{3}\displaystyle\sum_{k=0}^{\infty}\left(\dfrac{1}{12}\right)^k = 4 + \dfrac{4}{11}$ o'clock

39. In the general case of a ball which rebounds σh feet if dropped from a height of h feet $(0 < \sigma < 1)$, we find for a ball dropped initially from h_0 feet that the total distance traveled is given by

$$d = h_0 + 2\sigma h_0 + 2\sigma^2 h_0 + 2\sigma^3 h_0 + \cdots$$

$$= h_0 + 2h_0\sigma[1 + \sigma + \sigma^2 + \cdots]$$

$$= h_0 + 2h_0\sigma\sum_{k=0}^{\infty}\sigma^k = h_0 + \frac{2h_0\sigma}{1-\sigma}.$$

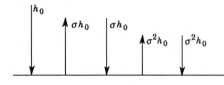

Here, $h_0 = 6$ ft and $\sigma = \frac{1}{3}$ so that distance $= 12$ ft.

41. A principal x deposited now at $r\%$ interest compounded annually will grow in k years to

$$x\left(1 + \frac{r}{100}\right)^k.$$

This means that in order to be able to withdraw n_k dollars after k years one must place

$$n_k\left(1 + \frac{r}{100}\right)^{-k}$$

dollars on deposit today. To extend this process in perpetuity as described in the text, the total deposit must be

$$\sum_{k=1}^{\infty} n_k\left(1 + \frac{r}{100}\right)^{-k}.$$

43. $s_n = \sum_{k=1}^{n} \ln \left(\frac{k+1}{k}\right) = [\ln(n+1) - \ln(n)] + [\ln n - \ln(n-1)] + \cdots + [\ln 2 - \ln 1] = \ln(n+1) \to \infty$

45. (a) $s_n = \sum_{k=1}^{n}(d_k - d_{k+1}) = d_1 - d_{n+1} \to d_1$

  (b) We use part (a).

  (i) $\sum_{k=1}^{\infty} \frac{\sqrt{k+1} - \sqrt{k}}{\sqrt{k(k+1)}} = \sum_{k=1}^{\infty} \left[\frac{1}{\sqrt{k}} - \frac{1}{\sqrt{k+1}}\right] = 1$

  (ii) $\sum_{k=1}^{\infty} \frac{2k+1}{2k^2(k+1)^2} = \sum_{k=1}^{\infty} \frac{1}{2}\left[\frac{1}{k^2} - \frac{1}{(k+1)^2}\right] = \frac{1}{2}$

SECTION 12.3

1. converges; basic comparison with $\sum \frac{1}{k^2}$ **3.** converges; basic comparison with $\sum \frac{1}{k^2}$

5. diverges; basic comparison with $\sum \frac{1}{k+1}$ **7.** diverges; limit comparison with $\sum \frac{1}{k}$

9. converges; integral test, $\int_{1}^{\infty} \frac{\tan^{-1} x}{1+x^2}\, dx = \lim_{b\to\infty} \left[\frac{1}{2}(\tan^{-1} x)^2\right]_{1}^{b} = \frac{3\pi^2}{32}$

11. diverges; p-series with $p = \frac{2}{3} \leq 1$ **13.** diverges; divergence test, $\left(\frac{3}{4}\right)^{-k} \not\to 0$

15. diverges; basic comparison with $\sum \frac{1}{k}$

17. diverges; divergence test, $\frac{1}{2+3^{-k}} \to \frac{1}{2} \neq 0$

19. converges; limit comparison with $\sum \frac{1}{k^2}$

21. diverges; integral test, $\int_{2}^{\infty} \frac{dx}{x \ln x} = \lim_{b\to\infty} \left[\ln(\ln x)\right]_{2}^{b} = \infty$

23. converges; limit comparison with $\sum \frac{1}{k^2}$ **25.** diverges; limit comparison with $\sum \frac{1}{k}$

27. converges; limit comparison with $\sum \frac{1}{k^{3/2}}$

29. converges; integral test, $\int_{1}^{\infty} x e^{-x^2}\, dx = \lim_{b\to\infty} \left[-\frac{1}{2} e^{-x^2}\right]_{1}^{b} = \frac{1}{2e}$

31. (a) If $a_k/b_k \to 0$, then $a_k/b_k < 1$ for all $k \geq K$ for some K. But then $a_k < b_k$ for all $k \geq K$

  and, since $\sum b_k$ converges, $\sum a_k$ converges. [The Basic Comparison Theorem 12.3.5.]

  (b) Similar to (a) except that this time we appeal to part (ii) of Theorem 12.3.5.

(c) $\displaystyle\sum a_k = \sum \frac{1}{k^2}$ converges, $\quad \displaystyle\sum b_k = \sum \frac{1}{k^{3/2}}$ converges, $\quad \displaystyle\frac{1/k^2}{1/k^{3/2}} = \frac{1}{\sqrt{k}} \to 0$

$\displaystyle\sum a_k = \sum \frac{1}{k^2}$ converges, $\quad \displaystyle\sum b_k = \sum \frac{1}{\sqrt{k}}$ diverges, $\quad \displaystyle\frac{1/k^2}{1/\sqrt{k}} = \frac{1}{k^{3/2}} \to 0$

(d) $\displaystyle\sum b_k = \sum \frac{1}{\sqrt{k}}$ diverges, $\quad \displaystyle\sum a_k = \sum \frac{1}{k^2}$ converges, $\quad \displaystyle\frac{1/k^2}{1/\sqrt{k}} = 1/k^{3/2} \to 0$

$\displaystyle\sum b_k = \sum \frac{1}{\sqrt{k}}$ diverges, $\quad \displaystyle\sum a_k = \sum \frac{1}{k}$ diverges, $\quad \displaystyle\frac{1/k}{1/\sqrt{k}} = \frac{1}{\sqrt{k}} \to 0$

33. (a) Set $f(x) = x^{1/4} - \ln x$. Then

$$f'(x) = \frac{1}{4}x^{-3/4} - \frac{1}{x} = \frac{1}{4x}(x^{1/4} - 4).$$

Since $f(e^{12}) = e^3 - 12 > 0$ and $f'(x) > 0$ for $x > e^{12}$, we have that

$$n^{1/4} > \ln n \quad \text{and therefore} \quad \frac{1}{n^{5/4}} > \frac{\ln n}{n^{3/2}}$$

for sufficiently large n. Since $\displaystyle\sum \frac{1}{n^{5/4}}$ is a convergent p-series, $\displaystyle\sum \frac{\ln n}{n^{3/2}}$ converges

by the basic comparison test.

(b) By L'Hospital's rule

$$\lim_{x \to \infty} \frac{(\ln x)/x^{3/2}}{1/x^{5/4}} = \lim_{x \to \infty} \frac{\ln x}{x^{1/4}} \overset{\star}{=} \lim_{x \to \infty} \frac{1/x}{\frac{1}{4}x^{-3/4}} = \lim_{x \to \infty} \frac{4}{x^{1/4}} = 0.$$

Thus, the limit comparison test does not apply.

SECTION 12.4

1. converges; ratio test, $\displaystyle\frac{a_{k+1}}{a_k} = \frac{10}{k+1} \to 0$ **3.** converges; root test, $(a_k)^{1/k} = \dfrac{1}{k} \to 0$

5. diverges; divergence test, $\displaystyle\frac{k!}{100^k} \to \infty$ **7.** diverges; limit comparison with $\displaystyle\sum \frac{1}{k}$

9. converges; root test, $(a_k)^{1/k} = \dfrac{2}{3}k^{1/k} \to \dfrac{2}{3}$ **11.** diverges; limit comparison with $\displaystyle\sum \frac{1}{\sqrt{k}}$

13. diverges; ratio test, $\displaystyle\frac{a_{k+1}}{a_k} = \frac{k+1}{10^4} \to \infty$ **15.** converges; basic comparison with $\displaystyle\sum \frac{1}{k^{3/2}}$

17. converges; basic comparison with $\displaystyle\sum \frac{1}{k^2}$

19. diverges; integral test, $\displaystyle\int_2^\infty \frac{1}{x}(\ln x)^{-1/2}dx = \lim_{b\to\infty}\left[2(\ln x)^{1/2}\right]_2^b = \infty$

21. diverges; divergence test, $\displaystyle\left(\frac{k}{k+100}\right)^k = \left(1+\frac{100}{k}\right)^{-k} \to e^{-100} \neq 0$

23. diverges; limit comparison with $\displaystyle\sum\frac{1}{k}$ **25.** converges; ratio test, $\displaystyle\frac{a_{k+1}}{a_k} = \frac{\ln(k+1)}{e\ln k} \to \frac{1}{e}$

27. converges; basic comparison with $\displaystyle\sum\frac{1}{k^{3/2}}$

29. converges; ratio test, $\displaystyle\frac{a_{k+1}}{a_k} = \frac{2(k+1)}{(2k+1)(2k+2)} \to 0$

31. converges; ratio test: $\displaystyle\frac{a_{k+1}}{a_k} = \frac{(k+1)(2k+1)(2k+2)}{(3k+1)(3k+2)(3k+3)} \to \frac{4}{27}$

33. By the hint

$$\sum_{k=1}^\infty k\left(\frac{1}{10}\right)^k = \frac{1}{10}\sum_{k=1}^\infty k\left(\frac{1}{10}\right)^{k-1} = \frac{1}{10}\left[\frac{1}{1-1/10}\right]^2 = \frac{10}{81}.$$

35. Set $b_k = a_k r^k$. If $(a_k)^{1/k} \to \rho$ and $\rho < \dfrac{1}{r}$, then

$$(b_k)^{1/k} = (a_k r^k)^{1/k} = (a_k)^{1/k}r \to \rho r < 1$$

and thus, by the root test, $\Sigma b_k = \Sigma a_k r^k$ converges.

SECTION 12.5

1. diverges; $a_k \not\to 0$ **3.** diverges; $\dfrac{k}{k+1} \to 1 \neq 0$

5. (a) does not converge absolutely; integral test,

$$\int_1^\infty \frac{\ln x}{x}dx = \lim_{b\to\infty}\left[\frac{1}{2}(\ln x)^2\right]_1^b = \infty$$

(b) converges conditionally; Theorem 12.5.3

7. diverges; limit comparison with $\displaystyle\sum\frac{1}{k}$

another approach: $\displaystyle\sum\left(\frac{1}{k}-\frac{1}{k!}\right) = \sum\frac{1}{k}-\sum\frac{1}{k!}$ diverges since $\displaystyle\sum\frac{1}{k}$ diverges and

$\displaystyle\sum\frac{1}{k!}$ converges

9. (a) does not converge absolutely; limit comparison with $\displaystyle\sum\frac{1}{k}$

(b) converges conditionally; Theorem 12.5.3

11. diverges; $a_k \not\to 0$

13. (a) does not converge absolutely;

$$(\sqrt{k+1} - \sqrt{k}) \cdot \frac{(\sqrt{k+1} + \sqrt{k})}{(\sqrt{k+1} + \sqrt{k})} = \frac{1}{\sqrt{k+1} + \sqrt{k}}$$

and

$$\sum \frac{1}{\sqrt{k} + \sqrt{k+1}} > \sum \frac{1}{2\sqrt{k+1}} = 2 \sum \frac{1}{\sqrt{k+1}} \qquad (\text{a } p\text{-series with } p < 1)$$

(b) converges conditionally;   Theorem 12.5.3

15. converges absolutely (terms already positive); basic comparison,

$$\sum \sin\left(\frac{\pi}{4k^2}\right) \le \sum \frac{\pi}{4k^2} = \frac{\pi}{4} \sum \frac{1}{k^2} \qquad (|\sin x| \le |x|)$$

17. converges absolutely;   ratio test, $\dfrac{a_{k+1}}{a_k} = \dfrac{k+1}{2k} \to \dfrac{1}{2}$

19. (a) does not converge absolutely;   limit comparison with $\sum \dfrac{1}{k}$

(b) converges conditionally;   Theorem 12.5.3

21. diverges; $a_k \not\to 0$

23. $\dfrac{10}{11}$;   geometric series with $a = 1$ and $r = -\dfrac{1}{10}$,   sum $= \dfrac{a}{1-r} = \dfrac{10}{11}$

25. Use (12.5.4).

(a) $n = 4$; $\dfrac{1}{(n+1)!} < 0.01 \implies 100 < (n+1)!$

(b) $n = 6$; $\dfrac{1}{(n+1)!} < 0.001 \implies 1000 < (n+1)!$

27. No.   For instance, set $a_{2k} = 2/k$ and $a_{2k+1} = 1/k$.

29. See the proof of Theorem 12.8.2.

SECTION 12.6

1. $-1 + x + \frac{1}{2}x^2 - \frac{1}{24}x^4$ **3.** $-\frac{1}{2}x^2 - \frac{1}{12}x^4$

5. $1 - x + x^2 - x^3 + x^4 - x^5$ **7.** $x + \frac{1}{3}x^3 + \frac{2}{15}x^5$

9. $P_0(x) = 1$, $P_1(x) = 1 - x$, $P_2(x) = 1 - x + 3x^2$, $P_3(x) = 1 - x + 3x^2 + 5x^3$

11. $\displaystyle\sum_{k=0}^{n} (-1)^k \frac{x^k}{k!}$ **13.** $\displaystyle\sum_{k=0}^{m} \frac{x^{2k}}{(2k)!}$ where $m = \dfrac{n}{2}$ and n is even

15. The Taylor polynomial

$$P_n(0.5) = 1 + (0.5) + \frac{(0.5)^2}{2!} + \cdots + \frac{(0.5)^n}{n!}$$

estimates $e^{0.5}$ within

$$|R_{n+1}(0.5)| \le e^{0.5} \frac{|0.5|^{n+1}}{(n+1)!} < 2\frac{(0.5)^{n+1}}{(n+1)!}.$$

Since

$$2\frac{(0.5)^4}{4!} = \frac{1}{8(24)} < 0.01,$$

we can take $n = 3$ and be sure that

$$P_3(0.5) = 1 + (0.5) + \frac{(0.5)^2}{2} + \frac{(0.5)^3}{6} = \frac{79}{48}$$

differs from $\sqrt{e}$ by less than 0.01. Our calculator gives

$$\tfrac{79}{48} \cong 1.645833 \quad \text{and} \quad \sqrt{e} \cong 1.6487213.$$

17. At $x = 1$ the sine series gives

$$\sin 1 = 1 - \frac{1}{3!} + \frac{1}{5!} - \frac{1}{7!} + \cdots.$$

This is a convergent alternating series with decreasing terms. The first term of magnitude less than 0.01 is $1/5! = 1/120$. Thus

$$1 - \frac{1}{3!} = 1 - \frac{1}{6} = \frac{5}{6}$$

differs from $\sin 1$ by less than 0.01. Our calculator gives

$$\tfrac{5}{6} \cong 0.8333333 \quad \text{and} \quad \sin 1 \cong 0.84114709.$$

The estimate

$$1 - \frac{1}{3!} + \frac{1}{5!} = \frac{101}{120}$$

is much more accurate:

$$\tfrac{101}{120} \cong 0.8416666.$$

19. At $x = 1$ the cosine series gives

$$\cos 1 = 1 - \frac{1}{2!} + \frac{1}{4!} - \frac{1}{6!} + \frac{1}{8!} + \cdots.$$

This is a convergent alternating series with decreasing terms. The first term of magnitude less than 0.01 is $1/6! = 1/720$. Thus

$$1 - \frac{1}{2!} + \frac{1}{4!} = 1 - \frac{1}{2} + \frac{1}{24} = \frac{13}{24}$$

differs from $\cos 1$ by less than 0.01. Our calculator gives

$$\tfrac{13}{24} \cong 0.5416666 \quad \text{and} \quad \cos 1 \cong 0.5403023.$$

21. By (12.6.7)
$$P_n(x) = x - \frac{x^2}{2} + \frac{x^3}{3} - \frac{x^4}{4} + \cdots + (-1)^{n+1}\frac{x^n}{n}.$$

For $0 \le x \le 1$ we know from (12.5.4) that
$$|P_n(x) - \ln(1+x)| < \frac{x^{n+1}}{n+1}.$$

(a) $n = 4; \dfrac{(0.5)^{n+1}}{n+1} \le 0.01 \quad\Longrightarrow\quad 100 \le (n+1)2^{n+1} \quad\Longrightarrow\quad n \ge 4$

(b) $n = 2; \dfrac{(0.3)^{n+1}}{n+1} \le 0.01 \quad\Longrightarrow\quad 100 \le (n+1)\left(\dfrac{10}{3}\right)^{n+1} \quad\Longrightarrow\quad n \ge 2$

(c) $n = 999; \dfrac{(1)^{n+1}}{n+1} \le 0.001 \quad\Longrightarrow\quad 1000 \le n+1 \quad\Longrightarrow\quad n \ge 999$

23. The result follows from the fact that $\quad P^{(k)}(0) = \begin{cases} k!a_k, & 0 \le k \le n \\ 0, & n < k \end{cases}$.

25. Set $t = ax$. Then, $e^{ax} = e^t = \displaystyle\sum_{k=0}^{\infty} \frac{t^k}{k!} = \sum_{k=0}^{\infty} a^k \frac{x^k}{k!}, \quad (-\infty, \infty).$

27. Set $t = ax$. Then, $\cos ax = \cos t = \displaystyle\sum_{k=0}^{\infty} \frac{(-1)^k}{(2k)!} t^{2k} = \sum_{k=0}^{\infty} \frac{(-1)^k a^{2k}}{(2k)!} x^{2k}, \quad (-\infty, \infty).$

29. By the hint
$$\ln(a+x) = \ln\left[a\left(1 + \frac{x}{a}\right)\right] = \ln a + \ln\left(1 + \frac{x}{a}\right) = \ln a + \sum_{k=1}^{\infty} \frac{(-1)^{k+1}}{ka^k} x^k.$$

By (12.6.7) the series converges for $-1 < \dfrac{x}{a} \le 1$; that is, $-a < x \le a$.

31. $\ln 2 = \ln\left(\dfrac{1 + 1/3}{1 - 1/3}\right) \cong 2\left[\dfrac{1}{3} + \dfrac{1}{3}\left(\dfrac{1}{3}\right)^3 + \dfrac{1}{5}\left(\dfrac{1}{3}\right)^5\right] = \dfrac{842}{1215}.$

Our calculator gives $\dfrac{842}{1215} \cong 0.6930041$ and $\ln 2 \cong 0.6931471.$

33. Set $u = (x-t)^k, \qquad dv = f^{(k+1)}(t)\, dt$
$\qquad du = -k(x-t)^{k-1}\, dt, \quad v = f^{(k)}(t).$

Then, $-\dfrac{1}{k!}\displaystyle\int_0^x f^{(k+1)}(t)(x-t)^k\, dt$
$$= -\frac{1}{k!}\left[(x-t)^k f^{(k)}(t)\right]_0^x - \frac{1}{k!}\int_0^x k(x-t)^{k-1} f^{(k)}(t)\, dt$$
$$= \frac{f^{(k)}(0)}{k!} x^k - \frac{1}{(k-1)!}\int_0^x f^{(k)}(t)(x-t)^{k-1}\, dt.$$

The given identity follows.

35. $f^{(0)}(0) = f(0) = 0$ by the definition

$$f'(0) = \lim_{x \to 0} \frac{f(x) - f(0)}{x - 0} = \lim_{x \to 0} \frac{e^{-1/x^2}}{x} = 0$$

$$f''(0) = \lim_{x \to 0} \frac{f'(x) - f'(0)}{x - 0} = \lim_{x \to 0} \frac{e^{-1/x^2}(2x^{-3})}{x} = 0$$

SECTION 12.7

1. $g(x) = 6 + 9(x - 1) + 7(x - 1)^2 + 3(x - 1)^3,$ $(-\infty, \infty)$

3. $g(x) = -3 + 5(x + 1) - 19(x + 1)^2 + 20(x + 1)^3 - 10(x + 1)^4 + 2(x + 1)^5,$ $(-\infty, \infty)$

5. $g(x) = \dfrac{1}{1 + x} = \dfrac{1}{2 + (x - 1)} = \dfrac{1}{2}\left[\dfrac{1}{1 + \left(\dfrac{x-1}{2}\right)}\right] = \dfrac{1}{2}\displaystyle\sum_{k=0}^{\infty}(-1)^k\left(\dfrac{x-1}{2}\right)^k$

(geometric series) ⟶

$$= \sum_{k=0}^{\infty}(-1)^k \frac{(x-1)^k}{2^{k+1}} \quad \text{for} \quad \left|\frac{x-1}{2}\right| < 1 \quad \text{and thus for} \quad -1 < x < 3$$

7. $g(x) = \dfrac{1}{1 - 2x} = \dfrac{1}{5 - 2(x + 2)} = \dfrac{1}{5}\left[\dfrac{1}{1 - \frac{2}{5}(x + 2)}\right] = \dfrac{1}{5}\displaystyle\sum_{k=0}^{\infty}\left[\dfrac{2}{5}(x+2)\right]^k$

(geometric series) ⟶

$$= \sum_{k=0}^{\infty} \frac{2^k}{5^{k+1}}(x + 2)^k \quad \text{for} \quad \left|\frac{2}{5}(x + 2)\right| < 1 \quad \text{and thus for} \quad -\frac{9}{2} < x < \frac{1}{2}$$

9. $g(x) = \sin x = \sin[(x - \pi) + \pi] = \sin(x - \pi)\cos\pi + \cos(x - \pi)\sin\pi$

$$= -\sin(x - \pi) = -\sum_{k=0}^{\infty}(-1)^k \frac{(x - \pi)^{2k+1}}{(2k + 1)!}$$

(12.6.5) ⟶

$$= \sum_{k=0}^{\infty}(-1)^{k+1}\frac{(x - \pi)^{2k+1}}{(2k + 1)!}, \quad (-\infty, \infty)$$

11. $g(x) = \cos x = \cos[(x - \pi) + \pi] = \cos(x - \pi)\cos\pi - \sin(x - \pi)\sin\pi$

$$= -\cos(x - \pi) = -\sum_{k=0}^{\infty}(-1)^k \frac{(x - \pi)^{2k}}{(2k)!} = \sum_{k=0}^{\infty}(-1)^{k+1}\frac{(x - \pi)^{2k}}{(2k)!}, \quad (-\infty, \infty)$$

(12.6.6) ⟶

13. $g(x) = \sin \frac{1}{2}\pi x = \sin\left[\frac{\pi}{2}(x-1) + \frac{\pi}{2}\right]$

$= \sin\left[\frac{\pi}{2}(x-1)\right]\cos\frac{\pi}{2} + \cos\left[\frac{\pi}{2}(x-1)\right]\sin\frac{\pi}{2}$

$= \cos\left[\frac{\pi}{2}(x-1)\right] = \sum_{k=0}^{\infty}(-1)^k\left(\frac{\pi}{2}\right)^{2k}\frac{(x-1)^{2k}}{(2k)!}, \quad (-\infty, \infty)$

$\underset{(12.6.6)}{\big\uparrow}$

15. $g(x) = \ln(1+2x) = \ln[3 + 2(x-1)] = \ln\left[3\left(1 + \frac{2}{3}(x-1)\right)\right]$

$= \ln 3 + \ln\left[1 + \frac{2}{3}(x-1)\right] = \ln 3 + \sum_{k=1}^{\infty}\frac{(-1)^{k+1}}{k}\left[\frac{2}{3}(x-1)\right]^k$

$\underset{(12.6.7)}{\big\uparrow}$

$= \ln 3 + \sum_{k=1}^{\infty}\frac{(-1)^{k+1}}{k}\left(\frac{2}{3}\right)^k(x-1)^k.$

This result holds if $\;-1 < \frac{2}{3}(x-1) \le 1,\;$ which is to say, if $\;-\frac{1}{2} < x \le \frac{5}{2}.$

17.
$$\begin{aligned} g(x) &= x\ln x \\ g'(x) &= 1 + \ln x \\ g''(x) &= x^{-1} \\ g'''(x) &= -x^{-2} \\ g^{(\text{iv})}(x) &= 2x^{-3} \\ &\vdots \\ g^{(k)}(x) &= (-1)^k(k-2)!\,x^{1-k}, \quad k \ge 2. \end{aligned}$$

Then, $g(2) = 2\ln 2, \; g'(2) = 1 + \ln 2, \quad$ and $\quad g^{(k)}(2) = \frac{(-1)^k(k-2)!}{2^{k-1}}, \quad k \ge 2.$

Thus, $g(x) = 2\ln 2 + (1 + \ln 2)(x-2) + \sum_{k=2}^{\infty}\frac{(-1)^k}{k(k-1)2^{k-1}}(x-2)^k.$

19. $g(x) = x\sin x = x\sum_{k=0}^{\infty}(-1)^k\frac{x^{2k+1}}{(2k+1)!} = \sum_{k=0}^{\infty}(-1)^k\frac{x^{2k+2}}{(2k+1)!}$

21.
$$\begin{aligned} g(x) &= (1-2x)^{-3} \\ g'(x) &= -2(-3)(1-2x)^{-4} \\ g''(x) &= (-2)^2(4\cdot 3)(1-2x)^{-5} \\ g'''(x) &= (-2)^3(-5\cdot 4\cdot 3)(1-2x)^{-6} \\ &\vdots \\ g^{(k)}(x) &= (-2)^k\left[(-1)^k\frac{(k+2)!}{2}\right](1-2x)^{-k-3}, \quad k \ge 0. \end{aligned}$$

Thus, $g^{(k)}(-2) = (-2)^k \left[(-1)^k \dfrac{(k+2)!}{2}\right] 5^{-k-3} = \dfrac{2^{k-1}}{5^{k+3}}(k+2)!$

and $g(x) = \displaystyle\sum_{k=0}^{\infty}(k+2)(k+1)\dfrac{2^{k-1}}{5^{k+3}}(x-2)^k.$

23. $g(x) = \cos^2 x = \dfrac{1+\cos 2x}{2} = \dfrac{1}{2} + \dfrac{1}{2}\cos\left[2(x-\pi)+2\pi\right]$

$\qquad = \dfrac{1}{2} + \dfrac{1}{2}\cos\left[2(x-\pi)\right] = \dfrac{1}{2} + \dfrac{1}{2}\displaystyle\sum_{k=0}^{\infty}(-1)^k\dfrac{[2(x-\pi)]^{2k}}{(2k)!}$

$\qquad = 1 + \displaystyle\sum_{k=1}^{\infty}\dfrac{(-1)^k 2^{2k-1}}{(2k)!}(x-\pi)^{2k}$

$\qquad\qquad\Big\uparrow$

$\qquad\qquad\quad (k=0 \text{ term is } \tfrac{1}{2})$

25.

$$g(x) = x^n$$
$$g'(x) = nx^{n-1}$$
$$g''(x) = n(n-1)x^{n-2}$$
$$g'''(x) = n(n-1)(n-2)x^{n-3}$$
$$\vdots$$
$$g^{(k)}(x) = n(n-1)\cdots(n-k+1)x^{n-k}, \qquad 0 \le k \le n$$
$$g^{(k)}(x) = 0, \qquad k > n.$$

Thus,

$$g^{(k)}(1) = \begin{cases} \dfrac{n!}{(n-k)!}, & 0 \le k \le n \\[2mm] 0, & k > n \end{cases} \quad\text{and}\quad g(x) = \sum_{k=0}^{n}\dfrac{n!}{(n-k)!k!}(x-1)^k.$$

27. (a) $\dfrac{e^x}{e^a} = e^{x-a} = \displaystyle\sum_{k=0}^{\infty}\dfrac{(x-a)^k}{k!}, \quad e^x = e^a\displaystyle\sum_{k=0}^{\infty}\dfrac{(x-a)^k}{k!}$

(b) $e^{a+(x-a)} = e^x = e^a\displaystyle\sum_{k=0}^{\infty}\dfrac{(x-a)^k}{k!}, \quad e^{x_1+x_2} = e^{x_1}\displaystyle\sum_{k=0}^{\infty}\dfrac{x_2^k}{k!} = e^{x_1}e^{x_2}$

(c) $e^{-a}\displaystyle\sum_{k=0}^{\infty}(-1)^k\dfrac{(x-a)^k}{k!}$

SECTION 12.8

1. $(-1, 1)$;   ratio test: $\dfrac{b_{k+1}}{b_k} = \dfrac{k+1}{k}|x| \to |x|,$ series converges for $|x| < 1.$

At the endpoints $x = 1$ and $x = -1$ the series diverges since at those points $b_k \not\to 0.$

3. $(-\infty, \infty)$;   ratio test: $\dfrac{b_{k+1}}{b_k} = \dfrac{|x|}{(2k+1)(2k+2)} \to 0,$ series converges all $x.$

5. Converges only at 0;   divergence test: $(-k)^{2k} x^{2k} \rightarrow 0$ only if $x = 0$, and series
    clearly converges at $x = 0$.

7. $[-2, 2)$;   root test: $(b_k)^{1/k} = \dfrac{|x|}{2k^{1/k}} \rightarrow \dfrac{|x|}{2}$,   series converges for $|x| < 2$.

    At $x = 2$ series becomes $\sum \dfrac{1}{k}$, the divergent harmonic series.

    At $x = -2$ series becomes $\sum (-1)^k \dfrac{1}{k}$, a convergent alternating series.

9. Converges only at 0;   divergence test: $\left(\dfrac{k}{100} \right)^k x^k \rightarrow 0$ only if $x = 0$, and series
    clearly converges at $x = 0$.

11. $\left[-\dfrac{1}{2}, \dfrac{1}{2} \right)$;   root test: $(b_k)^{1/k} = \dfrac{2|x|}{\sqrt{k^{1/k}}} \rightarrow 2|x|$,   series converges for $|x| < \dfrac{1}{2}$.

    At $x = \dfrac{1}{2}$ series becomes $\sum \dfrac{1}{\sqrt{k}}$, a divergent p-series.

    At $x = -\dfrac{1}{2}$ series becomes $\sum (-1)^k \dfrac{1}{\sqrt{k}}$, a convergent alternating series.

13. $(-1, 1)$;   ratio test: $\dfrac{b_{k+1}}{b_k} = \dfrac{k^2}{(k+1)(k-1)} |x| \rightarrow |x|$,   series converges for $|x| < 1$.

    At the endpoints $x = 1$ and $x = -1$ the series diverges since there $b_k \not\rightarrow 0$.

15. $(-10, 10)$;   root test: $(b_k)^{1/k} = \dfrac{k^{1/k}}{10} |x| \rightarrow \dfrac{|x|}{10}$,   series converges for $|x| < 10$.

    At the endpoints $x = 10$ and $x = -10$ the series diverges since there $b_k \not\rightarrow 0$.

17. $(-\infty, \infty)$;   root test: $(b_k)^{1/k} = \dfrac{|x|}{k} \rightarrow 0$,   series converges all x.

19. $(-\infty, \infty)$;   root test: $(b_k)^{1/k} = \dfrac{|x-2|}{k} \rightarrow 0$,   series converges all x.

21. $(0, 4)$;   ratio test: $\dfrac{b_{k+1}}{b_k} = \dfrac{\ln(k+1)}{\ln k} \dfrac{|x-2|}{2} \rightarrow \dfrac{|x-2|}{2}$,   series converges for $|x - 2| < 2$.

    At the endpoints $x = 0$ and $x = 4$ the series diverges since there $b_k \not\rightarrow 0$.

23. $\left(-\frac{5}{2}, \frac{1}{2}\right)$;   root test: $(b_k)^{1/k} = \frac{2}{3}|x+1| \to \frac{2}{3}|x+1|$,   series converges for $|x+1| < \frac{3}{2}$.

At the endpoints $x = -\frac{5}{2}$ and $x = \frac{1}{2}$ the series diverges since there $b_k \not\to 0$.

25.   Examine the convergence of $\sum |a_k x^k|$; for (a) use the root test and for (b) use the ratio rest.

SECTION 12.9

1.   Use the fact that $\dfrac{d}{dx}\left(\dfrac{1}{1-x}\right) = \dfrac{1}{(1-x)^2}$:

$$\frac{1}{(1-x)^2} = \frac{d}{dx}(1 + x + x^2 + x^3 + \cdots + x^n + \cdots) = 1 + 2x + 3x^2 + \cdots + nx^{n-1} + \cdots .$$

3.   Use the fact that $\dfrac{d^{(k-1)}}{dx^{(k-1)}}\left[\dfrac{1}{1-x}\right] = \dfrac{(k-1)!}{(1-x)^k}$:

$$\frac{1}{(1-x)^k} = \frac{1}{(k-1)!}\frac{d^{(k-1)}}{dx^{(k-1)}}\left[1 + x + \cdots + x^{k-1} + x^k + x^{k+1} + \cdots + x^{n+k-1} + \cdots\right]$$

$$= \frac{1}{(k-1)!}\frac{d^{(k-1)}}{dx^{(k-1)}}\left[x^{k-1} + x^k + x^{k+1} + \cdots + x^{n+k-1} + \cdots\right]$$

$$= 1 + kx + \frac{(k+1)k}{2}x^2 + \cdots + \frac{(n+k-1)(n+k-2)\cdots(n+1)}{(k-1)!}x^n + \cdots$$

$$= 1 + kx + \frac{(k+1)k}{2!}x^2 + \cdots + \frac{(n+k-1)!}{n!(k-1)!}x^n + \cdots .$$

5.   Use the fact that $\dfrac{d}{dx}[\ln(1-x^2)] = \dfrac{-2x}{1-x^2}$:

$$\frac{1}{1-x^2} = 1 + x^2 + x^4 + \cdots + x^{2n} + \cdots$$

$$\frac{-2x}{1-x^2} = -2x - 2x^3 - 2x^5 - \cdots - 2x^{2n+1} - \cdots .$$

By integration

$$\ln(1-x^2) = \left(-x^2 - \frac{1}{2}x^4 - \frac{1}{3}x^6 - \cdots - \frac{x^{2n+2}}{n+1} - \cdots\right) + C.$$

At $x = 0$, both $\ln(1-x^2)$ and the series are 0. Thus, $C = 0$ and

$$\ln(1-x^2) = -x^2 - \frac{1}{2}x^4 - \frac{1}{3}x^6 - \cdots - \frac{1}{n+1}x^{2n+2} - \cdots .$$

7. $\sec^2 x = \dfrac{d}{dx}(\tan x) = \dfrac{d}{dx}\left(x + \dfrac{1}{3}x^3 + \dfrac{2}{15}x^5 + \dfrac{17}{315}x^7 + \cdots\right) = 1 + x^2 + \dfrac{2}{3}x^4 + \dfrac{17}{45}x^6 + \cdots$

9.   On its interval of convergence a power series is the Taylor series of its sum. Thus,

$$f(x) = x^2 \sin^2 x = x^2\left(x - \dfrac{x^3}{3!} + \dfrac{x^5}{5!} - \dfrac{x^7}{7!} + \cdots\right)$$

$$= x^3 - \dfrac{x^5}{3!} + \dfrac{x^7}{5!} - \dfrac{x^9}{7!} + \cdots = \sum_{n=0}^{\infty} f^{(n)}(0)\dfrac{x^n}{n!}$$

implies $f^{(9)}(0) = -9!/7! = -72.$

11. $\sin x^2 = \displaystyle\sum_{k=0}^{\infty}(-1)^k \dfrac{(x^2)^{2k+1}}{(2k+1)!} = \sum_{k=0}^{\infty}(-1)^k \dfrac{x^{4k+2}}{(2k+1)!}$

13. $e^{3x^3} = \displaystyle\sum_{k=0}^{\infty}\dfrac{(3x^3)^k}{k!} = \sum_{k=0}^{\infty}\dfrac{3^k}{k!}x^{3k}$

15. $\dfrac{2x}{1-x^2} = 2x\left(\dfrac{1}{1-x^2}\right) = 2x\displaystyle\sum_{k=0}^{\infty}(x^2)^k = \sum_{k=0}^{\infty}2x^{2k+1}$

17. $\dfrac{1}{1-x} + e^x = \displaystyle\sum_{k=0}^{\infty}x^k + \sum_{k=0}^{\infty}\dfrac{x^k}{k!} = \sum_{k=0}^{\infty}\dfrac{(k!+1)}{k!}x^k$

19. $x\ln(1+x^3) = x\displaystyle\sum_{k=1}^{\infty}\dfrac{(-1)^{k+1}}{k}(x^3)^k = \sum_{k=1}^{\infty}\dfrac{(-1)^{k+1}}{k}x^{3k+1}$

$\uparrow$
(12.6.7) $\_\_\_|$

21. $x^3 e^{-x^3} = x^3\displaystyle\sum_{k=0}^{\infty}\dfrac{(-x^3)^k}{k!} = \sum_{k=0}^{\infty}\dfrac{(-1)^k}{k!}x^{3k+3}$

23. $0.804 \le I \le 0.808;$ $I = \displaystyle\int_0^1\left(1 - x^3 + \dfrac{x^6}{2!} - \dfrac{x^9}{3!} + \cdots\right)dx$

$$= \left[x - \dfrac{x^4}{4} + \dfrac{x^7}{14} - \dfrac{x^{10}}{60} + \dfrac{x^{13}}{(13)(24)} - \cdots\right]_0^1$$

$$= 1 - \tfrac{1}{4} + \tfrac{1}{14} - \tfrac{1}{60} + \tfrac{1}{312} - \cdots.$$

Since $\tfrac{1}{312} < 0.01,$ we can stop there:

$$1 - \tfrac{1}{4} + \tfrac{1}{14} - \tfrac{1}{60} \le I \le 1 - \tfrac{1}{4} + \tfrac{1}{14} - \tfrac{1}{60} + \tfrac{1}{312}$$

gives $0.804 \le I \le 0.808.$

25. $0.600 \leq I \leq 0.603$; $I = \displaystyle\int_0^1 \left(x^{1/2} - \dfrac{x^{3/2}}{3!} + \dfrac{x^{5/2}}{5!} - \cdots \right) dx$

$$= \left[\tfrac{2}{3}x^{3/2} - \tfrac{1}{15}x^{5/2} + \tfrac{1}{420}x^{7/2} - \cdots \right]_0^1$$

$$= \tfrac{2}{3} - \tfrac{1}{15} + \tfrac{1}{420} - \cdots .$$

Since $\tfrac{1}{420} < 0.01$,   we can stop there:

$$\tfrac{2}{3} - \tfrac{1}{15} \leq I \leq \tfrac{2}{3} - \tfrac{1}{15} + \tfrac{1}{420}$$

gives $0.600 \leq I \leq 0.603$.

27. $0.294 \leq I \leq 0.304$; $I = \displaystyle\int_0^1 \left(x^2 - \dfrac{x^6}{3} + \dfrac{x^{10}}{5} - \dfrac{x^{14}}{7} + \cdots \right) dx$

(12.9.6) —

$$= \left[\tfrac{1}{3}x^3 - \tfrac{1}{21}x^7 + \tfrac{1}{55}x^{11} - \tfrac{1}{105}x^{15} + \cdots \right]_0^1$$

$$= \tfrac{1}{3} - \tfrac{1}{21} + \tfrac{1}{55} - \tfrac{1}{105} + \cdots .$$

Since $\tfrac{1}{105} < 0.01$,   we can stop there:

$$\tfrac{1}{3} - \tfrac{1}{21} + \tfrac{1}{55} - \tfrac{1}{105} \leq I \leq \tfrac{1}{3} - \tfrac{1}{21} + \tfrac{1}{55}$$

gives $0.294 \leq I \leq 0.304$.

29. e^{x^3};   by (12.6.4)   **31.** $3x^2 e^{x^3} = \dfrac{d}{dx}(e^{x^3})$

33.   Let $f(x)$ be the sum of these series; a_k and b_k are both $\dfrac{f^{(k)}(0)}{k!}$.

35. $0.0352 \leq I \leq 0.0359$; $I = \displaystyle\int_0^{1/2} \left(x^2 - \dfrac{x^3}{2} + \dfrac{x^4}{3} - \dfrac{x^5}{4} + \cdots \right) dx$

$$= \left[\dfrac{x^3}{3} - \dfrac{x^4}{8} + \dfrac{x^5}{15} - \dfrac{x^6}{24} + \cdots \right]_0^{1/2}$$

$$= \dfrac{1}{3(2^3)} - \dfrac{1}{8(2^4)} + \dfrac{1}{15(2^5)} - \dfrac{1}{24(2^6)} + \cdots .$$

Since $\dfrac{1}{24(2^6)} = \dfrac{1}{1536} < 0.001$, we can stop there:

$$\frac{1}{3(2^3)} - \frac{1}{8(2^4)} + \frac{1}{15(2^5)} - \frac{1}{24(2^6)} \le I \le \frac{1}{3(2^3)} - \frac{1}{8(2^4)} + \frac{1}{15(2^5)}$$

gives $0.0352 \le I \le 0.0359.$ Direct integration gives

$$I = \int_0^{1/2} x \ln(1+x)\, dx = \left[\frac{1}{2}(x^2 - 1)\ln(1+x) - \frac{1}{4}x^2 + \frac{1}{2}x\right]_0^{1/2} = \frac{3}{16} - \frac{3}{8}\ln 1.5 \cong 0.0354505.$$

37. $0.2640 \le I \le 0.2643;$

$$I = \int_0^1 \left(x - x^2 + \frac{x^3}{2!} - \frac{x^4}{3!} + \frac{x^5}{4!} - \frac{x^6}{5!} + \frac{x^7}{6!} - \cdots \right) dx$$

$$= \left[\frac{x^2}{2} - \frac{x^3}{3} + \frac{x^4}{4(2!)} - \frac{x^5}{5(3!)} + \frac{x^6}{6(4!)} - \frac{x^7}{7(5!)} + \frac{x^8}{8(6!)} - \cdots\right]_0^1$$

$$= \frac{1}{2} - \frac{1}{3} + \frac{1}{4(2!)} - \frac{1}{5(3!)} + \frac{1}{6(4!)} - \frac{1}{7(5!)} + \frac{1}{8(6!)} - \cdots.$$

Note that $\dfrac{1}{8(6!)} = \dfrac{1}{5760} < 0.001.$ The integral lies between

$$\frac{1}{2} - \frac{1}{3} + \frac{1}{4(2!)} - \frac{1}{5(3!)} + \frac{1}{6(4!)} - \frac{1}{7(5!)}$$

and

$$\frac{1}{2} - \frac{1}{3} + \frac{1}{4(2!)} - \frac{1}{5(3!)} + \frac{1}{6(4!)} - \frac{1}{7(5!)} + \frac{1}{8(6!)}.$$

The first sum is greater than 0.2640 and the second sum is less than 0.2643.

Direct integration gives

$$\int_0^1 x e^{-x}\, dx = \left[-x e^{-x} - e^{-x}\right]_0^1 = 1 - 2/e \cong 0.2642411.$$

SECTION 12.10

1.   Take $\alpha = 1/2$ in (12.10.2) to obtain $1 + \frac{1}{2}x - \frac{1}{8}x^2 + \frac{1}{16}x^3 - \frac{5}{128}x^4.$

3.   In (12.10.2) replace x by x^2 and take $\alpha = 1/2$ to obtain $1 + \frac{1}{2}x^2 - \frac{1}{8}x^4.$

5.   Take $\alpha = -1/2$ in (12.10.2) to obtain $1 - \frac{1}{2}x + \frac{3}{8}x^2 - \frac{5}{16}x^3 + \frac{35}{128}x^4.$

7.   In (12.10.2) replace x by $-x$ and take $\alpha = 1/4$ to obtain $1 - \frac{1}{4}x - \frac{3}{32}x^2 - \frac{7}{128}x^3 - \frac{77}{2048}x^4$.

9.   9.8995; $\sqrt{98} = (100 - 2)^{1/2} = 10\left(1 - \frac{1}{50}\right)^{1/2} \cong 10\left[1 - \frac{1}{100} - \frac{1}{20000}\right] = 9.8995$

11.   2.0799; $\sqrt[3]{9} = (8 + 1)^{1/3} = 2\left(1 + \frac{1}{8}\right)^{1/3} \cong 2\left[1 + \frac{1}{24} - \frac{1}{576}\right] \cong 2.0799$

13.   0.4925; $17^{-1/4} = (16 + 1)^{-1/4} = \frac{1}{2}\left(1 + \frac{1}{16}\right)^{-1/4} \cong \frac{1}{2}\left[1 - \frac{1}{64} + \frac{5}{8192}\right] \cong 0.4925$

SECTION 12.11

1. $\frac{4}{3}$; $\displaystyle\sum_{k=0}^{\infty}\left(\frac{1}{4}\right)^k = \frac{1}{1 - 1/4} = \frac{4}{3}$ **3.** $\frac{2}{3}$; $\displaystyle\sum_{k=0}^{\infty}(-1)^k\left(\frac{1}{2}\right)^k = \frac{1}{1 - (-1/2)} = \frac{2}{3}$

5.   1; $s_n = \left(1 - \frac{1}{2}\right) + \left(\frac{1}{2} - \frac{1}{3}\right) + \cdots + \left(\frac{1}{n} - \frac{1}{n+1}\right) = 1 - \frac{1}{n+1} \to 1$

7. $\displaystyle\sum_{k=0}^{\infty} x^{5k+1} = x\sum_{k=0}^{\infty}(x^5)^k = \frac{x}{1 - x^5}$ for $|x^5| < 1$ and thus for $|x| < 1$

9. $\displaystyle\sum_{k=1}^{\infty} \frac{3}{2}x^{2k-1} = \frac{3x}{2}\sum_{k=1}^{\infty} x^{2k-2} = \frac{3x}{2}\sum_{k=0}^{\infty}(x^2)^k = \frac{3x}{2(1 - x^2)}$ for $|x^2| < 1$ and thus for $|x| < 1$

11. $\displaystyle\sum_{k=1}^{\infty} \frac{k^2}{k!} = 1 + \frac{4}{2!} + \frac{9}{3!} + \frac{16}{4!} + \frac{25}{5!} + \cdots$

$$= 1 + \frac{2}{1!} + \frac{3}{2!} + \frac{4}{3!} + \frac{5}{4!} + \cdots$$

$$= 1 + 2 + \left(\frac{2}{2!} + \frac{1}{2!}\right) + \left(\frac{3}{3!} + \frac{1}{3!}\right) + \left(\frac{4}{4!} + \frac{1}{4!}\right) + \cdots$$

$$= 4 + \frac{1}{2!} + \left(\frac{1}{2!} + \frac{1}{3!}\right) + \left(\frac{1}{3!} + \frac{1}{4!}\right) + \cdots$$

$$= 2\left(2 + \frac{1}{2!} + \frac{1}{3!} + \frac{1}{4!} + \cdots\right) = 2e$$

13.   diverges;   limit comparison with $\displaystyle\sum \frac{1}{k}$

15.   converges absolutely;   basic comparison with $\displaystyle\sum \frac{1}{k^2}$

17.   (a)   does not converge absolutely;   limit comparison with $\displaystyle\sum \frac{1}{k}$

       (b)   converges conditionally;   Theorem 12.5.3

19.   converges absolutely;   root test: $\left(\dfrac{k}{3^{k-1}}\right)^{1/k} \to \dfrac{1}{3}$

21. (a) does not converge absolutely; basic comparison with $\sum \frac{1}{k}$

(b) converges conditionally; Theorem 12.5.3

23. $\left[\frac{9}{5}, \frac{11}{5}\right)$; root test: $(b_k)^{1/k} = \frac{5|x-2|}{k^{1/k}} \to 5|x-2|$, series converges for $|x-2| < 1/5$.

At $x = 9/5$ series becomes $\sum \frac{(-1)^k}{k}$, a convergent alternating series.

At $x = 11/5$ series becomes $\sum \frac{1}{k}$, the divergent harmonic series.

25. $(0,2)$; ratio test: $\frac{b_{k+1}}{b_k} = \frac{k+2}{k}|x-1|^2 \to |x-1|^2$, series converges for $|x-1| < 1$.

At the endpoints $x = 0$ and $x = 2$ the series becomes $\sum k(k+1)$ which diverges since $b_k \not\to 0$.

27. $(-1,1)$; ratio test: $\frac{b_{k+1}}{b_k} = \frac{(2k+1)(k+1)}{k(2k+3)}x^2 \to x^2$, series converges for $|x| < 1$.

At the endpoints $x = \pm 1$ the series diverges since there $b_k \not\to 0$.

29. Converges only at -1; ratio test: $\frac{b_{k+1}}{b_k} = (k+1)|x+1| \to \infty$ unless $x = -1$,

and series clearly converges at $x = -1$.

31. $(-9,9)$; root test: $(b_k)^{1/k} = \frac{k^{1/k}}{9}|x| \to \frac{1}{9}|x|$, series converges for $|x| < 9$.

At the endpoints $x = \pm 9$ the series diverges since there $b_k \not\to 0$.

33. $(-3,7)$; root test: $(b_k)^{1/k} \to \frac{1}{5}|x-2|$, series converges for $|x-2| < 5$.

At the endpoints $x = -3$ and $x = 7$ the series diverges since there $b_k \not\to 0$.

35. $xe^{5x^2} = x\sum_{k=0}^{\infty} \frac{(5x^2)^k}{k!} = \sum_{k=0}^{\infty} \frac{5^k}{k!}x^{2k+1}$

37. $\sqrt{x}\tan^{-1}\sqrt{x} = \sqrt{x}\sum_{k=0}^{\infty} \frac{(-1)^k}{2k+1}(\sqrt{x})^{2k+1} = \sum_{k=0}^{\infty} \frac{(-1)^k}{2k+1}x^{k+1}$

$\uparrow$
$\llcorner$ (12.9.6)

39. $(x+x^2)\sin x^2 = (x+x^2)\sum_{k=0}^{\infty} \frac{(-1)^k}{(2k+1)!}(x^2)^{2k+1} = \sum_{k=0}^{\infty} \frac{(-1)^k}{(2k+1)!}(x^{4k+3} + x^{4k+4})$

$\uparrow$
$\llcorner$ (12.6.5)

41. We use (12.6.4) and (12.6.5) and collect powers of x up to x^4:

$$e^{\sin x} = \sum_{k=0}^{\infty} \frac{(\sin x)^k}{k!} = 1 + \sin x + \frac{\sin^2 x}{2} + \frac{\sin^3 x}{6} + \frac{\sin^4 x}{24} + \cdots$$

$$= 1 + \left(x - \frac{x^3}{6} + \cdots\right) + \frac{1}{2}\left(x - \frac{x^3}{6} + \cdots\right)^2 + \frac{1}{6}\left(x - \frac{x^3}{6} + \cdots\right)^3 + \frac{1}{24}\left(x - \frac{x^3}{6} + \cdots\right)^4$$

$$\cong 1 + \left(x - \frac{x^3}{6}\right) + \frac{1}{2}\left(x^2 - \frac{1}{3}x^4\right) + \frac{1}{6}(x^3) + \frac{1}{24}(x^4)$$

$$= 1 + x + \tfrac{1}{2}x^2 - \tfrac{1}{8}x^4.$$

43. In (12.10.2) replace x by $-x^2$ and take $\alpha = -1/2$ to obtain $1 + \tfrac{1}{2}x^2 + \tfrac{3}{8}x^4$.

45. $0.493 \le I \le 0.500$; $I = \displaystyle\int_0^{1/2} (1 - x^4 + x^8 - \cdots)\, dx = \left[x - \frac{1}{5}x^5 + \frac{1}{9}x^9 - \cdots\right]_0^{1/2} = \frac{1}{2} - \frac{1}{160} + \frac{1}{4608} - \cdots .$

Since $\frac{1}{160} < 0.01,$ we can stop there:

$$\tfrac{1}{2} - \tfrac{1}{160} \le I \le \tfrac{1}{2}$$

gives $0.493 \le I \le 0.500.$

47. $4.081 \le \sqrt[3]{68} \le 4.084;$

$$\sqrt[3]{68} = (64 + 4)^{1/3} = 4\left(1 + \tfrac{1}{16}\right)^{1/3}$$

$$= 4\left[1 + \frac{1}{48} - \frac{1}{9(16^2)} + \begin{array}{l}\text{convergent alternating series with}\\\text{declining terms; first term positive}\end{array}\right]$$

$$= 4 + \frac{1}{12} - \frac{1}{576} + \begin{array}{l}\text{convergent alternating series with}\\\text{declining terms; first term positive.}\end{array}$$

Since $\frac{1}{576} < 0.01,$ we can stop there:

$$4 + \tfrac{1}{12} - \tfrac{1}{576} \le \sqrt[3]{68} \le 4 + \tfrac{1}{12}$$

gives $4.081 \le \sqrt[3]{68} \le 4.084.$

49. $\displaystyle\sum_{k=1}^{\infty} a_k = \sum_{k=1}^{\infty} \int_k^{k+1} x e^{-x}\, dx = \int_1^{\infty} x e^{-x}\, dx = \lim_{b \to \infty} \int_1^b x e^{-x}\, dx$

(by parts)

$$\overset{\downarrow}{=} \lim_{b \to \infty} \left[(-x - 1)e^{-x}\right]_1^b = \lim_{b \to \infty} \left[\frac{2}{e} - \frac{b+1}{e^b}\right] = \frac{2}{e}$$

51. (a) diverges; $a_k = \dfrac{1}{1 - 1/k} = \dfrac{k}{k-1} \to 1 \neq 0$

     (b) diverges; $a_k = \dfrac{1/k}{1 - 1/k} = \dfrac{1}{k-1}$, $\displaystyle\sum_{k=2}^{\infty} a_k$ is the divergent harmonic series

     (c) converges to 1; $a_k = \dfrac{1/k^2}{1 - 1/k} = \dfrac{1}{k(k-1)} = \dfrac{1}{k-1} - \dfrac{1}{k}$;

     thus, $s_n = \left(1 - \dfrac{1}{2}\right) + \left(\dfrac{1}{2} - \dfrac{1}{3}\right) + \cdots + \left(\dfrac{1}{n-1} - \dfrac{1}{n}\right) = 1 - \dfrac{1}{n} \to 1$

53. Set

$$a_n = x_n - \frac{\epsilon}{4^n}, \quad b_n = x_n + \frac{\epsilon}{4^n}.$$

Then

$$\sum_{n=1}^{\infty} (b_n - a_n) = 2\epsilon \sum_{n=1}^{\infty} \left(\frac{1}{4}\right)^n = \frac{2}{3}\epsilon < \epsilon.$$

CHAPTER 13

SECTION 13.1

1.

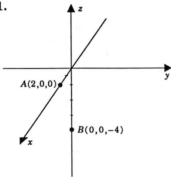

3.

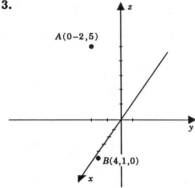

length $\overline{AB}$: $2\sqrt{5}$

midpoint: $(1, 0, -2)$

length $\overline{AB}$: $5\sqrt{2}$

midpoint: $\left(2, -\frac{1}{2}, \frac{5}{2}\right)$

5. $z = -2$ 7. $y = 1$ 9. $x = 3$

11. $x^2 + (y - 2)^2 + (z + 1)^2 = 9$ 13. $(x - 2)^2 + (y - 4)^2 + (z + 4)^2 = 36$

15. $(x - 3)^2 + (y - 2)^2 + (z - 2)^2 = 13$ 17. $(x - 2)^2 + (y - 3)^2 + (z + 4)^2 = 25$

19. $(2, 3, -5)$ 21. $(-2, 3, 5)$ 23. $(-2, 3, -5)$

25. $(-2, -3, -5)$ 27. $(2, -5, 5)$ 29. $(-2, 1, -3)$

31.   Each such sphere has an equation of the form

$$(x - a)^2 + (y - a)^2 + (z - a)^2 = a^2.$$

Substituting $x = 5$, $y = 1$, $z = 4$ we get

$$(5 - a)^2 + (1 - a)^2 + (4 - a)^2 = a^2.$$

This reduces to $a^2 - 10a + 21 = 0$ and gives $a = 3$ or $a = 7$.

33.   (a)   Take R as (x, y, z). Since

$$d(P, R) = t\,d(P, Q)$$

we conclude by similar triangles that

$$d(AR) = t\,d(B, Q)$$

and therefore

$$z - a_3 = t(b_3 - a_3).$$

Thus

$$z = a_3 + t(b_3 - a_3).$$

In similar fashion

$$x = a_1 + t(b_1 - a_1) \quad \text{and} \quad y = a_2 + t(b_2 - a_2).$$

(b)   The midpoint of PQ, $\left(\dfrac{a_1 + b_1}{2}, \dfrac{a_2 + b_2}{2}, \dfrac{a_3 + b_3}{2} \right)$,   occurs at $t = \dfrac{1}{2}$.

SECTION 13.3

1. $3\mathbf{i} - 4\mathbf{j} + 6\mathbf{k}$ 3. $-3\mathbf{i} - \mathbf{j} + 8\mathbf{k}$

5.   5   7.   3   9. $\sqrt{6}$

11.   (a) $\mathbf{a}$, $\mathbf{c}$, and $\mathbf{d}$ since $\mathbf{a} = \frac{1}{3}\mathbf{c} = -\frac{1}{2}\mathbf{d}$

(b) $\mathbf{a}$ and $\mathbf{c}$ since $\mathbf{a} = \frac{1}{3}\mathbf{c}$

(c) $\mathbf{a}$ and $\mathbf{c}$ both have direction opposite to $\mathbf{d}$

13.   (i) $\mathbf{a} + \mathbf{b}$ (ii) $-(\mathbf{a} + \mathbf{b})$ (iii) $\mathbf{a} - \mathbf{b}$ (iv) $\mathbf{b} - \mathbf{a}$

15.   (a) $\mathbf{a} - 3\mathbf{b} + 2\mathbf{c} + 4\mathbf{d} = (2\mathbf{i} - \mathbf{k}) - 3(\mathbf{i} + 3\mathbf{j} + 5\mathbf{k}) + 2(-\mathbf{i} + \mathbf{j} + \mathbf{k}) + 4(\mathbf{i} + \mathbf{j} + 6\mathbf{k})$

$$= \mathbf{i} - 3\mathbf{j} + 10\mathbf{k}$$

(b)   The vector equation

$$(1, 1, 6) = A(2, 0, -1) + B(1, 3, 5) + C(-1, 1, 1)$$

implies

$$
\begin{aligned}
1 &= 2A + B - C, \\
1 &= \phantom{2A + {}} 3B + C, \\
6 &= -A + 5B + C.
\end{aligned}
$$

Simultaneous solution gives $A = -2$, $B = \frac{3}{2}$, $C = -\frac{7}{2}$.

17. $\|3\mathbf{i} + \mathbf{j}\| = \|\alpha\mathbf{j} - \mathbf{k}\|$ $\implies$ $10 = \alpha^2 + 1$ so $\alpha = \pm 3$

19. $\|\alpha\mathbf{i} + (\alpha - 1)\mathbf{j} + (\alpha + 1)\mathbf{k}\| = 2$ $\implies$ $\alpha^2 + (\alpha - 1)^2 + (\alpha + 1)^2 = 4$

$$\implies \quad 3\alpha^2 = 2 \quad \text{so} \quad \alpha = \pm\tfrac{1}{3}\sqrt{6}$$

21. $\pm\frac{2}{13}\sqrt{13}\,(3\mathbf{j} + 2\mathbf{k})$ since $\|\alpha(3\mathbf{j} + 2\mathbf{k})\| = 2$ $\implies$ $\alpha = \pm\frac{2}{13}\sqrt{13}$

23.   (a)   Since $\|\mathbf{a} - \mathbf{b}\|$ and $\|\mathbf{a} + \mathbf{b}\|$ are the lengths of the diagonals of the parallelogram, the parallelogram must be a rectangle.

(b)   Simplify

$$\sqrt{(a_1 - b_1)^2 + (a_2 - b_2)^2 + (a_3 - b_3)^2} = \sqrt{(a_1 + b_1)^2 + (a_2 + b_2)^2 + (a_3 + b_3)^2}.$$

25.   (a) $\|\mathbf{r} - \mathbf{a}\| = 3$ where $\mathbf{a} = a_1\mathbf{i} + a_2\mathbf{j} + a_3\mathbf{k}$

       (b) $\|\mathbf{r}\| \leq 2$ (c) $\|\mathbf{r} - \mathbf{a}\| \leq 1$ where $\mathbf{a} = a_1\mathbf{i} + a_2\mathbf{j} + a_3\mathbf{k}$

       (d) $\|\mathbf{r} - \mathbf{a}\| = \|\mathbf{r} - \mathbf{b}\|$ (e) $\|\mathbf{r} - \mathbf{a}\| + \|\mathbf{r} - \mathbf{b}\| = k$

SECTION 13.4

1. $\mathbf{a} \cdot \mathbf{b}$ **3.** $(\mathbf{a} - \mathbf{b}) \cdot \mathbf{c} + \mathbf{b} \cdot (\mathbf{c} + \mathbf{a}) = \mathbf{a} \cdot \mathbf{c} - \mathbf{b} \cdot \mathbf{c} + \mathbf{b} \cdot \mathbf{c} + \mathbf{b} \cdot \mathbf{a} = \mathbf{a} \cdot (\mathbf{b} + \mathbf{c})$

5.   (a) $\mathbf{a} \cdot \mathbf{b} = (2)(3) + (1)(-1) + (0)(2) = 5$

           $\mathbf{a} \cdot \mathbf{c} = (2)(4) + (1)(0) + (0)(3) = 8$

           $\mathbf{b} \cdot \mathbf{c} = (3)(4) + (-1)(0) + (2)(3) = 18$

       (b) $\|\mathbf{a}\| = \sqrt{5}, \quad \|\mathbf{b}\| = \sqrt{14}, \quad \|\mathbf{c}\| = 5.$ Then,

$$\cos \sphericalangle(\mathbf{a}, \mathbf{b}) = \frac{\mathbf{a} \cdot \mathbf{b}}{\|\mathbf{a}\| \, \|\mathbf{b}\|} = \frac{5}{(\sqrt{5})(\sqrt{14})} = \frac{1}{14}\sqrt{70},$$

$$\cos \sphericalangle(\mathbf{a}, \mathbf{c}) = \frac{8}{(\sqrt{5})(5)} = \frac{8}{25}\sqrt{5},$$

$$\cos \sphericalangle(\mathbf{b}, \mathbf{c}) = \frac{18}{(\sqrt{14})(5)} = \frac{9}{35}\sqrt{14}.$$

       (c) $\mathbf{u_b} = \dfrac{1}{\sqrt{14}}(3\mathbf{i} - \mathbf{j} + 2\mathbf{k}), \quad \text{comp}_\mathbf{b}\, \mathbf{a} = \mathbf{a} \cdot \mathbf{u_b} = \dfrac{1}{\sqrt{14}}(6 - 1) = \dfrac{5}{14}\sqrt{14},$

           $\mathbf{u_c} = \frac{1}{5}(4\mathbf{i} + 3\mathbf{k}), \quad \text{comp}_\mathbf{c}\, \mathbf{a} = \mathbf{a} \cdot \mathbf{u_c} = \frac{8}{5}$

       (d) $\text{proj}_\mathbf{b}\, \mathbf{a} = (\text{comp}_\mathbf{b}\, \mathbf{a})\, \mathbf{u_b} = \frac{5}{14}(3\mathbf{i} - \mathbf{j} + 2\mathbf{k})$

           $\text{proj}_\mathbf{c}\, \mathbf{a} = (\text{comp}_\mathbf{c}\, \mathbf{a})\, \mathbf{u_c} = \frac{8}{25}(4\mathbf{i} + 3\mathbf{k})$

7. $\mathbf{u} = \cos\dfrac{\pi}{3}\mathbf{i} + \cos\dfrac{\pi}{4}\mathbf{j} + \cos\dfrac{2\pi}{3}\mathbf{k} = \dfrac{1}{2}\mathbf{i} + \dfrac{1}{2}\sqrt{2}\,\mathbf{j} - \dfrac{1}{2}\mathbf{k}$

9. $\cos\theta = \dfrac{(3\mathbf{i} - \mathbf{j} - 2\mathbf{k}) \cdot (\mathbf{i} + 2\mathbf{j} - 3\mathbf{k})}{\|3\mathbf{i} - \mathbf{j} - 2\mathbf{k}\| \, \|\mathbf{i} + 2\mathbf{j} - 3\mathbf{k}\|} = \dfrac{7}{\sqrt{14}\sqrt{14}} = \dfrac{1}{2}, \quad \theta = \dfrac{\pi}{3}$

11.   Since $\|\mathbf{i} - \mathbf{j} + \sqrt{2}\,\mathbf{k}\| = 2,$ we have $\cos\alpha = \frac{1}{2}, \quad \cos\beta = -\frac{1}{2}, \quad \cos\gamma = \frac{1}{2}\sqrt{2}.$

       The direction angles are $\frac{1}{3}\pi, \quad \frac{2}{3}\pi, \quad \frac{1}{4}\pi.$

13.   (a) $\text{proj}_\mathbf{b}\, \alpha\mathbf{a} = (\alpha\mathbf{a} \cdot \mathbf{u_b})\mathbf{u_b} = \alpha(\mathbf{a} \cdot \mathbf{u_b})\mathbf{u_b} = \alpha\, \text{proj}_\mathbf{b}\, \mathbf{a}$

       (b) $\text{proj}_\mathbf{b}\, (\mathbf{a} + \mathbf{c}) = [(\mathbf{a} + \mathbf{c}) \cdot \mathbf{u_b}]\mathbf{u_b}$

$$= (\mathbf{a} \cdot \mathbf{u_b} + \mathbf{c} \cdot \mathbf{u_b})\mathbf{u_b}$$

$$= (\mathbf{a} \cdot \mathbf{u_b})\mathbf{u_b} + (\mathbf{c} \cdot \mathbf{u_b})\mathbf{u_b} = \text{proj}_\mathbf{b}\, \mathbf{a} + \text{proj}_\mathbf{b}\, \mathbf{c}$$

15. (a) For $\mathbf{a} \neq \mathbf{0}$ the following statements are equivalent:

$$\mathbf{a} \cdot \mathbf{b} = \mathbf{a} \cdot \mathbf{c}, \quad \mathbf{b} \cdot \mathbf{a} = \mathbf{c} \cdot \mathbf{a},$$

$$\mathbf{b} \cdot \frac{\mathbf{a}}{\|\mathbf{a}\|} = \mathbf{c} \cdot \frac{\mathbf{a}}{\|\mathbf{a}\|}, \quad \mathbf{b} \cdot \mathbf{u_a} = \mathbf{c} \cdot \mathbf{u_a}$$

$$(\mathbf{b} \cdot \mathbf{u_a})\mathbf{u_a} = (\mathbf{c} \cdot \mathbf{u_a})\mathbf{u_a},$$

$$\operatorname{proj}_{\mathbf{a}} \mathbf{b} = \operatorname{proj}_{\mathbf{a}} \mathbf{c}.$$

$$\mathbf{a} \cdot \mathbf{b} = \mathbf{a} \cdot \mathbf{c} \quad \text{but} \quad \mathbf{b} \neq \mathbf{c}$$

(b) $\mathbf{b} = (\mathbf{b} \cdot \mathbf{i})\mathbf{i} + (\mathbf{b} \cdot \mathbf{j})\mathbf{j} + (\mathbf{b} \cdot \mathbf{k})\mathbf{k} = (\mathbf{c} \cdot \mathbf{i})\mathbf{i} + (\mathbf{c} \cdot \mathbf{j})\mathbf{j} + (\mathbf{c} \cdot \mathbf{k})\mathbf{k} = \mathbf{c}$

$\quad\quad\quad\llcorner$ (13.4.13) $\quad\quad\quad\quad\quad\quad\quad\quad\quad\quad\quad$ (13.4.13) $\lrcorner$

17. (a) $\|\mathbf{a}+\mathbf{b}\|^2 - \|\mathbf{a}-\mathbf{b}\|^2 = (\mathbf{a}+\mathbf{b}) \cdot (\mathbf{a}+\mathbf{b}) - (\mathbf{a}-\mathbf{b}) \cdot (\mathbf{a}-\mathbf{b})$

$$= [(\mathbf{a} \cdot \mathbf{a}) + 2(\mathbf{a} \cdot \mathbf{b}) + (\mathbf{b} \cdot \mathbf{b})] - [(\mathbf{a} \cdot \mathbf{a}) - 2(\mathbf{a} \cdot \mathbf{b}) + (\mathbf{b} \cdot \mathbf{b})] = 4(\mathbf{a} \cdot \mathbf{b})$$

(b) The following statements are equivalent:

$$\mathbf{a} \perp \mathbf{b}, \quad \mathbf{a} \cdot \mathbf{b} = 0, \quad \|\mathbf{a}+\mathbf{b}\|^2 - \|\mathbf{a}-\mathbf{b}\|^2 = 0, \quad \|\mathbf{a}+\mathbf{b}\| = \|\mathbf{a}-\mathbf{b}\|.$$

(c) By (b), the relation $\|\mathbf{a}+\mathbf{b}\| = \|\mathbf{a}-\mathbf{b}\|$ gives $\mathbf{a} \perp \mathbf{b}$. The relation $\mathbf{a}+\mathbf{b} \perp \mathbf{a}-\mathbf{b}$ gives

$$0 = (\mathbf{a}+\mathbf{b}) \cdot (\mathbf{a}-\mathbf{b}) = \|\mathbf{a}\|^2 - \|\mathbf{b}\|^2 \quad \text{and thus} \quad \|\mathbf{a}\| = \|\mathbf{b}\|.$$

The parallelogram is a square since it has two adjacent sides of equal length and these meet at right angles.

19. Let $\theta_1, \theta_2, \theta_3$ be the direction angles of $-\mathbf{a}$. Then

$$\theta_1 = \cos^{-1}\left[\frac{(-\mathbf{a} \cdot \mathbf{i})}{\|-\mathbf{a}\|}\right] = \cos^{-1}\left[-\frac{(\mathbf{a} \cdot \mathbf{i})}{\|\mathbf{a}\|}\right] = \cos^{-1}(-\cos\alpha) = \cos^{-1}(\pi - \alpha) = \pi - \alpha.$$

Similarly $\quad \theta_2 = \pi - \beta \quad$ and $\quad \theta_3 = \pi - \gamma$.

21. If $\mathbf{a} \perp \mathbf{b}$ and $\mathbf{a} \perp \mathbf{c}$, then

$$\mathbf{a} \cdot \mathbf{b} = 0, \quad \mathbf{a} \cdot \mathbf{c} = 0$$

$$\mathbf{a} \cdot (\alpha\mathbf{b} + \beta\mathbf{c}) = \alpha(\mathbf{a} \cdot \mathbf{b}) + \beta(\mathbf{a} \cdot \mathbf{c}) = 0$$

$$\mathbf{a} \perp (\alpha\mathbf{b} + \beta\mathbf{c}).$$

23. Existence of decomposition:

$$\mathbf{a} = (\mathbf{a} \cdot \mathbf{u_b})\mathbf{u_b} + [\mathbf{a} - (\mathbf{a} \cdot \mathbf{u_b})\mathbf{u_b}].$$

Uniqueness of decomposition: suppose that

$$\mathbf{a} = \mathbf{a}_{\|} + \mathbf{a}_{\perp} = \mathbf{A}_{\|} + \mathbf{A}_{\perp}.$$

Then the vector $\mathbf{a}_\parallel - \mathbf{A}_\parallel = \mathbf{A}_\perp - \mathbf{a}_\perp$ is both parallel to $\mathbf{b}$ and perpendicular to $\mathbf{b}$. (Exercises 21 and 22.) Therefore it is zero. Consequently $\mathbf{A}_\parallel = \mathbf{a}_\parallel$ and $\mathbf{A}_\perp = \mathbf{a}_\perp$.

25. $\cos\dfrac{\pi}{3} = \dfrac{\mathbf{c}\cdot\mathbf{d}}{\|\mathbf{c}\|\,\|\mathbf{d}\|}, \quad \dfrac{1}{2} = \dfrac{2x+1}{x^2+2}, \quad x^2 = 4x; \qquad x = 0, 4$

27.

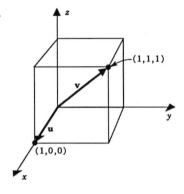

We take $\mathbf{u} = \mathbf{i}$ as an edge and $\mathbf{v} = \mathbf{i}+\mathbf{j}+\mathbf{k}$ as a diagonal of a cube. Then,

$$\cos\theta = \frac{\mathbf{u}\cdot\mathbf{v}}{\|\mathbf{u}\|\,\|\mathbf{v}\|} = \frac{1}{3}\sqrt{3},$$

$$\theta = \cos^{-1}\left(\tfrac{1}{3}\sqrt{3}\right) \cong 0.96 \text{ radians.}$$

29. (a) The direction angles of a vector always satisfy

$$\cos^2\alpha + \cos^2\beta + \cos^2\gamma = 1$$

and, as you can check,

$$\cos^2\tfrac{1}{4}\pi + \cos^2\tfrac{1}{6}\pi + \cos^2\tfrac{2}{3}\pi \neq 1.$$

(b) The relation

$$\cos^2\alpha + \cos^2\tfrac{1}{4}\pi + \cos^2\tfrac{1}{4}\pi = 1$$

gives

$$\cos^2\alpha + \tfrac{1}{2} + \tfrac{1}{2} = 1, \quad \cos\alpha = 0, \quad a_1 = \|\mathbf{a}\|\cos\alpha = 0.$$

31. Set $\mathbf{u} = a\mathbf{i} + b\mathbf{j} + c\mathbf{k}$.   The relations

$$(a\mathbf{i}+b\mathbf{j}+c\mathbf{k})\cdot(\mathbf{i}+2\mathbf{j}+\mathbf{k}) = 0 \quad \text{and} \quad (a\mathbf{i}+b\mathbf{j}+c\mathbf{k})\cdot(3\mathbf{i}-4\mathbf{j}+2\mathbf{k}) = 0$$

give

$$a + 2b + c = 0 \qquad 3a - 4b + 2c = 0$$

so that $b = \tfrac{1}{8}a$ and $c = -\tfrac{5}{4}a$.

Then, since $\mathbf{u}$ is a unit vector,

$$a^2 + b^2 + c^2 = 1, \quad a^2 + \left(\frac{a}{8}\right)^2 + \left(\frac{-5a}{4}\right)^2 = 1, \quad \frac{165}{64}a^2 = 1.$$

Thus, $a = \pm\dfrac{8}{165}\sqrt{165}$ and $\mathbf{u} = \pm\dfrac{\sqrt{165}}{165}(8\mathbf{i}+\mathbf{j}-10\mathbf{k}).$

33. Place center of sphere at the origin.

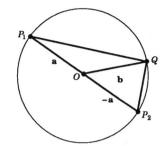

$$\overrightarrow{P_1Q} \cdot \overrightarrow{P_2Q} = (-\mathbf{a}+\mathbf{b}) \cdot (\mathbf{a}+\mathbf{b})$$

$$= -\|\mathbf{a}\|^2 + \|\mathbf{b}\|^2$$

$$= 0.$$

SECTION 13.5

1. -2

3. 0

5. 5

7. -6

9. no

11. yes

13. yes

15. $x = (ac - 1)/b$

17. $x = 0, \quad x = \frac{1}{2}(1 \pm \sqrt{5})$

19. $\begin{vmatrix} 4 & 5 & 6 \\ 1 & 2 & 3 \\ 7 & 8 & 9 \end{vmatrix} = - \begin{vmatrix} 1 & 2 & 3 \\ 4 & 5 & 6 \\ 7 & 8 & 9 \end{vmatrix}$ (interchanged two rows)

21. The second determinant is zero because two rows are the same.

23. $\begin{vmatrix} 1 & 2 & 3 \\ x & 2x & 3x \\ 4 & 5 & 6 \end{vmatrix} = x \begin{vmatrix} 1 & 2 & 3 \\ 1 & 2 & 3 \\ 4 & 5 & 6 \end{vmatrix} = 0$

    ⎿ two rows are the same

25. the expansion on the left can be written

$$\lambda^2 \begin{vmatrix} 1 & 1 \\ a & b \end{vmatrix} - 2\lambda \begin{vmatrix} 1 & 1 \\ a & b \end{vmatrix} + 1 \begin{vmatrix} 1 & 1 \\ a & b \end{vmatrix} = (\lambda^2 - 2\lambda + 1) \begin{vmatrix} 1 & 1 \\ a & b \end{vmatrix} = (\lambda - 1)^2 \begin{vmatrix} 1 & 1 \\ a & b \end{vmatrix}.$$

27. $\begin{vmatrix} a_1 & a_2 & a_3 \\ b_1 & b_2 & b_3 \\ c_1 & c_2 & c_3 \end{vmatrix} = - \begin{vmatrix} a_1 & a_2 & a_3 \\ c_1 & c_2 & c_3 \\ b_1 & b_2 & b_3 \end{vmatrix} = \begin{vmatrix} c_1 & c_2 & c_3 \\ a_1 & a_2 & a_3 \\ b_1 & b_2 & b_3 \end{vmatrix} = c_1 \begin{vmatrix} a_2 & a_3 \\ b_2 & b_3 \end{vmatrix} - c_2 \begin{vmatrix} a_1 & a_3 \\ b_1 & b_3 \end{vmatrix} + c_3 \begin{vmatrix} a_1 & a_2 \\ b_1 & b_2 \end{vmatrix}$

29. $\begin{vmatrix} a & a^2 \\ b & b^2 \end{vmatrix} = a \begin{vmatrix} 1 & a \\ b & b^2 \end{vmatrix} = ab \begin{vmatrix} 1 & a \\ 1 & b \end{vmatrix}$

31. $\begin{vmatrix} a_1 + b_1 & a_2 + b_2 & a_3 + b_3 \\ b_1 & b_2 & b_3 \\ c_1 & c_2 & c_3 \end{vmatrix} = \begin{vmatrix} a_1 & a_2 & a_3 \\ b_1 & b_2 & b_3 \\ c_1 & c_2 & c_3 \end{vmatrix} + \begin{vmatrix} b_1 & b_2 & b_3 \\ b_1 & b_2 & b_3 \\ c_1 & c_2 & c_3 \end{vmatrix}$

    by Exercise 28 ⎤

The second determinant is zero because two rows are the same.

SECTION 13.6

1. $(\mathbf{i}+\mathbf{j}) \times (\mathbf{i}-\mathbf{j}) = [\mathbf{i} \times (\mathbf{i}-\mathbf{j})] + [\mathbf{j} \times (\mathbf{i}-\mathbf{j})] = (\mathbf{0}-\mathbf{k}) + (-\mathbf{k}-\mathbf{0}) = -2\mathbf{k}$

3. $(\mathbf{i}-\mathbf{j}) \times (\mathbf{j}-\mathbf{k}) = [\mathbf{i} \times (\mathbf{j}-\mathbf{k})] - [\mathbf{j} \times (\mathbf{j}-\mathbf{k})] = (\mathbf{j}+\mathbf{k}) - (\mathbf{0}-\mathbf{i}) = \mathbf{i}+\mathbf{j}+\mathbf{k}$

5. $(2\mathbf{j}-\mathbf{k}) \times (\mathbf{i}-3\mathbf{j}) = [2\mathbf{j} \times (\mathbf{i}-3\mathbf{j})] - [\mathbf{k} \times (\mathbf{i}-3\mathbf{j})] = (-2\mathbf{k}) - (\mathbf{j}+3\mathbf{i}) = -3\mathbf{i}-\mathbf{j}-2\mathbf{k}$

7. $\mathbf{j} \cdot (\mathbf{i} \times \mathbf{k}) = \mathbf{j} \cdot (-\mathbf{j}) = -1$ 9. $(\mathbf{i} \times \mathbf{j}) \times \mathbf{k} = \mathbf{k} \times \mathbf{k} = \mathbf{0}$ 11. $\mathbf{j} \cdot (\mathbf{k} \times \mathbf{i}) = \mathbf{j} \cdot (\mathbf{j}) = 1$

13. $(\mathbf{i}+\mathbf{j}+\mathbf{k}) \times (2\mathbf{i}+\mathbf{k}) = [(\mathbf{i}+\mathbf{j}+\mathbf{k}) \times 2\mathbf{i}] + [(\mathbf{i}+\mathbf{j}+\mathbf{k}) \times \mathbf{k}]$

$$= (-2\mathbf{k}+2\mathbf{j}) + (-\mathbf{j}+\mathbf{i})$$
$$= \mathbf{i}+\mathbf{j}-2\mathbf{k}$$

15. $[2\mathbf{i}+\mathbf{j}] \cdot [(\mathbf{i}-3\mathbf{j}+\mathbf{k}) \times (4\mathbf{i}+\mathbf{k})] = [2\mathbf{i}+\mathbf{j}] \cdot [(\mathbf{i}-3\mathbf{j}+\mathbf{k}) \times 4\mathbf{i} + (\mathbf{i}-3\mathbf{j}+\mathbf{k}) \times \mathbf{k}]$

$$= [2\mathbf{i}+\mathbf{j}] \cdot [(12\mathbf{k}+4\mathbf{j}) + (-\mathbf{j}-3\mathbf{i})]$$
$$= (2\mathbf{i}+\mathbf{j}) \cdot (-3\mathbf{i}+3\mathbf{j}+12\mathbf{k}) = -3$$

17. $[(\mathbf{i}-\mathbf{j}) \times (\mathbf{j}-\mathbf{k})] \times [\mathbf{i}+5\mathbf{k}] = \{[\mathbf{i} \times (\mathbf{j}-\mathbf{k})] - [\mathbf{j} \times (\mathbf{j}-\mathbf{k})]\} \times [\mathbf{i}+5\mathbf{k}]$

$$= [(\mathbf{k}+\mathbf{j}) - (-\mathbf{i})] \times [\mathbf{i}+5\mathbf{k}]$$
$$= (\mathbf{i}+\mathbf{j}+\mathbf{k}) \times (\mathbf{i}+5\mathbf{k})$$
$$= [(\mathbf{i}+\mathbf{j}+\mathbf{k}) \times \mathbf{i}] + [(\mathbf{i}+\mathbf{j}+\mathbf{k}) \times 5\mathbf{k}]$$
$$= (-\mathbf{k}+\mathbf{j}) + (-5\mathbf{j}+5\mathbf{i})$$
$$= 5\mathbf{i}-4\mathbf{j}-\mathbf{k}$$

19. Set $\mathbf{a} = \overrightarrow{PQ} = -\mathbf{i}+2\mathbf{k}$ and $\mathbf{b} = \overrightarrow{PR} = 2\mathbf{i}-\mathbf{k}$ to obtain
$$A = \tfrac{1}{2}\|\mathbf{a} \times \mathbf{b}\| = \tfrac{1}{2}\|3\mathbf{j}\| = \tfrac{3}{2}.$$

21. $V = \left|[(\mathbf{i}+\mathbf{j}) \times (2\mathbf{i}-\mathbf{k})] \cdot (3\mathbf{j}+\mathbf{k})\right| = \left|(-\mathbf{i}+\mathbf{j}-2\mathbf{k}) \cdot (3\mathbf{j}+\mathbf{k})\right| = 1$

23. $(\mathbf{a}+\mathbf{b}) \times (\mathbf{a}-\mathbf{b}) = [\mathbf{a} \times (\mathbf{a}-\mathbf{b})] + [\mathbf{b} \times (\mathbf{a}-\mathbf{b})]$

$$= [\mathbf{a} \times (-\mathbf{b})] + [\mathbf{b} \times \mathbf{a}]$$
$$= -(\mathbf{a} \times \mathbf{b}) - (\mathbf{a} \times \mathbf{b}) = -2(\mathbf{a} \times \mathbf{b})$$

25. $\mathbf{a} \times \mathbf{i} = \mathbf{0}, \quad \mathbf{a} \times \mathbf{j} = \mathbf{0} \implies \mathbf{a}\|\mathbf{i}, \quad \mathbf{a}\|\mathbf{j} \implies \mathbf{a} = \mathbf{0}$

27. $(\alpha\mathbf{a}+\beta\mathbf{b}) \times (\gamma\mathbf{a}+\delta\mathbf{b}) = (\alpha\mathbf{a} \times \delta\mathbf{b}) + (\beta\mathbf{b} \times \gamma\mathbf{a})$

$$= \alpha\delta(\mathbf{a} \times \mathbf{b}) - \beta\gamma(\mathbf{a} \times \mathbf{b})$$
$$= (\alpha\delta - \beta\gamma)(\mathbf{a} \times \mathbf{b}) = \begin{vmatrix} \alpha & \beta \\ \gamma & \delta \end{vmatrix}(\mathbf{a} \times \mathbf{b})$$

29. $\mathbf{a} \cdot (\mathbf{b} \times \mathbf{c}) = (\mathbf{a} \times \mathbf{b}) \cdot \mathbf{c} = (\mathbf{c} \times \mathbf{a}) \cdot \mathbf{b} = (\mathbf{b} \times \mathbf{c}) \cdot \mathbf{a} = (\mathbf{a} \times -\mathbf{c}) \cdot \mathbf{b}$

$\mathbf{a} \cdot (\mathbf{c} \times \mathbf{b}) = \mathbf{c} \cdot (\mathbf{b} \times \mathbf{a}) = (-\mathbf{a} \times \mathbf{b}) \cdot \mathbf{c}$

31. $\mathbf{c} \times \mathbf{a} = (\mathbf{a} \times \mathbf{b}) \times \mathbf{a} = (\mathbf{a} \cdot \mathbf{a})\mathbf{b} - (\mathbf{a} \cdot \mathbf{b})\mathbf{a} = (\mathbf{a} \cdot \mathbf{a})\mathbf{b} = \|\mathbf{a}\|^2 \mathbf{b}$

$\underset{\text{Exercise 30(a)}}{\uparrow}\underset{\mathbf{a} \cdot \mathbf{b} = 0}{\uparrow}$

SECTION 13.7

1. P (when $t = 0$) and Q (when $t = -1$)

3. Take $\mathbf{r}_0 = \overrightarrow{OP} = 3\mathbf{i} + \mathbf{j}$ and $\mathbf{d} = \mathbf{k}$. Then, $\mathbf{r}(t) = (3\mathbf{i} + \mathbf{j}) + t\mathbf{k}$.

5. Take $\mathbf{r}_0 = \mathbf{0}$ and $\mathbf{d} = \overrightarrow{OQ}$. Then, $\mathbf{r}(t) = t(x_1\mathbf{i} + y_1\mathbf{j} + z_1\mathbf{k})$.

7. $\overrightarrow{PQ} = \mathbf{i} - \mathbf{j} + \mathbf{k}$ so direction numbers are $1, -1, 1$. Using P as a point on the line, we have

$$x(t) = 1 + t, \quad y(t) = -t, \quad z(t) = 3 + t.$$

9. The line is parallel to the y-axis so we can take $0, 1, 0$ as direction numbers. Therefore

$$x(t) = 2, \quad y(t) = -2 + t, \quad z(t) = 3.$$

11. Since the line $2(x + 1) = 4(y - 3) = z$ can be written

$$\frac{x + 1}{2} = \frac{y - 3}{1} = \frac{z}{4},$$

it has direction numbers $2, 1, 4$. The line through $P(-1, 2, -3)$ with direction vector

$2\mathbf{i} + \mathbf{j} + 4\mathbf{k}$ can be parametrized

$$\mathbf{r}(t) = (-\mathbf{i} + 2\mathbf{j} - 3\mathbf{k}) + t(2\mathbf{i} + \mathbf{j} + 4\mathbf{k}).$$

13. We set $\mathbf{r}_1(t) = \mathbf{r}_2(u)$ and solve for t and u:

$$\mathbf{i} + t\mathbf{j} = \mathbf{j} + u(\mathbf{i} + \mathbf{j}),$$

$$(1 - u)\mathbf{i} + (-1 - u + t)\mathbf{j} = \mathbf{0}.$$

Thus,

$$1 - u = 0 \quad \text{and} \quad -1 - u + t = 0.$$

The equation gives $u = 1$, $t = 2$. The point of intersection is $P(1, 2, 0)$.

As direction vectors for the lines we can take $\mathbf{u} = \mathbf{j}$ and $\mathbf{v} = \mathbf{i} + \mathbf{j}$. Thus

$$\cos\theta = \frac{\mathbf{u} \cdot \mathbf{v}}{\|\mathbf{u}\| \, \|\mathbf{v}\|} = \frac{1}{(1)(\sqrt{2})} = \frac{1}{2}\sqrt{2}.$$

The angle of intersection is $\frac{1}{4}\pi$ radians.

15. We solve the system

$$3 + t = 1, \quad 1 - t = 4 + u, \quad 5 + 2t = 2 + u$$

for t and u to find that $t = -2$, $u = -1$. The point of intersection is $(1, 3, 1)$.

Since $\quad \mathbf{i} - \mathbf{j} + 2\mathbf{k}\quad$ is a direction vector for l_1 and $\quad \mathbf{j} + \mathbf{k}\quad$ is a direction vector for l_2,

$$\cos \theta = \frac{(\mathbf{i} - \mathbf{j} + 2\mathbf{k}) \cdot (\mathbf{j} + \mathbf{k})}{\sqrt{6}\sqrt{2}} = \frac{1}{2\sqrt{3}} = \frac{1}{6}\sqrt{3} \quad \text{and} \quad \theta \cong 1.28 \text{ radians.}$$

17. $\left(x_0 - \dfrac{d_1}{d_3}z_0, \ \ y_0 - \dfrac{d_2}{d_3}z_0, \ \ 0 \right)$ **19.**   The lines are parallel.

21. $\mathbf{r}(t) = (2\mathbf{i} + 7\mathbf{j} - \mathbf{k}) + t(2\mathbf{i} - 5\mathbf{j} + 4\mathbf{k}), \quad 0 \le t \le 1$

23.   Set $\qquad \mathbf{u} = \dfrac{\overrightarrow{PQ}}{\|\overrightarrow{PQ}\|} = \dfrac{-4\mathbf{i} + 2\mathbf{j} + 4\mathbf{k}}{\| -4\mathbf{i} + 2\mathbf{j} + 4\mathbf{k}\|} = -\dfrac{2}{3}\mathbf{i} + \dfrac{1}{3}\mathbf{j} + \dfrac{2}{3}\mathbf{k}.$

Then $\qquad \mathbf{r}(t) = (6\mathbf{i} - 5\mathbf{j} + \mathbf{k}) + t\mathbf{u}$ is $\overrightarrow{OP}$ at $t = 9$ and it is $\overrightarrow{OQ}$ at $t = 15$. (Check this.)

Answer: $\mathbf{u} = -\frac{2}{3}\mathbf{i} + \frac{1}{3}\mathbf{j} + \frac{2}{3}\mathbf{k}, \quad 9 \le t \le 15.$

25.   The given line, call it l, has direction vector $2\mathbf{i} - 4\mathbf{j} + 6\mathbf{k}$.

If $a\mathbf{i} + b\mathbf{j} + c\mathbf{k}$ is a direction vector for a line perpendicular to l, then

$$(2\mathbf{i} - 4\mathbf{j} + 6\mathbf{k}) \cdot (a\mathbf{i} + b\mathbf{j} + c\mathbf{k}) = 2a - 4b + 6c = 0.$$

The lines through $P(3, -1, 8)$ perpendicular to l can be parametrized

$$X(u) = 3 + au, \quad Y(u) = -1 + bu, \quad Z(u) = 8 + cu$$

with $2a - 4b + 6c = 0$.

27. $d(P, l) = \dfrac{\|(\mathbf{i} + 2\mathbf{k}) \times (2\mathbf{i} - \mathbf{j} + 2\mathbf{k})\|}{\|2\mathbf{i} - \mathbf{j} + 2\mathbf{k}\|} = 1$

29.   The line contains the point $P_0(1, 0, 2)$. Therefore

$$d(P, l) = \frac{\|(2\mathbf{j} + \mathbf{k}) \times (\mathbf{i} - 2\mathbf{j} + 3\mathbf{k})\|}{\|\mathbf{i} - 2\mathbf{j} + 3\mathbf{k}\|} = \sqrt{\frac{69}{14}} \cong 2.22$$

31.   The line contains the point $P_0(2, -1, 0)$. Therefore

$$d(P, l) = \frac{\|(\mathbf{i} - \mathbf{j} - \mathbf{k}) \times (\mathbf{i} + \mathbf{j})\|}{\|\mathbf{i} + \mathbf{j}\|} = \sqrt{3} \cong 1.73.$$

33.   (a)   The line passes through $P(1, 1, 1)$ with direction vector $\mathbf{i} + \mathbf{j}$. Therefore

$$d(0, l) = \frac{\|(\mathbf{i} + \mathbf{j} + \mathbf{k}) \times (\mathbf{i} + \mathbf{j})\|}{\|\mathbf{i} + \mathbf{j}\|} = 1.$$

(b)   The distance from the origin to the line segment is $\sqrt{3}$.

Solution. The line segment can be parametrized

$$\mathbf{r}(t) = \mathbf{i} + \mathbf{j} + \mathbf{k} + t(\mathbf{i} + \mathbf{j}), \quad t \in [0, 1].$$

This is the set of all points $P(1+t, 1+t, 1)$ with $t \in [0, 1]$.

The distance from the origin to such a point is

$$f(t) = \sqrt{2(1+t^2)+1}.$$

The minimum value of this function is $f(0) = \sqrt{3}$.

Explanation. The point on the line through P and Q closest to the origin is not on the line

segment $\overline{PQ}$.

35. We begin with $\mathbf{r}(t) = \mathbf{j} - 2\mathbf{k} + t(\mathbf{i} - \mathbf{j} + 3\mathbf{k})$. The scalar t_0 for which $\mathbf{r}(t_0) \perp l$ can be found by solving the equation

$$[\mathbf{j} - 2\mathbf{k} + t_0(\mathbf{i} - \mathbf{j} + 3\mathbf{k})] \cdot [\mathbf{i} - \mathbf{j} + 3\mathbf{k}] = 0.$$

This equation gives $-7 + 11t_0 = 0$ and thus $t_0 = 7/11$. Therefore

$$\mathbf{r}(t_0) = \mathbf{j} - 2\mathbf{k} + \tfrac{7}{11}(\mathbf{i} - \mathbf{j} + 3\mathbf{k}) = \tfrac{7}{11}\mathbf{i} + \tfrac{4}{11}\mathbf{j} - \tfrac{1}{11}\mathbf{k}.$$

The vectors of norm 1 parallel to $\mathbf{i} - \mathbf{j} + 3\mathbf{k}$ are

$$\pm \frac{1}{\sqrt{11}}(\mathbf{i} - \mathbf{j} + 3\mathbf{k}).$$

The standard parametrizations are

$$\mathbf{R}(t) = \frac{7}{11}\mathbf{i} + \frac{4}{11}\mathbf{j} - \frac{1}{11}\mathbf{k} \pm \frac{t}{\sqrt{11}}(\mathbf{i} - \mathbf{j} + 3\mathbf{k})$$

$$= \frac{1}{11}(7\mathbf{i} + 4\mathbf{j} - \mathbf{k}) \pm t\left[\frac{\sqrt{11}}{11}(\mathbf{i} - \mathbf{j} + 3\mathbf{k})\right].$$

37. $0 < t < s$

By similar triangles, if $0 < s < 1$, the tip of $\overrightarrow{OA} + s\overrightarrow{AB} + s\overrightarrow{BC}$ falls on $\overline{AC}$. If $0 < t < s$, then the tip of $\overrightarrow{OA} + s\overrightarrow{AB} + t\overrightarrow{BC}$ falls short of $\overline{AC}$ and stays within the triangle. Clearly all points in the interior of the triangle can be reached in this manner.

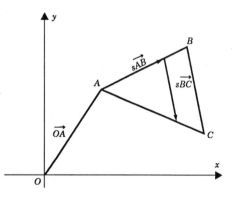

SECTION 13.8

1. Q

3. Since $\mathbf{i} - 4\mathbf{j} + 3\mathbf{k}$ is normal to the plane, we have

$$(x - 2) - 4(y - 3) + 3(z - 4) = 0 \quad \text{and thus} \quad x - 4y + 3z - 2 = 0.$$

5. The vector $3\mathbf{i} - 2\mathbf{j} + 5\mathbf{k}$ is normal to the given plane and thus to every parallel plane: the equation we want can be written

$$3(x - 2) - 2(y - 1) + 5(z - 1) = 0, \quad 3x - 2y + 5z - 9 = 0.$$

7. $\pm\dfrac{1}{62}\sqrt{62}\,(2\mathbf{i} - 3\mathbf{j} + 7\mathbf{k})$

9. $\dfrac{x}{15} + \dfrac{y}{12} + \dfrac{z}{(-10)} = 1$

11. $\mathbf{u_{N_1}} = \dfrac{\sqrt{38}}{38}\,(5\mathbf{i} - 3\mathbf{j} + 2\mathbf{k}), \quad \mathbf{u_{N_2}} = \dfrac{\sqrt{14}}{14}\,(\mathbf{i} + 3\mathbf{j} + 2\mathbf{k}), \quad \cos\theta = |\mathbf{u_{N_1}} \cdot \mathbf{u_{N_2}}| = 0.$

Therefore $\theta = \pi/2$ radians.

13. $\mathbf{u_{N_1}} = \dfrac{\sqrt{3}}{3}(\mathbf{i} - \mathbf{j} + \mathbf{k}), \quad \mathbf{u_{N_2}} = \dfrac{\sqrt{14}}{14}\,(2\mathbf{i} + \mathbf{j} + 3\mathbf{k}), \quad \cos\theta = |\mathbf{u_{N_1}} \cdot \mathbf{u_{N_2}}| = \dfrac{2}{21}\sqrt{42} \cong 0.617.$

Therefore $\theta \cong 0.91$ radians.

15. coplanar since $\quad 0(4\mathbf{j} - \mathbf{k}) + 0(3\mathbf{i} + \mathbf{j} + 2\mathbf{k}) + 1(\mathbf{0}) = 0$

17. We need to determine whether there exist scalars s, t, u not all zero such that

$$s(\mathbf{i} + \mathbf{j} + \mathbf{k}) + t(2\mathbf{i} - \mathbf{j}) + u(3\mathbf{i} - \mathbf{j} - \mathbf{k}) = \mathbf{0}$$

$$(s + 2t + 3u)\,\mathbf{i} + (s - t - u)\,\mathbf{j} + (s - u)\,\mathbf{k} = \mathbf{0}.$$

The only solution of the system

$$s + 2t + 3u = 0, \quad s - t - u = 0, \quad s - u = 0$$

is $\quad s = t = u = 0$. Thus, the vectors are not coplanar.

19. By (13.8.7), $\quad d(P, p) = \dfrac{|2(2) + 4(-1) - (3) + 1|}{\sqrt{4 + 16 + 1}} = \dfrac{2}{\sqrt{21}} = \dfrac{2}{21}\sqrt{21}.$

21. $\overrightarrow{P_1 P} = (x - 1)\mathbf{i} + y\mathbf{j} + (z - 1)\mathbf{k}, \quad \overrightarrow{P_1 P_2} = \mathbf{i} + \mathbf{j} - \mathbf{k}, \quad \overrightarrow{P_1 P_3} = \mathbf{j}.$

Therefore

$$(\overrightarrow{P_1 P_2} \times \overrightarrow{P_1 P_3}) = (\mathbf{i} + \mathbf{j} - \mathbf{k}) \times \mathbf{j} = \mathbf{i} + \mathbf{k}$$

and

$$\overrightarrow{P_1 P} \cdot (\overrightarrow{P_1 P_2} \times \overrightarrow{P_1 P_3}) = [(x - 1)\mathbf{i} + y\mathbf{j} + (z - 1)\mathbf{k}] \cdot [\mathbf{i} + \mathbf{k}] = x - 1 + z - 1.$$

The equation for the plane can be written $\quad x + z = 2$.

23. $\dfrac{x - x_0}{d_1} = \dfrac{y - y_0}{d_2}, \quad \dfrac{y - y_0}{d_2} = \dfrac{z - z_0}{d_3}$

25. Following the hint we take $x = 0$ and find that $P_0(0,0,0)$ lies on the line of intersection. As normals to the plane we use

$$\mathbf{N}_1 = \mathbf{i} + 2\mathbf{j} + 3\mathbf{k} \quad \text{and} \quad \mathbf{N}_2 = -3\mathbf{i} + 4\mathbf{j} + \mathbf{k}.$$

Note that

$$\mathbf{N}_1 \times \mathbf{N}_2 = (\mathbf{i} + 2\mathbf{j} + 3\mathbf{k}) \times (-3\mathbf{i} + 4\mathbf{j} + \mathbf{k}) = -10\mathbf{i} - 10\mathbf{j} + 10\mathbf{k}.$$

We take $-\frac{1}{10}(\mathbf{N}_1 \times \mathbf{N}_2) = \mathbf{i} + \mathbf{j} - \mathbf{k}$ as a direction vector for the line through $P_0(0,0,0)$. Then

$$x(t) = t, \quad y(t) = t, \quad z(t) = -t.$$

27. Straightforward computations give us

$$l: x(t) = 1 - 3t, \quad y(t) = -1 + 4t, \quad z(t) = 2 - t$$

and

$$p: x + 4y - z = 6.$$

Substitution of the scalar parametric equations for l in the equation for p gives

$$(1 - 3t) + 4(-1 + 4t) - (2 - t) = 6 \quad \text{and thus} \quad t = 11/14.$$

Using $t = 11/14$, we get $\quad x = -19/14, \quad y = 15/7, \quad z = 17/14.$

29. Let $\quad \mathbf{N} = A\mathbf{i} + B\mathbf{j} + C\mathbf{k} \quad$ be normal to the plane. Then

$$\mathbf{N} \cdot \mathbf{d} = (\mathbf{i} + B\mathbf{j} + C\mathbf{k}) \cdot (\mathbf{i} + 2\mathbf{j} + 4\mathbf{k}) = 1 + 2B + 4C = 0$$

and

$$\mathbf{N} \cdot \mathbf{d} = (\mathbf{i} + B\mathbf{j} + C\mathbf{k}) \cdot (-\mathbf{i} - \mathbf{j} + 3\mathbf{k}) = -1 - B + 3C = 0.$$

This gives $\quad B = -7/10 \quad$ and $\quad C = 1/10$. The equation for the plane can be written

$$1(x - 0) - \tfrac{7}{10}(y - 0) + \tfrac{1}{10}(z - 0) = 0, \quad \text{which simplifies to} \quad 10x - 7y + z = 0.$$

31. $\mathbf{N} + \overrightarrow{PQ}$ and $\mathbf{N} - \overrightarrow{PQ}$ are the diagonals of a rectangle with sides $\mathbf{N}$ and $\overrightarrow{PQ}$. Since the diagonals are perpendicular, the rectangle is a square; that is $\|\mathbf{N}\| = \|\overrightarrow{PQ}\|$. Thus, the points Q form a circle centered at P with radius $\|\mathbf{N}\|$.

33. If $\alpha > 0$, then P_1 lies on the same side of the plane as the tip of $\mathbf{N}$; if $\alpha < 0$, then P_1 and the tip of $\mathbf{N}$ lie on opposite sides of the plane.

To see this, suppose that the tip of $\mathbf{N}$ is at $P_0(x_0, y_0, z_0)$. Then

$$\mathbf{N} \cdot \overrightarrow{P_0 P_1} = A(x_1 - x_0) + B(y_1 - y_0) + C(z_1 - z_0) = Ax_1 + By_1 + Cz_1 + D = \alpha.$$

If $\alpha > 0$, $0 \leq \sphericalangle\left(\mathbf{N}, \overrightarrow{P_0 P_1}\right) < \pi/2$; if $\alpha < 0$, then $\pi/2 < \sphericalangle\left(\mathbf{N}, \overrightarrow{P_0 P_1}\right) < \pi$. Since $\mathbf{N}$ is perpendicular to the plane, the result follows.

SECTION 13.10

1. (a) $3\sqrt{10}$ (b) $(5, -\frac{3}{2}, \frac{3}{2})$ (c) $(11, -12, 9)$ or $(-1, 9, -6)$

    (d) a plane through the midpoint of $\overline{P_1P_2}$ with $\overrightarrow{P_1P_2}$ as a normal vector:

    $$4x - 7y + 5z - 38 = 0$$

    (e) $x(t) = 6 + 4t, \quad y(t) = 6 - 7t, \quad z(t) = -4 + 5t$

    (f) Let P be the foot of the perpendicular from P_3 to the line through P_1 and P_2.
    Since $\overrightarrow{P_1P_2} = 4\mathbf{i} - 7\mathbf{j} + 5\mathbf{k}$, we can take P as

    $$(3 + 4t, \quad 2 - 7t, \quad -1 + 5t). \qquad (t \text{ real})$$

    Then

    $$\overrightarrow{PP_3} = (3 - 4t)\mathbf{i} + (4 + 7t)\mathbf{j} + (-3 - 5t)\mathbf{k}$$

    and from $\overrightarrow{P_1P_2} \perp \overrightarrow{PP_3}$ we have

    $$4(3 - 4t) - 7(4 + 7t) + 5(-3 - 5t) = 0.$$

    Solving for t we get $t = -31/90$. Thus

    $$\overrightarrow{PP_3} = \tfrac{394}{90}\mathbf{i} + \tfrac{143}{90}\mathbf{j} - \tfrac{115}{90}\mathbf{k}.$$

    We take $394\mathbf{i} + 143\mathbf{j} - 115\mathbf{k}$ as a direction vector for the line through $P_3(6, 6, -4)$ perpendicular to the line through P_1 and P_2. The line is given by

    $$x(t) = 6 + 394t, \quad y(t) = 6 + 143t, \quad z(t) = -4 - 115t.$$

    (g) As a normal to the plane we take

    $$\overrightarrow{P_1P_2} \times \overrightarrow{P_1P_3} = (4\mathbf{i} - 7\mathbf{j} + 5\mathbf{k}) \times (3\mathbf{i} + 4\mathbf{j} - 3\mathbf{k})$$

    $$= \mathbf{i} + 27\mathbf{j} + 37\mathbf{k}.$$

    An equation for the plane is

    $$(x - 3) + 27(y - 2) + 37(z + 1) = 0 \quad \text{which gives} \quad x + 27y + 37z - 20 = 0.$$

3. Any point on l is of the form $(-1 + t, \ -2 + t, \ -1 + t)$ and $\overrightarrow{PQ} = (t - 4)\mathbf{i} + (t - 3)\mathbf{j} + (t + 1)\mathbf{k}$.
    The line l has direction vector $\mathbf{d} = \mathbf{i} + \mathbf{j} + \mathbf{k}$.   If $\overrightarrow{PQ} \perp l$, then

    $$\overrightarrow{PQ} \cdot \mathbf{d} = 0, \quad (t - 4) + (t - 3) + (t + 1) = 0, \quad t = 2.$$

    Thus Q is $(1, 0, 1)$.

5. 1   [using (13.8.7)]

7. The plane passes through $P(2, 1, -3)$ with normal $2\mathbf{i} + 3\mathbf{j} - 4\mathbf{k}$. The equation for the plane can be written $2(x - 2) + 3(y - 1) - 4(z + 3) = 0$. This simplifies to $2x + 3y - 4z - 19 = 0$.

9. The lines have direction vectors

$$\mathbf{d}_1 = 2\mathbf{i} + 3\mathbf{j} + \mathbf{k}, \quad \mathbf{d}_2 = \mathbf{i} - \mathbf{j} + 2\mathbf{k}.$$

The vector $\mathbf{d}_1 \times \mathbf{d}_2 = 7\mathbf{i} - 3\mathbf{j} - 5\mathbf{k}$ is normal to the plane. Since the point $(-1, 2, 1)$ lies on the plane, the equation of the plane can be written $7(x+1) - 3(y-2) - 5(z-1) = 0$. This simplifies to $7x - 3y - 5z + 18 = 0$.

11. Setting $z = 0$ we get

$$2x + y + 6 = 0 \quad \text{and} \quad x + 4y - 7 = 0.$$

These equations are satisfied by $x = -31/7$, $y = 20/7$. Thus the point $(-31/7,\ 20/7,\ 0)$ lies on both planes.

The vectors

$$\mathbf{N}_1 = 2\mathbf{i} + \mathbf{j} - 3\mathbf{k}, \quad \mathbf{N}_2 = \mathbf{i} + 4\mathbf{j} - 5\mathbf{k}$$

are normal to the planes. Since $\mathbf{N}_1 \times \mathbf{N}_2 = 7\mathbf{i} + 7\mathbf{j} + 7\mathbf{k}$, we can take $\mathbf{i} + \mathbf{j} + \mathbf{k}$ as a direction vector for the line. The line can be parametrized by

$$x(t) = t - \tfrac{31}{7}, \quad y(t) = t + \tfrac{20}{7}, \quad z(t) = t.$$

13. Every point on the line is of the form $(-2 + 3t,\ 1 + 2t, -6 + t)$. Such a point lies on the plane if and only if

$$2(-2 + 3t) + (1 + 2t) - 3(-6 + t) + 6 = 0.$$

Solving for t, we get $t = -21/5$. The line intersects the plane at $(-73/5,\ -37/5,\ -51/5)$.

15. Taking $z = 0$, we see that $P_1(1, -1, 0)$ lies on both planes; taking $z = 1$, we see that $P_2(6, -15, 1)$ also lies on both planes. The equation for the plane through $P(2, 1, -3)$, $P_1(1, -1, 0)$, $P_2(6, -15, 1)$ can be written $5x + 2y + 3z - 3 = 0$. (usual method)

17. 3 (using 13.7.6)

19.

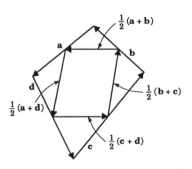

$$\mathbf{a} + \mathbf{b} + \mathbf{c} + \mathbf{d} = \mathbf{0}$$

gives

$$\tfrac{1}{2}(\mathbf{a} + \mathbf{b}) = -\tfrac{1}{2}(\mathbf{c} + \mathbf{d})$$

$$\tfrac{1}{2}(\mathbf{b} + \mathbf{c}) = -\tfrac{1}{2}(\mathbf{a} + \mathbf{d}).$$

21. $\alpha = -2$; $\quad \alpha(\mathbf{a} \times \mathbf{b}) = (\mathbf{a} + \mathbf{b}) \times (\mathbf{a} - \mathbf{b}) = (\mathbf{a} \times -\mathbf{b}) + (\mathbf{b} \times \mathbf{a}) = -2(\mathbf{a} \times \mathbf{b})$

23. $(\|\mathbf{b}\| \mathbf{a} - \|\mathbf{a}\| \mathbf{b}) \cdot (\|\mathbf{b}\| \mathbf{a} + \|\mathbf{a}\| \mathbf{b})$

$$= (\|\mathbf{b}\| \mathbf{a}) \cdot (\|\mathbf{b}\| \mathbf{a} + \|\mathbf{a}\| \mathbf{b}) \ - \ (\|\mathbf{a}\| \mathbf{b}) \cdot (\|\mathbf{b}\| \mathbf{a} + \|\mathbf{a}\| \mathbf{b})$$

$$= \|\mathbf{b}\|^2 \, \|\mathbf{a}\|^2 + \|\mathbf{a}\| \, \|\mathbf{b}\| (\mathbf{a} \cdot \mathbf{b}) - \|\mathbf{a}\| \, \|\mathbf{b}\| (\mathbf{b} \cdot \mathbf{a}) - \|\mathbf{a}\|^2 \, \|\mathbf{b}\|^2$$

$$= 0$$

25.   By Exercise 24 the plane has an equation of the form

$$\alpha \, (3x - 2y + 4z - 7) + \beta \, (x + 5y - 2z + 9) = 0.$$

Substituting $x = 1$, $y = 4$, $z = -3$ into the equation we find that $\beta = 2\alpha/3$. The equation of the plane can be written

$$\alpha \, (3x - 2y + 4z - 7) + \frac{2\alpha}{3} \, (x + 5y - 2z + 9) = 0.$$

Dividing through by α and simplifying we get

$$11x + 4y + 8z - 3 = 0.$$

CHAPTER 14

SECTION 14.1

1. $\mathbf{f}'(t) = 2\mathbf{i} - \mathbf{j} + 3\mathbf{k}$

3. $\mathbf{f}'(t) = -\dfrac{1}{2\sqrt{1-t}}\mathbf{i} + \dfrac{1}{2\sqrt{1+t}}\mathbf{j} + \dfrac{1}{(1-t)^2}\mathbf{k}$

5. $\mathbf{f}'(t) = \cos t\,\mathbf{i} - \sin t\,\mathbf{j} + \sec^2 t\,\mathbf{k}$

7. $\displaystyle\int_1^2 (\mathbf{i} + 2t\mathbf{j})\,dt = \left[t\mathbf{i} + t^2\mathbf{j}\right]_1^2 = \mathbf{i} + 3\mathbf{j}$

9. $\displaystyle\int_0^1 (e^t\mathbf{i} + e^{-t}\mathbf{k})\,dt = \left[e^t\mathbf{i} - e^{-t}\mathbf{k}\right]_0^1 = (e-1)\mathbf{i} + \left(1 - \dfrac{1}{e}\right)\mathbf{k}$

11. $\frac{1}{2}\mathbf{i} + \mathbf{j}$

13. (a) $\mathbf{f}(t) = 3\cos t\,\mathbf{i} + 2\sin t\,\mathbf{j}$ (b) $\mathbf{f}(t) = 3\cos t\,\mathbf{i} - 2\sin t\,\mathbf{j}$

15. (a) $\mathbf{f}(t) = t\mathbf{i} + t^2\mathbf{j}$ (b) $\mathbf{f}(t) = -t\mathbf{i} + t^2\mathbf{j}$

17. $\mathbf{f}(t) = (1+2t)\mathbf{i} + (4+5t)\mathbf{j} + (-2+8t)\mathbf{k}, \quad 0 \le t \le 1$

19. $\mathbf{f}'(t_0) = \mathbf{i} + m\mathbf{j}$,

$$\int_a^b \mathbf{f}(t)\,dt = \left[\tfrac{1}{2}t^2\mathbf{i}\right]_a^b + \left[\int_a^b f(t)\,dt\right]\mathbf{j} = \tfrac{1}{2}\left(b^2 - a^2\right)\mathbf{i} + A\mathbf{j},$$

$$\int_a^b \mathbf{f}'(t)\,dt = \left[t\mathbf{i} + f(t)\mathbf{j}\right]_a^b = (b-a)\mathbf{i} + (d-c)\mathbf{j}$$

21.
$$\mathbf{f}'(t) = \mathbf{i} + t^2\mathbf{j}$$
$$\mathbf{f}(t) = (t + C_1)\mathbf{i} + \left(\tfrac{1}{3}t^3 + C_2\right)\mathbf{j} + C_3\mathbf{k}$$
$$\mathbf{f}(0) = \mathbf{j} - \mathbf{k} \implies C_1 = 0, \quad C_2 = 1, \quad C_3 = -1$$
$$\mathbf{f}(t) = t\mathbf{i} + \left(\tfrac{1}{3}t^3 + 1\right)\mathbf{j} - \mathbf{k}$$

23. $\mathbf{f}'(t) = \alpha\,\mathbf{f}(t) \implies \mathbf{f}(t) = e^{\alpha t}\,\mathbf{f}(0) = e^{\alpha t}\mathbf{c}$

25. (a) If $\mathbf{f}'(t) = \mathbf{0}$ on an interval, then the derivative of each component is 0 on that interval, each component is constant on that interval, and therefore $\mathbf{f}$ itself is constant on that interval.

    (b) Set $\mathbf{h}(t) = \mathbf{f}(t) - \mathbf{g}(t)$ and apply part (a).

27. If $\mathbf{f}$ is differentiable at t, then each component is differentiable at t, each component is continuous at t, and therefore $\mathbf{f}$ is continuous at t.

29. no; as a counter-example, set $\mathbf{f}(t) = \mathbf{i} = \mathbf{g}(t)$.

SECTION 14.2

1. $\mathbf{f}'(t) = \mathbf{b}, \quad \mathbf{f}''(t) = \mathbf{0}$

3. $\mathbf{f}'(t) = 2e^{2t}\mathbf{i} - \cos t\,\mathbf{j}, \quad \mathbf{f}''(t) = 4e^{2t}\mathbf{i} + \sin t\,\mathbf{j}$

5. $$\mathbf{f}'(t) = \left[(e^t\mathbf{i} + t\mathbf{k}) \times \frac{d}{dt}(t\mathbf{j} + e^{-t}\mathbf{k})\right] + \left[\frac{d}{dt}(e^t\mathbf{i} + t\mathbf{k}) \times (t\mathbf{j} + e^{-t}\mathbf{k})\right]$$

$$= [(e^t\mathbf{i} + t\mathbf{k}) \times (\mathbf{j} - e^{-t}\mathbf{k})] + [(e^t\mathbf{i} + \mathbf{k}) \times (t\mathbf{j} + e^{-t}\mathbf{k})]$$

$$= (-t\mathbf{i} + \mathbf{j} + e^t\mathbf{k}) + (-t\mathbf{i} - \mathbf{j} + te^t\mathbf{k})$$

$$= -2t\mathbf{i} + e^t(t+1)\mathbf{k}$$

$$\mathbf{f}''(t) = -2\mathbf{i} + e^t(t+2)\mathbf{k}$$

7. $\mathbf{f}'(t) = \dfrac{1}{2}\sqrt{t}\,\mathbf{g}'(\sqrt{t}) + \mathbf{g}(\sqrt{t}), \quad \mathbf{f}''(t) = \dfrac{1}{4}\mathbf{g}''(\sqrt{t}) + \dfrac{3}{4\sqrt{t}}\mathbf{g}'(\sqrt{t})$

9. $-\sin t\, e^{\cos t}\,\mathbf{i} + \cos t\, e^{\sin t}\,\mathbf{j}$

11. $\dfrac{d}{dt}[(\mathbf{a} + t\mathbf{b}) \times (\mathbf{c} + t\mathbf{d})] = [(\mathbf{a} + t\mathbf{b}) \times \mathbf{d}] + [\mathbf{b} \times (\mathbf{c} + t\mathbf{d})] = (\mathbf{a} \times \mathbf{d}) + (\mathbf{b} \times \mathbf{c}) + 2t\,(\mathbf{b} \times \mathbf{d})$

13. $\dfrac{d}{dt}[(\mathbf{a} + t\mathbf{b}) \cdot (\mathbf{c} + t\mathbf{d})] = [(\mathbf{a} + t\mathbf{b}) \cdot \mathbf{d}] + [\mathbf{b} \cdot (\mathbf{c} + t\mathbf{d})] = (\mathbf{a} \cdot \mathbf{d}) + (\mathbf{b} \cdot \mathbf{c}) + 2t\,(\mathbf{b} \cdot \mathbf{d})$

15. $\mathbf{r}(t) = \mathbf{a} + t\mathbf{b}$ 17. $\mathbf{r}(t) = \frac{1}{2}t^2\mathbf{a} + \frac{1}{6}t^3\mathbf{b} + t\mathbf{c} + \mathbf{d}$

19. $$\mathbf{r}(t) \cdot \mathbf{r}'(t) = (\cos t\,\mathbf{i} + \sin t\,\mathbf{j}) \cdot (-\sin t\,\mathbf{i} + \cos t\,\mathbf{j}) = 0$$

$$\mathbf{r}(t) \times \mathbf{r}'(t) = (\cos t\,\mathbf{i} + \sin t\,\mathbf{j}) \times (-\sin t\,\mathbf{i} + \cos t\,\mathbf{j})$$

$$= \cos^2 t\,\mathbf{k} + \sin^2 t\,\mathbf{k} = (\cos^2 t + \sin^2 t)\,\mathbf{k} = \mathbf{k}$$

21. $\dfrac{d}{dt}[\mathbf{f}(t) \times \mathbf{f}'(t)] = [\mathbf{f}(t) \times \mathbf{f}''(t)] + \underbrace{[\mathbf{f}'(t) \times \mathbf{f}'(t)]}_{\mathbf{0}} = \mathbf{f}(t) \times \mathbf{f}''(t).$

23. $\|\mathbf{r}(t)\|$ is constant $\iff \|\mathbf{r}(t)\|^2 = \mathbf{r}(t) \cdot \mathbf{r}(t)$ is constant

$\iff \dfrac{d}{dt}[\mathbf{r}(t) \cdot \mathbf{r}(t)] = 2[\mathbf{r}(t) \cdot \mathbf{r}'(t)] = 0$ identically

$\iff \mathbf{r}(t) \cdot \mathbf{r}'(t) = 0$ identically

25. Write

$$\frac{[\mathbf{f}(t+h) \times \mathbf{g}(t+h)] - [\mathbf{f}(t) \times \mathbf{g}(t)]}{h}$$

as

$$\left(\mathbf{f}(t+h) \times \left[\frac{\mathbf{g}(t+h) - \mathbf{g}(t)}{h}\right]\right) + \left(\left[\frac{\mathbf{f}(t+h) - \mathbf{f}(t)}{h}\right] \times \mathbf{g}(t)\right)$$

and take the limit as $h \to 0$. (Appeal to Theorem 14.1.3.)

SECTION 14.3

1. $\mathbf{r}'(t) = -\pi \sin \pi t\, \mathbf{i} + \pi \cos \pi t\, \mathbf{j} + \mathbf{k}, \quad \mathbf{r}'(2) = \pi \mathbf{j} + \mathbf{k}$

    $\mathbf{R}(u) = (\mathbf{i} + 2\mathbf{k}) + u(\pi \mathbf{j} + \mathbf{k})$

3. $\mathbf{r}'(t) = \mathbf{b} + 2t\mathbf{c}, \quad \mathbf{r}'(-1) = \mathbf{b} - 2\mathbf{c}, \quad \mathbf{R}(u) = (\mathbf{a} - \mathbf{b} + \mathbf{c}) + u(\mathbf{b} - 2\mathbf{c})$

5. $\mathbf{r}'(t) = 4t\mathbf{i} - \mathbf{j} + 4t\mathbf{k}, \quad P$ is tip of $\mathbf{r}(1), \quad \mathbf{r}'(1) = 4\mathbf{i} - \mathbf{j} + 4\mathbf{k}$

    $\mathbf{R}(u) = (2\mathbf{i} + 5\mathbf{k}) + u(4\mathbf{i} - \mathbf{j} + 4\mathbf{k})$

7. The scalar components $x(t) = at$ and $y(t) = bt^2$ satisfy the equation

    $$a^2 y(t) = a^2(bt^2) = b(a^2 t^2) = b[x(t)]^2$$

    and generate the parabola $\quad a^2 y = bx^2$.

9. $$\mathbf{r}(t) = t\mathbf{i} + (1 + t^2)\mathbf{j}, \qquad \mathbf{r}'(t) = \mathbf{i} + 2t\mathbf{j}$$

    (a) $\mathbf{r}(t) \perp \mathbf{r}'(t) \implies \mathbf{r}(t) \cdot \mathbf{r}'(t) = [t\mathbf{i} + (1 + t^2)\mathbf{j}] \cdot (\mathbf{i} + 2t\mathbf{j})$

    $$= t(2t^2 + 3) = 0 \implies t = 0$$

    $\mathbf{r}(t)$ and $\mathbf{r}'(t)$ are perpendicular at $(0, 1)$.

    (b)   and (c) $\mathbf{r}(t) = \alpha \mathbf{r}'(t)$ with $\alpha \neq 0 \implies t = \alpha$ and $1 + t^2 = 2t\alpha \implies t = \pm 1$.

    If $\alpha > 0$, then $t = 1$. $\mathbf{r}(t)$ and $\mathbf{r}'(t)$ have the same direction at $(1, 2)$.

    If $\alpha < 0$, then $t = -1$. $\mathbf{r}(t)$ and $\mathbf{r}'(t)$ have opposite directions at $(-1, 2)$.

11. The tangent line at $t = t_0$ has the form $\mathbf{R}(u) = \mathbf{r}(t_0) + u\mathbf{r}'(t_0)$. If $\mathbf{r}'(t_0) = \alpha \mathbf{r}(t_0)$, then

    $$\mathbf{R}(u) = \mathbf{r}(t_0) + u\,\alpha \mathbf{r}(t_0) = (1 + u\alpha)\mathbf{r}(t_0).$$

    The tangent line passes through the origin at $u = -1/\alpha$.

13. $\mathbf{r}_1(t) = \mathbf{r}_2(u)$ implies

    $$\left\{ \begin{array}{c} e^t = u \\ 2\sin\left(t + \tfrac{1}{2}\pi\right) = 2 \\ t^2 - 2 = u^2 - 3 \end{array} \right\} \quad \text{so that} \quad t = 0, \quad u = 1.$$

    The point of intersection is $(1, 2, -2)$.

$$r_1'(t) = e^t i + 2\cos\left(t + \frac{\pi}{2}\right)j + 2tk, \quad r_1'(0) = i$$

$$r_2'(u) = i + 2uk, \quad r_2'(1) = i + 2k$$

$$\cos\theta = \frac{r_1'(0) \cdot r_2'(1)}{\|r_1'(0)\|\,\|r_2'(1)\|} = \frac{1}{5}\sqrt{5} \cong 0.447, \quad \theta \cong 1.11 \text{ radians}$$

15. (a) $r(t) = a\cos t\, i + b\sin t\, j$ (b) $r(t) = a\cos t\, i - b\sin t\, j$

    (c) $r(t) = a\cos 2t\, i + b\sin 2t\, j$ (d) $r(t) = a\cos 3t\, i - b\sin 3t\, j$

17. $r(t) = (t^2 + 1)i + tj, \quad t \geq 1; \quad \text{or,} \quad r(t) = \sec^2 t\, i + \tan t\, j, \quad t \in [\frac{1}{4}\pi, \frac{1}{2}\pi)$

19. $r(t) = \cos t \sin 3t\, i = \sin t \sin 3t\, j, \quad t \in [0, \pi]$

21. $y^3 = x^2$

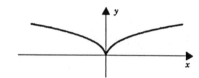

There is no tangent vector at the origin.

23. We substitute $x = t, \; y = t^2, \; z = t^3$ in the plane equation to obtain

$$4t + 2t^2 + t^3 = 24, \quad (t - 2)(t^2 + 4t + 12) = 0, \quad t = 2.$$

The twisted cubic intersects the plane at the tip of $r(2)$, the point $(2, 4, 8)$.

The angle between the curve and the normal line at the pont of intersection is the angle between the tangent vector $r'(2) = i + 4j + 12k$ and the normal $N = 4i + 2j + k$:

$$\cos\theta = \frac{(i + 4j + 12k) \cdot (4i + 2j + k)}{\|i + 4j + 12k\|\,\|4i + 2j + k\|} = \frac{24}{\sqrt{161}\sqrt{21}} \cong 0.412, \quad \theta \cong 1.15 \text{ radians.}$$

25.
$$r'(t) = -2\sin 2t\, i + 2\cos 2t\, j + k, \quad \|r'(t)\| = \sqrt{5}$$

$$T(t) = \frac{r'(t)}{\|r'(t)\|} = \frac{1}{5}\sqrt{5}\,(-2\sin 2t\, i + 2\cos 2t\, j + k)$$

$$T'(t) = -\tfrac{4}{5}\sqrt{5}\,(\cos 2t\, i + \sin 2t\, j), \quad \|T'(t)\| = \tfrac{4}{5}\sqrt{5}$$

$$N(t) = \frac{T'(t)}{\|T'(t)\|} = -(\cos 2t\, i + \sin 2t\, j)$$

at $t = \pi/4$:   tip of $r = (0, 1, \pi/4), \; T = \frac{1}{5}\sqrt{5}\,(-2i + k), \; N = -j$

normal for osculating plane:

$$\mathbf{T} \times \mathbf{N} = \tfrac{1}{5}\sqrt{5}\,(-2\mathbf{i}+\mathbf{k}) \times (-\mathbf{j}) = \tfrac{1}{5}\sqrt{5}\,\mathbf{i} + \tfrac{2}{5}\sqrt{5}\,\mathbf{k}$$

equation for osculating plane:

$$\tfrac{1}{5}\sqrt{5}\,(x-0) + \tfrac{2}{5}\sqrt{5}\left(z - \frac{\pi}{4}\right) = 0, \quad \text{which gives} \quad x + 2z = \frac{\pi}{2}$$

27.
$$\mathbf{r}'(t) = \mathbf{i} + 2t\mathbf{j} + 3t^2\mathbf{k}, \quad \|\mathbf{r}'(t)\| = \sqrt{1 + 4t^2 + 9t^4}$$

$$\mathbf{T}(t) = \frac{\mathbf{r}'(t)}{\|\mathbf{r}'(t)\|} = \frac{1}{\sqrt{1+4t^2+9t^4}}\,(\mathbf{i} + 2t\mathbf{j} + 3t^2\mathbf{k}),$$

$$\mathbf{T}'(t) = \frac{1}{(1+4t^2+9t^4)^{3/2}}\left[\left(-4t - 18t^3\right)\mathbf{i} + \left(2 - 18t^4\right)\mathbf{j} + \left(6t + 12t^3\right)\mathbf{k}\right]$$

at $t = 1$: tip of $\mathbf{r} = (1,1,1)$, $\mathbf{T} = \dfrac{1}{\sqrt{14}}\,(\mathbf{i} + 2\mathbf{j} + 3\mathbf{k})$,

$$\mathbf{T}' = \frac{1}{7\sqrt{14}}\,(-11\mathbf{i} - 8\mathbf{j} + 9\mathbf{k}), \quad \|\mathbf{T}'\| = \frac{\sqrt{266}}{7\sqrt{14}}, \quad \mathbf{N} = \frac{1}{\sqrt{266}}\,(-11\mathbf{i} - 8\mathbf{j} + 9\mathbf{k})$$

normal for osculating plane:

$$\mathbf{T} \times \mathbf{N} = \frac{1}{\sqrt{14}}\,(\mathbf{i} + 2\mathbf{j} + 3\mathbf{k}) \times \frac{1}{\sqrt{266}}\,(-11\mathbf{i} - 8\mathbf{j} + 9\mathbf{k}) = \frac{\sqrt{19}}{19}\,(3\mathbf{i} - 3\mathbf{j} + \mathbf{k})$$

equation for osculating plane:

$$3(x-1) - 3(y-1) + (z-1) = 0, \quad \text{which gives} \quad 3x - 3y + z = 1$$

29. $\mathbf{T}_1 = \dfrac{\mathbf{R}'(u)}{\|\mathbf{R}'(u)\|} = -\dfrac{\mathbf{r}'(a+b-u)}{\|\mathbf{r}'(a+b-u)\|} = -\mathbf{T}.$

Therefore $\mathbf{T}_1'(u) = \mathbf{T}'(a+b-u)$ and $\mathbf{N}_1 = \mathbf{N}.$

31. Let $\mathbf{T}$ be the unit tangent at the tip of $\mathbf{R}(u) = \mathbf{r}(\phi(u))$ as calculated from the parametrization $\mathbf{r}$ and let $\mathbf{T}_1$ be the unit tangent at the same point as calculated from the parametrization $\mathbf{R}$. Then

$$\mathbf{T}_1 = \frac{\mathbf{R}'(u)}{\|\mathbf{R}'(u)\|} = \frac{\mathbf{r}'(\phi(u))\,\phi'(u)}{\|\mathbf{r}'(\phi(u))\,\phi'(u)\|} \underset{\underset{\phi'(u)>0}{\uparrow}}{=} \frac{\mathbf{r}'(\phi(u))}{\|\mathbf{r}'(\phi(u))\|} = \mathbf{T}.$$

This shows the invariance of the unit tangent.

The invariance of the principal normal and the osculating plane follows directly from the invariance of the unit tangent.

SECTION 14.4

1. $r'(t) = -a \sin t \, i + a \cos t \, j + b k, \quad \|r'(t)\| = \sqrt{a^2 + b^2}$

$$L = \int_0^{2\pi} \sqrt{a^2 + b^2} \, dt = 2\pi \sqrt{a^2 + b^2}$$

3. $r'(t) = i + \tan t \, j, \quad \|r'(t)\| = \sqrt{1 + \tan^2 t} = |\sec t|$

$$L = \int_0^{\pi/4} |\sec t| \, dt = \int_0^{\pi/4} \sec t \, dt = [\ln|\sec t + \tan t|]_0^{\pi/4} = \ln(1 + \sqrt{2})$$

5. $r'(t) = 3t^2 i + 2t j, \quad \|r'(t)\| = \sqrt{9t^4 + 4t^2} = |t|\sqrt{4 + 9t^2}$

$$L = \int_0^1 |t\sqrt{4 + 9t2}| \, dt = \int_0^1 t\sqrt{4 + 9t^2} \, dt = \left[\frac{1}{27} (4 + 9t^2)^{3/2} \right]_0^1 = \frac{1}{27} \left(13\sqrt{13} - 8 \right)$$

7. $r'(t) = (\cos t - \sin t)e^t i + (\sin t + \cos t)e^t j, \quad \|r'(t)\| = \sqrt{2} \, e^t$

$$L = \int_0^\pi \sqrt{2} \, e^t \, dt = \sqrt{2} \, (e^\pi - 1)$$

9. $r'(t) = 2i + 2t j - 2t k, \quad \|r'(t)\| = 2\sqrt{1 + 2t^2}$

$$L = \int_0^2 2\sqrt{1 + 2t^2} \, dt = \sqrt{2} \int_0^{\tan^{-1}(2\sqrt{2})} \sec^3 u \, du$$
$$\underset{\llcorner (t\sqrt{2} = \tan u)}{\uparrow}$$

$$= \tfrac{1}{2}\sqrt{2} \, [\sec u \tan u + \ln|\sec u + \tan u|]_0^{\tan^{-1}(2\sqrt{2})} = 6 + \tfrac{1}{2}\sqrt{2} \ln(3 + 2\sqrt{2})$$

11.
$$s = s(t) = \int_a^t \|r'(u)\| \, du$$

$$s'(t) = \|r'(t)\| = \|x'(t) \, i + y'(t) \, j + z'(t) \, k\|$$

$$= \sqrt{[x'(t)]^2 + [y'(t)]^2 + [z'(t)]^2}.$$

In the Leibniz notation this translates to

$$\frac{ds}{dt} = \sqrt{\left(\frac{dx}{dt} \right)^2 + \left(\frac{dy}{dt} \right)^2 + \left(\frac{dz}{dt} \right)^2}.$$

13.
$$s = s(x) = \int_a^x \sqrt{1 + [f'(t)]^2}\, dt$$

$$s'(x) = \sqrt{1 + [f'(x)]^2}.$$

In the Leibniz notation this translates to

$$\frac{ds}{dx} = \sqrt{1 + \left(\frac{dy}{dx}\right)^2}.$$

15. Let L be the length as computed from $\mathbf{r}$ and L^* the length as computed from $\mathbf{R}$.   Then

$$L^* = \int_c^d \|\mathbf{R}'(u)\|\, du = \int_c^d \|\mathbf{r}'(\phi\,(u))\|\,\|\phi'(u)\,du \underset{t = \phi\,(u)}{=} \int_a^b \|\mathbf{r}'(t)\|\, dt = L.$$

17.
$$L = \int_0^{1/2} \|\mathbf{r}'(t)\|\, dt = \int_0^{1/2} (1 + t^4)^{1/2}\, dt$$

$$\cong \int_0^{1/2} \left(1 + \frac{1}{2}\, t^4 - \frac{1}{8}\, t^8 + \cdots\right) dt$$

$$= \left[t + \tfrac{1}{10}\, t^5 - \tfrac{1}{72}\, t^9 + \cdots\right]_0^{1/2}$$

$$= \tfrac{1}{2} + \tfrac{1}{320} - \tfrac{1}{36864} + \cdots .$$

This goes on as a convergent alternating series with decreasing terms. Therefore

$$0.5030 \le \tfrac{1}{2} + \tfrac{1}{320} - \tfrac{1}{36864} \le L \le \tfrac{1}{2} + \tfrac{1}{320} \le 0.5032.$$

We can take $L \cong 0.503$.

SECTION 14.5

1.
$$\mathbf{r}(t) = r[\cos\theta(t)\,\mathbf{i} + \sin\theta(t)\,\mathbf{j}]$$

$$\mathbf{r}'(t) = r[-\sin\theta(t)\,\mathbf{i} + \cos\theta(t)\,\mathbf{j})]\theta'(t)$$

$$\|\mathbf{r}'(t)\| = v \implies r|\theta'(t)| = v \implies |\theta'(t)| = v/r$$

$$\mathbf{r}''(t) = r[-\cos\theta(t)\,\mathbf{i} - \sin\theta(t)\,\mathbf{j}]\,[\theta'(t)]^2$$

$$\|\mathbf{r}''(t)\| = r[\theta'(t)]^2 = v^2/r$$

3.
$$\mathbf{r}(t) = at\mathbf{i} + b\sin at\,\mathbf{j}$$

$$\mathbf{r}'(t) = a\mathbf{i} + ab\cos at\,\mathbf{j}$$

$$\mathbf{r}''(t) = -a^2 b\sin at\,\mathbf{j}$$

$$\|\mathbf{r}''(t)\| = a^2\,|b\sin at|$$

$$= a^2\,|y(t)|$$

5. $y = \cos \pi x, \quad 0 \leq x \leq 2$

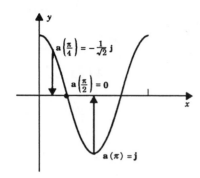

7. $x = \sqrt{1 + y^2}, \quad y \geq -1$

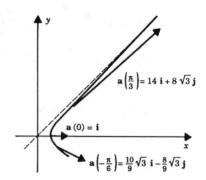

9. (a) initial position is tip of $r(0) = x_0 \mathbf{i} + y_0 \mathbf{j} + z_0 \mathbf{k}$

(b) $r'(t) = (\alpha \cos \theta)\mathbf{j} + (\alpha \sin \theta - 32t)\mathbf{k}, \quad r'(0) = (\alpha \cos \theta)\mathbf{j} + (\alpha \sin \theta)\mathbf{k}$

(c) $|r'(0)| = |\alpha|$ \qquad\qquad\qquad (d) $r''(t) = -32\mathbf{k}$

(e) a parabolic arc from the parabola

$$z = z_0 + (\tan \theta)(y - y_0) - 16 \frac{(y - y_0)^2}{\alpha^2 \cos^2 \theta}$$

in the plane $x = x_0$

11. (a) $r'(t) = \dfrac{a\omega}{2} \left(e^{\omega t} - e^{-\omega t} \right) \mathbf{i} + \dfrac{b\omega}{2} \left(e^{\omega t} + e^{-\omega t} \right) \mathbf{j}, \quad r'(0) = b\omega \mathbf{j}$

(b) $r''(t) = \dfrac{a\omega^2}{2} \left(e^{\omega t} + e^{-\omega t} \right) \mathbf{i} + \dfrac{b\omega^2}{2} \left(e^{\omega t} - e^{-\omega t} \right) \mathbf{j} = \omega^2 r(t)$

(c) The torque τ is $\mathbf{0}$: $\quad \tau(t) = r(t) \times m a(t) = r(t) \times m\omega^2 r(t) = \mathbf{0}.$

The angular momentum $\mathbf{L}(t)$ is constant since $\mathbf{L}'(t) = \tau(t) = \mathbf{0}$.

13. We begin with the force equation $\mathbf{F}(t) = \alpha \mathbf{k}$. In general, $\mathbf{F}(t) = m a(t)$, so that here

$$a(t) = \frac{\alpha}{m} \mathbf{k}.$$

Integration gives

$$v(t) = C_1 \mathbf{i} + C_2 \mathbf{j} + \left(\frac{\alpha}{m} t + C_3 \right) \mathbf{k}.$$

Since $v(0) = 2\mathbf{j}$, we can conclude that $C_1 = 0$, $C_2 = 2$, $C_3 = 0$. Thus

$$v(t) = 2\mathbf{j} + \frac{\alpha}{m} t\mathbf{k}.$$

Another integration gives

$$r(t) = D_1 \mathbf{i} + (2t + D_2)\mathbf{j} + \left(\frac{\alpha}{2m} t^2 + D_3 \right) \mathbf{k}.$$

Since $\mathbf{r}(0) = y_0\mathbf{j} + z_0\mathbf{k}$, we have $D_1 = 0$, $D_2 = y_0$, $D_3 = z_0$, and therefore

$$\mathbf{r}(t) = (2t + y_0)\mathbf{j} + \left(\frac{\alpha}{2m}t^2 + z_0\right)\mathbf{k}.$$

The conditions of the problem require that t be restricted to nonnegative values.
To obtain an equation for the path in Cartesian coordinates, we write out the components

$$x(t) = 0, \quad y(t) = 2t + y_0, \quad z(t) = \frac{\alpha}{2m}t^2 + z_0. \qquad (t \geq 0)$$

From the second equation we have

$$t = \tfrac{1}{2}[y(t) - y_0]. \qquad (y(t) \geq y_0)$$

Substituting this into the third equation, we get

$$z(t) = \frac{\alpha}{8m}[y(t) - y_0]^2 + z_0. \qquad (y(t) \geq y_0)$$

Eliminating t altogether, we have

$$z = \frac{\alpha}{8m}(y - y_0)^2 + z_0. \qquad (y \geq y_0)$$

Since $x = 0$, the path of the object is a parabolic arc in the yz-plane.

Answers to (a) through (d):

(a)   velocity: $\mathbf{v}(t) = 2\mathbf{j} + \dfrac{\alpha}{m}t\mathbf{k}.$ (b)   speed: $v(t) = \dfrac{1}{m}\sqrt{4m^2 + \alpha^2 t^2}.$

(c)   momentum: $\mathbf{p}(t) = 2m\mathbf{j} + \alpha t\mathbf{k}.$

(d)   path in vector form: $\mathbf{r}(t) = (2t + y_0)\mathbf{j} + \left(\dfrac{\alpha}{2m}t^2 + z_0\right)\mathbf{k}, \quad t \geq 0.$

path in Cartesian coordinates: $z = \dfrac{\alpha}{8m}(y - y_0)^2 + z_0, \quad y \geq y_0, \quad x = 0.$

15. $\mathbf{F}(t) = m\,\mathbf{a}(t) = m\,\mathbf{r}''(t) = 2m\mathbf{k}$

17.   From $\mathbf{F}(t) = m\,\mathbf{a}(t)$ we obtain

$$\mathbf{a}(t) = \pi^2[a\cos\pi t\,\mathbf{i} + b\sin\pi t\,\mathbf{j}].$$

By direct calculation using $\mathbf{v}(0) = -\pi b\mathbf{j} + \mathbf{k}$ and $\mathbf{r}(0) = b\mathbf{j}$ we obtain

$$\mathbf{v}(t) = a\pi\sin\pi t\,\mathbf{i} - b\pi\cos\pi t\,\mathbf{j} + \mathbf{k}$$

$$\mathbf{r}(t) = a(1 - \cos\pi t)\mathbf{i} + b(1 - \sin\pi t)\mathbf{j} + t\mathbf{k}.$$

(a) $\mathbf{v}(1) = b\pi\mathbf{j} + \mathbf{k}$ (b) $\|\mathbf{v}(1)\| = \sqrt{\pi^2 b^2 + 1}$

(c) $\mathbf{a}(1) = -\pi^2 a\mathbf{i}$ (d) $m\,\mathbf{v}(1) = m\,(\pi b\mathbf{j} + \mathbf{k})$

(e) $\mathbf{L}(1) = \mathbf{r}(1) \times m\,\mathbf{v}(1) = [2a\mathbf{i} + b\mathbf{j} + \mathbf{k}] \times [m\,(b\pi\mathbf{j} + \mathbf{k})]$

$\qquad = m\,[b(1 - \pi)\mathbf{i} - 2a\mathbf{j} + 2ab\pi\mathbf{k}]$

(f) $\boldsymbol{\tau}(1) = \mathbf{r}(1) \times \mathbf{F}(1) = [2a\mathbf{i} + b\mathbf{j} + \mathbf{k}] \times [-m\pi^2 a\mathbf{i}] = -m\pi^2 a\,[\mathbf{j} - b\mathbf{k}]$

19. We have $mv = mv_1 + mv_2$ and $\frac{1}{2}mv^2 = \frac{1}{2}mv_1{}^2 + \frac{1}{2}mv_2{}^2$.

Therefore $\mathbf{v} = \mathbf{v}_1 + \mathbf{v}_2$ and $v^2 = v_1{}^2 + v_2{}^2$.

Since $v^2 = \mathbf{v} \cdot \mathbf{v} = (\mathbf{v}_1 + \mathbf{v}_2) \cdot (\mathbf{v}_1 + \mathbf{v}_2) = v_1{}^2 + v_2{}^2 + 2(\mathbf{v}_1 \cdot \mathbf{v}_2)$,

we have $\mathbf{v}_1 \cdot \mathbf{v}_2 = 0$ and $\mathbf{v}_1 \perp \mathbf{v}_2$.

21. $\mathbf{r}''(t) = \mathbf{a}, \quad \mathbf{r}'(t) = \mathbf{v}(0) + t\mathbf{a}, \quad \mathbf{r}(t) = \mathbf{r}(0) + t\mathbf{v}(0) + \frac{1}{2}t^2\mathbf{a}.$

If neither $\mathbf{v}(0)$ nor $\mathbf{a}$ is zero, the displacement $\mathbf{r}(t) - \mathbf{r}(0)$ is a linear combination of $\mathbf{v}(0)$ and $\mathbf{a}$ and thus remains on the plane determined by these vectors. The equation of this plane can be written

$$[\mathbf{a} \times \mathbf{v}(0)] \cdot [\mathbf{r} - \mathbf{r}(0)] = 0.$$

(If either $\mathbf{v}(0)$ or $\mathbf{a}$ is zero, the motion is restricted to a straight line; if both of these vectors are zero, the particle remains at its initial position $\mathbf{r}(0)$.)

23. $\mathbf{r}(t) = \mathbf{i} + t\mathbf{j} + \left(\dfrac{qE_0}{2m}\right)t^2\mathbf{k}$ **25.** $\mathbf{r}(t) = \left(1 + \dfrac{t^3}{6m}\right)\mathbf{i} + \dfrac{t^4}{12m}\mathbf{j} + t\mathbf{k}$

27.

$(14.2.3)$ ⟶

$$\frac{d}{dt}\left(\frac{1}{2}mv^2\right) = mv\frac{dv}{dt} = m\left(\mathbf{v} \cdot \frac{d\mathbf{v}}{dt}\right) = m\frac{d\mathbf{v}}{dt} \cdot \mathbf{v} = \mathbf{F} \cdot \frac{d\mathbf{r}}{dt}$$

$$= 4r^2\left(\mathbf{r} \cdot \frac{d\mathbf{r}}{dt}\right) = 4r^2\left(r\frac{dr}{dt}\right) = 4r^3\frac{dr}{dt} = \frac{d}{dt}\left(r^4\right).$$

Therefore $d/dt\left(\frac{1}{2}mv^2 - r^4\right) = 0$ and $\frac{1}{2}mv^2 - r^4$ is a constant E. Evaluating E from $t = 0$, we find that $E = 2m$.

Thus $\frac{1}{2}mv^2 - r^4 = 2m$ and $v = \sqrt{4 + (2/m)\,r^4}$.

SECTION * 14.6

1. On Earth: year of length T, average distance from sun d.

On Venus: year of length αT, average distance from sun $0.72d$.

Therefore

$$\frac{(\alpha T)^2}{T^2} = \frac{(0.72d)^3}{d^3}.$$

This gives $\alpha^2 = (0.72)^3 \cong 0.372$ and $\alpha \cong 0.615$. Answer: about 61.5% of an Earth year.

3.
$$\left(\frac{dx}{dt}\right)^2 + \left(\frac{dy}{dt}\right)^2 = \left[\frac{d}{dt}(r\cos\theta)\right]^2 + \left[\frac{d}{dt}(r\sin\theta)\right]^2$$

$$= \left[r(-\sin\theta)\frac{d\theta}{dt} + \frac{dr}{dt}\cos\theta\right]^2 + \left[r\cos\theta\frac{d\theta}{dt} + \frac{dr}{dt}\sin\theta\right]^2$$

$$= r^2\sin^2\theta\left(\frac{d\theta}{dt}\right)^2 + \left(\frac{dr}{dt}\right)^2\cos^2\theta + r^2\cos^2\theta\left(\frac{d\theta}{dt}\right)^2 + \left(\frac{dr}{dt}\right)^2\sin^2\theta$$

$$= \left(\frac{dr}{dt}\right)^2 + r^2\left(\frac{d\theta}{dt}\right)^2$$

5.   Substitute
$$r = \frac{a}{1 + e\cos\theta}, \quad \left(\frac{dr}{d\theta}\right)^2 = \left[\frac{-a}{(1+e\cos\theta)^2}\cdot(-e\sin\theta)\right]^2 = \frac{(ae\sin\theta)^2}{(1+e\cos\theta)^4}$$

into the right side of the equation and you will see that, with a and e^2 as given, the expression reduces to E.

SECTION 14.7

1. $k = \dfrac{e^{-x}}{(1 + e^{-2x})^{3/2}}$

3. $k = \dfrac{\sec^2 x}{(1 + \tan^2 x)^{3/2}} = |\cos x|$

5. $k = \dfrac{|\sin x|}{(1 + \cos^2 x)^{3/2}}$

7. $k = \dfrac{|x|}{(1 + x^4/4)^{3/2}}$;   at $\left(2, \dfrac{4}{3}\right)$, $\rho = \dfrac{5}{2}\sqrt{5}$

9. $k = \dfrac{|-1/y^3|}{(1 + 1/y^2)^{3/2}} = \dfrac{1}{(1 + y^2)^{3/2}}$;   at $(2,2)$, $\rho = 5\sqrt{5}$

11. $k(x) = \dfrac{|-1/x^2|}{(1 + 1/x^2)^{3/2}} = \dfrac{x}{(x^2 + 1)^{3/2}}$, $x > 0$

$k'(x) = \dfrac{(1 - 2x^2)}{(x^2 + 1)^{5/2}}$, $k'(x) = 0 \implies x = \dfrac{1}{2}\sqrt{2}$

Since k increases on $\left(0, \frac{1}{2}\sqrt{2}\right]$ and decreases on $\left[\frac{1}{2}\sqrt{2}, \infty\right)$, k is maximal at $\left(\frac{1}{2}\sqrt{2}, \frac{1}{2}\ln\frac{1}{2}\right)$.

13. $k = \dfrac{1}{(1 + t^2)^{3/2}}$

15. $k = \dfrac{\left|2e^{2t}\cos t\,(\cos t - \sin t) + 2e^{2t}\sin t\,(\cos t + \sin t)\right|}{[e^{2t}(\cos t - \sin t)^2 + e^{2t}(\cos t + \sin t)^2]^{3/2}} = \dfrac{2e^{2t}}{(2e^{2t})^{3/2}} = \dfrac{1}{2}\sqrt{2}\,e^{-t}$

17. $k = \dfrac{\left|2/x^3\right|}{[1 + 1/x^4]^{3/2}} = \dfrac{2|x^3|}{(x^4 + 1)^{3/2}}; \quad$ at $x = \pm 1$, $\rho = \dfrac{2^{3/2}}{2} = \sqrt{2}$

19. We use (14.7.3) and the hint to obtain

$$k = \frac{\left|ab\sinh^2 t - ab\cosh^2 t\right|}{[a^2\sinh^2 t + b^2\cosh^2 t]^{3/2}} = \frac{\left|\dfrac{a}{b}y^2 - \dfrac{b}{a}x^2\right|}{\left[\left(\dfrac{ay}{b}\right)^2 + \left(\dfrac{bx}{a}\right)^2\right]^{3/2}}$$

$$= \frac{a^3 b^3\left|\dfrac{a}{b}y^2 - \dfrac{b}{a}x^2\right|}{[a^4 y^2 + b^4 x^2]^{3/2}} = \frac{a^4 b^4}{[a^4 y^2 + b^4 x^2]^{3/2}}.$$

21. By the hint and the fact that $\|\mathbf{T} \times \mathbf{N}\| = 1$,

$$\frac{\|\mathbf{v} \times \mathbf{a}\|}{(ds/dt)^3} = \frac{\left\| \left(\dfrac{ds}{dt}\mathbf{T}\right) \times \left(\dfrac{d^2 s}{dt^2}\mathbf{T} + k\left(\dfrac{ds}{dt}\right)^2 \mathbf{N}\right) \right\|}{(ds/dt)^3}$$

$$\underset{\mathbf{T} \times \mathbf{T} = 0 \ \uparrow}{=} \frac{\left\| k\,(ds/dt)^3\,(\mathbf{T} \times \mathbf{N}) \right\|}{(ds/dt)^3} = k.$$

23. $\mathbf{r}'(t) = e^t(\cos t - \sin t)\mathbf{i} + e^t(\sin t + \cos t)\mathbf{j} + e^t\mathbf{k}$

$\dfrac{ds}{dt} = \|\mathbf{r}'(t)\| = \sqrt{3}\,e^t, \quad \dfrac{d^2 s}{dt^2} = \sqrt{3}\,e^t$

$\mathbf{T}(t) = \dfrac{\mathbf{r}'(t)}{\|\mathbf{r}'(t)\|} = \dfrac{1}{\sqrt{3}}[(\cos t - \sin t)\mathbf{i} + (\sin t + \cos t)\mathbf{j} + \mathbf{k}]$

$\mathbf{T}'(t) = \dfrac{1}{\sqrt{3}}[(-\sin t - \cos t)\mathbf{i} + (\cos t - \sin t)\mathbf{j}]$

Then,

$$k = \frac{\|\mathbf{T}'(t)\|}{ds/dt} = \frac{\sqrt{2/3}}{\sqrt{3}\,e^t} = \frac{1}{3}\sqrt{2}\,e^{-t},$$

$$\mathbf{a_T} = \frac{d^2 s}{dt^2} = \sqrt{3}\,e^t, \quad \mathbf{a_N} = k\left(\frac{ds}{dt}\right)^2 = \sqrt{2}\,e^t.$$

25. $\mathbf{r}'(t) = \mathbf{i} + t\mathbf{j} + t^2\mathbf{k}, \quad \dfrac{ds}{dt} = \|\mathbf{r}'(t)\| = \sqrt{t^4 + t^2 + 1}, \quad \dfrac{d^2s}{dt^2} = \dfrac{2t^3 + t}{\sqrt{t^4 + t^2 + 1}}$

$$\mathbf{T}(t) = \frac{\mathbf{r}'(t)}{\|\mathbf{r}'(t)\|} = \frac{1}{\sqrt{t^4 + t^2 + 1}}\left(\mathbf{i} + t\mathbf{j} + t^2\mathbf{k}\right),$$

$$\mathbf{T}'(t) = \frac{1}{(t^4 + t^2 + 1)^{3/2}}\left[-t\left(2t^2 + 1\right)\mathbf{i} + \left(1 - t^4\right)\mathbf{j} + t\left(t^2 + 2\right)\mathbf{k}\right].$$

Then,

$$k = \frac{\|\mathbf{T}'(t)\|}{ds/dt} = \frac{\sqrt{t^2\left(2t^2 + 1\right)^2 + \left(1 + t^4\right)^2 + t^2\left(t^2 + 2\right)^2}}{(t^4 + t^2 + 1)^2}$$

$$= \frac{\sqrt{\left(t^4 + 4t^2 + 1\right)\left(t^4 + t^2 + 1\right)}}{(t^4 + t^2 + 1)^2} = \frac{\sqrt{t^4 + 4t^2 + 1}}{(t^4 + t^2 + 1)^{3/2}},$$

$$\mathbf{a_T} = \frac{d^2s}{dt^2} = \frac{2t^3 + t}{\sqrt{t^4 + t^2 + 1}}, \quad \mathbf{a_N} = k\left(\frac{ds}{dt}\right)^2 = \frac{\sqrt{t^4 + 4t^2 + 1}}{\sqrt{t^4 + t^2 + 1}}.$$

27. By Exercise 26

$$k = \frac{\left|\left(e^{a\theta}\right)^2 + 2\left(ae^{a\theta}\right)^2 - \left(e^{a\theta}\right)\left(a^2 e^{a\theta}\right)\right|}{\left[\left(e^{a\theta}\right)^2 + \left(ae^{a\theta}\right)^2\right]^{3/2}} = \frac{e^{-a\theta}}{\sqrt{1 + a^2}}.$$

29. By Exercise 26

$$k = \frac{\left|a^2(1 - \cos\theta)^2 + 2a^2\sin^2\theta - a^2(1 - \cos\theta)(\cos\theta)\right|}{\left[a^2(1 - \cos\theta)^2 + a^2\sin^2\theta\right]^{3/2}}$$

$$= \frac{3a^2(1 - \cos\theta)}{\left[2a^2(1 - \cos\theta)\right]^{3/2}} = \frac{3ar}{[2ar]^{3/2}} = \frac{3}{2\sqrt{2ar}}.$$

31. (a) For $0 \leq \theta \leq \pi$

$$s(\theta) = \int_\theta^\pi \sqrt{[x'(t)]^2 + [y'(t)]^2}\, dt = \int_\theta^\pi \sqrt{R^2(1 - \cos t)^2 + R^2\sin^2 t}\, dt$$

$$= \int_\theta^\pi R\sqrt{2(1 - \cos t)}\, dt = \int_\theta^\pi 2R\sin\frac{1}{2}t\, dt = 4R\cos\frac{1}{2}\theta = 4R\left|\cos\frac{1}{2}\theta\right|.$$

For $\pi \leq \theta \leq 2\pi$

$$s(\theta) = \int_\pi^\theta 2R\sin\frac{1}{2}t\, dt = -4R\cos\frac{1}{2}\theta = 4R\left|\cos\frac{1}{2}\theta\right|.$$

(b) $k(\theta) = \dfrac{|x'(\theta)y''(\theta) - y'(\theta)x''(\theta)|}{\{[x'(\theta)]^2 + [y'(\theta)]\}^{3/2}} = \dfrac{|R(1 - \cos\theta)R\cos\theta - R\sin\theta\,(R\sin\theta)|}{8R^3\sin^3\frac{1}{2}\theta}.$

This reduces to $k(\theta) = 1/(4R\sin\frac{1}{2}\theta)$ and gives $\rho(\theta) = 4R\,\sin\frac{1}{2}\theta.$

(c) $\rho^2 + s^2 = 16R^2$

33.   Straightforward calculation gives

$$s(\theta) = 4a\left|\cos\tfrac{1}{2}\theta\right| \quad \text{and} \quad \rho(\theta) = \tfrac{4}{3}a\sin\tfrac{1}{2}\theta.$$

Therefore

$$9\rho^2 + s^2 = 16a^2.$$

CHAPTER 15

SECTION 15.1

1. dom (f) = the first and third quadrants, including the axes;   ran $(f) = [0, \infty)$

3. dom (f) = the set of all points (x, y) except those on the line $y = -x$;   ran $(f) = (-\infty, 0) \cup (0, \infty)$

5. dom (f) = the entire plane;   ran $(f) = (-1, 1)$ since

$$\frac{e^x - e^y}{e^x + e^y} = \frac{e^x + e^y - 2e^y}{e^x + e^y} = 1 - \frac{2}{e^{x-y} + 1}$$

and the last quotient takes on all values between 0 and 2.

7. dom (f) = the first and third quadrants, excluding the axes;   ran $(f) = (-\infty, \infty)$

9. dom (f) = the set of all points (x, y) with $x^2 < y$ —in other words, the set of all points of the plane above the parabola $y = x^2$;   ran $(f) = (0, \infty)$

11. dom (f) = the set of all points (x, y) with $-3 \le x \le 3$, $-2 \le y \le 2$ (a rectangle);   ran $(f) = [-2, 3]$

13. dom (f) = the set of all points (x, y, z) not on the plane $x + y + z = 0$;   ran $(f) = \{-1, 1\}$

15. dom (f) = the set of all points (x, y, z) with $|y| < |x|$;   ran $(f) = (-\infty, 0]$

17. dom (f) = the set of all points (x, y) with $x^2 + y^2 < 9$ —in other words, the set of all points of the plane inside the circle $x^2 + y^2 = 9$;   ran $(f) = [2/3, \infty)$

19. dom (f) = the set of all points (x, y, z) with $x + 2y + 3z > 0$ —in other words, the set of all points in space that lie on the same side of the plane $x + 2y + 3z = 0$ as the point $(1, 1, 1)$;   ran $(f) = (-\infty, \infty)$

21. dom (f) = all of space;   ran $(f) = (0, \infty)$

23. dom (f) = the set of all points (x, y, z) with $-3 \le x \le 3$, $-2 \le y \le 2$, $-1 \le z \le 1$ (a rectangular solid);   ran $(f) = [-2, 3]$

25. f assigns the number $\sqrt{x}$ to each real number $x \ge 0$; g assigns the number $\sqrt{x}$ to each point (x, y) of the plane with $x \ge 0$; h assigns the number $\sqrt{x}$ to each point (x, y, z) of space with $x \ge 0$.

SECTION 15.2

1. a quadric cone

3. a parabolic cylinder

5. a hyperboloid of one sheet

7. a sphere

9. an elliptic paraboloid

11. a hyperbolic paraboloid

13.
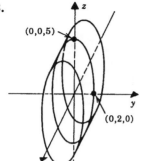
(0,0,5)

(0,2,0)

15.
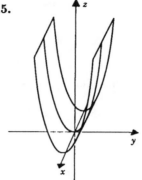

17.
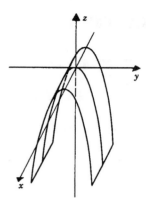

19. elliptic paraboloid
xy-trace: the origin
xz-trace: the parabola $x^2 = 4z$
yz-trace: the parabola $y^2 = 9z$
surface has the form of Figure 15.2.5

21. quadric cone
xy-trace: the origin
xz-trace: the lines $x = \pm 2z$
yz-trace: the lines $y = \pm 3z$
surface has the form of Figure 15.2.4

23. hyperboloid of two sheets
xy-trace: none
xz-trace: the hyperbola $4z^2 - x^2 = 4$
yz-trace: the hyperbola $9z^2 - y^2 = 9$
surface has the form of Figure 15.2.3

25. (a) an elliptic paraboloid (vertex down if A and B are both positive, vertex up if A and B are both negative)

(b) a hyperbolic paraboloid

(c) the xy-plane if A and B are both zero; otherwise a parabolic cylinder

27. $x^2 + y^2 - 4z = 0$ (paraboloid of revolution)

29. (a) a circle

(b) (i) $\sqrt{x^2 + y^2} = -3z$ (ii) $\sqrt{x^2 + z^2} = \frac{1}{3}y$

31. $x + 2y + 3\left(\dfrac{x + y - 6}{2}\right) = 6$ or $5x + 7y = 30$,   a line

33. $\left.\begin{array}{c} x^2 + y^2 + (z - 1)^2 = \frac{3}{2} \\ x^2 + y^2 - z^2 = 1 \end{array}\right\}$ $(z^2 + 1) + (z - 1)^2 = \dfrac{3}{2}$; $(2z - 1)^2 = 0$, $z = \dfrac{1}{2}$ so that $x^2 + y^2 = \dfrac{5}{4}$

35. $x^2 + y^2 + (x^2 + 3y^2) = 4$ or $x^2 + 2y^2 = 2$,   an ellipse

37. $x^2 + y^2 = (2 - y)^2$ or $x^2 = -4(y - 1)$,   a parabola

SECTION 15.3

1. lines of slope 1: $y = x - c$

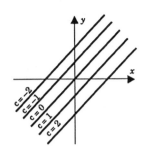

3. parabolas: $y = x^2 - c$

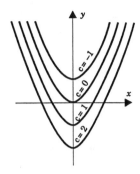

5. the y-axis and the lines $y = \left(\dfrac{1-c}{c}\right) x$

with the origin omitted throughout

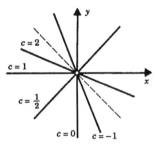

7. the cubics $y = x^3 - c$

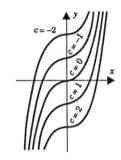

9. the lines $y = \pm x$ and the hyperbolas $x^2 - y^2 = c$

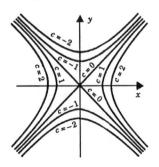

11. pairs of horizontal lines $y = \pm\sqrt{c}$ and the x-axis

13. the circles $x^2 + y^2 = e^c$, c real

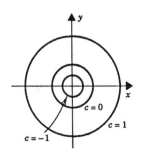

15. the curves $y = e^{cx^2}$ with the point $(0,1)$ omitted

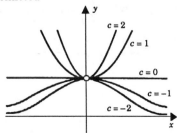

17. the coordinate axes and pairs of lines

$$y = \pm \frac{\sqrt{1-c}}{\sqrt{c}}\, x, \text{ the origin}$$

omitted throughout

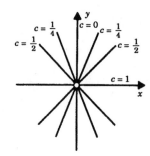

19. $x + 2y + 3z = 0$, plane through the origin

21. $z = \sqrt{x^2 + y^2}$, the upper nappe of the circular cone $z^2 = x^2 + y^2$ (Figure 15.2.4)

23. the elliptic paraboloid $\dfrac{x^2}{(1/2)^2} + \dfrac{y^2}{(1/3)^2} = 72z$ (Figure 15.2.5)

25.   (i)   hyperboloid of two sheets   (Figure 15.2.3)

    (ii)   circular cone   (Figure 15.2.4)

    (iii)   hyperboloid of one sheet   (Figure 15.2.2)

SECTION 15.4

1. $\dfrac{\partial f}{\partial x} = 6x - y, \quad \dfrac{\partial f}{\partial y} = 1 - x$

3. $\dfrac{\partial \rho}{\partial \phi} = \cos \phi \cos \theta, \quad \dfrac{\partial \rho}{\partial \theta} = -\sin \phi \sin \theta$

5. $\dfrac{\partial f}{\partial x} = e^{x-y} + e^{y-x}, \quad \dfrac{\partial f}{\partial y} = -e^{x-y} - e^{y-x}$

7. $\dfrac{\partial g}{\partial x} = \dfrac{(AD - BC)y}{(Cx + Dy)^2}, \quad \dfrac{\partial g}{\partial y} = \dfrac{(BC - AD)x}{(Cx + Dy)^2}$

9. $\dfrac{\partial u}{\partial x} = y + z, \quad \dfrac{\partial u}{\partial y} = x + z, \quad \dfrac{\partial u}{\partial z} = x + y$

11. $\dfrac{\partial f}{\partial x} = z \cos (x - y), \quad \dfrac{\partial f}{\partial y} = -z \cos (x - y), \quad \dfrac{\partial f}{\partial z} = \sin (x - y)$

13. $\dfrac{\partial \rho}{\partial \theta} = e^{\theta + \phi} [\cos (\theta - \phi) - \sin (\theta - \phi)], \quad \dfrac{\partial \rho}{\partial \phi} = e^{\theta + \phi} [\cos (\theta - \phi) + \sin (\theta - \phi)]$

15. $\dfrac{\partial h}{\partial x} = 2f(x)f'(x)g(y), \quad \dfrac{\partial h}{\partial y} = [f(x)]^2 g'(y)$

17. $\dfrac{\partial f}{\partial x} = (y^2 \ln z)z^{xy^2}, \quad \dfrac{\partial f}{\partial y} = (2xy \ln z)z^{xy^2}, \quad \dfrac{\partial f}{\partial z} = xy^2 z^{xy^2 - 1}$

19. $f_x(x, y) = e^x \ln y, \quad f_x(0, e) = 1; \quad f_y(x, y) = \dfrac{1}{y} e^x, \quad f_y(0, e) = e^{-1}$

21. $f_x(x,y) = \dfrac{y}{(x+y)^2}, \quad f_x(1,2) = \dfrac{2}{9}; \quad f_y(x,y) = \dfrac{-x}{(x+y)^2}, \quad f_y(1,2) = -\dfrac{1}{9}$

23. $f_x(x,y) = \lim\limits_{h\to 0} \dfrac{(x+h)^2 y - x^2 y}{h} = \lim\limits_{h\to 0} y\left(\dfrac{2xh + h^2}{h}\right) = y \lim\limits_{h\to 0}(2x+h) = 2xy$

       $f_x(x,y) = \lim\limits_{h\to 0} \dfrac{x^2(y+h) - x^2 y}{h} = \lim\limits_{h\to 0} \dfrac{x^2 h}{h} = \lim\limits_{h\to 0} x^2 = x^2$

25. $f_x(x,y) = \lim\limits_{h\to 0} \dfrac{\ln\left(y\,(x+h)^2\right) - \ln x^2 y}{h} = \lim\limits_{h\to 0} \dfrac{\ln y + 2\ln(x+h) - 2\ln x - \ln y}{h}$

$$= 2\lim\limits_{h\to 0} \dfrac{\ln(x+h) - \ln x}{h} = 2\dfrac{d}{dx}(\ln x) = \dfrac{2}{x}$$

       $f_y(x,y) = \lim\limits_{h\to 0} \dfrac{\ln\left(x^2(y+h)\right) - \ln x^2 y}{h} = \lim\limits_{h\to 0} \dfrac{\ln x^2 + \ln(y+h) - \ln x^2 - \ln y}{h}$

$$= \lim\limits_{h\to 0} \dfrac{\ln(y+h) - \ln y}{h} = \dfrac{d}{dy}(\ln y) = \dfrac{1}{y}$$

27. $f_x(x,y) = \lim\limits_{h\to 0} \dfrac{1}{h}\left\{\dfrac{1}{(x+h) - y} - \dfrac{1}{x - y}\right\} = \lim\limits_{h\to 0} \dfrac{1}{h}\left\{\dfrac{-h}{(x+h-y)(x-y)}\right\}$

$$= \lim\limits_{h\to 0} \dfrac{-1}{(x+h-y)(x-y)} = \dfrac{-1}{(x-y)^2}$$

       $f_x(x,y) = \lim\limits_{h\to 0} \dfrac{1}{h}\left\{\dfrac{1}{x - (y+h)} - \dfrac{1}{x - y}\right\} = \lim\limits_{h\to 0} \dfrac{1}{h}\left\{\dfrac{h}{(x-y-h)(x-y)}\right\}$

$$= \lim\limits_{h\to 0} \dfrac{1}{(x-y-h)(x-y)} = \dfrac{1}{(x-y)^2}$$

29. $f_x(x,y,z) = \lim\limits_{h\to 0} \dfrac{(x+h)y^2 z - xy^2 z}{h} = \lim\limits_{h\to 0} y^2 z = y^2 z$

       $f_y(x,y,z) = \lim\limits_{h\to 0} \dfrac{x(y+h)^2 z - xy^2 z}{h} = \lim\limits_{h\to 0} \dfrac{xz(2yh + h^2)}{h}$

$$= \lim\limits_{h\to 0} xz(2y+h) = 2xyz$$

       $f_z(x,y,z) = \lim\limits_{h\to 0} \dfrac{xy^2(x+h) - xy^2 z}{h} = \lim\limits_{h\to 0} xy^2 = xy^2$

31.
$$x\frac{\partial u}{\partial x} + y\frac{\partial u}{\partial y} = x\left(4Ax^3 + 4Bxy^2\right) + y\left(4Bx^2y + 4Cy^3\right)$$

$$= 4\left(Ax^4 + 2Bx^2y^2 + Cy^4\right) = 4u$$

33. $\dfrac{\partial x}{\partial r}\dfrac{\partial y}{\partial \theta} - \dfrac{\partial x}{\partial \theta}\dfrac{\partial y}{\partial r} = (\cos\theta)(r\cos\theta) - (-r\sin\theta)(\sin\theta) = r$

35. $V\dfrac{\partial P}{\partial V} = V\left(-\dfrac{kT}{V^2}\right) = -\dfrac{kT}{V} = -P$

$V\dfrac{\partial P}{\partial V} + T\dfrac{\partial P}{\partial T} = V\left(-\dfrac{kT}{V^2}\right) + T\left(\dfrac{k}{V}\right) = -\dfrac{kT}{V} + \dfrac{kT}{V} = 0$

37. (a) $50\sqrt{3}$ in.^{2}

  (b) $\dfrac{\partial A}{\partial b} = \dfrac{1}{2}c\sin\theta$;   at time t_0, $\dfrac{\partial A}{\partial b} = 5\sqrt{3}$

  (c) $\dfrac{\partial A}{\partial \theta} = \dfrac{1}{2}bc\cos\theta$;   at time t_0, $\dfrac{\partial A}{\partial \theta} = 50$

  (d) with $h = \dfrac{\pi}{180}$, $A(b,c,\theta+h) - A(b,c,\theta) \cong h\dfrac{\partial A}{\partial \theta} = \dfrac{\pi}{180}(50) = \dfrac{5\pi}{18}$ in.^{2}

  (e) $0 = \dfrac{1}{2}\sin\theta\left(b\dfrac{\partial c}{\partial b} + c\right)$;   at time t_0, $\dfrac{\partial c}{\partial b} = \dfrac{-c}{b} = -2$

39. (a) y_0-section: $\mathbf{r}(x) = x\mathbf{i} + y_0\mathbf{j} + f(x,y_0)\mathbf{k}$

   tangent line: $\mathbf{R}(t) = [x_0\mathbf{i} + y_0\mathbf{j} + f(x_0,y_0)\mathbf{k}] + t\left[\mathbf{i} + \dfrac{\partial f}{\partial x}(x_0,y_0)\mathbf{k}\right]$

  (b) x_0-section: $\mathbf{r}(y) = x_0\mathbf{i} + y\mathbf{j} + f(x_0,y)\mathbf{k}$

   tangent line: $\mathbf{R}(t) = [x_0\mathbf{i} + y_0\mathbf{j} + f(x_0,y_0)\mathbf{k}] + t\left[\mathbf{j} + \dfrac{\partial f}{\partial y}(x_0,y_0)\mathbf{k}\right]$

  (c) For (x,y,z) in the plane

   $[(x-x_0)\mathbf{i} + (y-y_0)\mathbf{j} + (z-f(x_0,y_0))\mathbf{k}] \cdot \left[\left(\mathbf{i} + \dfrac{\partial f}{\partial x}(x_0,y_0)\mathbf{k}\right) \times \left(\mathbf{j} + \dfrac{\partial f}{\partial y}(x_0,y_0)\mathbf{k}\right)\right] = 0.$

   From this it follows that
   $$z - f(x_0,y_0) = (x-x_0)\dfrac{\partial f}{\partial x}(x_0,y_0) + (y-y_0)\dfrac{\partial f}{\partial y}(x_0,y_0).$$

41. (a) Set $u = ax + by$. Then

   $$b\dfrac{\partial w}{\partial x} - a\dfrac{\partial w}{\partial y} = b(a\,g'(u)) - a(b\,g'(u)) = 0.$$

  (b) Set $u = x^m y^n$. Then

   $$nx\dfrac{\partial w}{\partial x} - my\dfrac{\partial w}{\partial y} = nx\left[mx^{m-1}y^n g'(u)\right] - my\left[nx^m y^{n-1}g'(u)\right] = 0.$$

SECTION 15.5

1. interior $= \{(x,y) : 2 < x < 4, \quad 1 < y < 3\}$ (the
    inside of the rectangle), boundary $=$ the union of
    the four boundary line segments; set is closed.

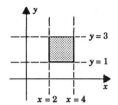

3. interior $=$ the entire set (region between the two
    concentric circles), boundary $=$ the two circles, one
    of radius 1, the other of radius 2; set is open.

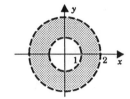

5. interior $= \{(x,y) : 1 < x^2 < 4\} =$
    $\{(x,y) : -2 < x < -1\} \cup \{(x,y) : 1 < x < 2\}$
    (two vertical strips without the boundary lines),
    boundary $= \{(x,y) : x = -2, x = -1, x = 1, \text{ or }$
    $x = 2\}$ (four vertical lines);   set is neither open
    nor closed.

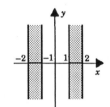

7. interior $=$ region below the parabola $y = x^2$,
    boundary $=$ the parabola $y = x^2$;   the set is closed.

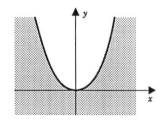

9. interior $= \{(x,y,z) : x^2 + y^2 < 1, 0 < z \le 4\}$
    (the inside of the cylinder),   boundary $=$ the total
    surface of the cylinder (the curved part,
    the top, and the bottom);   the set is closed.

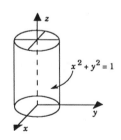

11. (a) ϕ (b) S (c) closed

13. interior $= \{x : 1 < x < 3\}$, boundary $= \{1,3\}$;   set is closed.

15. interior $=$ the entire set, boundary $= \{1\}$;   set is open.

17. interior $= \{x : |x| > 1\}$, boundary $= \{1,-1\}$;   set is neither open nor closed.

19. interior $= \phi$, boundary $= \{\text{the entire set}\} \cup \{0\}$;   the set is neither open nor closed.

SECTION 15.6

1. $\dfrac{\partial^2 f}{\partial x^2} = 2A, \quad \dfrac{\partial^2 f}{\partial y^2} = 2C, \quad \dfrac{\partial^2 f}{\partial y \partial x} = \dfrac{\partial^2 f}{\partial x \partial y} = 2B$

3. $\dfrac{\partial^2 f}{\partial x^2} = Cy^2 e^{xy}, \quad \dfrac{\partial^2 f}{\partial y^2} = Cx^2 e^{xy}, \quad \dfrac{\partial^2 f}{\partial y \partial x} = \dfrac{\partial^2 f}{\partial x \partial y} = Ce^{xy}(xy + 1)$

5. $\dfrac{\partial^2 f}{\partial x^2} = 2, \quad \dfrac{\partial^2 f}{\partial y^2} = 4(x + 3y^2 + z^3), \quad \dfrac{\partial^2 f}{\partial z^2} = 6z(2x + 2y^2 + 5z^3)$

   $\dfrac{\partial^2 f}{\partial x \partial y} = \dfrac{\partial^2 f}{\partial y \partial x} = 4y, \quad \dfrac{\partial^2 f}{\partial z \partial x} = \dfrac{\partial^2 f}{\partial x \partial z} = 6z^2, \quad \dfrac{\partial^2 f}{\partial z \partial y} = \dfrac{\partial^2 f}{\partial y \partial z} = 12yz^2$

7. $\dfrac{\partial^2 f}{\partial x^2} = \dfrac{1}{(x + y)^2} - \dfrac{1}{x^2}, \quad \dfrac{\partial^2 f}{\partial y^2} = \dfrac{1}{(x + y)^2}, \quad \dfrac{\partial^2 f}{\partial y \partial x} = \dfrac{\partial^2 f}{\partial x \partial y} = \dfrac{1}{(x + y)^2}$

9. $\dfrac{\partial^2 f}{\partial x^2} = 2(y + z), \quad \dfrac{\partial^2 f}{\partial y^2} = 2(x + z), \quad \dfrac{\partial^2 f}{\partial z^2} = 2(x + y)$

   all the second mixed partials are $2(x + y + z)$

11. $\dfrac{\partial^2 f}{\partial x^2} = y(y - 1)x^{y-2}, \quad \dfrac{\partial^2 f}{\partial y^2} = (\ln x)^2 x^y, \quad \dfrac{\partial^2 f}{\partial y \partial x} = \dfrac{\partial^2 f}{\partial x \partial y} = x^{y-1}(1 + y \ln x)$

13. $x^2 \dfrac{\partial^2 u}{\partial x^2} + 2xy \dfrac{\partial^2 u}{\partial x \partial y} + y^2 \dfrac{\partial^2 u}{\partial y^2} = x^2 \left(\dfrac{-2y^2}{(x + y)^3} \right) + 2xy \left(\dfrac{2xy}{(x + y)^3} \right) + y^2 \left(\dfrac{-2x^2}{(x + y)^3} \right) = 0$

15. (a) $\dfrac{\partial^2 f}{\partial t^2} - c^2 \dfrac{\partial^2 f}{\partial x^2} = (0) - c^2(0) = 0$

    (b) $\dfrac{\partial^2 f}{\partial t^2} - c^2 \dfrac{\partial^2 f}{\partial x^2} = [c^2 k^2 f(x, t)] - c^2 [k^2 f(x, t)] = 0$

    (c) Set $u = x - ct$. By Exercise 40 Section 15.4 we obtain

    $$\dfrac{\partial^2 f}{\partial t^2} - c^2 \dfrac{\partial^2 f}{\partial x^2} = [c^2 g''(u)] - c^2 [g''(u)] = 0.$$

17. (a) mixed partials are 0

    (b) mixed partials are $g'(x) h'(y)$

    (c) by the hint mixed partials for each term $x^m y^n$ are $mn x^{m-1} y^{n-1}$

19. (a) no, since $\dfrac{\partial^2 f}{\partial y \partial x} \neq \dfrac{\partial^2 f}{\partial x \partial y}$ (b) no, since $\dfrac{\partial^2 f}{\partial y \partial x} \neq \dfrac{\partial^2 f}{\partial x \partial y}$ for $x \neq y$

21. $\dfrac{\partial^3 f}{\partial x^2 \partial y} = \underset{\text{by definition}}{\dfrac{\partial}{\partial x}\left(\dfrac{\partial^2 f}{\partial x \partial y}\right)} = \underset{(15.6.5)}{\dfrac{\partial}{\partial x}\left(\dfrac{\partial^2 f}{\partial y \partial x}\right)} = \underset{\text{by definition}}{\dfrac{\partial^2 f}{\partial x \partial y}\left(\dfrac{\partial f}{\partial x}\right)} = \underset{(15.6.5)}{\dfrac{\partial^2}{\partial y \partial x}\left(\dfrac{\partial f}{\partial x}\right)} = \underset{\text{by def.}}{\dfrac{\partial}{\partial y}\left(\dfrac{\partial^2 f}{\partial x^2}\right)} = \underset{\text{by def.}}{\dfrac{\partial^3 f}{\partial y \partial x^2}}$

23. (a) $\displaystyle\lim_{x \to 0} \dfrac{(x)(0)}{x^2 + 0} = \lim_{x \to 0} 0 = 0$ (b) $\displaystyle\lim_{y \to 0} \dfrac{(0)(y)}{0 + y^2} = \lim_{y \to 0} 0 = 0$

(c) $\displaystyle\lim_{x \to 0} \dfrac{(x)(mx)}{x^2 + (mx)^2} = \lim_{x \to 0} \dfrac{m}{1 + m^2} = \dfrac{m}{1 + m^2}$

(d) $\displaystyle\lim_{\theta \to 0^+} \dfrac{(\theta \cos\theta)(\theta \sin\theta)}{(\theta \cos\theta)^2 + (\theta \sin\theta)^2} = \lim_{\theta \to 0^+} \cos\theta \sin\theta = 0$

(e) By L'Hospital's rule $\displaystyle\lim_{x \to 0} \dfrac{f(x)}{x} = \lim_{x \to 0} f'(x) = f'(0).$ Thus

$$\lim_{x \to 0} \dfrac{xf(x)}{x^2 + [f(x)]^2} = \lim_{x \to 0} \dfrac{f(x)/x}{1 + [f(x)/x]^2} = \dfrac{f'(0)}{1 + [f'(0)]^2}.$$

(f) $\displaystyle\lim_{\theta \to (\pi/3)^-} = \dfrac{(\cos\theta \sin 3\theta)(\sin\theta \sin 3\theta)}{(\cos\theta \sin 3\theta)^2 + (\sin\theta \sin 3\theta)^2} = \lim_{\theta \to (\pi/3)^-} \cos\theta \sin\theta = \dfrac{1}{4}\sqrt{3}$

(g) $\displaystyle\lim_{t \to \infty} \dfrac{(1/t)(\sin t)/t}{1/t^2 + (\sin^2 t)/t^2} = \lim_{t \to \infty} \dfrac{\sin t}{1 + \sin^2 t};$ does not exist

25. (a) $\dfrac{\partial g}{\partial x}(0,0) = \displaystyle\lim_{h \to 0} \dfrac{g(h,0) - g(0,0)}{h} = \lim_{h \to 0} 0 = 0,$

$\dfrac{\partial g}{\partial y}(0,0) = \displaystyle\lim_{h \to 0} \dfrac{g(0,h) - g(0,0)}{h} = \lim_{h \to 0} 0 = 0$

(b) as (x,y) tends to $(0,0)$ along the x-axis, $g(x,y) = g(x,0) = 0$ tends to 0;

as (x,y) tends to $(0,0)$ along the line $y = x$, $g(x,y) = g(x,x) = \frac{1}{2}$ tends to $\frac{1}{2}$

27. For $y \neq 0$, $\dfrac{\partial f}{\partial x}(0,y) = \displaystyle\lim_{h \to 0} \dfrac{f(h,y) - f(0,y)}{h} = \lim_{h \to 0} \dfrac{y(y^2 - h^2)}{h^2 + y^2} = y.$

Since $\dfrac{\partial f}{\partial x}(0,0) = \displaystyle\lim_{h \to 0} \dfrac{f(h,0) - f(0,0)}{h} = \lim_{h \to 0} 0 = 0,$

we have $\dfrac{\partial f}{\partial x}(0,y) = y$ for all y.

For $x \neq 0$, $\qquad \dfrac{\partial f}{\partial y}(x,0) = \lim\limits_{h \to 0} \dfrac{f(x,h) - f(x,0)}{h} = \lim\limits_{h \to 0} \dfrac{x(h^2 - x^2)}{x^2 + h^2} = -x.$

Since $\qquad \dfrac{\partial f}{\partial y}(0,0) = \lim\limits_{h \to 0} \dfrac{f(0,h) - f(0,0)}{h} = \lim\limits_{h \to 0} 0 = 0,$

we have $\qquad \dfrac{\partial f}{\partial y}(x,0) = -x \quad$ for all x.

Therefore $\qquad \dfrac{\partial^2 f}{\partial y \partial x}(0,y) = 1 \ $ for all $y \quad$ and $\quad \dfrac{\partial^2 f}{\partial x \partial y}(x,0) = -1$ for all x.

In particular $\qquad \dfrac{\partial^2 f}{\partial y \partial x}(0,0) = 1 \quad$ while $\quad \dfrac{\partial^2 f}{\partial x \partial y}(0,0) = -1.$

CHAPTER 16

SECTION 16.1

1. $\nabla f = e^{xy}[(xy+1)\mathbf{i} + x^2\mathbf{j}]$ 3. $\nabla f = (6x - y)\mathbf{i} + (1-x)\mathbf{j}$

5. $\nabla f = 2xy^{-2}\mathbf{i} - 2x^2y^{-3}\mathbf{j}$

7. $\nabla f = z\cos(x-y)\mathbf{i} - z\cos(x-y)\mathbf{j} + \sin(x-y)\mathbf{k}$

9. $\nabla f = (y+z)\mathbf{i} + (x+z)\mathbf{j} + (x+y)\mathbf{k}$ 11. $\nabla f = e^{x-y}[(1+x+y)\mathbf{i} + (1-x-y)\mathbf{j}]$

13. $\nabla f = e^x[\ln y\,\mathbf{i} + y^{-1}\mathbf{j}]$ 15. $\nabla f = \dfrac{AD - BC}{(Cx + Dy)^2}[y\mathbf{i} - x\mathbf{j}]$

17. $\nabla f = (4x - 3y)\mathbf{i} + (8y - 3x)\mathbf{j}$;   at $(2,3)$, $\nabla f = -\mathbf{i} + 18\mathbf{j}$

19. $\nabla f = -e^{-x}\sin(z+2y)\mathbf{i} + 2e^{-x}\cos(z+2y)\mathbf{j} + e^{-x}\cos(z+2y)\mathbf{k}$;

     at $(0, \pi/4, \pi/4)$, $\nabla f = -\tfrac{1}{2}\sqrt{2}\,(\mathbf{i} + 2\mathbf{j} + \mathbf{k})$

21.   With $r = (x^2 + y^2 + z^2)^{1/2}$ we have

$$\frac{\partial r}{\partial x} = \frac{x}{r}, \quad \frac{\partial r}{\partial y} = \frac{y}{r}, \quad \frac{\partial r}{\partial z} = \frac{z}{r}.$$

    (a)

$$\nabla(\ln r) = \frac{\partial}{\partial x}(\ln r)\mathbf{i} + \frac{\partial}{\partial y}(\ln r)\mathbf{j} + \frac{\partial}{\partial z}(\ln r)\mathbf{k}$$

$$= \frac{1}{r}\frac{\partial r}{\partial x}\mathbf{i} + \frac{1}{r}\frac{\partial r}{\partial y}\mathbf{j} + \frac{1}{r}\frac{\partial r}{\partial z}\mathbf{k}$$

$$= \frac{x}{r^2}\mathbf{i} + \frac{y}{r^2}\mathbf{j} + \frac{z}{r^2}\mathbf{k} = \frac{\mathbf{r}}{r^2}$$

    (b)

$$\nabla(\sin r) = \frac{\partial}{\partial x}(\sin r)\mathbf{i} + \frac{\partial}{\partial y}(\sin r)\mathbf{j} + \frac{\partial}{\partial z}(\sin r)\mathbf{k}$$

$$= \cos r\frac{\partial r}{\partial x}\mathbf{i} + \cos r\frac{\partial r}{\partial y}\mathbf{j} + \cos r\frac{\partial r}{\partial z}\mathbf{k}$$

$$= (\cos r)\frac{x}{r}\mathbf{i} + (\cos r)\frac{y}{r}\mathbf{j} + (\cos r)\frac{z}{r}\mathbf{k}$$

$$= \left(\frac{\cos r}{r}\right)\mathbf{r}$$

    (c) $\nabla e^r = \left(\dfrac{e^r}{r}\right)\mathbf{r}$ [same method as in (a) and (b)]

23. (a) Let $\mathbf{c} = c_1\mathbf{i} + c_2\mathbf{j} + c_3\mathbf{k}$. First, we take $\mathbf{h} = h\mathbf{i}$. Since $\mathbf{c} \cdot \mathbf{h}$ is $o(\mathbf{h})$,

$$0 = \lim_{h \to 0} \frac{\mathbf{c} \cdot \mathbf{h}}{\|\mathbf{h}\|} = \lim_{h \to 0} \frac{c_1 h}{h} = c_1.$$

Similarly, $c_2 = 0$ and $c_3 = 0$.

(b) $(\mathbf{y} - \mathbf{z}) \cdot \mathbf{h} = [f(\mathbf{x} + \mathbf{h}) - f(\mathbf{x}) - \mathbf{z} \cdot \mathbf{h}] + [\mathbf{y} \cdot \mathbf{h} - f(\mathbf{x} + \mathbf{h}) + f(\mathbf{x})] = o(\mathbf{h}) + o(\mathbf{h}) = o(\mathbf{h})$,
so that, by part (a), $\mathbf{y} - \mathbf{z} = \mathbf{0}$.

25. (a) In Section 15.6 we showed that f was not continuous at $(0,0)$. It is therefore not differentiable at $(0,0)$.

(b) For $(x, y) \neq (0, 0)$, $\dfrac{\partial f}{\partial x} = \dfrac{2y(y^2 - x^2)}{(x^2 + y^2)^2}$. As (x, y) tends to $(0, 0)$ along the positive y-axis,

$$\frac{\partial f}{\partial x} = \frac{2y^3}{y^4} = \frac{2}{y} \quad \text{tends to } \infty.$$

SECTION 16.2

1. $\nabla f = 2x\mathbf{i} + 6y\mathbf{j}, \quad \nabla f(1,1) = 2\mathbf{i} + 6\mathbf{j}, \quad \mathbf{u} = \frac{1}{2}\sqrt{2}\,(\mathbf{i} - \mathbf{j})$,

$f'_{\mathbf{u}}(1,1) = \nabla f(1,1) \cdot \mathbf{u} = -2\sqrt{2}$

3. $\nabla f\,(e^y - ye^x)\mathbf{i} + (xe^y - e^x)\mathbf{j}, \quad \nabla f(1,0) = \mathbf{i} + (1 - e)\mathbf{j}, \quad \mathbf{u} = \frac{1}{5}(3\mathbf{i} + 4\mathbf{j})$,

$f'_{\mathbf{u}}(1,0) = \nabla f(1,0) \cdot \mathbf{u} = \frac{1}{5}(7 - 4e)$

5. $\nabla f = \dfrac{(a - b)y}{(x + y)^2}\mathbf{i} + \dfrac{(b - a)x}{(x + y)^2}\mathbf{j}, \quad \nabla f(1,1) = \dfrac{a - b}{4}(\mathbf{i} - \mathbf{j}), \quad \mathbf{u} = \dfrac{1}{2}\sqrt{2}\,(\mathbf{i} - \mathbf{j})$,

$f'_{\mathbf{u}}(1,1) = \nabla f(1,1) \cdot \mathbf{u} = \frac{1}{4}\sqrt{2}\,(a - b)$

7. $\nabla f = (y + z)\mathbf{i} + (x + z)\mathbf{j} + (y + x)\mathbf{k}, \quad \nabla f(1,-1,1) = 2\mathbf{j}, \quad \mathbf{u} = \frac{1}{6}\sqrt{6}\,(\mathbf{i} + 2\mathbf{j} + \mathbf{k})$,

$f'_{\mathbf{u}}(1,-1,1) = \nabla f(1,-1,1) \cdot \mathbf{u} = \frac{2}{3}\sqrt{6}$

9. $\nabla f = 2\left(x + y^2 + z^2\right)\left(\mathbf{i} + 2y\mathbf{j} + 3z^2\mathbf{k}\right), \quad \nabla f(1,-1,1) = 6(\mathbf{i} - 2\mathbf{j} + 3\mathbf{k}), \quad \mathbf{u} = \frac{1}{2}\sqrt{2}\,(\mathbf{i} + \mathbf{j})$,

$f'_{\mathbf{u}}(1,-1,1) = \nabla f(1,-1,1) \cdot \mathbf{u} = -3\sqrt{2}$

11. $\nabla f = \dfrac{x}{x^2 + y^2}\mathbf{i} + \dfrac{y}{x^2 + y^2}\mathbf{j}, \quad \mathbf{u} = \dfrac{1}{\sqrt{x^2 + y^2}}(-x\mathbf{i} - y\mathbf{j}), \quad f'_{\mathbf{u}}(x,y) = \nabla f \cdot \mathbf{u} = -\dfrac{1}{\sqrt{x^2 + y^2}}$

13. $\nabla f = (2Ax + 2By)\mathbf{i} + (2Bx + 2Cy)\mathbf{j}, \quad \nabla f(a,b) = (2aA + 2bB)\mathbf{i} + (2aB + 2bC)\mathbf{j}$

  (a) $\mathbf{u} = \frac{1}{2}\sqrt{2}(-\mathbf{i} + \mathbf{j}), \quad f'_{\mathbf{u}}(a,b) = \nabla f(a,b) \cdot \mathbf{u} = \sqrt{2}\,[a(B-A) + b(C-B)]$

  (b) $\mathbf{u} = \frac{1}{2}\sqrt{2}(\mathbf{i} - \mathbf{j}), \quad f'_{\mathbf{u}}(a,b) = \nabla f(a,b) \cdot \mathbf{u} = \sqrt{2}\,[a(A-B) + b(B-C)]$

15. $\nabla f = e^{y^2 - z^2}(\mathbf{i} + 2xy\mathbf{j} - 2xz\mathbf{k}), \quad \nabla f(1,2,-2) = \mathbf{i} + 4\mathbf{j} + 4\mathbf{k}, \quad \mathbf{r}'(t) = \mathbf{i} - 2\sin(t-1)\mathbf{j} - 2e^{t-1}\mathbf{k},$

  at $(1,2,-2)$ $t = 1, \quad \mathbf{r}'(1) = \mathbf{i} - 2\mathbf{k}, \quad \mathbf{u} = \frac{1}{5}\sqrt{5}\,(\mathbf{i} - 2\mathbf{k}), \quad f'_{\mathbf{u}}(1,2,-2) = \nabla f(1,2,-2) \cdot \mathbf{u} = -\frac{7}{5}\sqrt{5}$

17. $\nabla f = f'(x_0)\mathbf{i}$. If $f'(x_0) \neq 0$, the gradient points in the direction in which f increases: to the right if $f'(x_0) > 0$, to the left if $f'(x_0) < 0$.

19. (a) $\displaystyle\lim_{h \to 0} \frac{f(h,0) - f(0,0)}{h} = \lim_{h \to 0} \frac{\sqrt{h^2}}{h} = \lim_{h \to 0} \frac{|h|}{h}$ does not exist

  (b) no; by Theorem 16.2.5 f cannot be differentiable at $(0,0)$

21. $\nabla \lambda(x,y) = -\frac{8}{3}x\mathbf{i} - 6y\mathbf{j}$

  (a) $\nabla \lambda(1,-1) = -\frac{8}{3}\mathbf{i} = 6\mathbf{j}, \quad \mathbf{u} = \dfrac{-\nabla\lambda(1,-1)}{\|\nabla\lambda(1,-1)\|} = \dfrac{\frac{8}{3}\mathbf{i} - 6\mathbf{j}}{\frac{2}{3}\sqrt{97}}, \quad \lambda'_{\mathbf{u}}(1,-1) = \nabla\lambda(1,-1) \cdot \mathbf{u} = -\frac{2}{3}\sqrt{97}$

  (b) $\mathbf{u} = \mathbf{i}, \quad \lambda'_{\mathbf{u}}(1,2) = \nabla\lambda(1,2) \cdot \mathbf{u} = \left(-\frac{8}{3}\mathbf{i} - 12\mathbf{j}\right) \cdot \mathbf{i} = -\frac{8}{3}$

  (c) $\mathbf{u} = \frac{1}{2}\sqrt{2}\,(\mathbf{i} + \mathbf{j}), \quad \lambda'_{\mathbf{u}}(2,2) = \nabla\lambda(2,2) \cdot \mathbf{u} = \left(-\frac{16}{3}\mathbf{i} - 12\mathbf{j}\right) \cdot \left[\frac{1}{2}\sqrt{2}\,(\mathbf{i} + \mathbf{j})\right] = -\frac{26}{3}\sqrt{2}$

23. (a) The projection of the path onto the xy-plane is the curve

$$C\colon \mathbf{r}(t) = x(t)\mathbf{i} + y(t)\mathbf{j}$$

which begins at $(1,1)$ and at each point has its tangent vector in the direction of $-\nabla f$. Since

$$\nabla f = 2x\mathbf{i} + 6y\mathbf{j},$$

we have the initial-value problems

$$x'(t) = -2x(t), \quad x(0) = 1 \qquad \text{and} \qquad y'(t) = -6y(t), \quad y(0) = 1.$$

From Theorem 7.6.1 we find that

$$x(t) = e^{-2t} \qquad \text{and} \qquad y(t) = e^{-6t}.$$

Eliminating the parameter t, we find that C is the curve $y = x^3$ from $(1,1)$ to $(0,0)$.

  (b) Here

$$x'(t) = -2x(t), \quad x(0) = 1 \qquad \text{and} \qquad y'(t) = -6y(t), \quad y(0) = -2$$

so that

$$x(t) = e^{-2t} \qquad \text{and} \qquad y(t) = -2e^{-6t}.$$

Eliminating the parameter t, we find that the projection of the path onto the xy-plane is the curve $y = -2x^3$ from $(1, -2)$ to $(0, 0)$.

25.   The projection of the path onto the xy-plane is the curve

$$C: \mathbf{r}(t) = x(t)\mathbf{i} + y(t)\mathbf{j}$$

which begins at (a^2, b^2) and at each point has its tangent vector in the direction of

$-\nabla f = -\left(2a^2 x \mathbf{i} + 2b^2 y \mathbf{j}\right)$.   Thus,

$$x'(t) = -2a^2 x(t), \quad x(0) = a^2 \qquad \text{and} \qquad y'(t) = -2b^2 y(t), \quad y(0) = b^2$$

so that

$$x(t) = a^2 e^{-2a^2 t} \qquad \text{and} \qquad y(t) = b^2 e^{-2b^2 t}.$$

Since

$$\left[\frac{x}{a^2}\right]^{b^2} = \left(e^{-2a^2 t}\right)^{b^2} = \left[\frac{y}{b^2}\right]^{a^2},$$

C is the curve $\left(b^2\right)^{a^2} x^{b^2} = \left(a^2\right)^{b^2} y^{a^2}$ from (a^2, b^2) to $(0, 0)$.

27.   We want the curve

$$C: \mathbf{r}(t) = x(t)\mathbf{i} + y(t)\mathbf{j}$$

which begins at $(\pi/4, 0)$ and at each point has its tangent vector in the direction of

$$\nabla T = -\sqrt{2}\, e^{-y} \sin x \, \mathbf{i} - \sqrt{2}\, e^{-y} \cos x \, \mathbf{j}.$$

From

$$x'(t) = -\sqrt{2}\, e^{-y} \sin x \qquad \text{and} \qquad y'(t) = -\sqrt{2}\, e^{-y} \cos x$$

we obtain

$$\frac{dy}{dx} = \frac{y'(t)}{x'(t)} = \cot x$$

so that

$$y = \ln|\sin x| + C.$$

Since $y = 0$ when $x = \pi/4$, we get $C = \ln \sqrt{2}$ and $y = \ln|\sqrt{2} \sin x|$. As $\nabla T(\pi/4, 0) = -\mathbf{i} - \mathbf{j}$, the curve $y = \ln|\sqrt{2} \sin x|$ is followed in the direction of decreasing x.

29. (a) $\displaystyle \lim_{h \to 0} \frac{f(2+h, (2+h)^2) - f(2, 4)}{h} = \lim_{h \to 0} \frac{3(2+h)^2 + (2+h)^2 - 16}{h}$

$$= \lim_{h \to 0} 4 \left[\frac{4h + h^2}{h}\right] = \lim_{h \to 0} 4(4 + h) = 16$$

(b)
$$\lim_{h \to 0} \frac{f\left(\dfrac{h+8}{4}, 4+h\right) - f(2,4)}{h} = \lim_{h \to 0} \frac{3\left(\dfrac{h+8}{4}\right)^2 + (4+h) - 16}{h}$$

$$= \lim_{h \to 0} \frac{\frac{3}{16}h^2 + 3h + 12 + 4 + h - 16}{h}$$

$$= \lim_{h \to 0} \left(\tfrac{3}{16}h + 4\right) = 4$$

(c) $\mathbf{u} = \frac{1}{17}\sqrt{17}\,(\mathbf{i} + 4\mathbf{j}), \quad \nabla f(2,4) = 12\mathbf{i} + \mathbf{j}; \quad f'_{\mathbf{u}}(2,4) = \nabla f(2,4) \cdot \mathbf{u} = \frac{16}{17}\sqrt{17}$

(d)   The limits computed in (a) and (b) are not directional derivatives. In (a) and (b) we have, in essence, computed $\nabla f(2,4) \cdot \mathbf{r}_0$ taking $\mathbf{r}_0 = \mathbf{i} + 4\mathbf{j}$ in (a) and $\mathbf{r}_0 = \frac{1}{4}\mathbf{i} + \mathbf{j}$ in (b). In neither case is $\mathbf{r}_0$ a unit vector.

SECTION 16.3

1.   (a) $f(x, y, z) = a_1 x + a_2 y + a_3 z + C$ (b) $f(x, y, z) = g(x, y, z) + a_1 x + a_2 y + a_3 z + C$

3.   (a) U is not connected

     (b)   (i) $g(\mathbf{x}) = f(\mathbf{x}) - 1$ (ii) $g(\mathbf{x}) = -f(\mathbf{x})$

     (c) U is not connected

5.   Since f is continuous at $\mathbf{a}$ and $f(\mathbf{a}) = A$, there exists $\delta > 0$ such that

$$\text{if} \quad \|\mathbf{x} - \mathbf{a}\| < \delta \quad \text{and} \quad x \in \Omega, \quad \text{then} \quad |f(\mathbf{x}) - A| < \epsilon.$$

Whether $\mathbf{a}$ is on the boundary of Ω or in the interior of Ω, there exists $\mathbf{x}_1$ in the interior of Ω within δ of $\mathbf{a}$. That implies that $|f(\mathbf{x}_1) - A| < \epsilon$ and thus that $f(\mathbf{x}_1) < A + \epsilon$. A similar argument shows the existence of $\mathbf{x}_2$ with the desired property.

SECTION 16.4

1. $\nabla f = 2xy\mathbf{i} + x^2\mathbf{j};$

     $\nabla f(\mathbf{r}(t)) \cdot \mathbf{r}'(t) = \left(2\mathbf{i} + e^{2t}\mathbf{j}\right) \cdot \left(e^t\mathbf{i} - e^{-t}\mathbf{j}\right) = e^t$

3. $\nabla f = y\mathbf{i} + (x - z)\mathbf{j} - y\mathbf{k};$

     $\nabla f(\mathbf{r}(t)) \cdot \mathbf{r}'(t) = \left(t^2\mathbf{i} + \left(t - t^3\right)\mathbf{j} - t^2\mathbf{k}\right) \cdot \left(\mathbf{i} + 2t\mathbf{j} + 3t^2\mathbf{k}\right) = 3t^2 - 5t^4$

5. $\nabla f = 2x\mathbf{i} + 2y\mathbf{j} + \mathbf{k};$

     $\nabla f(\mathbf{r}(t)) \cdot \mathbf{r}'(t) = (2a\cos\omega t\,\mathbf{i} + 2b\sin\omega t\,\mathbf{j} + \mathbf{k}) \cdot (-a\omega\sin\omega t\,\mathbf{i} + b\omega\cos\omega t\,\mathbf{j} + b\omega\mathbf{k})$

     $\qquad\qquad = 2\omega\left(b^2 - a^2\right)\sin\omega t\,\cos\omega t + b\omega$

7. $\dfrac{du}{dt} = \dfrac{\partial u}{\partial x}\dfrac{dx}{dt} + \dfrac{\partial u}{\partial y}\dfrac{dy}{dt} = (2x - 3y)(-\sin t) + (4y - 3x)(\cos t)$

$\qquad = 2\cos t \sin t + 3\sin^2 t - 3\cos^2 t = \sin 2t - 3\cos 2t$

9. $\dfrac{du}{dt} = \dfrac{\partial u}{\partial x}\dfrac{dx}{dt} + \dfrac{\partial u}{\partial y}\dfrac{dy}{dt}$

$\qquad = (e^x \sin y + e^y \cos x)\left(\tfrac{1}{2}\right) + (e^x \cos y + e^y \sin x)(2)$

$\qquad = e^{t/2}\left(\tfrac{1}{2}\sin 2t + 2\cos 2t\right) + e^{2t}\left(\tfrac{1}{2}\cos \tfrac{1}{2}t + 2\sin \tfrac{1}{2}t\right)$

11. $\dfrac{du}{dt} = \dfrac{\partial u}{\partial x}\dfrac{dx}{dt} + \dfrac{\partial u}{\partial y}\dfrac{dy}{dt} + \dfrac{\partial u}{\partial z}\dfrac{dz}{dt}$

$\qquad = (y + z)(2t) + (x + z)(1 - 2t) + (y + x)(2t - 2)$

$\qquad = (1 - t)(2t) + (2t^2 - 2t + 1)(1 - 2t) + t(2t - 2)$

$\qquad = 1 - 4t + 6t^2 - 4t^3$

13. $V = \dfrac{1}{3}\pi r^2 h, \quad \dfrac{dV}{dt} = \dfrac{\partial V}{\partial r}\dfrac{dr}{dt} + \dfrac{\partial V}{\partial h}\dfrac{dh}{dt} = \left(\dfrac{2}{3}\pi rh\right)\dfrac{dr}{dt} + \left(\dfrac{1}{3}\pi r^2\right)\dfrac{dh}{dt}.$

At the given instant,

$$\dfrac{dV}{dt} = \dfrac{2}{3}\pi(280)(3) + \dfrac{1}{3}\pi(196)(-2) = \dfrac{1288}{3}\pi.$$

The volume is increasing at the rate of $\dfrac{1288}{3}\pi$ in.^{3}/ sec .

15. $\dfrac{\partial u}{\partial s} = \dfrac{\partial u}{\partial x}\dfrac{\partial x}{\partial s} + \dfrac{\partial u}{\partial y}\dfrac{\partial y}{\partial s} = (2x - y)(\cos t) + (-x)(t\cos s)$

$\qquad = 2s\cos^2 t - t\sin s \cos t - st\cos s \cos t$

$\quad \dfrac{\partial u}{\partial t} = \dfrac{\partial u}{\partial x}\dfrac{\partial x}{\partial t} + \dfrac{\partial u}{\partial y}\dfrac{\partial y}{\partial t} = (2x - y)(-s\sin t) + (-x)(\sin s)$

$\qquad = -2s^2 \cos t \sin t + st\sin s \sin t - s\cos t \sin s$

17. $\dfrac{\partial u}{\partial s} = \dfrac{\partial u}{\partial x}\dfrac{\partial x}{\partial s} + \dfrac{\partial u}{\partial y}\dfrac{\partial y}{\partial s} + \dfrac{\partial u}{\partial z}\dfrac{\partial z}{\partial s}$

$\qquad = (2x - y)(\cos t) + (-x)(-\cos (t - s)) + 2z(t\cos s)$

$\qquad = 2s\cos^2 t - \sin (t - s)\cos t + s\cos t \cos (t - s) + 2t^2 \sin s \cos s$

$\quad \dfrac{\partial u}{\partial t} = \dfrac{\partial u}{\partial x}\dfrac{\partial x}{\partial t} + \dfrac{\partial u}{\partial y}\dfrac{\partial y}{\partial t} + \dfrac{\partial u}{\partial z}\dfrac{\partial z}{\partial t}$

$\qquad = (2x - y)(-s\sin t) + (-x)(\cos (t - s)) + 2z(\sin s)$

$\qquad = -2s^2 \cos t \sin t + s\sin (t - s)\sin t - s\cos t \cos (t - s) + 2t\sin^2 s$

19.
$$\frac{d}{dt}[f(\mathbf{r}(t))] = \left[\boldsymbol{\nabla} f(\mathbf{r}(t)) \cdot \frac{\mathbf{r}'(t)}{\|\mathbf{r}'(t)\|}\right]\|\mathbf{r}'(t)\|$$

$$= f'_{\mathbf{u}(t)}(\mathbf{r}(t))\|\mathbf{r}'(t)\| \quad \text{where} \quad \mathbf{u}(t) = \frac{\mathbf{r}'(t)}{\|\mathbf{r}'(t)\|}$$

21. (a) $(\cos r)\dfrac{\mathbf{r}}{r}$ (b) $(r\cos r + \sin r)\dfrac{\mathbf{r}}{r}$

23. (a) $(r\cos r - \sin r)\dfrac{\mathbf{r}}{r^3}$ (b) $\left(\dfrac{\sin r - r\cos r}{\sin^2 r}\right)\dfrac{\mathbf{r}}{r}$

25. (a) $\dfrac{\partial u}{\partial r} = \dfrac{\partial u}{\partial x}\dfrac{\partial x}{\partial r} + \dfrac{\partial u}{\partial y}\dfrac{\partial y}{\partial r} = \dfrac{\partial u}{\partial x}\cos\theta + \dfrac{\partial u}{\partial y}\sin\theta$

$\qquad\quad \dfrac{\partial u}{\partial \theta} = \dfrac{\partial u}{\partial x}\dfrac{\partial x}{\partial \theta} + \dfrac{\partial u}{\partial y}\dfrac{\partial y}{\partial \theta} = \dfrac{\partial u}{\partial x}(-r\sin\theta) + \dfrac{\partial u}{\partial y}(r\cos\theta)$

(b) $\left(\dfrac{\partial u}{\partial r}\right)^2 = \left(\dfrac{\partial u}{\partial x}\right)^2\cos^2\theta + 2\dfrac{\partial u}{\partial x}\dfrac{\partial u}{\partial y}\cos\theta\sin\theta + \left(\dfrac{\partial u}{\partial y}\right)^2\sin^2\theta,$

$\qquad \dfrac{1}{r^2}\left(\dfrac{\partial u}{\partial \theta}\right)^2 = \left(\dfrac{\partial u}{\partial x}\right)^2\sin^2\theta - 2\dfrac{\partial u}{\partial x}\dfrac{\partial u}{\partial y}\cos\theta\sin\theta + \left(\dfrac{\partial u}{\partial y}\right)^2\cos^2\theta,$

$\qquad \left(\dfrac{\partial u}{\partial r}\right)^2 + \dfrac{1}{r^2}\left(\dfrac{\partial u}{\partial \theta}\right)^2 = \left(\dfrac{\partial u}{\partial x}\right)^2\left(\cos^2\theta + \sin^2\theta\right) + \left(\dfrac{\partial u}{\partial y}\right)^2\left(\sin^2\theta + \cos^2\theta\right) = \left(\dfrac{\partial u}{\partial x}\right)^2 + \left(\dfrac{\partial u}{\partial y}\right)^2$

27. Set $u = z^4 + x^2 z^3 + y^2 + xy - 2.$ Then

$$\frac{\partial u}{\partial z} = 4z^3 + 3x^2 z^2 = z^2(4z + 3x^2), \qquad \frac{\partial z}{\partial x} = -\frac{\partial u/\partial x}{\partial u/\partial z} = -\frac{2xz^3 + y}{z^2(4z + 3x^2)},$$

$$\frac{\partial z}{\partial y} = -\frac{\partial u/\partial y}{\partial u/\partial z} = -\frac{2y + x}{z^2(4z + 3x^2)}.$$

29. (a)

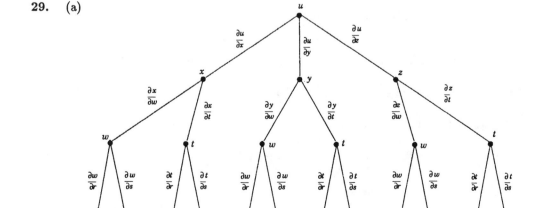

(b) $\quad \dfrac{\partial u}{\partial r} = \dfrac{\partial u}{\partial x}\left(\dfrac{\partial x}{\partial w}\dfrac{\partial w}{\partial r} + \dfrac{\partial x}{\partial t}\dfrac{\partial t}{\partial r}\right) + \dfrac{\partial u}{\partial y}\left(\dfrac{\partial y}{\partial w}\dfrac{\partial w}{\partial r} + \dfrac{\partial y}{\partial t}\dfrac{\partial t}{\partial r}\right) + \dfrac{\partial u}{\partial z}\left(\dfrac{\partial z}{\partial w}\dfrac{\partial w}{\partial r} + \dfrac{\partial z}{\partial t}\dfrac{\partial t}{\partial r}\right).$

To obtain $\partial u/\partial s$, replace each r by s.

31. (a) We write $u = u(x,y); \quad x = t, \ y = y(t).$

By (16.4.3)

$$\frac{du}{dt} = \frac{\partial u}{\partial x}\frac{dx}{dt} + \frac{\partial u}{\partial y}\frac{dy}{dt}.$$

Since

$$u(t, y(t)) = 0 \quad \text{for all } t, \quad \frac{du}{dt} = 0.$$

Since

$$x = t, \qquad \frac{dx}{dt} = 1 \quad \text{and} \quad \frac{dy}{dt} = \frac{dy}{dx}.$$

Thus we have

$$\frac{\partial u}{\partial x} + \frac{\partial u}{\partial y}\frac{dy}{dx} = 0.$$

Since

$$\partial u/\partial y \neq 0, \quad \frac{dy}{dx} = -\frac{\partial u/\partial x}{\partial u/\partial y}.$$

(b) $\quad \dfrac{dy}{dx} = -\dfrac{2xy^4 + \cos(x+y)}{4x^2 y^3 + \cos(x+y)}$

33. $\quad \dfrac{\partial u}{\partial s} = \dfrac{\partial u}{\partial x}\dfrac{\partial x}{\partial s} + \dfrac{\partial u}{\partial y}\dfrac{\partial y}{\partial s}, \quad \dfrac{\partial u}{\partial t} = \dfrac{\partial u}{\partial x}\dfrac{\partial x}{\partial t} + \dfrac{\partial u}{\partial y}\dfrac{\partial y}{\partial t}$

SECTION 16.5

1. Set $f(x,y) = x^2 + xy + y^2.$ Then,

$$\nabla f = (2x+y)\mathbf{i} + (x+2y)\mathbf{j}, \quad \nabla f(-1,-1) = -3\mathbf{i} - 3\mathbf{j}.$$

normal vector $\mathbf{i} + \mathbf{j}$; tangent vector $\mathbf{i} - \mathbf{j}$

tangent line $x + y + 2 = 0$; normal line $x - y = 0$

3. Set $f(x,y) = \left(x^2 + y^2\right)^2 - 9\left(x^2 - y^2\right).$ Then,

$$\nabla f = [4x(x^2+y^2) - 18x]\mathbf{i} + \left[4y\left(x^2+y^2\right) + 18y\right]\mathbf{j}, \quad \nabla f\left(\sqrt{2},1\right) = -6\sqrt{2}\,\mathbf{i} + 30\mathbf{j}.$$

normal vector $\sqrt{2}\,\mathbf{i} - 5\mathbf{j}$; tangent vector $5\mathbf{i} + \sqrt{2}\,\mathbf{j}$

tangent line $\sqrt{2}x - 5y + 3 = 0$; normal line $5x + \sqrt{2}\,y - 6\sqrt{2} = 0$

5. Set $f(x, y) = xy^2 - 2x^2 + y + 5x$.   Then,

$$\nabla f = (y^2 - 4x + 5)\mathbf{i} + (2xy + 1)\mathbf{j}, \quad \nabla f(4, 2) = -7\mathbf{i} + 17\mathbf{j}.$$

normal vector $7\mathbf{i} - 17\mathbf{j}$; tangent vector $17\mathbf{i} + 7\mathbf{j}$

tangent line $7x - 17y + 6 = 0$; normal line $17x + 7y - 82 = 0$

7. Set $f(x, y) = 2x^3 - x^2y^2 - 3x + y$.   Then,

$$\nabla f = (6x^2 - 2xy^2 - 3)\mathbf{i} + (-2x^2y + 1)\mathbf{j}, \quad \nabla f(1, -2) = -5\mathbf{i} + 5\mathbf{j}.$$

normal vector $\mathbf{i} - \mathbf{j}$; tangent vector $\mathbf{i} + \mathbf{j}$

tangent line $x - y - 3 = 0$; normal line $x + y + 1 = 0$

9. Set $f(x, y) = x^2y + a^2y$.   By (16.5.4)

$$m = -\frac{\partial f/\partial x}{\partial f/\partial y} = -\frac{2xy}{x^2 + a^2}.$$

At $(0, a)$ the slope is 0.

11. Set $f(x, y, z) = x^3 + y^3 - 3xyz$.   Then,

$$\nabla f = (3x^2 - 3yz)\mathbf{i} + (3y^2 - 3xz)\mathbf{j} - 3xy\mathbf{k}, \quad \nabla f\left(1, 2, \tfrac{3}{2}\right) = -6\mathbf{i} + \tfrac{15}{2}\mathbf{j} - 6\mathbf{k};$$

tangent plane at $\left(1, 2, \tfrac{3}{2}\right)$: $-6(x - 1) + \tfrac{15}{2}(y - 2) - 6\left(z - \tfrac{3}{2}\right) = 0$, which reduces to $4x - 5y + 4z = 0$.

13. Set $z = g(x, y) = axy$.   Then, $\nabla g = ay\mathbf{i} + ax\mathbf{j}$, $\nabla g\left(1, \dfrac{1}{a}\right) = \mathbf{i} + a\mathbf{j}$.

tangent plane at $\left(1, \dfrac{1}{a}, 1\right)$: $z - 1 = 1(x - 1) + a\left(y - \dfrac{1}{a}\right)$, which reduces to $x + ay - z - 1 = 0$

15. Set $z = g(x, y) = \sin x + \sin y + \sin(x + y)$.   Then,

$$\nabla g = [\cos x + \cos(x + y)]\mathbf{i} + [\cos y + \cos(x + y)]\mathbf{j}, \quad \nabla g(0, 0) = 2\mathbf{i} + 2\mathbf{j};$$

tangent plane at $(0, 0, 0)$: $z - 0 = 2(x - 0) + 2(y - 0)$, $\quad 2x + 2y - z = 0$.

17. Set $f(x, y, z) = b^2c^2x^2 - a^2c^2y^2 - a^2b^2z^2$.   Then,

$$\nabla f(x_0, y_0, z_0) = 2b^2c^2x_0\mathbf{i} - 2a^2c^2y_0\mathbf{j} - 2a^2b^2z_0\mathbf{k};$$

tangent plane at (x_0, y_0, z_0):

$$2b^2c^2x_0(x - x_0) - 2a^2c^2y_0(y - y_0) - 2a^2b^2z_0(z - z_0) = 0,$$

which can be rewritten as follows:

$$b^2c^2x_0x - a^2c^2y_0y - a^2b^2z_0z = b^2c^2x_0{}^2 - a^2c^2y_0{}^2 - a^2b^2z_0{}^2 = f(x_0, y_0, z_0) = a^2 + b^2 + c^2.$$

19. Set $z = g(x, y) = xy + a^3x^{-1} + b^3y^{-1}$.

$$\nabla g = (y - a^3x^{-2})\mathbf{i} + (x - b^3y^{-2})\mathbf{j}, \qquad \nabla g = 0 \implies y = a^3x^{-2} \quad \text{and} \quad x = b^3y^{-2}.$$

Thus,

$$y = a^3b^{-6}y^4, \quad y^3 = b^6a^{-3}, \quad y = b^2/a, \quad x = b^3y^{-2} = a^2/b \quad \text{and} \quad g(a^2/b, \ b^2/a) = 3ab.$$

The tangent plane is horizontal at $(a^2/b, \ b^2/a, \ 3ab)$.

21. Set $z = g(x, y) = xy$.   Then, $\nabla g = y\mathbf{i} + x\mathbf{j}$.

$$\nabla g = 0 \implies x = y = 0.$$

The tangent plane is horizontal at $(0, 0; 0)$.

23. Set $z = g(x, y) = 2x^2 + 2xy - y^2 - 5x + 3y - 2$.   Then,

$$\nabla g = (4x + 2y - 5)\mathbf{i} + (2x - 2y + 3)\mathbf{j}.$$

$$\nabla g = 0 \implies 4x + 2y - 5 = 0 = 2x - 2y + 3 \implies x = \tfrac{1}{3}, \quad y = \tfrac{11}{6}.$$

The tangent plane is horizontal at $\left(\tfrac{1}{3}, \ \tfrac{11}{6}, \ -\tfrac{1}{12}\right)$.

25.
$$\frac{x - x_0}{(\partial f/\partial x)(x_0, y_0, z_0)} = \frac{y - y_0}{(\partial f/\partial y)(x_0, y_0, z_0)} = \frac{z - z_0}{(\partial f/\partial z)(x_0, y_0, z_0)}$$

27. Since the tangent planes meet at right angles, the normals ∇F and ∇G meet at right angles:

$$\frac{\partial F}{\partial x}\frac{\partial G}{\partial x} + \frac{\partial F}{\partial y}\frac{\partial G}{\partial y} + \frac{\partial F}{\partial z}\frac{\partial G}{\partial z} = 0.$$

29. The tangent plane at an arbitrary point (x_0, y_0, z_0) has equation

$$y_0z_0(x - x_0) + x_0z_0(y - y_0) + x_0y_0(z - z_0) = 0,$$

which simplifies to

$$y_0z_0x + x_0z_0y + x_0y_0z = 3x_0y_0z_0 \quad \text{and thus to} \quad \frac{x}{3x_0} + \frac{y}{3y_0} + \frac{z}{3z_0} = 1.$$

The volume of the pyramid is

$$V = \frac{1}{3}Bh = \frac{1}{3}\left[\frac{(3x_0)(3y_0)}{2}\right](3z_0) = \frac{9}{2}x_0y_0z_0 = \frac{9}{2}a^3.$$

31.   The point $(2, 3, -2)$ is the tip of $\mathbf{r}(1)$.

Since $\mathbf{r}'(t) = 2\mathbf{i} - \dfrac{3}{t^2}\mathbf{j} - 4t\mathbf{k}$,   we have $\mathbf{r}'(1) = 2\mathbf{i} - 3\mathbf{j} - 4\mathbf{k}$.

Now set $f(x, y, z) = x^2 + y^2 + 3z^2 - 25$.   The function has gradient $2x\mathbf{i} + 2y\mathbf{j} + 6z\mathbf{k}$.

At the point $(2, 3, -2)$,

$$\nabla f = 2(2\mathbf{i} + 3\mathbf{j} - 6\mathbf{k}).$$

The angle θ between $\mathbf{r}'(1)$ and the gradient gives

$$\cos\theta = \frac{(2\mathbf{i} - 3\mathbf{j} - 4\mathbf{k})}{\sqrt{29}} \cdot \frac{(2\mathbf{i} + 3\mathbf{j} - 6\mathbf{k})}{7} = \frac{19}{7\sqrt{29}} \cong 0.504.$$

Therefore $\theta \cong 1.043$ radians. The angle between the curve and the plane is

$$\frac{\pi}{2} - \theta \cong 1.571 - 1.043 \cong 0.528 \text{ radians.}$$

33.   Set $f(x, y, z) = x^2 y^2 + 2x + z^3$.   Then,

$$\nabla f = (2xy^2 + 2)\mathbf{i} + 2x^2 y\mathbf{j} + 3z^2\mathbf{k}, \quad \nabla f(2, 1, 2) = 6\mathbf{i} + 8\mathbf{j} + 12\mathbf{k}.$$

The plane tangent to $f(x, y, z) = 16$ at $(2, 1, 2)$ has equation

$$6(x - 2) + 8(y - 1) + 12(z - 2) = 0, \quad 3x + 4y + 6z = 22.$$

Next, set $g(x, y, z) = 3x^2 + y^2 - 2z$.   Then,

$$\nabla g = 6x\mathbf{i} + 2y\mathbf{j} - 2\mathbf{k}, \quad \nabla g(2, 1, 2) = 12\mathbf{i} + 2\mathbf{j} - 2\mathbf{k}.$$

The plane tangent to $g(x, y, z) = 9$ at $(2, 1, 2)$ is

$$12(x - 2) + 2(y - 1) - 2(z - 2) = 0, \quad 6x + y - z = 11.$$

35.   The gradient to the sphere at $(1, 1, 2)$ is

$$2x\mathbf{i} + (2y - 4)\mathbf{j} + (2z - 2)\mathbf{k} = 2\mathbf{i} - 2\mathbf{j} + 2\mathbf{k}.$$

The gradient to the paraboloid at $(1, 1, 2)$ is

$$6x\mathbf{i} + 4y\mathbf{j} - 2\mathbf{k} = 6\mathbf{i} + 4\mathbf{j} - 2\mathbf{k}.$$

Since

$$(2\mathbf{i} - 2\mathbf{j} + 2\mathbf{k}) \cdot (6\mathbf{i} + 4\mathbf{j} - 2\mathbf{k}) = 0,$$

the surfaces intersect at right angles.

37. (a) $3x + 4y + 6 = 0$ since plane p is vertical.   (b) $y = -\frac{1}{4}(3x + 6) = -\frac{1}{4}[3(4t - 2) + 6] = -3t$

$z = x^2 + 3y^2 + 2 = (4t - 2)^2 + 3(-3t)^2 + 2 = 43t^2 - 16t + 6$

$\mathbf{r}(t) = (4t - 2)\mathbf{i} - 3t\mathbf{j} + (43t^2 - 16t + 6)\mathbf{k}$

(c) From part (b) the tip of $\mathbf{r}(1)$ is $(2, -3, 33)$.   We take

$\mathbf{r}'(1) = 4\mathbf{i} - 3\mathbf{j} + 70\mathbf{j}$ as $\mathbf{d}$ to write

$$\mathbf{R}(s) = (2\mathbf{i} - 3\mathbf{j} + 33\mathbf{k}) + s(4\mathbf{i} - 3\mathbf{j} + 70\mathbf{k}).$$

(d) Set $g(x, y) = x^2 + 3y^2 + 2$.   Then,

$$\nabla g = 2x\mathbf{i} + 6y\mathbf{j} \quad \text{and} \quad \nabla g(2, -3) = 4\mathbf{i} - 18\mathbf{j}.$$

An equation for the plane tangent to $z = g(x, y)$ at $(2, -3, 33)$ is

$$z - 33 = 4(x - 2) - 18(y + 3) \quad \text{which reduces to} \quad 4x - 18y - z = 29.$$

(e) Substituting t for x in the equations for p and p_1, we obtain

$$3t + 4y + 6 = 0 \quad \text{and} \quad 4t - 18y - z = 29.$$

From the first equation

$$y = -\tfrac{3}{4}(t + 2)$$

and then from the second equation

$$z = 4t - 18\left[-\tfrac{3}{4}(t + 2)\right] - 29 = \tfrac{35}{2}t - 2.$$

Thus,

$$(*) \quad \mathbf{r}(t) = t\mathbf{i} - (\tfrac{3}{4}t + \tfrac{3}{2})\mathbf{j} + (\tfrac{35}{2}t - 2)\mathbf{k}.$$

Lines l and l' are the same. To see this, consider how l and l' are formed; to assure yourself, replace t in $(*)$ by $4s + 2$ to obtain $\mathbf{R}(s)$ found in part (c).

SECTION 16.6

1. $\nabla f(x, y) = (2 - 2x)\mathbf{i} - 2y\mathbf{j} = \mathbf{0}$ only at $(1, 0)$.

The difference

$$f(1 + h, k) - f(1, 0) = \left[2(1 + h) - (1 + h)^2 - k^2\right] - 1 = -h^2 - k^2$$

is negative for all small h and k; there is a local maximum of 1 at $(1, 0)$.

3. $\nabla f(x, y) = (2 - 2x)\mathbf{i} + (2 + 2y)\mathbf{j} = \mathbf{0}$ only at $(1, -1)$.

The difference

$$f(1 + h, -1 + k) - f(1, -1)$$

$$= [2(1 + h) + 2(-1 + k) - (1 + h)^2 + (-1 + k)^2 + 5] - 5 = -h^2 + k^2$$

does not keep a constant sign for small h and k; $(1, -1)$ is a saddle point.

5. $\nabla f(x,y) = (2x + y + 3)\mathbf{i} + (x + 2y)\mathbf{j} = \mathbf{0}$ only at $(-2,1)$.

The difference

$f(-2 + h,\ 1 + k) - f(-2,1)$

$\qquad = [(-2 + h)^2 + (-2 + h)(1 + k) + (1 + k)^2 + 3(-2 + h) + 1] - (-2) = h^2 + hk + k^2$

is positive for all small h and k. To see this, note that

$$h^2 + hk + k^2 \geq h^2 + k^2 - |h|\,|k| > 0;$$

there is a local minimum of -2 at $(-2,1)$.

7. $\nabla f = (3x^2 - 3)\mathbf{i} + \mathbf{j}$ is never $\mathbf{0}$;   there are no stationary points and no local extreme values.

9. $\nabla f = (2x + y - 3)\mathbf{i} + (x + 2y - 3)\mathbf{j} = \mathbf{0}$ only at $(1,1)$.

The difference

$f(1 + h,\ 1 + k) - f(1,1) = [(1+h)^2 + (1+h)(1+k) + (1+k)^2 - 3(1+h) - 3(1+k)] - (-3) = h^2 + hk + k^2$

is positive for all small h and k. (See solution to Exercise 5 for details.) There is a local minimum of -3 at $(1,1)$.

11. $\nabla f = (2x + y + 2)\mathbf{i} + (x + 2)\mathbf{j} = \mathbf{0}$ only at $(-2,2)$.

The difference

$$f(-2 + h,\ 2 + k) - f(-2,2) = [(-2 + h)^2 + (-2 + h)(2 + k) + 2(-2 + h) + 2(2 + k) + 1] - 1$$

$$= h^2 + hk = h(h + k)$$

does not keep a constant sign for all small h and k. To see this, suppose $h > 0$. Then, $h(h + k)$ is positive when $k > -h$ and negative when $k < -h$. The point $(-2,2)$ is a saddle point.

13. $\nabla f = (12x^2 y - 4y^3)\mathbf{i} + (4x^3 - 12xy^2)\mathbf{j} = \mathbf{0}$ only at $(0,0)$.

The difference

$$f(h,k) - f(0,0) = 4h^3 k - 4hk^3 = 4hk(h^2 - k^2)$$

does not keep a constant sign for all small h and k. To see this, suppose h and k are positive. Then, $4hk(h^2 - k^2)$ is positive when $h > k$ and negative when $h < k$. The origin is a saddle point.

15. $\nabla f = \dfrac{1}{(x^2 + y^2)^{3/2}}(-x\mathbf{i} - y\mathbf{j})$ is never $\mathbf{0}$ on D. Note that $f(x,y)$ is the reciprocal of the distance of (x,y) from the origin. The point of D closest to the origin (draw a figure) is $(1,1)$. Therefore $f(1,1) = 1/\sqrt{2}$ is the maximum value of f. The point of D furthest from the origin is $(3,4)$. Therefore $f(3,4) = 1/5$ is the least value taken on by f.

17. $\nabla f = 8x\mathbf{i} - 18y\mathbf{j} = \mathbf{0}$ only at $(0,0)$.   Since the difference

$$f(h,k) - f(0,0) = 4h^2 - 9k^2$$

does not keep a constant sign for all small h and k, the point $(0,0)$ is a saddle point. It follows that f takes on no extreme values on the interior of D.

The boundary of D consists of the vertical lines $x = -1$ and $x = 1$. Note that

$$f(-1,y) = 4 - 9y^2 \text{ and } f(1,y) = 4 - 9y^2.$$

It is clear then that f takes on a maximum value of 4 [at $(-1,0)$ and $(1,0)$] but no minimum value.

19. $\nabla f = 2(x-1)\mathbf{i} + 2(y-1)\mathbf{j} = \mathbf{0}$ only at $(1,1)$. As the sum of two squares, $f(x,y) \geq 0$. Thus, $f(1,1) = 0$ is a minimum. To examine the behavior of f on the boundary of D, we note that f represents the square of the distance between (x,y) and $(1,1)$. Thus, f is maximal at the point of the boundary furthest from $(1,1)$. This is the point $(-\sqrt{2}, -\sqrt{2})$; the maximum value of f is $f\left(-\sqrt{2}, -\sqrt{2}\right) = 6 + 4\sqrt{2}$.

21. $\nabla f = 2(x-y)\mathbf{i} - 2(x-y)\mathbf{j} = \mathbf{0}$ at each point of the line segment $y = x$ from $(0,0)$ to $(4,4)$. Since $f(x,x) = 0$ and $f(x,y) \geq 0$, f takes on its minimum of 0 at each of these points.

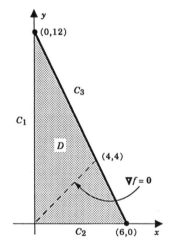

Next we consider the boundary of D.   We parametrize each side of the triangle:

$$C_1: \mathbf{r}_1(t) = t\mathbf{j}, \quad t \in [0,12]$$
$$C_2: \mathbf{r}_2(t) = t\mathbf{i}, \quad t \in [0,6]$$
$$C_3: \mathbf{r}_3(t) = t\mathbf{i} + (12 - 2t)\mathbf{j}, \quad t \in [0,6]$$

and observe from

$$f(\mathbf{r}_1(t)) = t^2, \quad t \in [0,12]$$
$$f(\mathbf{r}_2(t)) = t^2, \quad t \in [0,6]$$
$$f(\mathbf{r}_3(t)) = (3t - 12)^2, \quad t \in [0,6]$$

that f takes on its maximum of 144 at the point $(0,12)$.

23. $\nabla f = 2(x-4)\mathbf{i} + 2y\mathbf{j}$ is never $\mathbf{0}$ at an interior point
of D. Next we examine f on the boundary of D:

$$C_1 : \ \mathbf{r}_1(t) = t\mathbf{i} + 4t\mathbf{j}, \quad t \in [0,2,],$$
$$C_2 : \ \mathbf{r}_2(t) = t\mathbf{i} + t^3\mathbf{j}, \quad t \in [0,2].$$

Note that

$$f_1(t) = f(\mathbf{r}_1(t)) = 17t^2 - 8t + 16,$$
$$f_2(t) = f(\mathbf{r}_2(t)) = (t-4)^2 + t^6.$$

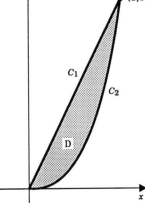

Next

$$f_1'(t) = 34t - 8 = 0 \quad \Longrightarrow \quad t = 4/17 \quad \text{and gives} \quad x = 4/17, \ \ y = 16/17$$

and

$$f_2'(t) = 6t^5 + 2t - 8 = 0 \quad \Longrightarrow \quad t = 1 \quad \text{and gives} \quad x = 1, \ \ y = 1.$$

The extreme values of f can be culled from the following list:

$$f(0,0) = 16, \quad f(2,8) = 68, \quad f\left(\tfrac{4}{17}, \tfrac{16}{17}\right) = \tfrac{256}{17}, \quad f(1,1) = 10.$$

We see that $f(1,1) = 10$ is the absolute minimum and $f(2,8)$ is the absolute maximum.

25. (a) $\nabla f = \tfrac{1}{2}x\mathbf{i} - \tfrac{2}{9}y\mathbf{j} = \mathbf{0}$ only at $(0,0)$.

(b) The difference

$$f(h,k) - f(0,0) = \tfrac{1}{4}h^2 - \tfrac{1}{9}k^2$$

does not keep a constant sign for all small h and k; $(0,0)$ is a saddle point. The function has
no local extreme values.

(c) Being the difference of two squares, f can be maximized by maximizing $\tfrac{1}{4}x^2$ and minimizing
$\tfrac{1}{9}y^2$; $(1,0)$ and $(-1,0)$ give absolute maximum value $\tfrac{1}{4}$. Similarly, $(0,1)$ and $(0,-1)$ give
absolute minimum value $-\tfrac{1}{9}$.

27.
$$f(x,y) = \sum_{i=1}^{3} \left[(x - x_i)^2 + (y - y_i)^2 \right]$$

$$\nabla f(x,y) = 2\left[(3x - x_1 - x_2 - x_3)\mathbf{i} + (3y - y_1 - y_2 - y_3)\mathbf{j} \right]$$

$$\nabla f = \mathbf{0} \quad \text{only at} \quad \left(\frac{x_1 + x_2 + x_3}{3}, \frac{y_1 + y_2 + y_3}{3} \right) = (x_0, y_0).$$

The difference $f(x_0 + h, \, y_0 + k) - f(x_0, y_0)$

$$= \sum_{i=1}^{3} \left[(x_0 + h - x_i)^2 + (y_0 + k - y_i)^2 - (x_0 - x_i)^2 - (y_0 - y_i)^2 \right]$$

$$= \sum_{i=1}^{3} \left[2h(x_0 - x_i) + h^2 + 2k(y_0 - y_i) + k^2 \right]$$

$$= 2h(3x_0 - x_1 - x_2 - x_3) + 2k(3y_0 - y_1 - y_2 - y_3) + h^2 + k^2$$

$$= h^2 + h^2$$

is positive for all small h and k. Thus, f has its absolute minimum at (x_0, y_0).

29.   From the geometry of the situation we recognize that we want to find the distance between $\left(\frac{1}{2}, \frac{1}{2}, \frac{1}{2}\right)$ and the point on the sphere at the end of the radius through the point $\left(\frac{1}{2}, \frac{1}{2}, \frac{1}{2}\right)$. This means the distance we want is the radius diminished by the distance between the origin (center) and $\left(\frac{1}{2}, \frac{1}{2}, \frac{1}{2}\right)$:

$$1 - \sqrt{\left(\tfrac{1}{2}\right)^2 + \left(\tfrac{1}{2}\right)^2 + \left(\tfrac{1}{2}\right)^2} = 1 - \tfrac{1}{2}\sqrt{3}.$$

31. $P(x, y) = 1000[(x - 80)(320 - x) + (y - 60)(140 - y)], \quad 80 \le x \le 320, \quad 60 \le y \le 140.$

$\nabla P = 1000[(400 - 2x)\mathbf{i} + (200 - 2y)\mathbf{j}] = \mathbf{0}$ only at $(200, 100)$.

The difference

$P(200 + h, \, 100 + k) - P(200, 100)$

$$= 1000[(120 + h)(120 - h) + (40 + k)(40 - k) - (120)^2 - (40)^2]$$

$$= -1000(h^2 + k^2)$$

is negative for all small h and k. The profit is maximized by setting the selling prices at \$2 per razor, \$1 per dozen blades.

SECTION 16.7

1. $\nabla f = (2x + y - 6)\mathbf{i} + (x + 2y)\mathbf{j} = \mathbf{0}$ only at $(4, -2)$.

$f_{xx} = 2, \quad f_{xy} = 1, \quad f_{yy} = 2.$

At $(4, -2)$, $D = -3 < 0$ and $A = 2 > 0$ so we have a local min; the value is -10.

3. $\nabla f = (4 - 2x + y)\mathbf{i} + (2 + x - 2y)\mathbf{j} = \mathbf{0}$ only at $\left(\frac{10}{3}, \frac{8}{3}\right)$.

$f_{xx} = -2, \quad f_{xy} = 1, \quad f_{yy} = -2.$

At $\left(\frac{10}{3}, \frac{8}{3}\right)$, $D = -3 < 0$ and $A = -2 < 0$ so we have a local max; the value is $\frac{28}{3}$.

5. $\nabla f = (3x^2 - 6y)\mathbf{i} + (3y^2 - 6x)\mathbf{j} = \mathbf{0}$ at $(2,2)$ and $(0,0)$.

     $f_{xx} = 6x, \quad f_{xy} = -6, \quad f_{yy} = 6y, \quad D = 36 - 36xy.$

     At $(2,2)$, $D = -108 < 0$ and $A = 12 > 0$ so we have a local min; the value is -8.

     At $(0,0)$, $D = 36 > 0$ so we have a saddle point.

7. $\nabla f = (3x^2 - 6y + 6)\mathbf{i} + (2y - 6x + 3)\mathbf{j} = \mathbf{0}$ at $\left(5, \frac{27}{2}\right)$ and $\left(1, \frac{3}{2}\right)$.

     $f_{xx} = 6x, \quad f_{xy} = -6, \quad f_{yy} = 2, \quad D = 36 - 12x.$

     At $\left(5, \frac{27}{2}\right)$, $D = -24 < 0$ and $A = 30 > 0$ so we have a local min; the value is $-\frac{117}{4}$.

     At $\left(1, \frac{3}{2}\right)$, $D = 24 > 0$ so we have a saddle point.

9. $\nabla f = e^x \cos y\,\mathbf{i} - e^x \sin y\,\mathbf{j}$ is never $\mathbf{0}$; there are no stationary points and no local extreme values.

11. $\nabla f = \sin y\,\mathbf{i} + x \cos y\,\mathbf{j} = \mathbf{0}$ at $(0, n\pi)$ for all integral n.

     $f_{xx} = 0, \quad f_{xy} = \cos y, \quad f_{yy} = -x \sin y, \quad D = \cos^2 y.$

     Since $D = \cos^2 n\pi = 1 > 0$, each stationary point is a saddle point.

13. $\nabla f = (2xy + 1 + y^2)\mathbf{i} + (x^2 + 2xy + 1)\mathbf{j} = \mathbf{0}$ at $(1,-1)$ and $(-1,1)$.

     $f_{xx} = 2y, \quad f_{xy} = 2x + 2y, \quad f_{yy} = 2x, \quad D = 4(x+y)^2 - 4xy.$

     At both $(1,-1)$ and $(-1,1)$ we have saddle points since $D = 4 > 0$.

15. $\nabla f = (y - x^{-2})\mathbf{i} + (x - 8y^{-2})\mathbf{j} = \mathbf{0}$ only at $\left(\frac{1}{2}, 4\right)$.

     $f_{xx} = 2x^{-3}, \quad f_{xy} = 1, \quad f_{yy} = 16y^{-3}, \quad D = 1 - 32x^{-3}y^{-3}.$

     At $\left(\frac{1}{2}, 4\right)$, $D = -3 < 0$ and $A = 16 > 0$ so we have a local min; the value is 6.

17. $\nabla f = (y - x^{-2})\mathbf{i} + (x - y^{-2})\mathbf{j} = \mathbf{0}$ only at $(1,1)$.

     $f_{xx} = 2x^{-3}, \quad f_{xy} = 1, \quad f_{yy} = 2y^{-3}, \quad D = 1 - 4x^{-3}y^{-3}.$

     At $(1,1)$, $D = -3 < 0$ and $A = 2 > 0$ so we have a local min; the value is 3.

19. $f(x,y) = xy(1 - x - y), \quad 0 \le x \le 1, \quad 0 \le y \le 1 - x.$

     [dom(f) is the triangle with vertices $(0,0)$, $(1,0)$, $(0,1)$.]

     $\nabla f = (y - 2xy - y^2)\mathbf{i} + (x - 2xy - x^2)\mathbf{j} = \mathbf{0} \quad \Longrightarrow \quad x = y = \frac{1}{3}.$

     (Note that $[0,0]$ is not an interior point of the domain of f.)

     $f_{xx} = -2x, \quad f_{xy} = 1 - 2x - 2y, \quad f_{yy} = -2x, \quad D = (1 - 2x - 2y)^2 - 4xy.$

     At $\left(\frac{1}{3}, \frac{1}{3}\right)$, $D = -\frac{1}{3} < 0$ and $A > 0$ so we have a local max; the value is $1/27$.

Since $f(x, y) = 0$ at each point on the boundary of the domain, the local max of $1/27$ is also the absolute max.

21. (a) $f(x, y) = 0$ along the plane curve $y = x^{2/3}$.
Since $f(x, y)$ is positive for points below the
curve and is negative for points above the
curve, there is a saddle point at the origin.

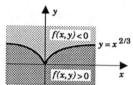

(b) $\nabla f = (ye^{xy} - 2\sin(x + y))\mathbf{i} + (xe^{xy} - 2\sin(x + y))\mathbf{j}$

$$f_{xx} = -2\cos(x + y) + y^2 e^{xy},$$

$$f_{xy} = -2\cos(x + y) + e^{xy}(1 + xy),$$

$$f_{yy} = -2\cos(x + y) + x^2 e^{xy}.$$

At the origin $A = -2$, $B = -1$, $C = -2$, and $D = -3$. We have a local max; the value is 3.

23.

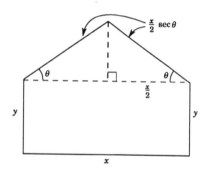

$$A = xy + \frac{1}{2}x\left(\frac{x}{2}\tan\theta\right),$$

$$P = x + 2y + 2\left(\frac{x}{2}\sec\theta\right),$$

$$0 < \theta < \frac{1}{2}\pi, \quad 0 < x < \frac{P}{1 + \sec\theta}.$$

$$A(x, \theta) = \tfrac{1}{2}x(P - x - x\sec\theta) + \tfrac{1}{4}x^2\tan\theta,$$

$$\nabla A = \left(\frac{P}{2} - x - x\sec\theta + \frac{x}{2}\tan\theta\right)\mathbf{i} + \left(\frac{x^2}{4}\sec^2\theta - \frac{x^2}{2}\sec\theta\tan\theta\right)\mathbf{j},$$

$$\nabla A = \frac{1}{2}[P + x(\tan\theta - 2\sec\theta - 2)]\mathbf{i} + \frac{x^2}{4}\sec\theta\,(\sec\theta - 2\tan\theta)\mathbf{j}.$$

From $\dfrac{\partial A}{\partial \theta} = 0$ we get $\theta = \tfrac{1}{6}\pi$ and then from $\dfrac{\partial A}{\partial x} = 0$ we get

$$P + x\left(\tfrac{1}{3}\sqrt{3} - \tfrac{4}{3}\sqrt{3} - 2\right) = 0 \quad \text{so that} \quad x = (2 - \sqrt{3})P.$$

Next,

$$A_{xx} = \tfrac{1}{2}(\tan\theta - 2\sec\theta - 2),$$

$$A_{x\theta} = \frac{x}{2}\sec\theta\,(\sec\theta - 2\tan\theta),$$

$$A_{\theta\theta} = \frac{x^2}{2}\sec\theta\,(\sec\theta\tan\theta - \sec^2\theta - \tan^2\theta).$$

By the second-partials test

$$A = -\tfrac{1}{2}(2 + \sqrt{3}), \quad B = 0, \quad C = -\tfrac{1}{3}P^2\sqrt{3}(2 - \sqrt{3})^2, \quad D < 0.$$

The area is a maximum when $\theta = \tfrac{1}{6}\pi$, $x = (2 - \sqrt{3})P$ and $y = \tfrac{1}{6}(3 - \sqrt{3})P$.

25. From $\qquad\qquad x = \tfrac{1}{2}y = \tfrac{1}{3}z = t \quad$ and $\quad x = y - 2 = z = s$

we take $\qquad\qquad\qquad (t, 2t, 3t) \quad$ and $\quad (s, 2 + s, s)$

as arbitrary points on the lines. It suffices to minimize the square of the distance between these points:

$$f(t, s) = (t - s)^2 + (2t - 2 - s)^2 + (3t - s)^2$$

$$= 14t^2 - 12ts + 3s^2 - 8t + 4s + 4, \qquad t, s \text{ real.}$$

$$\nabla f = (28t - 12s - 8)\mathbf{i} + (-12t + 6s + 4)\mathbf{j}; \qquad \nabla f = 0 \implies t = 0, \ s = -2/3.$$

$$f_{tt} = 28, \quad f_{ts} = -12, \quad f_{ss} = 6, \quad D = (-12)^2 - 6(28) = -24 < 0.$$

By the second-partials test, the distance is a minimum when $t = 0$, $s = -2/3$; the nature of the problem tells us the minimum is absolute. The distance is $\sqrt{f(0, 2/3)} = \tfrac{2}{3}\sqrt{6}$.

27.

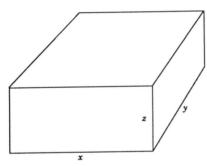

$$96 = xyz,$$

$$C = 30xy + 10(2xz + 2yz)$$

$$= 30xy + 20(x + y)\frac{96}{xy}.$$

$$C(x, y) = 30\left[xy + \frac{64}{x} + \frac{64}{y}\right],$$

$$\nabla C = 30(y - 64x^{-2})\mathbf{i} + 30(x - 64y^{-2})\mathbf{j} = 0 \implies x = y = 4.$$

$$C_{xx} = 128x^{-3}, \quad C_{xy} = 1, \quad C_{yy} = 128y^{-3}.$$

When $x = y = 4$, we have $D = -3 < 0$ and $A = 2 > 0$ so the cost is minimized by making the dimensions of the crate $4 \times 4 \times 6$ meters.

29. (a) $f(m, b) = [2 - b]^2 + [-5 - (m + b)]^2 + [4 - (2m + b)]^2.$

$f_m = 10m + 6b - 6, \quad f_b = 6m + 6b - 2; \qquad f_m = f_b = 0 \implies m = 1, \ b = -\tfrac{2}{3}.$

$f_{mm} = 10, \quad f_{mb} = 6, \quad f_{bb} = 6, \quad D = -24 < 0 \implies$ a min.

Answer: the line $y = x - \tfrac{2}{3}$.

(b) $f(\alpha, \beta) = [2 - \beta]^2 + [-5 - (\alpha + \beta)]^2 + [4 - (4\alpha + \beta)]^2.$

$f_\alpha = 34\alpha + 10\beta - 22, \quad f_\beta = 10\alpha + 6\beta - 2; \qquad f_\alpha = f_\beta = 0 \quad \Longrightarrow \quad \left\{ \begin{array}{l} \alpha = \frac{14}{13} \\ \\ \beta = -\frac{19}{13} \end{array} \right].$

$f_{\alpha\alpha} = 34, \quad f_{\alpha\beta} = 10, \quad f_{\beta\beta} = 6, \quad D = -104 < 0 \quad \Longrightarrow \quad \text{a min.}$

Answer: the parabola $y = \frac{1}{13}\left(14x^2 - 19\right).$

SECTION 16.8

1.
$$f(x, y) = x^2 + y^2, \qquad\qquad g(x, y) = xy - 1$$
$$\nabla f = 2x\mathbf{i} + 2y\mathbf{j}, \qquad\qquad \nabla g = y\mathbf{i} + x\mathbf{j}.$$

$$\nabla f = \lambda \nabla g \quad \Longrightarrow \quad 2x = \lambda y \text{ and } 2y = \lambda x.$$

Multiplying the first equation by x and the second equation by y, we get

$$2x^2 = \lambda xy = 2y^2.$$

Thus, $x = \pm y$. From $g(x, y) = 0$ we conclude that $x = y = \pm 1$. The points $(1, 1)$ and $(-1, -1)$ clearly give a minimum, since f represents the square of the distance of a point on the hyperbola from the origin. The minimum is 2.

3.
$$f(x, y) = xy, \qquad\qquad g(x, y) = b^2 x^2 + a^2 y^2 - a^2 b^2$$
$$\nabla f = y\mathbf{i} + x\mathbf{j}, \qquad\qquad \nabla g = 2b^2 x\mathbf{i} + 2a^2 y\mathbf{j}.$$

$$\nabla f = \lambda \nabla g \quad \Longrightarrow \quad y = 2\lambda b^2 x \text{ and } x = 2\lambda a^2 y.$$

Multiplying the first equation by $a^2 y$ and the second equation by $b^2 x$, we get

$$a^2 y^2 = 2\lambda a^2 b^2 xy = b^2 x^2.$$

Thus, $ay = \pm bx$. From $g(x, y) = 0$ we conclude that $x = \pm\frac{1}{2}a\sqrt{2}$ and $y = \pm\frac{1}{2}b\sqrt{2}.$

Since f is continuous and the ellipse is closed and bounded, the minimum exists. It occurs at $\left(\frac{1}{2}a\sqrt{2}, -\frac{1}{2}b\sqrt{2}\right)$ and $\left(-\frac{1}{2}a\sqrt{2}, \frac{1}{2}b\sqrt{2}\right)$; the minimum is $-\frac{1}{2}ab.$

5.   Since f is continuous and the ellipse is closed and bounded, the maximum exists.

$$f(x, y) = xy^2, \qquad\qquad g(x, y) = b^2 x^2 + a^2 y^2 - a^2 b^2$$
$$\nabla f = y^2\mathbf{i} + 2xy\mathbf{j}, \qquad\qquad \nabla g = 2b^2 x\mathbf{i} + 2a^2 y\mathbf{j}.$$

$$\nabla f = \lambda \nabla g \quad \Longrightarrow \quad y^2 = 2\lambda b^2 x \text{ and } 2xy = 2\lambda a^2 y.$$

Multiplying the first equation by $a^2 y$ and the second equation by $b^2 x$, we get

$$a^2 y^3 = 2\lambda a^2 b^2 xy = 2b^2 x^2 y.$$

We can exclude $y = 0$; it clearly cannot produce the maximum. Thus,

$$a^2 y^2 = 2b^2 x^2 \quad \text{and, from } g(x,y) = 0, \quad 3b^2 x^2 = a^2 b^2.$$

This gives us $x = \pm \frac{1}{3}\sqrt{3}\, a$ and $y = \pm \frac{1}{3}\sqrt{6}\, b$. This maximum occurs at $x = \frac{1}{3}\sqrt{3}\, a$, $y = \pm \frac{1}{3}\sqrt{6}\, b$; the value there is $\frac{2}{9}\sqrt{3}\, ab^2$.

7. The given curve is closed and bounded. Since $x^2 + y^2$ represents the square of the distance from points on this curve to the origin, the maximum exists.

$$f(x,y) = x^2 + y^2, \qquad\qquad g(x,y) = x^4 + 7x^2 y^2 + y^4 - 1$$

$$\nabla f = 2x\mathbf{i} + 2y\mathbf{j}, \qquad\qquad \nabla g = \left(4x^3 + 14xy^2\right)\mathbf{i} + \left(4y^3 + 14x^2 y\right)\mathbf{j}.$$

We use the cross-product equation (16.8.4):

$$2x(4y^3 + 14x^2 y) - 2y(4x^3 + 14xy^2) = 0,$$

$$20x^3 y - 20xy^3 = 0,$$

$$xy(x^2 - y^2) = 0.$$

Thus, $x = 0$, $y = 0$, or $x = \pm y$. From $g(x,y) = 0$ we conclude that the points to examine are

$$(0, \pm 1), \quad (\pm 1, 0), \quad \left(\pm \tfrac{1}{3}\sqrt{3}, \pm \tfrac{1}{3}\sqrt{3}\right).$$

The value of f at each of the first four points is 1; the value at the last four points is $2/3$. The maximum is 1.

9. The maximum exists since xyz is continuous and the ellipsoid is closed and bounded.

$$f(x,y,z) = xyz, \qquad\qquad g(x,y,z) = \frac{x^2}{a^2} + \frac{y^2}{b^2} + \frac{z^2}{c^2} - 1$$

$$\nabla f = yz\mathbf{i} + xz\mathbf{j} + xy\mathbf{k}, \qquad\qquad \nabla g = \frac{2x}{a^2}\mathbf{i} + \frac{2y}{b^2}\mathbf{j} + \frac{2z}{c^2}\mathbf{k}.$$

$$\nabla f = \lambda \nabla g \implies yz = \frac{2x}{a^2}\lambda, \quad xz = \frac{2y}{b^2}\lambda, \quad xy = \frac{2z}{c^2}\lambda.$$

We can assume x, y, z are non-zero, for otherwise $f(x,y,z) = 0$, which is clearly not a maximum. Then from the first two equations

$$\frac{yza^2}{x} = 2\lambda = \frac{xzb^2}{y} \quad \text{so that} \quad a^2 y^2 = b^2 x^2 \quad \text{or} \quad x^2 = \frac{a^2 y^2}{b^2}.$$

Similarly from the second and third equations we get

$$b^2 z^2 = c^2 y^2 \quad \text{or} \quad z^2 = \frac{c^2 y^2}{b^2}.$$

Substituting these expressions for x^2 and z^2 in $g(x,y,z) = 0$, we obtain

$$\frac{1}{a^2}\left[\frac{a^2 y^2}{b^2}\right] + \frac{y^2}{b^2} + \frac{1}{c^2}\left[\frac{c^2 y^2}{b^2}\right] - 1 = 0, \quad \frac{3y^2}{b^2} = 1, \quad y = \pm\frac{1}{3}b\sqrt{3}.$$

Then, $x = \pm\frac{1}{3}a\sqrt{3}$ and $z = \pm\frac{1}{3}c\sqrt{3}$. The maximum value is $\frac{1}{9}\sqrt{3}\,abc$.

11. Since the sphere is closed and bounded and $2x + 3y + 5z$ is continuous, the maximum exists.

$$f(x,y,z) = 2x + 3y + 5z, \qquad g(x,y,z) = x^2 + y^2 + z^2 - 19$$

$$\nabla f = 2\mathbf{i} + 3\mathbf{j} + 5\mathbf{k}, \qquad \nabla g = 2x\mathbf{i} + 2y\mathbf{j} + 2z\mathbf{k}.$$

$$\nabla f = \lambda\nabla g \implies 2 = 2\lambda x, \quad 3 = 2\lambda y, \quad 5 = 2\lambda z.$$

Since $\lambda \neq 0$ here, we solve the equations for x, y and z:

$$x = \frac{1}{\lambda}, \quad y = \frac{3}{2\lambda}, \quad z = \frac{5}{2\lambda},$$

and substitute these results in $g(x,y,z) = 0$ to obtain

$$\frac{1}{\lambda^2} + \frac{9}{4\lambda^2} + \frac{25}{4\lambda^2} - 19 = 0, \quad \frac{38}{4\lambda^2} - 19 = 0, \quad \lambda = \pm\frac{1}{2}\sqrt{2}.$$

The positive value of λ will produce positive values for x, y, z and thus the maximum for f. We get $x = \sqrt{2}$, $y = \frac{3}{2}\sqrt{2}$, $z = \frac{5}{2}\sqrt{2}$, and $2x + 3y + 5z = 19\sqrt{2}$.

13.

$$f(x,y,z) = xyz, \qquad g(x,y,z) = \frac{x}{a} + \frac{y}{b} + \frac{z}{c} - 1$$

$$\nabla f = yz\mathbf{i} + xz\mathbf{j} + xy\mathbf{k}, \qquad \nabla g = \frac{1}{a}\mathbf{i} + \frac{1}{b}\mathbf{j} + \frac{1}{c}\mathbf{k}.$$

$$\nabla f = \lambda\nabla g \implies yz = \frac{\lambda}{a}, \quad xz = \frac{\lambda}{b}, \quad xy = \frac{\lambda}{c}.$$

Multiplying these equations by x, y, z respectively, we obtain

$$xyz = \frac{\lambda x}{a}, \quad xyz = \frac{\lambda y}{b}, \quad xyz = \frac{\lambda z}{c}.$$

Adding these equations and using the fact that $g(x,y,z) = 0$, we have

$$3xyz = \lambda\left(\frac{x}{a} + \frac{y}{b} + \frac{z}{c}\right) = \lambda.$$

Since x, y, z are non-zero,

$$yz = \frac{\lambda}{a} = \frac{3xyz}{a}, \quad 1 = \frac{3x}{a}, \quad x = \frac{a}{3}.$$

Similarly, $y = \frac{b}{3}$ and $z = \frac{c}{3}$. The maximum is $\frac{1}{27}abc$.

15. It suffices to minimize the square of the distance from $(0,1)$ to the parabola. Clearly, the minimum exists.

$$f(x,y) = x^2 + (y-1)^2, \qquad g(x,y) = x^2 - 4y$$

$$\nabla f = 2x\mathbf{i} + 2(y-1)\mathbf{j}, \qquad \nabla g = 2x\mathbf{i} - 4\mathbf{j}.$$

We use the cross-product equation (16.8.4):

$$2x(-4) - 2x(2y-2) = 0, \quad 4x + 4xy = 0, \quad x(y+1) = 0.$$

Since $y \geq 0$, we have $x = 0$ and thus $y = 0$. The minimum is 1.

17. It's easier to work with the square of the distance; the minimum certainly exists.

$$f(x,y,z) = x^2 + y^2 + z^2, \qquad g(x,y,z) = Ax + By + Cz + D$$

$$\nabla f = 2x\mathbf{i} + 2y\mathbf{j} + 2z\mathbf{k}, \qquad \nabla g = A\mathbf{i} + B\mathbf{j} + C\mathbf{k}.$$

$$\nabla f = \lambda \nabla g \implies 2x = A\lambda, \quad 2y = B\lambda, \quad 2z = C\lambda.$$

Substituting these equations in $g(x,y,z) = 0$, we have

$$\frac{1}{2}\lambda \left(A^2 + B^2 + C^2\right) + D = 0, \quad \lambda \frac{-2D}{A^2 + B^2 + C^2}.$$

Thus, in turn,

$$x = \frac{-DA}{A^2 + B^2 + C^2}, \quad y = \frac{-DB}{A^2 + B^2 + C^2}, \quad z = \frac{-DC}{A^2 + B^2 + C^2}$$

so the minimum value of $\sqrt{x^2 + y^2 + z^2}$ is $|D|\left(A^2 + B^2 + C^2\right)^{-1/2}$.

19.

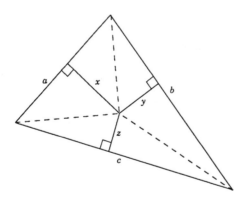

area $A = \frac{1}{2}ax + \frac{1}{2}by + \frac{1}{2}cz$.

The geometry suggests that

$$x^2 + y^2 + z^2$$

has a minimum.

$$f(x,y,z) = x^2 + y^2 + z^2, \qquad g(x,y,z) = ax + by + cz - 2A$$

$$\nabla f = 2x\mathbf{i} + 2y\mathbf{j} + 2z\mathbf{k}, \qquad \nabla g = a\mathbf{i} + b\mathbf{j} + c\mathbf{k}.$$

$$\nabla f = \lambda \nabla g \implies 2x = a\lambda, \quad 2y = b\lambda, \quad 2z = c\lambda.$$

Solving these equations for x, y, z and substituting the results in $g(x, y, z) = 0$, we have

$$\frac{a^2\lambda}{2} + \frac{b^2\lambda}{2} + \frac{c^2\lambda}{2} - 2A = 0, \quad \lambda = \frac{4A}{a^2 + b^2 + c^2}$$

and thus

$$x = \frac{2aA}{a^2 + b^2 + c^2}, \quad y = \frac{2bA}{a^2 + b^2 + c^2}, \quad z = \frac{2cA}{a^2 + b^2 + c^2}.$$

The minimum is $4A^2(a^2 + b^2 + c^2)^{-1}$.

21. Since the curve is asymptotic to the line $y = x$ as $x \to -\infty$ and as $x \to \infty$, the maximum exists. The distance between the point (x, y) and the line $y - x = 0$ is given by

$$\frac{|y - x|}{\sqrt{1 + 1}} = \frac{1}{2}\sqrt{2}\,|y - x|. \qquad \text{(by Theorem 9.2.2)}$$

Since the points on the curve are below the line $y = x$, we can replace $|y - x|$ by $x - y$. To simplify the work we drop the constant factor $\frac{1}{2}\sqrt{2}$.

$$f(x, y) = x - y, \qquad g(x, y) = x^3 - y^3 - 1$$

$$\nabla f = \mathbf{i} - \mathbf{j}, \qquad \nabla g = 3x^2\mathbf{i} - 3y^2\mathbf{j}.$$

We use the cross-product equation (16.8.4):

$$1\left(-3y^2\right) - \left(3x^2\right)(-1) = 0, \quad 3x^2 - 3y^2 = 0, \quad x = -y \ (x \neq y).$$

Now $g(x, y) = 0$ gives us

$$x^3 - (-x)^3 - 1 = 0, \quad 2x^3 = 1, \quad x = 2^{-1/3}.$$

The point is $\left(2^{-1/3}, -2^{-1/3}\right)$.

23. It suffices to show that the square of the area is a maximum when $a = b = c$.

$$f(a, b, c) = s(s - a)(s - b)(s - c), \quad g(a, b, c) = a + b + c - 2s$$

$$\nabla f = -s(s - b)(s - c)\mathbf{i} - s(s - a)(s - c)\mathbf{j} - s(s - a)(s - b)\mathbf{k}, \quad \nabla g = \mathbf{i} + \mathbf{j} + \mathbf{k}.$$

$$\nabla f = \lambda \nabla g \implies -s(s - b)(s - c) = -s(s - a)(s - c) = -s(s - a)(s - b) = \lambda.$$

Thus, $s - b = s - a = s - c$ so that $a = b = c$. This gives us the maximum, as no minimum exists. [The area can be made arbitrarily small by taking a close to s.]

25. To simplify notation we set $x = Q_1, \quad y = Q_2, \quad z = Q_3$.

$$f(x, y, z) = 2x + 8y + 24z, \qquad g(x, y, z) = x^2 + 2y^2 + 4z^2 - 4,500,000,000$$

$$\nabla f = 2\mathbf{i} + 8\mathbf{j} + 24\mathbf{k}, \qquad \nabla g = 2x\mathbf{i} + 4y\mathbf{j} + 8z\mathbf{k}.$$

$$\nabla f = \lambda \nabla g \implies 2 = 2\lambda x, \quad 8 = 4\lambda y, \quad 24 = 8\lambda z.$$

Since $\lambda \neq 0$ here, we solve the equations for x, y, z:

$$x = \frac{1}{\lambda}, \quad y = \frac{2}{\lambda}, \quad z = \frac{3}{\lambda},$$

and substitute these results in $g(x, y, z) = 0$ to obtain

$$\frac{1}{\lambda^2} + 2\left(\frac{4}{\lambda^2}\right) + 4\left(\frac{9}{\lambda^2}\right) - 45 \times 10^8 = 0, \quad \frac{45}{\lambda^2} = 45 \times 10^8, \quad \lambda = \pm 10^{-4}.$$

Since x, y, z are non-negative, $\lambda = 10^{-4}$ and

$$x = 10^4 = Q_1, \quad y = 2 \times 10^4 = Q_2, \quad z = 3 \times 10^4 = Q_3.$$

SECTION 16.9

1. $df = \left(3x^2y - 2xy^2\right)\Delta x + \left(x^3 - 2x^2y\right)\Delta y$

3. $df = (\cos y + y \sin x)\,\Delta x - (x \sin y + \cos x)\,\Delta y$

5. $df = \Delta x - (\tan z)\,\Delta y - \left(y \sec^2 z\right)\Delta z$

7.
$$\Delta u = \left[(x+\Delta x)^2 - 3(x+\Delta x)(y+\Delta y) + 2(y+\Delta y)^2\right] - \left(x^2 - 3xy + 2y^2\right)$$

$$= \left[(1.7)^2 - 3(1.7)(-2.8) + 2(-2.8)^2\right] - \left(2^2 - 3(2)(-3) + 2(-3)^2\right)$$

$$= (2.89 + 14.28 + 15.68) - 40 = -7.15$$

$$du = (2x - 3y)\,\Delta x + (-3x + 4y)\,\Delta y$$

$$= (4 + 9)(-0.3) + (-18)(0.2) = -7.50$$

9. $f(x, y) = x^{1/2}y^{1/4}; \quad x = 121, \quad y = 16, \quad \Delta x = 4, \quad \Delta y = 1$

$$f(x + \Delta x, y + \Delta y) \cong f(x, y) + df$$

$$= x^{1/2}\,y^{1/4} + \tfrac{1}{2}x^{-1/2}\,y^{1/4}\,\Delta x + \tfrac{1}{4}x^{1/2}\,y^{-3/4}\,\Delta y$$

$$\sqrt{125}\,\sqrt[4]{17} \cong \sqrt{121}\,\sqrt[4]{16} + \tfrac{1}{2}(121)^{-1/2}\,(16)^{1/4}\,(4) + \tfrac{1}{4}(121)^{1/2}\,(16)^{-3/4}\,(1)$$

$$= 11(2) + \tfrac{1}{2}\left(\tfrac{1}{11}\right)(2)(4) + \tfrac{1}{4}(11)\left(\tfrac{1}{8}\right)$$

$$= 22 + \tfrac{4}{11} + \tfrac{11}{32} = 22\,\tfrac{249}{352} \cong 22.71$$

11.
$$f(x, y) = \sin x \cos y; \quad x = \pi, \quad y = \frac{\pi}{4}, \quad \Delta x = -\frac{\pi}{7}, \quad \Delta y = -\frac{\pi}{20}$$

$$df = \cos x \cos y\,\Delta x - \sin x \sin y\,\Delta y$$

$$f(x + \Delta x, y + \Delta y) \cong f(x, y) + df$$

$$\sin \frac{6}{7}\pi \cos \frac{1}{5}\pi \cong \sin \pi \cos \frac{\pi}{4} + \left(\cos \pi \cos \frac{\pi}{4}\right)\left(-\frac{\pi}{7}\right) - \left(\sin \pi \sin \frac{\pi}{4}\right)\left(-\frac{\pi}{20}\right)$$

$$= 0 + \left(\tfrac{1}{2}\sqrt{2}\right)\left(\frac{\pi}{7}\right) + 0 = \frac{\pi\sqrt{2}}{14} \cong 0.32$$

13. $df = \dfrac{\partial z}{\partial x}\Delta x + \dfrac{\partial z}{\partial y}\Delta y = \dfrac{2y}{(x+y)^2}\Delta x - \dfrac{2x}{(x+y)^2}\Delta y$

With $x = 4,\ y = 2,\ \Delta x = 0.1,\ \Delta y = 0.1,$ we get

$$df = \tfrac{4}{36}(0.1) - \tfrac{8}{36}(0.1) = -\tfrac{1}{90}.$$

The exact change is $\dfrac{4.1 - 2.1}{4.1 + 2.1} - \dfrac{4 - 2}{4 + 2} = \dfrac{2}{6.2} - \dfrac{1}{3} = -\dfrac{1}{93}.$

15. $S = 2\pi r^2 + 2\pi rh;\quad r = 8,\ h = 12,\ \Delta r = -0.3,\ \Delta h = 0.2$

$$dS = \dfrac{\partial S}{\partial r}\Delta r + \dfrac{\partial S}{\partial h}\Delta h = (4\pi r + 2\pi h)\,\Delta r + (2\pi r)\,\Delta h$$

$$= 56\pi(-0.3) + 16\pi(0.2) = -13.6\pi.$$

The area decreases about 13.6π in.^{2}.

17. (a) $\pi r^2 h = \pi(r + \Delta r)^2(h + \Delta h) \implies \Delta h\dfrac{r^2 h}{(r + \Delta r)^2} - h = -\dfrac{(2r + \Delta r)h}{(r + \Delta r)^2}\Delta r.$

$$df = (2\pi rh)\,\Delta r + \pi r^2\,\Delta h, \qquad df = 0 \implies \Delta h = \dfrac{-2h}{r}\Delta r.$$

(b) $2\pi r^2 + 2\pi rh = 2\pi(r + \Delta r)^2 + 2\pi(r + \Delta r)(h + \Delta h).$

Solving for Δh,

$$\Delta h = \dfrac{r^2 + rh - (r + \Delta r)^2}{r + \Delta r} - h = -\dfrac{(2r + h + \Delta r)}{r + \Delta r}\Delta r.$$

$$df = (4\pi r + 2\pi h)\,\Delta r + 2\pi r\,\Delta h, \qquad df = 0 \implies \Delta h = -\left(\dfrac{2r + h}{r}\right)\Delta r.$$

19. $s = \dfrac{A}{A - W};\quad ds = \dfrac{-W}{(A - W)^2}\Delta A + \dfrac{A}{(A - W)^2}\Delta W.$ By the triangle inequality

$$|ds| \leq \dfrac{W\,|\Delta A| + A\,|\Delta W|}{(A - W)^2}.$$

Taking $A = 9,\ W = 5,\ |\Delta A| \leq 0.01,$ and $|\Delta W| \leq 0.02,$ we have

$$2.23 \leq s \pm |\Delta s| \leq 2.27.$$

SECTION 16.10

1. $\dfrac{\partial f}{\partial x} = xy^2,\quad f(x, y) = \tfrac{1}{2}x^2 y^2 + \phi(y),\quad \dfrac{\partial f}{\partial y} = x^2 y + \phi'(y) = x^2 y.$

Thus, $\phi'(y) = 0,\ \phi(y) = C,$ and $f(x, y) = \tfrac{1}{2}x^2 y^2 + C.$

3. $\dfrac{\partial f}{\partial x} = y,\quad f(x, y) = xy + \phi(y),\quad \dfrac{\partial f}{\partial y} = x + \phi'(y) = x.$

Thus, $\phi'(y) = 0,\ \phi(y) = C,$ and $f(x, y) = xy + C.$

5.   No; $\dfrac{\partial}{\partial y}\,(y^3 + x) = 3y^2$ whereas $\dfrac{\partial}{\partial x}\,(x^2 + y) = 2x.$

7. $\dfrac{\partial f}{\partial x} = \cos x - y\sin x,$ $f(x,y) = \sin x + y\cos x + \phi(y),$ $\dfrac{\partial f}{\partial y} = \cos x + \phi'(y) = \cos x.$

    Thus, $\phi'(y) = 0,$ $\phi(y) = C,$ and $f(x,y) = \sin x + y\cos x + C.$

9. $\dfrac{\partial f}{\partial x} = e^x\cos y^2,$ $f(x,y) = e^x\cos y^2 + \phi(y),$ $\dfrac{\partial f}{\partial y} = -2ye^x\sin y^2 + \phi'(y) = -2ye^x\sin y^2.$

    Thus, $\phi'(y) = 0,$ $\phi(y) = C,$ and $f(x,y) = e^x\cos y^2 + C.$

11. $\dfrac{\partial f}{\partial y} = xe^x - e^{-y},$ $f(x,y) = xye^x + e^{-y} + \phi(x),$ $\dfrac{\partial f}{\partial x} = ye^x + xye^x + \phi'(x) = ye^x(1 + x).$

    Thus, $\phi'(x) = 0,$ $\phi(x) = C,$ and $f(x,y) = xye^x + e^{-y} + C.$

13. $\dfrac{\partial f}{\partial x} = \dfrac{x}{\sqrt{x^2 + y^2}},$ $f(x,y) = \sqrt{x^2 + y^2} + \phi(y),$ $\dfrac{\partial f}{\partial y} = \dfrac{y}{\sqrt{x^2 + y^2}} + \phi'(y) = \dfrac{y}{\sqrt{x^2 + y^2}}.$

    Thus, $\phi'(y) = 0,$ $\phi(y) = C,$ and $f(x,y) = \sqrt{x^2 + y^2} + C.$

15. $\dfrac{\partial f}{\partial x} = x^2\sin^{-1} y,$ $f(x,y) = \tfrac{1}{3}x^3\sin^{-1} y + \phi(y),$ $\dfrac{\partial f}{\partial y} = \dfrac{x^3}{3\sqrt{1 - y^2}} + \phi'(y) = \dfrac{x^3}{3\sqrt{1 - y^2}} - \ln y.$

    Thus, $\phi'(y) = -\ln y,$ $\phi(y) = y - y\ln y + C,$ and

$$f(x,y) = \dfrac{x^3}{3\sqrt{1 - y^2}} + y - y\ln y + C.$$

17. $\dfrac{\partial f}{\partial x} = f(x,y),$ $\dfrac{\partial f/\partial x}{f(x,y)} = 1,$ $\ln f(x,y) = x + \phi(y),$ $\dfrac{\partial f/\partial y}{f(x,y)} = 0 + \phi'(y),$ $\dfrac{\partial f}{\partial y} = f(x,y).$

    Thus, $\phi'(y) = 1,$ $\phi(y) = y + K,$ and $f(x,y) = e^{x+y+K} = Ce^{x+y}.$

19.   (a) $P = 2x,$ $Q = z,$ $R = y;$ $\dfrac{\partial P}{\partial y} = 0 = \dfrac{\partial Q}{\partial x},$ $\dfrac{\partial P}{\partial z} = 0 = \dfrac{\partial R}{\partial x},$ $\dfrac{\partial Q}{\partial z} = 1 = \dfrac{\partial R}{\partial y}$

    (b), (c), and (d)

$$\dfrac{\partial f}{\partial x} = 2x, \quad f(x,y,z) = x^2 + g(y,z).$$

$$\dfrac{\partial f}{\partial y} = 0 + \dfrac{\partial g}{\partial y} \quad \text{with} \quad \dfrac{\partial f}{\partial y} = z \quad \Longrightarrow \quad \dfrac{\partial g}{\partial y} = z.$$

Then,

$$g(y, z) = yz + h(z),$$

$$f(x, y, z) = x^2 + yz + h(z),$$

$$\frac{\partial f}{\partial z} = 0 + y + h'(z) \quad \text{and} \quad \frac{\partial f}{\partial z} = y \quad \Longrightarrow \quad h'(z) = 0.$$

Thus, $h(z) = C$ and $f(x, y, z) = x^2 + yz + C.$

21. The function is a gradient by the test stated before Exercise 19.
Take $P = 2x + y$, $Q = 2y + x + z$, $R = y - 2z$. Then

$$\frac{\partial P}{\partial y} = 1 = \frac{\partial Q}{\partial x}, \quad \frac{\partial P}{\partial z} = 0 = \frac{\partial R}{\partial x}, \quad \frac{\partial Q}{\partial z} = 1 = \frac{\partial R}{\partial y}.$$

Next, we find f where $\nabla f = P\mathbf{i} + Q\mathbf{j} + R\mathbf{k}.$

$$\frac{\partial f}{\partial x} = 2x + y,$$

$$f(x, y, z) = x^2 + xy + g(y, z).$$

$$\frac{\partial f}{\partial y} = x + \frac{\partial g}{\partial y} \quad \text{with} \quad \frac{\partial f}{\partial y} = 2y + x + z \quad \Longrightarrow \quad \frac{\partial g}{\partial y} = 2y + z.$$

Then,

$$g(y, z) = y^2 + yz + h(z),$$

$$f(x, y, z) = x^2 + xy + y^2 + yz + h(z).$$

$$\frac{\partial f}{\partial z} = y + h'(z) = y - 2z \quad \Longrightarrow \quad h'(z) = -2z.$$

Thus, $h(z) = -z^2 + C$ and $f(x, y, z) = x^2 + xy + y^2 + yz - z^2 + C.$

23. $\mathbf{F}(\mathbf{r}) = \nabla \left(\dfrac{GmM}{r} \right)$

CHAPTER 17

SECTION 17.1

1. $\displaystyle\sum_{i=1}^{m} \Delta x_i = \Delta x_1 + \Delta x_2 + \cdots + \Delta x_n = (x_1 - x_0) + (x_2 - x_1) + \cdots + (x_n - x_{n-1})$

$$= x_n - x_0 = a_2 - a_1$$

3. $\displaystyle\sum_{i=1}^{m}\sum_{j=1}^{n} \Delta x_i \, \Delta y_j = \left(\sum_{i=1}^{m} \Delta x_i\right)\left(\sum_{j=1}^{n} \Delta y_j\right) = (a_2 - a_1)(b_2 - b_1)$

5. $\displaystyle\sum_{i=1}^{m} (x_i + x_{i-1}) \, \Delta x_i = \sum_{i=1}^{m} (x_i + x_{i-1})(x_i - x_{i-1}) = \sum_{i=1}^{m} \left(x_i^2 - x_{i-1}^2\right)$

$$= x_m^2 - x_0^2 = a_2^2 - a_1^2$$

7. $\displaystyle\sum_{i=1}^{m}\sum_{j=1}^{n} (x_i + x_{i-1}) \, \Delta x_i \, \Delta y_j = \left(\sum_{i=1}^{m} (x_i + x_{i-1}) \, \Delta x_i\right)\left(\sum_{j=1}^{n} \Delta y_j\right)$

(Exercise 5) $\stackrel{\downarrow}{=} \left(a_2^2 - a_1^2\right)(b_2 - b_1)$

9. $\displaystyle\sum_{i=1}^{m}\sum_{j=1}^{n} (2\Delta x_i - 3\Delta y_j) = 2\left(\sum_{i=1}^{m} \Delta x_i\right)\left(\sum_{j=1}^{n} 1\right) - 3\left(\sum_{i=1}^{m} 1\right)\left(\sum_{j=1}^{n} \Delta y_j\right)$

$$= 2n(a_2 - a_1) - 3m(b_2 - b_1)$$

11. $\displaystyle\sum_{i=1}^{m}\sum_{j=1}^{n}\sum_{k=1}^{q} \Delta x_i \, \Delta y_j \, \Delta z_k = \left(\sum_{i=1}^{m} \Delta x_i\right)\left(\sum_{j=1}^{n} \Delta y_j\right)\left(\sum_{k=1}^{q} \Delta z_k\right)$

$$= (a_2 - a_1)(b_2 - b_1)(c_2 - c_1)$$

13. $\displaystyle\sum_{i=1}^{n}\sum_{j=1}^{n}\sum_{k=1}^{n} \delta_{ijk} a_{ijk} = a_{111} + a_{222} + \cdots + a_{nnn} = \sum_{p=1}^{n} a_{ppp}$

SECTION 17.2

1. $L_f(P) = 2\frac{1}{4}$, $U_f(P) = 5\frac{3}{4}$

3. (a) $L_f(P) = \sum\limits_{i=1}^{m} \sum\limits_{j=1}^{n} (x_{i-1} + 2y_{j-1}) \, \Delta x_i \, \Delta y_j$, $U_f(P) = \sum\limits_{i=1}^{m} \sum\limits_{j=1}^{n} (x_i + 2y_j) \, \Delta x_i \, \Delta y_j$

(b) $L_f(P) \leq \sum\limits_{i=1}^{m} \sum\limits_{j=1}^{n} \left[\dfrac{x_{i-1} + x_i}{2} + 2\left(\dfrac{y_{j-1} + y_j}{2} \right) \right] \Delta x_i \, \Delta y_j \leq U_f(P)$.

The middle expression can be written

$$\sum\limits_{i=1}^{m} \sum\limits_{j=1}^{n} \frac{1}{2} \left(x_i^2 - x_{i-1}^2 \right) \Delta y_j + \sum\limits_{i=1}^{m} \sum\limits_{j=1}^{n} \left(y_j^2 - y_{j-1}^2 \right) \Delta x_i.$$

The first double sum reduces to

$$\sum\limits_{i=1}^{m} \sum\limits_{j=1}^{n} \frac{1}{2} \left(x_i^2 - x_{i-1}^2 \right) \Delta y_j = \frac{1}{2} \left(\sum\limits_{i=1}^{m} \left(x_i^2 - x_{i-1}^2 \right) \right) \left(\sum\limits_{j=1}^{n} \Delta y_j \right) = \frac{1}{2} (4-0)(1-0) = 2.$$

In like manner the second double sum also reduces to 2.   Thus, $I = 4$;   the volume of the prism bounded above by the plane $z = x + 2y$ and below by R.

5. $L_f(P) = -7/24$, $U_f(P) = 7/24$

7. (a) $L_f(P) = \sum\limits_{i=1}^{m} \sum\limits_{j=1}^{n} (4x_{i-1} \, y_{j-1}) \, \Delta x_i \, \Delta y_j$, $U_f(P) = \sum\limits_{i=1}^{m} \sum\limits_{j=1}^{n} (4x_i \, y_j) \, \Delta x_i \, \Delta y_j$

(b) $L_f(P) \leq \sum\limits_{i=1}^{m} \sum\limits_{j=1}^{n} (x_i + x_{i-1})(y_j + y_{j-1}) \, \Delta x_1 \, \Delta y_j \leq U_f(P)$.

The middle expression can be written

$$\sum\limits_{i=1}^{m} \sum\limits_{j=1}^{n} \left(x_i^2 - x_{i-1}^2 \right) \left(y_j^2 - y_{j-1}^2 \right) = \left(\sum\limits_{i=1}^{m} x_i^2 - x_{i-1}^2 \right) \left(\sum\limits_{j=1}^{n} y_j^2 - y_{j-1}^2 \right)$$

$$\text{by } (17.1.5) \uparrow$$

$$= \left(b^2 - 0^2 \right) \left(d^2 - 0^2 \right) = b^2 d^2.$$

It follows that $I = b^2 d^2$.

9. **(a)** $L_f(P) = \sum\limits_{i=1}^{m} \sum\limits_{j=1}^{n} 3\left(x_{i-1}^2 - y_j^2\right) \Delta x_i \, \Delta y_j, \quad U_f(P) = \sum\limits_{i=1}^{m} \sum\limits_{j=1}^{n} 3\left(x_i^2 - y_{j-1}^2\right) \Delta x_i \, \Delta y_j$

(b) $L_f(P) \le \sum\limits_{i=1}^{m} \sum\limits_{j=1}^{n} \left[\left(x_i^2 + x_i x_{i-1} + x_{i-1}^2\right) - \left(y_j^2 + y_j y_{j-1} + y_{j-1}^2\right)\right] \Delta x_i \, \Delta y_j \le U_f(P).$

Since in general $\left(A^2 + AB + B^2\right)\left(A - B\right) = A^3 - B^3$,   the middle expression can be written

$$\sum\limits_{i=1}^{m} \sum\limits_{j=1}^{n} \left(x_i^3 - x_{i-1}^3\right) \Delta y_j - \sum\limits_{i=1}^{m} \sum\limits_{j=1}^{n} \left(y_j^3 - y_{j-1}^3\right) \Delta x_i,$$

which reduces to

$$\left(\sum\limits_{i=1}^{m} x_i^3 - x_{i-1}^3\right) \left(\sum\limits_{j=1}^{n} \Delta y_j\right) - \left(\sum\limits_{i=1}^{m} \Delta x_i\right) \left(\sum\limits_{j=1}^{n} y_j^3 - y_{j-1}^3\right).$$

This can be evaluated as $b^3 d - bd^3 = bd\left(b^2 - d^2\right).$ It follows that $I = bd\left(b^2 - d^2\right).$

SECTION 17.3

1. $\displaystyle\iint\limits_{\Omega} dx\,dy = \int_a^b \phi(x)\,dx$

SECTION 17.4

1. $\displaystyle\int_0^1 \int_0^3 x^2 \, dy \, dx = \int_0^1 3x^2 \, dx = 1$

3. $\displaystyle\int_0^1 \int_0^3 xy^2 \, dy \, dx = \int_0^1 x \left[\frac{1}{3}y^3\right]_0^3 dx = \int_0^1 9x \, dx = \frac{9}{2}$

5. $\displaystyle\int_0^1 \int_0^x xy^3 \, dy \, dx = \int_0^1 x \left[\frac{1}{4}y^4\right]_0^x dx = \int_0^1 \frac{1}{4}x^5 \, dx = \frac{1}{24}$

7. $\displaystyle\int_0^{\pi/2} \int_0^{\pi/2} \sin(x+y) \, dy \, dx = \int_0^{\pi/2} \left[-\cos(x+y)\right]_0^{\pi/2} dx = \int_0^{\pi/2} \left[\cos x - \cos\left(x + \frac{\pi}{2}\right)\right] dx = 2$

9. $\displaystyle\int_0^{\pi/2} \int_0^{\pi/2} (1+xy) \, dy \, dx = \int_0^{\pi/2} \left[y + \frac{1}{2}xy^2\right]_0^{\pi/2} dx = \int_0^{\pi/2} \left(\frac{1}{2}\pi + \frac{1}{8}\pi^2 x\right) dx = \frac{1}{4}\pi^2 + \frac{1}{64}\pi^4$

11. $\displaystyle\int_0^1 \int_{y^2}^y \sqrt{xy} \, dx \, dy = \int_0^1 \sqrt{y} \left[\frac{2}{3}x^{3/2}\right]_{y^2}^y dy = \int_0^1 \frac{2}{3}\left(y^2 - y^{7/2}\right) dy = \frac{2}{27}$

13. $\displaystyle\int_{-2}^2 \int_{\frac{1}{2}y^2}^{4-\frac{1}{2}y^2} \left(4 - y^2\right) dx \, dy = \int_{-2}^2 \left(4 - y^2\right) \left[\left(4 - \frac{1}{2}y^2\right) - \left(\frac{1}{2}y^2\right)\right] dy$

$$= 2 \int_0^2 \left(16 - 8y^2 + y^4\right) dy = \frac{512}{15}$$

15. 0 by symmetry (integrand odd in y, Ω symmetric about x-axis)

17. $\displaystyle\int_0^2 \int_0^{x/2} e^{x^2} \, dy \, dx = \int_0^2 \frac{1}{2} x e^{x^2} \, dx = \left[\frac{1}{4}e^{x^2}\right]_0^2 = \frac{1}{4}\left(e^4 - 1\right)$

19.

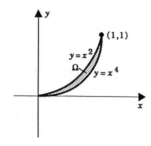

$$\int_0^1 \int_{y^{1/2}}^{y^{1/4}} f(x,y)\, dx\, dy$$

21.

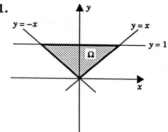

$$\int_{-1}^0 \int_{-x}^1 f(x,y)\, dy\, dx + \int_0^1 \int_x^1 f(x,y)\, dy\, dx$$

23.

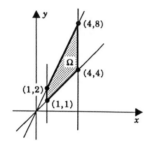

$$\int_1^2 \int_1^y f(x,y)\, dx\, dy + \int_2^4 \int_{y/2}^y f(x,y)\, dx\, dy$$

$$+ \int_4^8 \int_{y/2}^4 f(x,y)\, dx\, dy$$

25. $\displaystyle \int_{-2}^4 \int_{1/4x^2}^{\frac{1}{2}x+2} dy\, dx = \int_{-2}^4 \left[\frac{1}{2}x + 2 - \frac{1}{4}x^2 \right] dx = 9$

27. $\displaystyle \int_0^{1/4} \int_{2y^{3/2}}^y dx\, dy = \int_0^{1/4} \left[y - 2y^{3/2} \right] dy = \frac{1}{160}$

29.

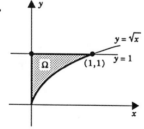

$$\int_0^1 \int_0^{y^2} \sin\left(\frac{y^3+1}{2} \right) dx\, dy = \int_0^1 y^2 \sin\left(\frac{y^3+1}{2} \right) dy$$

$$= \left[-\frac{2}{3} \cos\left(\frac{y^3+1}{2} \right) \right]_0^1$$

$$= \tfrac{2}{3}\left(\cos\tfrac{1}{2} - \cos 1 \right)$$

31.

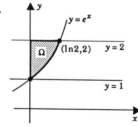

$$\int_0^{\ln 2} \int_{e^x}^2 e^{-x}\, dy\, dx = \int_0^{\ln 2} e^{-x}\left(2 - e^x \right) dx$$

$$= \left[-2e^{-x} - x \right]_0^{\ln 2} = 1 - \ln 2$$

33. $\displaystyle\int_1^2 \int_{y-1}^{2/y} dx\, dy = \int_1^2 \left[\frac{2}{y} - (y-1)\right] dy = \ln 4 - \frac{1}{2}$

35. $\displaystyle\int_0^2 \int_0^{3-\frac{3}{2}x} \left(4 - 2x - \frac{4}{3}y\right) dy\, dx = \int_0^3 \int_0^{2-\frac{2}{3}y} \left(4 - 2x - \frac{4}{3}y\right) dx\, dy = 4$

37. $\displaystyle\int_0^2 \int_0^{1-\frac{1}{2}x} x^3 y\, dy\, dx = \int_0^1 \int_0^{2-2y} x^3 y\, dx\, dy = \frac{2}{15}$

39.
$$\int_0^2 \int_{-\sqrt{2x-x^2}}^{\sqrt{2x-x^2}} (2x+1)\, dy\, dx = \int_{-1}^1 \int_{1-\sqrt{1-y^2}}^{1+\sqrt{1-y^2}} (2x+1)\, dx\, dy$$

$$= \int_{-1}^1 \left[x^2 + x\right]_{1-\sqrt{1-y^2}}^{1+\sqrt{1-y^2}} dy$$

$$= 6\int_{-1}^1 \sqrt{1-y^2}\, dy = 6\left(\frac{\pi}{2}\right) = 3\pi$$

41. $\displaystyle\int_0^1 \int_y^1 e^{y/x}\, dx\, dy = \int_0^1 \int_0^x e^{y/x}\, dy\, dx = \int_0^1 \left[x e^{y/x}\right]_0^x dx = \int_0^1 x(e-1)\, dx = \frac{1}{2}(e-1)$

43. $\displaystyle\int_0^1 \int_x^1 x^2 e^{y^4}\, dy\, dx = \int_0^1 \int_0^y x^2 e^{y^4}\, dx\, dy = \int_0^1 \left[\frac{1}{3}x^3 e^{y^4}\right]_0^y dy = \frac{1}{3}\int_0^1 y^3 e^{y^4}\, dy = \frac{1}{12}(e-1)$

45.
$$\iint_R f(x)g(y)\, dx dy = \int_c^d \int_a^b f(x)g(y)\, dx\, dy = \int_c^d \left(\int_a^b f(x)g(y)\, dx\right) dy$$

$$= \int_c^d g(y)\left(\int_a^b f(x)\, dx\right) dy = \left(\int_a^b f(x)\, dx\right)\left(\int_c^d g(y)\, dy\right)$$

47. We have $R:\ -a \le x \le a,\quad c \le y \le d.$ Set $f(x,y) = g_y(x).$ For each fixed $y \in [c,d]$, g_y is an odd function. Thus

$$\int_{-a}^a g_y(x)\, dx = 0. \tag{5.8.2}$$

Therefore

$$\iint_R f(x,y)\, dx dy = \int_c^d \int_{-a}^a f(x,y)\, dx\, dy$$

$$= \int_c^d \int_{-a}^a g_y(x)\, dx\, dy = \int_c^d 0\, dy = 0.$$

49. Note that $\Omega = \{(x,y): 0 \leq x \leq y,\ \ 0 \leq y \leq 1\}.$

Set $\Omega' = \{(x,y): 0 \leq y \leq x,\ \ 0 \leq x \leq 1\}.$

$$\iint\limits_{\Omega} f(x)f(y)\,dx\,dy = \int_0^1 \int_0^y f(x)f(y)\,dx\,dy$$

$$\underset{\underset{x \text{ and } y \text{ are dummy variables}}{\uparrow}}{=} \int_0^1 \int_0^x f(y)f(x)\,dy\,dx$$

$$= \int_0^1 \int_0^x f(x)f(y)\,dy\,dx = \iint\limits_{\Omega'} f(x)f(y)\,dx\,dy.$$

Note that Ω and Ω' don't overlap and their union is the unit square

$$R = \{(x,y): 0 \leq x \leq 1,\ \ 0 \leq y \leq 1\}.$$

If $\displaystyle\int_0^1 f(x)\,dx = 0,$ then

$$0 = \left(\int_0^1 f(x)\,dx\right)\left(\int_0^1 f(y)\,dy\right) \underset{\underset{\text{by Exercise 45}}{\uparrow}}{=} \iint\limits_{R} f(x)f(y)\,dx\,dy$$

$$= \iint\limits_{\Omega} f(x)f(y)\,dx\,dy + \iint\limits_{\Omega'} f(x)f(y)\,dx\,dy$$

$$= 2\iint\limits_{\Omega} f(x)f(y)\,dx\,dy$$

and therefore $\displaystyle\iint\limits_{\Omega} f(x)f(y)\,dx\,dy = 0.$

51. Let M be the maximum value of $|f(x,y)|$ on Ω.

$$\int_{\phi_1(x+h)}^{\phi_2(x+h)} = \int_{\phi_1(x+h)}^{\phi_1(x)} + \int_{\phi_1(x)}^{\phi_2(x)} + \int_{\phi_2(x)}^{\phi_2(x+h)}$$

$$|F(x+h) - F(x)| = \left| \int_{\phi_1(x+h)}^{\phi_2(x+h)} f(x,y)\,dy - \int_{\phi_1(x)}^{\phi_2(x)} f(x,y)\,dy \right|$$

$$= \left| \int_{\phi_1(x+h)}^{\phi_1(x)} f(x,y)\,dy + \int_{\phi_2(x)}^{\phi_2(x+h)} f(x,y)\,dy \right|$$

$$\leq \left| \int_{\phi_1(x+h)}^{\phi_1(x)} f(x,y)\,dy \right| + \left| \int_{\phi_2(x)}^{\phi_2(x+h)} f(x,y)\,dy \right|$$

$$\leq |\phi_1(x) - \phi_1(x+h)|\,M + |\phi_2(x+h) - \phi_2(x)|\,M.$$

The expression on the right tends to 0 as h tends to 0 since ϕ_1 and ϕ_2 are continuous.

SECTION 17.5

1. (a) $\Gamma: 0 \le \theta \le 2\pi, \quad 0 \le r \le 1$

$$\iint_{\Gamma} (\cos r^2) r \, dr d\theta = \int_0^{2\pi} \int_0^1 (\cos r^2) r \, dr \, d\theta$$

$$= 2\pi \int_0^1 r \cos r^2 \, dr = \pi \sin 1 \cong 0.84\,\pi$$

(b) $\Gamma: 0 \le \theta \le 2\pi, \quad 1 \le r \le 2$

$$\iint_{\Gamma} (\cos r^2) r \, dr d\theta = \int_0^{2\pi} \int_1^2 (\cos r^2) r \, dr \, d\theta$$

$$= 2\pi \int_1^2 r \cos r^2 \, dr = \pi(\sin 2 - \sin 1) \cong 0.07\pi$$

3. (a) $\Gamma: 0 \le \theta \le \pi/2, \quad 0 \le r \le 1$

$$\iint_{\Gamma} (r\cos\theta + r\sin\theta) r \, dr d\theta = \int_0^{\pi/2} \int_0^1 r^2(\cos\theta + \sin\theta) \, dr \, d\theta$$

$$= \left(\int_0^{\pi/2} (\cos\theta + \sin\theta) \, d\theta \right) \left(\int_0^1 r^2 \, dr \right) = 2\left(\frac{1}{3}\right) = \frac{2}{3}$$

(b) $\Gamma: 0 \le \theta \le \pi/2, \quad 1 \le r \le 2$

$$\iint_{\Gamma} (r\cos\theta + r\sin\theta) r \, dr d\theta = \int_0^{\pi/2} \int_1^2 r^2(\cos\theta + \sin\theta) \, dr \, d\theta$$

$$= \left(\int_0^{\pi/2} (\cos\theta + \sin\theta) \, d\theta \right) \left(\int_1^2 r^2 \, dr \right) = 2\left(\frac{7}{3}\right) = \frac{14}{3}$$

5. $\displaystyle \int_{-\pi/2}^{\pi/2} \int_0^1 r^2 \, dr \, d\theta = \frac{1}{3}\,\pi$

7. $\displaystyle \int_0^{\pi/3} \int_0^1 r^4 \, dr \, d\theta = \frac{1}{15}\,\pi$

9. $$\int_0^{2\pi} \int_0^b (r^2 \sin\theta + br) \, dr \, d\theta = \int_0^{2\pi} \left[\frac{1}{3}r^3 \sin\theta + \frac{b}{2}r^2 \right]_0^b \, d\theta$$

$$= b^3 \int_0^{2\pi} \left(\frac{1}{3}\sin\theta + \frac{1}{2} \right) \, d\theta = b^3 \pi$$

11. $$8\int_0^{\pi/2} \int_0^2 \frac{r}{2}\sqrt{12 - 3r^2} \, dr \, d\theta = 8 \int_0^{\pi/2} \left[-\frac{1}{18} (12 - 3r^2)^{3/2} \right]_0^2 \, d\theta$$

$$= 8 \int_0^{\pi/2} \frac{4}{3}\sqrt{3} \, d\theta = \frac{16}{3}\sqrt{3}\,\pi$$

13. $\displaystyle\int_0^{2\pi}\int_0^1 r\sqrt{4-r^2}\,dr\,d\theta = \int_0^{2\pi}\left[-\frac{1}{3}\left(4-r^2\right)^{3/2}\right]_0^1 d\theta$

$$= \int_0^{2\pi}\left(\frac{8}{3}-\sqrt{3}\right)d\theta = \frac{2}{3}(8-3\sqrt{3})\pi$$

15. $\displaystyle\int_{-\pi/2}^{\pi/2}\int_0^{2\cos\theta} 2r^2\cos\theta\,dr\,d\theta = \int_{-\pi/2}^{\pi/2}\left[\frac{2}{3}r^3\cos\theta\right]_0^{2\cos\theta} d\theta$

$$= \int_{-\pi/2}^{\pi/2}\frac{16}{3}\cos^4\theta\,d\theta = \frac{32}{3}\int_0^{\pi/2}\cos^4\theta\,d\theta = \frac{32}{3}\left(\frac{3}{16}\pi\right) = 2\pi$$

Section 8.3 ↑

17. $\displaystyle\frac{b}{a}\int_0^{\pi}\int_0^{a\sin\theta} r\sqrt{a^2-r^2}\,dr\,d\theta = \frac{b}{a}\int_0^{\pi}\left[-\frac{1}{3}\left(a^2-r^2\right)^{3/2}\right]_0^{a\sin\theta} d\theta$

$$= \frac{1}{3}a^2 b\int_0^{\pi}\left(1-\cos^3\theta\right)d\theta = \frac{1}{3}\pi a^2 b$$

SECTION 17.6

1. $\Omega:\ -L/2\le x\le L/2,\quad -W/2\le y\le W/2$

$$I_x = \iint_\Omega \frac{M}{LW}y^2\,dx dy = \frac{4M}{LW}\int_0^{W/2}\int_0^{L/2} y^2\,dx\,dy = \frac{1}{12}MW^2$$

symmetry ↑⌋

$$I_y = \iint_\Omega \frac{M}{LW}x^2\,dx dy = \frac{1}{12}ML^2$$

$$I_z = \iint_\Omega \frac{M}{LW}\left(x^2+y^2\right)dx dy = \frac{1}{12}M\left(L^2+W^2\right)$$

$$K_x = \sqrt{I_x/M} = W/2\sqrt{3}$$

$$K_y = \sqrt{I_y/M} = L/2\sqrt{3}$$

$$K_z = \sqrt{I_z/M} = \sqrt{L^2+W^2}\Big/2\sqrt{3}$$

3.

$$M = \iint_{\Omega} k\left(x + \frac{L}{2}\right) dx\,dy = \iint_{\Omega} \frac{1}{2}kL\,dx\,dy = \frac{1}{2}kL(\text{ area of }\Omega) = \frac{1}{2}kL^2 W$$

$$\uparrow$$
$$\text{symmetry}$$

$$x_M M = \iint_{\Omega} x\left[k\left(x + \frac{L}{2}\right)\right] dx\,dy = \iint_{\Omega} \left(kx^2 + \frac{1}{2}Lx\right) dx\,dy$$

$$= \iint_{\Omega} kx^2\,dx\,dy = 4k\int_0^{W/2}\int_0^{L/2} x^2\,dx\,dy = \frac{1}{12}kWL^3$$

$$\uparrow \qquad \uparrow$$
$$\text{symmetry} \quad \text{symmetry}$$

$$= \tfrac{1}{6}\left(\tfrac{1}{2}kL^2 W\right) L = \tfrac{1}{6}ML; \quad x_M = \tfrac{1}{6}L$$

$$y_M M = \iint_{\Omega} y\left[k\left(x + \frac{L}{2}\right)\right] dx\,dy = 0; \quad y_M = 0$$

$$\uparrow$$
$$\text{by symmetry}$$

5.

$$I_x = \iint_{\Omega} \frac{4M}{\pi R^2}y^2\,dx\,dy = \frac{4M}{\pi R^2}\int_0^{\pi/2}\int_0^R r^3\sin^2\theta\,dr\,d\theta$$

$$= \frac{4M}{\pi R^2}\left(\int_0^{\pi/2}\sin^2\theta\,d\theta\right)\left(\int_0^R r^3\,dr\right) = \frac{4M}{\pi R^2}\left(\frac{\pi}{4}\right)\left(\frac{1}{4}R^4\right) = \frac{1}{4}MR^2$$

$$I_y = \tfrac{1}{4}MR^2, \quad I_z = \tfrac{1}{2}MR^2$$

$$K_x = K_y = \tfrac{1}{2}R, \quad K_z = R/\sqrt{2}$$

7. I_M, the moment of inertia about the vertical line through the center of mass, is

$$\iint_{\Omega} \frac{M}{\pi R^2}\left(x^2 + y^2\right) dx\,dy$$

where Ω is the disc of radius R centered at the origin. Therefore

$$I_M = \frac{M}{\pi R^2}\int_0^{2\pi}\int_0^R r^3\,dr\,d\theta = \frac{1}{2}MR^2.$$

We need $I_0 = \frac{1}{2}MR^2 + d^2 M$ where d is the distance from the center of the disc to the origin. Solving this equation for d, we have $d = \sqrt{I - \frac{1}{2}MR^2}\Big/\sqrt{M}$.

9. $\Omega:\ 0 \le x \le a,\quad 0 \le y \le b$

$$I_x = \iint_\Omega \frac{4M}{\pi ab} y^2 \, dx\, dy = \frac{4M}{\pi ab} \int_0^a \int_0^{\frac{b}{a}\sqrt{a^2-x^2}} y^2 \, dy\, dx = \frac{1}{4} M b^2$$

$$I_y = \iint_\Omega \frac{4M}{\pi ab} x^2 \, dx\, dy = \frac{4M}{\pi ab} \int_0^a \int_0^{\frac{b}{a}\sqrt{a^2-x^2}} x^2 \, dy\, dx = \frac{1}{4} M a^2$$

$$I_z = \tfrac{1}{4} M \left(a^2 + b^2\right)$$

11. $\Omega:\ r_1{}^2 \le x^2 + y^2 \le r_2{}^2,\quad A = \pi \left(r_2{}^2 - r_1{}^2\right)$

(a)   Place the diameter on the x-axis.

$$I_x = \iint_\Omega \frac{M}{A} y^2 \, dx\, dy = \frac{M}{A} \int_0^{2\pi} \int_{r_1}^{r_2} \left(r^2 \sin^2 \theta\right) r \, dr\, d\theta = \frac{1}{4} M \left(r_2{}^2 + r_1{}^2\right)$$

(b) $\tfrac{1}{4} M \left(r_2{}^2 + r_1{}^2\right) + M r_1{}^2 = \tfrac{1}{4} M \left(r_2{}^2 + 5 r_1{}^2\right)$ (parallel axis theorem)

(c) $\tfrac{1}{4} M \left(r_2{}^2 + r_1{}^2\right) + M r_2{}^2 = \tfrac{1}{4} M \left(5 r_2{}^2 + r_1{}^2\right)$

13. $\Omega:\ r_1{}^2 \le x^2 + y^2 \le r_2{}^2,\quad A = \pi \left(r_2{}^2 - r_1{}^2\right)$

$$I = \iint_\Omega \frac{M}{A} \left(x^2 + y^2\right) \, dx\, dy = \frac{M}{A} \int_0^{2\pi} \int_{r_1}^{r_2} r^3 \, dr\, d\theta = \frac{1}{2} M (r_2{}^2 + r_1{}^2)$$

15.

$$M = \iint_\Omega k \left(R - \sqrt{x^2 + y^2}\right) \, dx\, dy = k \int_0^\pi \int_0^R \left(Rr - r^2\right) \, dr\, d\theta = \frac{1}{6} k \pi R^3$$

$x_M = 0$ by symmetry

$$y_M M = \iint_\Omega y \left[k \left(R - \sqrt{x^2 + y^2}\right)\right] \, dx\, dy = k \int_0^\pi \int_0^R \left(Rr^2 - r^3\right) \sin\theta \, dr\, d\theta = \frac{1}{6} k R^4$$

$y_M = R/\pi$

17.   Place P at the origin.

$$M = \iint_\Omega k\sqrt{x^2 + y^2} \, dx\, dy$$

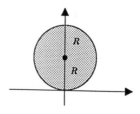

$$= k \int_0^\pi \int_0^{2R\sin\theta} r^2 \, dr\, d\theta = \frac{32}{9} k R^3$$

$x_M = 0$ by symmetry

$$y_M M = \iint_\Omega y \left(k\sqrt{x^2 + y^2}\right) \, dx\, dy = k \int_0^\pi \int_0^{2R\sin\theta} r^3 \sin\theta \, dr\, d\theta = \frac{64}{15} k R^4$$

$y_M = 6R/5$

Answer: the center of mass lies on the diameter through P at a distance $6R/5$ from P.

19. Suppose Ω, a basic region of area A, is broken up into n basic regions $\Omega_1, \cdots, \Omega_n$ with areas $A_1, \cdots, A_n$. Then

$$\bar{x}A = \iint\limits_{\Omega} x\,dx\,dy = \sum_{i=1}^{n} \left(\iint\limits_{\Omega_i} x\,dx\,dy \right) = \sum_{i=1}^{n} \bar{x}_i\,A_i = \bar{x}_1\,A_1 + \cdots + \bar{x}_n\,A_n.$$

The second formula can be derived in a similar manner.

SECTION 17.7

1. They are equal; they both give the volume of T.

3. Let $P_1 = \{x_0, \cdots, x_m\}$, $P_2 = \{y_0, \cdots, y_n\}$, $P_3 = \{z_0, \cdots, z_q\}$ be partitions of $[0,a]$, $[0,b]$, $[0,c]$ respectively and let $P = P_1 \times P_2 \times P_3$. Note that

$$x_{i-1}y_{j-1} \leq \left(\frac{x_i + x_{i-1}}{2} \right) \left(\frac{y_j + y_{j-1}}{2} \right) \leq x_i y_j$$

and therefore

$$x_{i-1}y_{j-1}\,\Delta x_i\,\Delta y_j\,\Delta z_k \leq \tfrac{1}{8} \left(x_i^2 - x_{i-1}^2 \right) \left(y_j^2 - y_{j-1}^2 \right) \Delta z_k \leq x_i y_j\,\Delta x_i\,\Delta y_j\,\Delta z_k.$$

It follows that

$$L_f(P) \leq \frac{1}{8} \sum_{i=1}^{m}\sum_{j=1}^{n}\sum_{k=1}^{q} \left(x_i^2 - x_{i-1}^2 \right)\left(y_j^2 - y_{j-1}^2 \right)\Delta z_k \leq U_f(P).$$

The middle term can be written

$$\frac{1}{8} \left(\sum_{i=1}^{m} x_i^2 - x_{i-1}^2 \right) \left(\sum_{j=1}^{n} y_j^2 - y_{j-1}^2 \right) \left(\sum_{k=1}^{q} \Delta z_k \right) = \frac{1}{8}\,a^2 b^2 c.$$

5. $\bar{x}_1 = a$, $\bar{y}_1 = b$, $\bar{z}_1 = c$; $\bar{x}_0 = A$, $\bar{y}_0 = B$, $\bar{z}_0 = C$

$$\bar{x}_1 V_1 + \bar{x}V = \bar{x}_0 V_0 \quad\Longrightarrow\quad a^2 bc + (ABC - abc)\bar{x} = A^2 BC$$

$$\Longrightarrow \quad \bar{x} = \frac{A^2 BC - a^2 bc}{ABC - abc}$$

similarly

$$\bar{y} = \frac{AB^2 C - ab^2 c}{ABC - abc}, \quad \bar{z} = \frac{ABC^2 - abc^2}{ABC - abc}$$

7. $M = \iiint\limits_{\Pi} Kz\,dx\,dy\,dz$

Let $P_1 = \{x_0, \cdots, x_m\}$, $P_2 = \{y_0, \cdots, y_n\}$, $P_3 = \{z_0, \cdots, z_q\}$ be partitions of $[0,a]$ and let $P = P_1 \times P_2 \times P_3$.   Note that

$$z_{k-1} \leq \tfrac{1}{2}\left(z_k + z_{k-1} \right) \leq z_k$$

hoffset .4in

and therefore

$$Kz_{k-1}\,\Delta x_i\,\Delta y_j\,\Delta z_k \le \tfrac{1}{2}K\,\Delta x_i\,\Delta y_j\left(z_k{}^2 - z_{k-1}^2\right) \le Kz_k\,\Delta x_i\,\Delta y_j\,\Delta z_k.$$

It follows that

$$L_f(P) \le \frac{1}{2}K\sum_{i=1}^{m}\sum_{j=1}^{n}\sum_{k=1}^{q}\Delta x_i\,\Delta y_j\left(z_k{}^2 - z_{k-1}^2\right) \le U_f(P).$$

The middle term can be written

$$\frac{1}{2}K\left(\sum_{i=1}^{m}\Delta x_i\right)\left(\sum_{j=1}^{n}\Delta y_j\right)\left(\sum_{k=1}^{q}z_k{}^2 - z_{k-1}^2\right) = \frac{1}{2}K(a)(a)(a^2) = \frac{1}{2}Ka^4.$$

$M = \tfrac{1}{2}Ka^4$ where K is the constant of proportionality for the density function.

9.
$$I_z = \iiint\limits_{\Pi} Kz\left(x^2 + y^2\right)\,dx\,dy\,dz$$

$$= \underbrace{\iiint\limits_{\Pi} Kx^2z\,dx\,dy\,dz}_{I_1} + \underbrace{\iiint\limits_{\Pi} Ky^2z\,dx\,dy\,dz}_{I_2}.$$

We will calculate I_1 using the partitions we used in doing Exercise 7.   Note that

$$x_{i-1}^2 z_{k-1} \le \left(\frac{x_i{}^2 + x_i x_{i-1} + x_{i-1}^2}{3}\right)\left(\frac{z_k + z_{k-1}}{2}\right) \le x_i{}^2 z_k$$

and therefore

$$Kx_{i-1}^2 z_{k-1}\,\Delta x_i\,\Delta y_j\,\Delta z_k \le \tfrac{1}{6}K\left(x_i{}^3 - x_{i-1}^3\right)\Delta y_j\left(z_k{}^2 - z_{k-1}^2\right) \le Kx_i{}^2 z_k{}^2\,\Delta x_i\,\Delta y_j\,\Delta z_k.$$

It follows that

$$L_f(P) \le \frac{1}{6}K\sum_{i=1}^{m}\sum_{j=1}^{n}\sum_{k=1}^{q}\left(x_i{}^3 - x_{i-1}^3\right)\Delta y_j\left(z_k{}^2 - z_{k-1}^2\right) \le U_f(P).$$

The middle term can be written

$$\frac{1}{6}K\left(\sum_{i=1}^{m}x_i{}^3 - x_{i-1}^3\right)\left(\sum_{j=1}^{n}\Delta y_j\right)\left(\sum_{k=1}^{q}z_k{}^2 - z_{k-1}^2\right) = \frac{1}{6}Ka^3(a)(a^2) = \frac{1}{6}Ka^6.$$

Similarly $I_2 = \tfrac{1}{6}Ka^6$ and therefore

by Exercise 7 ⤵

$$I_z = \tfrac{1}{3}Ka^6 = \tfrac{2}{3}\left(\tfrac{1}{2}Ka^4\right)a^2 = \tfrac{2}{3}Ma^2.$$

SECTION 17.8

1. $$\int_0^a\int_0^b\int_0^c dx\,dy\,dz = \int_0^a\int_0^b c\,dy\,dz = \int_0^a bc\,dz = abc$$

3. $\int_0^1 \int_1^{2y} \int_0^x (x+2z)\,dz\,dx\,dy = \int_0^1 \int_1^{2y} \left[xz + z^2\right]_0^x dx\,dy$

$= \int_0^1 \int_1^{2y} 2x^2\,dx\,dy = \int_0^1 \left[\frac{2}{3}x^3\right]_1^{2y} dy = \int_0^1 \left(\frac{16}{3}y^3 - \frac{2}{3}\right) dy = \frac{2}{3}$

5. $\iiint_\Pi f(x)g(y)h(z)\,dxdydz = \int_{c_1}^{c_2}\left[\int_{b_1}^{b_2}\left(\int_{a_1}^{a_2} f(x)g(y)h(z)\,dx\right)dy\right]dz$

$= \int_{c_1}^{c_2}\left[\int_{b_1}^{b_2} g(y)h(z)\left(\int_{a_1}^{a_2} f(x)\,dx\right)dy\right]dz$

$= \int_{c_1}^{c_2}\left[h(z)\left(\int_{a_1}^{a_2} f(x)\,dx\right)\left(\int_{b_1}^{b_2} g(y)\,dy\right)dz\right]$

$= \left(\int_{a_1}^{a_2} f(x)\,dx\right)\left(\int_{b_1}^{b_2} g(y)\,dy\right)\left(\int_{c_1}^{c_2} h(z)\,dz\right)$

7. $\left(\int_0^1 x^2\,dx\right)\left(\int_0^2 y^2\,dy\right)\left(\int_0^3 z^2\,dz\right) = \left(\frac{1}{3}\right)\left(\frac{8}{3}\right)\left(\frac{27}{3}\right) = 8$

9. $x_M M = \iiint_\Pi kx^2yz\,dxdydz = k\left(\int_0^a x^2\,dx\right)\left(\int_0^b y\,dy\right)\left(\int_0^c z\,dz\right)$

$= k\left(\frac{1}{3}a^3\right)\left(\frac{1}{2}b^2\right)\left(\frac{1}{2}c^2\right) = \frac{1}{12}ka^3b^2c^2.$

By Exercise 8, $M = \frac{1}{8}ka^2b^2c^2.$ Therefore $\bar{x} = \frac{2}{3}a.$ Similarly, $\bar{y} = \frac{2}{3}b$ and $\bar{z} = \frac{2}{3}c.$

11.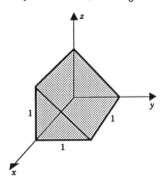

13. center of mass is the centroid

$\bar{x} = \frac{1}{2}$ by symmetry

$\bar{y}V = \iiint_T y\,dxdydz = \int_0^1 \int_0^1 \int_0^{1-y} y\,dz\,dy\,dx = \int_0^1 \int_0^1 (y - y^2)\,dy\,dx$

$= \int_0^1 \left[\frac{1}{2}y^2 - \frac{1}{3}y^3\right]_0^1 dx = \int_0^1 \frac{1}{6}\,dx = \frac{1}{6}$

$$\bar{z}V = \iiint_T z\,dxdydz = \int_0^1 \int_0^1 \int_0^{1-y} z\,dz\,dy\,dx = \int_0^1 \int_0^1 \frac{1}{2}(1-y)^2\,dy\,dx$$

$$= \frac{1}{2}\int_0^1 \int_0^1 (1-2y+y^2)\,dy\,dx = \frac{1}{2}\int_0^1 \left[y - y^2 + \frac{1}{3}y^3\right]_0^1 dx = \frac{1}{2}\int_0^1 \frac{1}{3}\,dx = \frac{1}{6}$$

$V = \frac{1}{2}$ (by Exercise 12); $\bar{y} = \frac{1}{3},\quad \bar{z} = \frac{1}{3}$

15. $\displaystyle\int_{-r}^{r} \int_{-\phi(x)}^{\phi(x)} \int_{-\psi(x,y)}^{\psi(x,y)} k\left(r - \sqrt{x^2 + y^2 + z^2}\right) dz\,dy\,dx$ with $\phi(x) = \sqrt{r^2 - x^2}$,

$\psi(x,y) = \sqrt{r^2 - (x^2 + y^2)}$, k the constant of proportionality

17. $\displaystyle\int_0^1 \int_{-\sqrt{x-x^2}}^{\sqrt{x-x^2}} \int_{-2x-3y-10}^{1-y^2} dz\,dy\,dx$

19. $\displaystyle\int_{-1}^{1} \int_{-2\sqrt{2-2x^2}}^{2\sqrt{2-2x^2}} \int_{3x^2+y^2/4}^{4-x^2-y^2/4} k\left(z - 3x^2 - \frac{1}{4}y^2\right) dz\,dy\,dx$

21. $\displaystyle V = \int_0^2 \int_{x^2}^{x+2} \int_0^x dz\,dy\,dx = \int_0^2 \int_{x^2}^{x+2} x\,dy\,dx = \int_0^2 (x^2 + 2x - x^3)\,dx = \frac{8}{3}$

$$\bar{x}V = \int_0^2 \int_{x^2}^{x+2} \int_0^x x\,dz\,dy\,dx = \int_0^2 \int_{x^2}^{x+2} x^2\,dy\,dx = \int_0^2 (x^3 + 2x^2 - x^4)\,dx = \frac{44}{15}$$

$$\bar{y}V = \int_0^2 \int_{x^2}^{x+2} \int_0^x y\,dz\,dy\,dx = \int_0^2 \int_{x^2}^{x+2} xy\,dy\,dx = \int_0^2 \frac{1}{2}(x^3 + 4x^2 + 4x - x^5)\,dx = 6$$

$$\bar{z}V = \int_0^2 \int_{x^2}^{x+2} \int_0^x z\,dz\,dy\,dx = \int_0^2 \int_{x^2}^{x+2} \frac{1}{2}x^2\,dy\,dx = \int_0^2 \frac{1}{2}(x^3 + 2x^2 - x^4)\,dx = \frac{22}{15}$$

$\bar{x} = \frac{11}{10}, \quad \bar{y} = \frac{9}{4}, \quad \bar{z} = \frac{11}{20}$

23. $\displaystyle V = \int_{-1}^2 \int_0^3 \int_{2-x}^{4-x^2} dz\,dy\,dx = \frac{27}{2};\quad (\bar{x}, \bar{y}, \bar{z}) = \left(\frac{1}{2}, \frac{3}{2}, \frac{12}{5}\right)$

25. $\displaystyle V = \int_0^a \int_0^{\phi(x)} \int_0^{\psi(x,y)} dz\,dy\,dx = \frac{1}{6}abc$ with $\phi(x) = b\left(1 - \frac{x}{a}\right),\quad \psi(x,y) = c\left(1 - \frac{x}{a} - \frac{y}{b}\right)$

$(\bar{x}, \bar{y}, \bar{z}) = \left(\frac{1}{4}a, \frac{1}{4}b, \frac{1}{4}c\right)$

27. $\Pi: 0 \le x \le a, \quad 0 \le y \le b, \quad 0 \le z \le c$

(a) $\displaystyle I_z = \int_0^a \int_0^b \int_0^c \frac{M}{abc}(x^2 + y^2)\,dz\,dy\,dx = \frac{1}{3}M(a^2 + b^2)$

(b) $I_M = I_z - d^2 M = \frac{1}{3} M (a^2 + b^2) - \frac{1}{4}(a^2 + b^2) M = \frac{1}{12} M (a^2 + b^2)$

       ⌐ parallel axis theorem (17.6.7)

(c) $I = I_M + d^2 M = \frac{1}{12} M (a^2 + b^2) + \frac{1}{4} a^2 M = \frac{1}{3} M a^2 + \frac{1}{12} M b^2$

       ⌐ parallel axis theorem (17.6.7)

29.
$$M = \int_0^1 \int_0^1 \int_0^y k \left(x^2 + y^2 + z^2\right) dz\, dy\, dx = \int_0^1 \int_0^1 k \left(x^2 y + y^3 + \frac{1}{3} y^3\right) dy\, dx$$
$$= \int_0^1 k \left(\frac{1}{2} x^2 + \frac{1}{3}\right) dx = \frac{1}{2} k$$

$(x_M, y_M, z_M) = \left(\frac{7}{12}, \frac{34}{45}, \frac{37}{90}\right)$

31. (a) 0 by symmetry

(b) $\underset{T}{\iiint} (a_1 x + a_2 y + a_3 z + a_4)\, dx\, dy\, dz = \underset{T}{\iiint} a_4\, dx\, dy\, dz = a_4$ (volume of ball) $= \frac{4}{3} \pi a_4$

                by symmetry ⌐

33.
$$M = \int_{-2}^2 \int_{-\sqrt{4-x^2}/2}^{\sqrt{4-x^2}/2} \int_{x^2+3y^2}^{4-y^2} k|x|\, dz\, dy\, dx = 4 \int_0^2 \int_0^{\sqrt{4-x^2}/2} \int_{x^2+3y^2}^{4-y^2} kx\, dz\, dy\, dx$$
$$= 4k \int_0^2 \int_0^{\sqrt{4-x^2}/2} \left(4x - x^3 - 4xy^2\right) dy\, dx = \frac{4}{3} k \int_0^2 x \left(4 - x^2\right)^{3/2} dx = \frac{128}{15} k$$

35. $M = \int_{-1}^2 \int_0^3 \int_{2-x}^{4-x^2} k(1+y)\, dz\, dy\, dx = \frac{135}{4} k; \quad (x_M, y_M, z_M) = \left(\frac{1}{2}, \frac{9}{5}, \frac{12}{5}\right)$

37. (a) $V = \int_0^6 \int_{z/2}^3 \int_x^{6-x} dy\, dx\, dz$

(b) $V = \int_0^3 \int_0^{2x} \int_x^{6-x} dy\, dz\, dx$

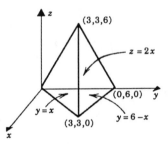

(c) $V = \int_0^6 \int_{z/2}^3 \int_{z/2}^y dx\, dy\, dz + \int_0^6 \int_3^{(12-z)/2} \int_{z/2}^{6-y} dx\, dy\, dz$

39. (a) $V = \underset{\Omega_{yz}}{\iint} 2y\, dy\, dz$

                                     (b) $V = \underset{\Omega_{yz}}{\iint} \left(\int_{-y}^y dx\right) dy\, dz$

(c) $V = \int_0^4 \int_{-\sqrt{4-y}}^{\sqrt{4-y}} \int_{-y}^y dx\, dz\, dy$

                                     (d) $V = \int_{-2}^2 \int_0^{4-z^2} \int_{-y}^y dx\, dy\, dz$

SECTION 17.9

1. Set the lower base of the cylinder on the xy-plane so that the axis of the cylinder coincides with the z-axis. Assume that the density varies directly as the distance from the lower base.

$$M = \int_0^{2\pi} \int_0^R \int_0^h kzr \, dz \, dr \, d\theta = \frac{1}{2} k\pi R^2 h^2$$

3.
$$I = I_z = k \int_0^{2\pi} \int_0^R \int_0^h zr^3 \, dr \, d\theta \, dz$$

$$= \frac{1}{4} k\pi R^4 h^2 = \frac{1}{2} \left(\frac{1}{2} k\pi R^2 h^2 \right) R^2 = \frac{1}{2} MR^2$$

from Exercise 1

5. Inverting the cone and placing the vertex at the origin, we have

$$V = \int_0^h \int_0^{2\pi} \int_0^{(R/h)z} r \, dr \, d\theta \, dz = \frac{1}{3} \pi R^2 h.$$

7. $\quad I = \dfrac{M}{V} \displaystyle\int_0^h \int_0^{2\pi} \int_0^{(R/h)z} r^3 \, dr \, d\theta \, dz = \dfrac{3}{10} MR^2$

9. $\quad V = \displaystyle\int_0^{2\pi} \int_0^1 \int_0^{1-r^2} r \, dz \, dr \, d\theta = \dfrac{1}{2}\pi$

11. $\quad M = \displaystyle\int_0^{2\pi} \int_0^1 \int_0^{1-r^2} k(r^2 + z^2) r \, dz \, dr \, d\theta = \dfrac{1}{4}k\pi$

13. $\displaystyle\int_0^{\pi/2} \int_0^1 \int_0^{\sqrt{4-r^2}} r \, dz \, dr \, d\theta = \int_0^{\pi/2} \int_0^1 r\sqrt{4 - r^2} \, dr \, d\theta = \int_0^{\pi/2} \left(\frac{8}{3} - \sqrt{3} \right) d\theta = \frac{1}{6} \left(8 - 3\sqrt{3} \right) \pi$

15.
$$V = \int_{-\pi/2}^{\pi/2} \int_0^{2a \cos \theta} \int_0^r r \, dz \, dr \, d\theta = \int_{-\pi/2}^{\pi/2} \int_0^{2a \cos \theta} r^2 \, dr \, d\theta$$

$$= \int_{-\pi/2}^{\pi/2} \frac{8}{3} a^3 \cos^3 \theta \, d\theta = \frac{32}{9} a^3$$

17.
$$V = \int_{-\pi/2}^{\pi/2} \int_0^{a \cos \theta} \int_0^{a-r} r \, dz \, dr \, d\theta = \int_{-\pi/2}^{\pi/2} \int_0^{a \cos \theta} r(a - r) \, dr \, d\theta$$

$$= \int_{-\pi/2}^{\pi/2} a^3 \left(\frac{1}{2} \cos^2 \theta - \frac{1}{3} \cos^3 \theta \right) d\theta = \frac{1}{36} a^2 (9\pi - 16)$$

19.
$$V = \int_{-\pi/2}^{\pi/2} \int_0^{\cos\theta} \int_{r^2}^{r\cos\theta} r\, dz\, dr\, d\theta = \int_{-\pi/2}^{\pi/2} \int_0^{\cos\theta} \left(r^2 \cos\theta - r^3\right) dr\, d\theta$$

$$= \int_{-\pi/2}^{\pi/2} \frac{1}{12} \cos^4\theta\, d\theta = \frac{1}{32}\pi$$

21.
$$V = \int_0^{2\pi} \int_0^{1/2} \int_{r\sqrt{3}}^{\sqrt{1-r^2}} r\, dz\, dr\, d\theta = \int_0^{2\pi} \int_0^{1/2} \left(r\sqrt{1-r^2} - r^2\sqrt{3}\right) dr\, d\theta = \frac{1}{3}\pi\left(2 - \sqrt{3}\right)$$

SECTION 17.10

1. $\left(\sqrt{3},\ \frac{1}{4}\pi,\ \cos^{-1}\left[\frac{1}{3}\sqrt{3}\right]\right)$ **3.** $\left(\frac{3}{4},\ \frac{3}{4}\sqrt{3},\ \frac{3}{2}\sqrt{3}\right)$

5. The circular cylinder $x^2 + y^2 = 1$;   the radius of the cylinder is 1 and the axis is the z-axis.

7. The lower nappe of the circular cone $z^2 = x^2 + y^2$.

9. Horizontal plane one unit above the xy-plane.

11. $V = \int_0^{2\pi} \int_0^{\pi} \int_0^{R} \rho^2 \sin\phi\, d\rho\, d\phi\, d\theta = \frac{4}{3}\pi R^3$

13. $V = \int_0^{\alpha} \int_0^{\pi} \int_0^{R} \rho^2 \sin\phi\, d\rho\, d\phi\, d\theta = \frac{2}{3}\alpha R^3$

15.
$$M = \int_0^{2\pi} \int_0^{\tan^{-1}(r/h)} \int_0^{h\sec\phi} k\rho^3 \sin\phi\, d\rho\, d\phi\, d\theta$$

$$= \int_0^{2\pi} \int_0^{\tan^{-1}(r/h)} \frac{kh^4}{4} \tan\phi \sec^3\phi\, d\phi\, d\theta$$

$$= \frac{kh^4}{4} \int_0^{2\pi} \frac{1}{3} \left[\sec^3\phi\right]_0^{\tan^{-1}(r/h)} d\theta = \frac{kh^4}{4} \int_0^{2\pi} \frac{1}{3}\left[\left(\frac{\sqrt{r^2+h^2}}{h}\right)^3 - 1\right] d\theta$$

$$= \frac{1}{6}k\pi h\left[(r^2+h^2)^{3/2} - h^3\right]$$

17. center ball at origin;   density $= \dfrac{M}{V} = \dfrac{3M}{4\pi R^3}$

   (a) $I = \dfrac{3M}{4\pi R^3} \displaystyle\int_0^{2\pi} \int_0^{\pi} \int_0^{R} \rho^4 \sin^3\phi\, d\rho\, d\phi\, d\theta = \dfrac{2}{5}MR^2$

   (b) $I = \frac{2}{5}MR^2 + R^2 M = \frac{7}{5}MR^2$ (parallel axis theorem)

19. center balls at origin;   density $= \dfrac{M}{V} = \dfrac{3M}{4\pi\left(R_2^{\ 3} - R_1^{\ 3}\right)}$

   (a) $I = \dfrac{3M}{4\pi\left(R_2^{\ 3} - R_1^{\ 3}\right)} \displaystyle\int_0^{2\pi} \int_0^{\pi} \int_{R_1}^{R_2} \rho^4 \sin^3\phi\, d\rho\, d\phi\, d\theta = \dfrac{2}{5}M\left(\dfrac{R_2^{\ 5} - R_1^{\ 5}}{R_2^{\ 3} - R_1^{\ 3}}\right)$

This result can be derived from Exercise 17 without further integration.   View the solid as a ball of mass M_2 from which is cut out a core of mass M_1.

$$M_2 = \frac{M}{V}V_2 = \frac{3M}{4\pi(R_2{}^3 - R_1{}^3)}\left(\frac{4}{3}\pi R_2{}^3\right) = \frac{MR_2{}^3}{R_2{}^3 - R_1{}^3}; \quad \text{similarly} \quad M_1 = \frac{MR_1{}^3}{R_2{}^3 - R_1{}^3}.$$

Then

$$I = I_2 - I_1 = \frac{2}{5}M_2R_2{}^2 - \frac{2}{5}M_1R_1{}^2 = \frac{2}{5}\left(\frac{MR_2{}^3}{R_2{}^3 - R_1{}^3}\right)R_2{}^2 - \frac{2}{5}\left(\frac{MR_1{}^3}{R_2{}^3 - R_1{}^3}\right)R_1{}^2$$

$$= \frac{2}{5}M\left(\frac{R_2{}^5 - R_1{}^5}{R_2{}^3 - R_1{}^3}\right).$$

(b)   Outer radius R and inner radius R_1 gives

$$\text{moment of inertia} = \frac{2}{5}M\left(\frac{R^5 - R_1{}^5}{R^3 - R_1{}^3}\right). \qquad \text{[part }(a)\text{]}$$

As $R_1 \to R$,

$$\frac{R^5 - R_1{}^5}{R^3 - R_1{}^3} = \frac{R^4 + R^3R_1 + R^2R_1{}^2 + RR_1{}^3 + R_1{}^4}{R^2 + RR_1 + R_1{}^2} \longrightarrow \frac{5R^4}{3R^2} = \frac{5}{3}R^2.$$

Thus the moment of inertia of spherical shell of radius R is

$$\tfrac{2}{5}M\left(\tfrac{5}{3}R^2\right) = \tfrac{2}{3}MR^2.$$

(c) $I = \frac{2}{3}MR^2 + R^2M = \frac{5}{3}MR^2$ (parallel axis theorem)

21. $V = \displaystyle\int_0^{2\pi}\int_0^{\alpha}\int_0^a \rho^2 \sin\phi \, d\rho \, d\phi \, d\theta = \frac{2}{3}\pi\left(1 - \cos\alpha\right)a^3$

23.   (a)   Substituting $x = \rho\sin\phi\cos\theta, \quad y = \rho\sin\phi\sin\theta, \quad z = \rho\cos\phi$

into $x^2 + y^2 + (z - R)^2 = R^2$

we have $\rho^2\sin^2\phi + (\rho\cos\phi - R)^2 = R^2,$

which simplifies to $\rho = 2R\cos\phi.$

(b) $0 \leq \theta \leq 2\pi, \quad 0 \leq \phi \leq \pi/4, \quad R\sec\phi \leq \rho \leq 2R\cos\phi$

25. $$V = \int_0^{2\pi}\int_0^{\pi/4}\int_0^{2}\rho^2\sin\phi\,d\rho\,d\phi\,d\theta + \int_0^{2\pi}\int_{\pi/4}^{\pi/2}\int_0^{2\sqrt{2}\cos\phi}\rho^2\sin\phi\,d\rho\,d\phi\,d\theta$$

$$= \frac{1}{3}\left(16 - 6\sqrt{2}\right)\pi$$

27.   Encase T in a spherical wedge W. W has spherical coordinates in a box Π that contains S.   Define f to be zero outside of T.   Then

$$F(\rho, \theta, \phi) = f\left(\rho\sin\phi\cos\theta, \; \rho\sin\phi\sin\theta, \; \rho\cos\phi\right)$$

is zero outside of S and

$$\iiint_T f(x,y,z)\,dxdydz = \iiint_W f(x,y,z)\,dxdydz$$

$$= \iiint_\Pi F(\rho,\theta,\phi)\,\rho^2 \sin\phi\,d\rho d\theta d\phi$$

$$= \iiint_S F(\rho,\theta,\phi)\,\rho^2 \sin\phi\,d\rho d\theta d\phi.$$

29. T is the set of all (x,y,z) with spherical coordinates (ρ,θ,ϕ) in the set

$$S: \quad 0 \le \theta \le 2\pi, \quad 0 \le \phi \le \pi/4, \quad R\sec\phi \le \rho \le 2R\cos\phi.$$

T has volume $V = \frac{2}{3}\pi R^3$. By symmetry the $\mathbf{i}, \mathbf{j}$ components of force are zero and

$$\mathbf{F} = \left\{ \frac{3GmM}{2\pi R^3} \iiint_T \frac{z}{(x^2+y^2+z^2)^{3/2}}\,dxdydz \right\}\mathbf{k}$$

$$= \left\{ \frac{3GmM}{2\pi R^3} \iiint_S \left(\frac{\rho\cos\phi}{\rho^3}\right)\rho^2 \sin\phi\,d\rho d\theta d\phi \right\}\mathbf{k}$$

$$= \left\{ \frac{3GmM}{2\pi R^3} \int_0^{2\pi}\int_0^{\pi/4}\int_{R\sec\phi}^{2R\cos\phi} \cos\phi \sin\phi\,d\rho\,d\phi\,d\theta \right\}\mathbf{k}$$

$$= \frac{GmM}{R^2}\left(\sqrt{2}-1\right)\mathbf{k}.$$

SECTION 17.11

1. $ad - bc$

3. $2\left(v^2 - u^2\right)$

5. $u^2v^2 - 4uv$

7. abc

9. r

11. $w\left(1 + w\cos v\right)$

13.   Set $u = x + y$, $v = x - y$.   Then

$$x = \frac{u+v}{2}, \quad y = \frac{u-v}{2} \quad \text{and} \quad J(u,v) = -\frac{1}{2}.$$

Ω is the set of all (x,y) with uv-coordinates in

$$\Gamma: \quad 0 \le u \le 1, \quad 0 \le v \le 2.$$

Then

$$\iint_\Omega \left(x^2 - y^2\right)\,dxdy = \iint_\Gamma \frac{1}{2}uv\,dudv = \frac{1}{2}\int_0^1\int_0^2 uv\,dv\,du$$

$$= \frac{1}{2}\left(\int_0^1 u\,du\right)\left(\int_0^2 v\,dv\right) = \frac{1}{2}\left(\frac{1}{2}\right)(2) = \frac{1}{2}.$$

15. $\dfrac{1}{2}\displaystyle\int_0^1\int_0^2 u\cos(\pi v)\,dv\,du = \dfrac{1}{2}\left(\displaystyle\int_0^1 u\,du\right)\left(\displaystyle\int_0^2\cos(\pi v)\,dv\right) = \dfrac{1}{2}\left(\dfrac{1}{2}\right)(0) = 0$

17. Set $u = x - y$, $v = x + 2y$. Then

$$x = \frac{2u+v}{3}, \quad y = \frac{v-u}{3}, \quad \text{and} \quad J(u,v) = \frac{1}{3}.$$

Ω is the set of all (x,y) with uv-coordinates in the set

$$\Gamma:\ 0 \le u \le \pi, \quad 0 \le v \le \pi/2.$$

Therefore

$$\iint_\Omega \sin(x-y)\cos(x+2y)\,dx\,dy = \iint_\Gamma \frac{1}{3}\sin u\cos v\,du\,dv = \frac{1}{3}\int_0^\pi\int_0^{\pi/2}\sin u\,\cos v\,dv\,du$$

$$= \frac{1}{3}\left(\int_0^\pi \sin u\,du\right)\left(\int_0^{\pi/2}\cos v\,dv\right) = \frac{1}{3}(2)(1) = \frac{2}{3}.$$

19. Set $u = xy$, $v = y$. Then

$$x = u/v, \quad y = v \quad \text{and} \quad J(u,v) = 1/v.$$

$$xy = 1, \qquad xy = 4 \implies u = 1, \qquad u = 4$$

$$y = x, \qquad y = 4x \implies u/v = v, \qquad 4u/v = v \implies v^2 = u, \qquad v^2 = 4u$$

Ω is the set of all (x,y) with uv-coordinates in the set

$$\Gamma:\ 1 \le u \le 4, \quad \sqrt{u} \le v \le 2\sqrt{u}.$$

(a) $A = \displaystyle\iint_\Gamma \frac{1}{v}\,du\,dv = \int_1^4\int_{\sqrt{u}}^{2\sqrt{u}}\frac{1}{v}\,dv\,du = \int_1^4 \ln 2\,du = 3\ln 2$

(b) $\bar{x}A = \displaystyle\int_1^4\int_{\sqrt{u}}^{2\sqrt{u}}\frac{u}{v^2}\,dv\,du = \frac{7}{3};\quad \bar{x} = \frac{7}{9\ln 2}$

$\bar{y}A = \displaystyle\int_1^4\int_{\sqrt{u}}^{2\sqrt{u}}\,dv\,du = \frac{14}{3};\quad \bar{y} = \frac{14}{9\ln 2}$

21. Set $u = x + y$, $v = 3x - 2y$. Then

$$x = \frac{2u+v}{5}, \quad y = \frac{3u-v}{5} \quad \text{and} \quad J(u,v) = -\frac{1}{5}.$$

With $\Gamma:\ 0 \le u \le 1$, $0 \le v \le 2$

$$M = \int_0^1\int_0^2 \frac{1}{5}\lambda\,dv\,du = \frac{2}{5}\lambda \quad \text{where} \quad \lambda \text{ is the density.}$$

Then

$$I_x = \int_0^1\int_0^2\left(\frac{3u-v}{5}\right)^2\frac{1}{5}\lambda\,dv\,du = \frac{8\lambda}{375} = \frac{4}{75}\left(\frac{2}{5}\lambda\right) = \frac{4}{75}M,$$

$$I_y = \int_0^1 \int_0^2 \left(\frac{2u+v}{5}\right)^2 \frac{1}{5} \lambda \, dv \, du = \frac{28\lambda}{375} = \frac{14}{75}\left(\frac{2}{5}\lambda\right) = \frac{14}{75} M,$$

$$I_z = I_x + I_y = \frac{18}{75} M.$$

23.   Set $u = x - 2y$, $v = 2x + y$.   Then

$$x = \frac{u+2v}{5}, \quad y = \frac{v-2u}{5} \quad \text{and} \quad J(u,v) = \frac{1}{5}.$$

Γ is the region between the parabola $v = u^2 - 1$ and the line $v = 2u + 2$.   A sketch of the curves shows that

$$\Gamma: \quad -1 \le u \le 3, \quad u^2 - 1 \le v \le 2u + 2.$$

Then

$$A = \frac{1}{5}(\text{area of } \Gamma) = \frac{1}{5}\int_{-1}^3 \left[(2u+2) - (u^2-1)\right] du = \frac{32}{15}.$$

25.   The choice $\theta = \pi/6$ reduces the equation to $13u^2 + 5v^2 = 1$.   This is an ellipse in the uv-plane with area $\pi ab = \pi/\sqrt{65}$.   Since $J(u,v) = 1$, the area of Ω is also $\pi/\sqrt{65}$.

27. $J = ab\alpha r \cos^{\alpha-1}\theta \sin^{\alpha-1}\theta$

29. $J = abc\rho^2 \sin\phi$; $V = \int_0^{2\pi} \int_0^\pi \int_0^1 abc\rho^2 \sin\phi \, d\rho \, d\phi \, d\theta = \frac{4}{3}\pi abc$

31.
$$V = \frac{2}{3}\pi abc, \quad \lambda = \frac{M}{V} = \frac{3M}{2\pi abc}$$

$$I_x = \frac{3M}{2\pi abc} \int_0^{2\pi} \int_0^{\pi/2} \int_0^1 \left(b^2\rho^2 \sin^2\phi \sin^2\theta + c^2\rho^2 \cos^2\phi\right)abc\rho^2 \sin\phi \, d\rho \, d\phi \, d\theta$$

$$= \frac{1}{5}M\left(b^2 + c^2\right)$$

$$I_y = \frac{1}{5}M\left(a^2 + c^2\right), \quad I_z = \frac{1}{5}M\left(a^2 + b^2\right)$$

33.
$$\iint_{S_a} \frac{e^{-(x-y)^2}}{1+(x+y)^2} \, dx \, dy = \frac{1}{2} \iint_\Gamma \frac{e^{-u^2}}{1+v^2} \, du \, dv$$

where Γ is the square in the uv-plane with vertices $(-2a, 0)$, $(0, -2a)$, $(2a, 0)$, $(0, 2a)$.
Γ contains the square $-a \le u \le a$, $-a \le v \le a$ and is contained in the square
$-2a \le u \le 2a$, $-2a \le v \le 2a$.   Therefore

$$\frac{1}{2}\int_{-a}^a \int_{-a}^a \frac{e^{-u^2}}{1+v^2} \, du \, dv \le \frac{1}{2}\iint_\Gamma \frac{e^{-u^2}}{1+v^2} \, du \, dv \le \frac{1}{2}\int_{-2a}^{2a} \int_{-2a}^{2a} \frac{e^{-u^2}}{1+v^2} \, du \, dv.$$

The two extremes can be written

$$\frac{1}{2}\left(\int_{-a}^{a} e^{-u^2}\,du\right)\left(\int_{-a}^{a} \frac{1}{1+v^2}\,dv\right) \quad \text{and} \quad \frac{1}{2}\left(\int_{-2a}^{2a} e^{-u^2}\,du\right)\left(\int_{-2a}^{2a} \frac{1}{1+v^2}\,dv\right).$$

As $a \to \infty$ both expressions tend to $\frac{1}{2}\left(\sqrt{\pi}\right)(\pi) = \frac{1}{2}\pi^{3/2}$. It follows that

$$\int_{-\infty}^{\infty}\int_{-\infty}^{\infty} \frac{e^{-(x-y)^2}}{1+(x+y)^2}\,dx\,dy = \frac{1}{2}\pi^{3/2}.$$

CHAPTER 18

SECTION 18.1

1. (a) $\mathbf{h}(x,y) = y\mathbf{i} + x\mathbf{j};$ $\mathbf{r}(u) = u\mathbf{i} + u^2\mathbf{j},$ $u \in [0,1]$

    $x(u) = u,$ $y(u) = u^2;$ $x'(u) = 1,$ $y'(u) = 2u$

    $\mathbf{h}(\mathbf{r}(u)) \cdot \mathbf{r}'(u) = y(u)\,x'(u) + x(u)\,y'(u) = u^2(1) + u(2u) = 3u^2$

    $\displaystyle \int_C \mathbf{h}(\mathbf{r}) \cdot d\mathbf{r} = \int_0^1 3u^2\,du = 1$

    (b) $h(x,y) = y\mathbf{i} + x\mathbf{j};$ $\mathbf{r}(u) = u^3\mathbf{i} - 2u\mathbf{j},$ $u \in [0,1]$

    $x(u) = u^3,$ $y(u) = -2u;$ $x'(u) = 3u^2,$ $y'(u) = -2$

    $\mathbf{h}(\mathbf{r}(u)) \cdot \mathbf{r}'(u) = y(u)\,x'(u) + x(u)\,y'(u) = (-2u)(3u^2) + u^3(-2) = -8u^3$

    $\displaystyle \int_C \mathbf{h}(\mathbf{r}) \cdot d\mathbf{r} = \int_0^1 -8u^3\,du = -2$

3. $h(x,y) = y\mathbf{i} + x\mathbf{j};$ $\mathbf{r}(u) = \cos u\,\mathbf{i} - \sin u\,\mathbf{j},$ $u \in [0, 2\pi]$

    $x(u) = \cos u,$ $y(u) = -\sin u;$ $x'(u) = -\sin u,$ $y'(u) = -\cos u$

    $\mathbf{h}(\mathbf{r}(u)) \cdot \mathbf{r}'(u) = y(u)\,x'(u) + x(u)\,y'(u) = \sin^2 u - \cos^2 u$

    $\displaystyle \int_C \mathbf{h}(\mathbf{r}) \cdot d\mathbf{r} = \int_0^{2\pi} (\sin^2 u - \cos^2 u)\,du = 0$

5. (a) $\mathbf{r}(u) = (2 - u)\mathbf{i} + (3 - u)\mathbf{j},$ $u \in [0,1]$

    $\displaystyle \int_C \mathbf{h}(\mathbf{r}) \cdot d\mathbf{r} = \int_0^1 (-5 + 5u - u^2)\,du = -\frac{17}{6}$

    (b) $\mathbf{r}(u) = (1 + u)\mathbf{i} + (2 + u)\mathbf{j},$ $u \in [0,1]$

    $\displaystyle \int_C \mathbf{h}(\mathbf{r}) \cdot d\mathbf{r} = \int_0^1 (1 + 3u + u^2)\,du = \frac{17}{6}$

7. $C = C_1 \cup C_2 \cup C_3$ where

    $C_1 : \mathbf{r}(u) = (1 - u)(-2\mathbf{i}) + u(2\mathbf{i}) = (4u - 2)\mathbf{i},$ $u \in [0,1]$

    $C_2 : \mathbf{r}(u) = (1 - u)(2\mathbf{i}) + u(2\mathbf{j}) = (2 - 2u)\mathbf{i} + 2u\mathbf{j},$ $u \in [0,1]$

    $C_3 : \mathbf{r}(u) = (1 - u)(2\mathbf{j}) + u(-2\mathbf{i}) = -2u\mathbf{i} + (2 - 2u)\mathbf{j},$ $u \in [0,1]$

    $\displaystyle \int_C = \int_{C_1} + \int_{C_2} + \int_{C_3} = 0 + (-4) + (-4) = -8$

9.

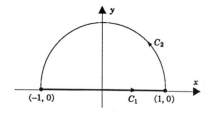

$C_1 : \mathbf{r}(u) = (-1 + 2u)\mathbf{i}, \quad u \in [0, 1]$

$C_2 : \mathbf{r}(u) = \cos u\,\mathbf{i} + \sin u\,\mathbf{j}, \quad u \in [0, \pi]$

$$\int_C = \int_{C_1} + \int_{C_2} = 0 + (-\pi) = -\pi$$

11. (a) $\mathbf{r}(u) = u\mathbf{i} + u\mathbf{j} + u\mathbf{k}, \quad u \in [0, 1]$

$$\int_C \mathbf{h}(\mathbf{r}) \cdot d\mathbf{r} = \int_0^1 3u^2 \, du = 1$$

(b) $\displaystyle \int_C \mathbf{h}(\mathbf{r}) \cdot d\mathbf{r} = \int_0^1 (2u^3 + u^5 + 3u^6) \, du = \frac{23}{21}$

13. $\mathbf{r}(u) = (1 - u)(\mathbf{j} + 4\mathbf{k}) + u(\mathbf{i} - 4\mathbf{k})$

$$= u\mathbf{i} + (1 - u)\mathbf{j} + (4 - 8u)\mathbf{k}, \quad u \in [0, 1]$$

$$\int_C \mathbf{F}(\mathbf{r}) \cdot d\mathbf{r} = \int_0^1 (-32u + 97u^2 - 64u^3) \, du = \frac{1}{3}$$

15. $$\int_C \mathbf{q} \cdot d\mathbf{r} = \int_a^b [\mathbf{q} \cdot \mathbf{r}'(u)] \, du = \int_a^b \frac{d}{du} [\mathbf{q} \cdot \mathbf{r}(u)] \, du$$

$$= [\mathbf{q} \cdot \mathbf{r}(b)] - [\mathbf{q} \cdot \mathbf{r}(a)]$$

$$= \mathbf{q} \cdot [\mathbf{r}(b) - \mathbf{r}(a)]$$

$$\int_C \mathbf{r} \cdot d\mathbf{r} = \int_a^b [\mathbf{r}(u) \cdot \mathbf{r}'(u)] \, du$$

$$= \frac{1}{2} \int_a^b \frac{d}{du} [\mathbf{r}(u) \cdot \mathbf{r}(u)] \, du$$

$$= \frac{1}{2} \int_a^b \frac{d}{du} \left(\|\mathbf{r}(u)\|^2 \right) \, du = \frac{1}{2} \left(\|\mathbf{r}(b)\|^2 - \|\mathbf{r}(a)\|^2 \right)$$

17. $\displaystyle \int_C \mathbf{f}(\mathbf{r}) \cdot d\mathbf{r} = \int_a^b [\mathbf{f}(\mathbf{r}(u)) \cdot \mathbf{r}'(u)] \, du = \int_a^b [f(u)\mathbf{i} \cdot \mathbf{i}] \, du = \int_a^b f(u) \, du$

19. $E : \mathbf{r}(u) = a \cos u\,\mathbf{i} + b \sin u\,\mathbf{j}, \quad u \in [0, 2\pi]$

$$W = \int_0^{2\pi} \left[\left(-\frac{1}{2}b \sin u \right) (-a \sin u) + \left(\frac{1}{2}a \cos u \right) (b \cos u) \right] \, du = \int_0^{2\pi} ab \, du = \pi ab$$

If the ellipse is traversed in the opposite direction, then $W = -\pi ab$. In both cases $|W| = \pi ab =$ area of the ellipse.

21. $\mathbf{r}(t) = \alpha t \mathbf{i} + \beta t^2 \mathbf{j} + \gamma t^3 \mathbf{k}$

$\mathbf{r}'(t) = \alpha \mathbf{i} + 2\beta t \mathbf{j} + 3\gamma t^2 \mathbf{k}$

force at time $t = m\mathbf{r}''(t) = m(2\beta \mathbf{j} + 6\gamma t \mathbf{k})$

$$W = \int_0^1 [m(2\beta \mathbf{j} + 6\gamma t \mathbf{k}) \cdot (\alpha \mathbf{i} + 2\beta t \mathbf{j} + 3\gamma t^2 \mathbf{k})]\, dt$$

$$= m \int_0^1 (4\beta^2 t + 18\gamma^2 t^3)\, dt = \left(2\beta^2 + \frac{9}{2}\gamma^2\right) m$$

23.   Take $C : \mathbf{r}(t) = r\cos t\,\mathbf{i} + r\sin t\,\mathbf{j}, \quad t \in [0, 2\pi]$

$$\int_C \mathbf{v}(\mathbf{r}) \cdot d\mathbf{r} = \int_0^{2\pi} [\mathbf{v}(\mathbf{r}(t)) \cdot \mathbf{r}'(t)]\, dt$$

$$= \int_0^{2\pi} [f(x(t), y(t))\,\mathbf{r}(t) \cdot \mathbf{r}'(t)]\, dt$$

$$= \int_0^{2\pi} f(x(t), y(t))\,[\mathbf{r}(t) \cdot \mathbf{r}'(t)]\, dt = 0$$

since for the circle $\mathbf{r}(t) \cdot \mathbf{r}'(t) = 0$ identically. The circulation is zero.

SECTION 18.2

1. $\mathbf{h}(x, y) = \nabla f(x, y)$ where $f(x, y) = \frac{1}{2}(x^2 + y^2)$

C is closed $\implies$ $\displaystyle\int_C \mathbf{h}(\mathbf{r}) \cdot d\mathbf{r} = 0$

3. $\mathbf{h}(x, y) = \nabla f(x, y)$ where $f(x, y) = x\cos \pi y;$ $\mathbf{r}(0) = \mathbf{0}, \quad \mathbf{r}(1) = \mathbf{i} - \mathbf{j}$

$$\int_C \mathbf{h}(\mathbf{r}) \cdot d\mathbf{r} = \int_C \nabla f(\mathbf{r}) \cdot d\mathbf{r} = f(\mathbf{r}(1)) - f(\mathbf{r}(0)) = f(1, -1) - f(0, 0) = -1$$

5. $\mathbf{h}(x, y) = \nabla f(x, y)$ where $f(x, y) = \frac{1}{2}x^2 y^2;$ $\mathbf{r}(0) = \mathbf{j}, \quad \mathbf{r}(1) = -\mathbf{j}$

$$\int_C \mathbf{h}(\mathbf{r}) \cdot d\mathbf{r} = \int_C \nabla f(\mathbf{r}) \cdot d\mathbf{r} = f(\mathbf{r}(1)) - f(\mathbf{r}(0)) = f(0, -1) - f(0, 1) = 0 - 0 = 0$$

7. $\mathbf{h}(x, y) = \nabla f(x, y)$ where $f(x, y) = x^2 y - xy^2;$ $\mathbf{r}(0) = \mathbf{i}, \ \mathbf{r}(\pi) = -\mathbf{i}$

$$\int_C \mathbf{h}(\mathbf{r}) \cdot d\mathbf{r} = \int_C \nabla f(\mathbf{r}) \cdot d\mathbf{r} = f(\mathbf{r}(\pi)) - f(\mathbf{r}(0)) = f(-1, 0) - f(1, 0) = 0 - 0 = 0$$

9. $\mathbf{h}(x, y) = \nabla f(x, y)$ where $f(x, y) = (x^2 + y^4)^{3/2}$

$$\int_C \mathbf{h}(\mathbf{r}) \cdot d\mathbf{r} = \int_C \nabla f(\mathbf{r}) \cdot d\mathbf{r} = f(1, 0) - f(-1, 0) = 1 - 1 = 0$$

11. $\mathbf{h}(x, y)$ is not a gradient, but part of it,

$$2x\cosh y\,\mathbf{i} + (x^2 \sinh y - y)\mathbf{j},$$

is a gradient. Since we are integrating over a closed curve, the contribution of the gradient part is 0. Thus

$$\int_C \mathbf{h}(\mathbf{r}) \cdot d\mathbf{r} = \int_C (-y\mathbf{i}) \cdot d\mathbf{r}.$$

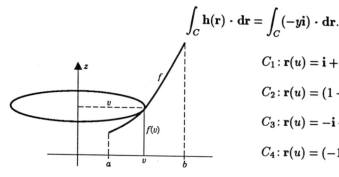

$$C_1 : \mathbf{r}(u) = \mathbf{i} + (-1 + 2u)\mathbf{j}, \quad u \in [0,1]$$

$$C_2 : \mathbf{r}(u) = (1 - 2u)\mathbf{i} + \mathbf{j}, \quad u \in [0,1]$$

$$C_3 : \mathbf{r}(u) = -\mathbf{i} + (1 - 2u)\mathbf{j}, \quad u \in [0,1]$$

$$C_4 : \mathbf{r}(u) = (-1 + 2u)\mathbf{i} - \mathbf{j}, \quad u \in [0,1]$$

$$\int_C \mathbf{h}(\mathbf{r}) \cdot d\mathbf{r} = \int_{C_1} (-y\mathbf{i}) \cdot d\mathbf{r} + \int_{C_2} (-y\mathbf{i}) \cdot d\mathbf{r} + \int_{C_3} (-y\mathbf{i}) \cdot d\mathbf{r} + \int_{C_4} (-y\mathbf{i}) \cdot d\mathbf{r}$$

$$= \quad 0 \quad + \int_0^1 -\mathbf{i} \cdot (-2\mathbf{i}) \, du + \quad 0 \quad + \int_0^1 \mathbf{i} \cdot (2\mathbf{i}) \, du$$

$$= \quad 0 \quad + \int_0^1 2 \, du \quad + \quad 0 \quad + \int_0^1 2 \, du$$

$$= \quad 4$$

13. Set $f(x,y,z) = g(x)$ and $C : \mathbf{r}(u) = u\mathbf{i}, \quad u \in [a,b]$.

In this case

$$\nabla f(\mathbf{r}(u)) = g'(x(u))\mathbf{i} = g'(u)\mathbf{i} \quad \text{and} \quad \mathbf{r}'(u) = \mathbf{i},$$

so that

$$\int_C \nabla f(\mathbf{r}) \cdot d\mathbf{r} = \int_a^b [\nabla f(\mathbf{r}(u)) \cdot \mathbf{r}'(u)] \, du = \int_a^b g'(u) \, du.$$

Since $\quad f(\mathbf{r}(b)) - f(\mathbf{r}(a)) = g(b) - g(a),$

$$\int_C \nabla f(\mathbf{r}) \cdot d\mathbf{r} = f(\mathbf{r}(b)) - f(\mathbf{r}(a)) \quad \text{gives} \quad \int_a^b g'(u) \, du = g(b) - g(a).$$

15. $\mathbf{F}(\mathbf{r}) = \nabla \left(\dfrac{mG}{r} \right); \quad W = \int_C \mathbf{F}(\mathbf{r}) \cdot d\mathbf{r} = mG \left(\dfrac{1}{r_2} - \dfrac{1}{r_1} \right)$

SECTION * 18.3

1. If f is continuous, then $-f$ is continuous and has antiderivatives u. The scalar fields $U(x,y,z) = u(x)$ are potential functions for $\mathbf{F}$:

$$\nabla U = \frac{\partial U}{\partial x}\mathbf{i} + \frac{\partial U}{\partial y}\mathbf{j} + \frac{\partial U}{\partial z}\mathbf{k} = \frac{du}{dx}\mathbf{i} = -f\mathbf{i} = -\mathbf{F}.$$

3. The scalar field $U(x, y, z) = cz + d$ is a potential energy function for **F**. We know that the total mechanical energy remains constant. Thus, for any times t_1 and t_2,

$$\tfrac{1}{2}m[\mathbf{v}(t_1)]^2 + U(\mathbf{r}(t_1)) = \tfrac{1}{2}m[\mathbf{v}(t_2)]^2 + U(\mathbf{r}(t_2)).$$

This gives

$$\tfrac{1}{2}m[\mathbf{v}(t_1)]^2 + cz(t_1) + d = \tfrac{1}{2}m[\mathbf{v}(t_2)]^2 + cz(t_2) + d.$$

Solve this equation for $\mathbf{v}(t_2)$ and you have the desired formula.

5. (a) We know that $-\nabla U$ points in the direction of maximum decrease of U. Thus $\mathbf{F} = -\nabla U$ attempts to drive objects toward a region where U has lower values.

(b) At a point where u has a minimum, $\nabla U = \mathbf{0}$ and therefore $\mathbf{F} = \mathbf{0}$.

7. (a) By conservation of energy $\tfrac{1}{2}mv^2 + U = E$. Since E is constant and U is constant, v is constant.

(b) ∇U is perpendicular to any surface where U is constant. Obviously so is $\mathbf{F} = -\nabla U$.

SECTION 18.4

1. $\mathbf{r}(u) = u\mathbf{i} + 2u\mathbf{j}, \quad u \in [0, 1]$

$$\int_C (x - 2y)\, dx + 2x\, dy = \int_0^1 \{[x(u) - 2y(u)]x'(u) + 2x(u)\, y'(u)\}\, du = \int_0^1 u\, du = \frac{1}{2}$$

3. $C = C_1 \cup C_2$

$C_1 : \mathbf{r}(u) = u\mathbf{i}, \quad u \in [0, 1]; \qquad C_2 : \mathbf{r}(u) = \mathbf{i} + 2u\mathbf{j}, \quad u \in [0, 1]$

$$\int_{C_1} (x - 2y)\, dx + 2x\, dy = \int_{C_1} x\, dx = \int_0^1 x(u)\, x'(u)\, du = \int_0^1 u\, du = \frac{1}{2}$$

$$\int_{C_2} (x - 2y)\, dx + 2x\, dy = \int_{C_2} 2x\, dy = \int_0^1 4\, du = 4$$

$$\int_C = \int_{C_1} + \int_{C_2} = 4\tfrac{1}{2}$$

5. $\mathbf{r}(u) = 2u^2\mathbf{i} + u\mathbf{j}, \quad u \in [0, 1]$

$$\int_C y\, dx + xy\, dy = \int_0^1 [y(u)\, x'(u) + x(u)\, y(u)\, y'(u)]\, du = \int_0^1 (4u^2 + 2u^3)\, du = \frac{11}{6}$$

7. $C = C_1 \cup C_2$

$C_1 : \mathbf{r}(u) = u\mathbf{j}, \quad u \in [0, 1]; \qquad C_2 : \mathbf{r}(u) = 2u\mathbf{i} + \mathbf{j}, \quad u \in [0, 1]$

$$\int_{C_1} y\,dx + xy\,dy = 0$$

$$\int_{C_2} y\,dx + xy\,dy = \int_{C_2} y\,dx = \int_0^1 y(u)\,x'(u)\,du = \int_0^1 2\,du = 2$$

$$\int_C = \int_{C_1} + \int_{C_2} = 2$$

9. $\mathbf{r}(u) = u\mathbf{i} + u\mathbf{j} + u\mathbf{k}, \quad u \in [0,1]$

$$\int_C y\,dx + 2z\,dy + x\,dz = \int_0^1 [y(u)\,x'(u) + 2z(u)\,y'(u) + x(u)\,z'(u)]\,du = \int_0^1 4u\,du = 2$$

11. $C = C_1 \cup C_2 \cup C_3$

$C_1 : \mathbf{r}(u) = u\mathbf{k}, \quad u \in [0,1]; \quad C_2 : \mathbf{r}(u) = u\mathbf{j} + \mathbf{k}, \quad u \in [0,1]; \quad C_3 : \mathbf{r}(u) = u\mathbf{i} + \mathbf{j} + \mathbf{k}, \quad u \in [0,1]$

$$\int_{C_1} y\,dx + 2z\,dy + x\,dz = 0$$

$$\int_{C_2} y\,dx + 2z\,dy + x\,dz = \int_{C_2} 2z\,dy = \int_0^1 2z(u)\,y'(u)\,du = \int_0^1 2\,du = 2$$

$$\int_{C_3} y\,dx + 2z\,dy + x\,dz = \int_{C_3} y\,dx = \int_0^1 y(u)\,x'(u)\,du = \int_0^1 du = 1$$

$$\int_C = \int_{C_1} + \int_{C_2} + \int_{C_3} = 3$$

13. $s'(u) = \sqrt{[x'(u)]^2 + [y'(u)]^2} = a$

(a) $$M = \int_C k(x+y)\,ds = k\int_0^{\pi/2} [x(u) + y(u)]\,s'(u)\,du = ka^2 \int_0^{\pi/2} (\cos u + \sin u)\,du = 2ka^2$$

(b)

$$x_M M = \int_C kx(x+y)\,ds = k\int_0^{\pi/2} x(u)\,[x(u) + y(u)]\,s'(u)\,du$$

$$= ka^3 \int_0^{\pi/2} (\cos^2 u + \cos u \sin u)\,du = \frac{1}{4}ka^3(\pi + 2)$$

$$y_M M = \int_C ky(x+y)\,ds = k\int_0^{\pi/2} y(u)\,[x(u) + y(u)]\,s'(u)\,du$$

$$= ka^3 \int_0^{\pi/2} (\sin u \cos u + \sin^2 u)\,du = \frac{1}{4}ka^3(\pi + 2)$$

$$x_M = y_M = \tfrac{1}{8}a(\pi + 2)$$

15. (a) $I_z = \int_C k(x+y)a^2\,ds = a^2 \int_C k(x+y)\,ds = a^2 M = Ma^2$

(b) The distance from the point (x,y) to the line $y=x$ is $|y-x|/\sqrt{2}$.   (See 9.2.2.) Therefore

$$I = \int_C k(x+y)\left[\frac{1}{2}(y-x)^2\right]ds = \frac{1}{2}k \int_0^{\pi/2} (a\cos u + a\sin u)(a\sin u - a\cos u)^2 a\,du$$

$$= \frac{1}{2}ka^4 \int_0^{\pi/2} (\sin u - \cos u)^2 \frac{d}{du}(\sin u - \cos u)\,du$$

$$= \frac{1}{2}ka^4 \left[\frac{1}{3}(\sin u - \cos u)^3\right]_0^{\pi/2} = \frac{1}{3}ka^4.$$

From Exercise 13, $M = 2ka^2$.   Therefore

$$I = \tfrac{1}{6}(2ka^2)a^2 = \tfrac{1}{6}Ma^2.$$

17. (a) $s'(u) = \sqrt{a^2 + b^2}$

$$L = \int_C ds = \int_0^{2\pi} \sqrt{a^2 + b^2}\,du = 2\pi\sqrt{a^2 + b^2}$$

(b) $x_M = 0,\quad y_M = 0$ (by symmetry)

$$z_M = \frac{1}{L}\int_C z\,ds = \frac{1}{2\pi\sqrt{a^2+b^2}}\int_0^{2\pi} bu\sqrt{a^2+b^2}\,du = b\pi$$

(c) $I_x = \int_C \frac{M}{L}(y^2 + z^2)\,ds = \frac{M}{2\pi}\int_0^{2\pi}(a^2\sin^2 u + b^2 u^2)\,du = \frac{1}{6}M(3a^2\pi + 8b^2\pi^2)$

$I_y = \frac{1}{6}M(3a^2\pi + 8b^2\pi^2)$ similarly

$I_z = Ma^2$ (all the mass is at distance a from the z-axis)

19. $$M = \int_C k(x^2 + y^2 + z^2)\,ds$$

$$= k\sqrt{a^2 + b^2}\int_0^{2\pi}(a^2 + b^2 u^2)\,du = \frac{2}{3}\pi k\sqrt{a^2+b^2}\,(3a^2 + 4\pi^2 b^2)$$

SECTION 18.5

1. $\oint_C 3y\,dx + 5x\,dy = \iint_\Omega (5-3)\,dxdy = 2A = 2\pi$

3. $\oint_C x^2\,dy = \iint_\Omega 2x\,dxdy = 2\overline{x}A = 2\left(\frac{a}{2}\right)(ab) = a^2 b$

5. $\oint_C (3xy + y^2)\, dx + (2xy + 5x^2)\, dy = \iint_\Omega [(2y + 10x) - (3x + 2y)]\, dx\, dy$

$$= \iint_\Omega 7x\, dx\, dy = 7\,\overline{x}A = 7(1)(\pi) = 7\pi$$

7. $\oint_C (2x^2 + xy - y^2)\, dx + (3x^2 - xy + 2y^2)\, dy = \iint_\Omega [(6x - y) - (x - 2y)]\, dx\, dy$

$$= \iint_\Omega (5x + y)\, dx\, dy = (5\overline{x} + \overline{y})A = (5a + 0)(\pi r^2) = 5a\pi r^2$$

9. $\oint_C e^x \sin y\, dx + e^x \cos y\, dy = \iint_\Omega [e^x \cos y - e^x \cos y]\, dx\, dy = 0$

11. $C : \mathbf{r}(u) = a \cos u\, \mathbf{i} + b \sin u\, \mathbf{j}, \quad u \in [0, 2\pi]$

$$A = \oint_C -y\, dx = \int_0^{2\pi} (-b \sin u)(-a \sin u)\, du = ab \int_0^{2\pi} \sin^2 u\, du = \pi ab$$

13. $\oint_C (ay + b)\, dx + (cx + d)\, dy = \iint_\Omega (c - a)\, dx\, dy = (c - a)A$

15. We take the arch from $x = 0$ to $x = 2\pi R$. (Figure 10.11.1) Let C_1 be the line segment from $(0,0)$ to $(2\pi R, 0)$ and let C_2 be the cycloidal arch from $(2\pi R, 0)$ back to $(0,0)$. Letting $C = C_1 \cup C_2$, we have

$$A = \oint_C x\, dy = \int_{C_1} x\, dy + \int_{C_2} x\, dy = 0 + \int_{C_2} x\, dy$$

$$= \int_{2\pi}^0 R(\theta - \sin\theta)(R \sin\theta)\, d\theta$$

$$= R^2 \int_0^{2\pi} (\sin^2\theta - \theta \sin\theta)\, d\theta$$

$$= R^2 \left[\frac{\theta}{2} - \frac{\sin 2\theta}{4} + \theta \cos\theta - \sin\theta \right]_0^{2\pi} = 3\pi R^2.$$

17. Taking Ω to be of type II (see Figure 18.5.2), we have

$$\iint_\Omega \frac{\partial Q}{\partial x}(x, y)\, dx\, dy = \int_c^d \int_{\phi_3(y)}^{\phi_4(y)} \frac{\partial Q}{\partial x}(x, y)\, dx\, dy$$

$$= \int_c^d \{Q[\phi_4(y), y] - Q[\phi_3(y), y]\}\, dy$$

$$(*) = \int_c^d Q[\phi_4(y), y]\, dy - \int_c^d Q[\phi_3(y), y]\, dy.$$

The graph of $x = \phi_4(y)$ from $x = c$ to $x = d$ is the curve

$$C_4 : r_4(u) = \phi_4(u)i + uj, \qquad u \in [c,d].$$

The graph of $x = \phi_3(y)$ from $x = c$ to $x = d$ is the curve

$$C_3 : r_3(u) = \phi_3(u)i + uj, \qquad u \in [c,d].$$

Then

$$\oint_C Q(x,y)\,dy = \int_{C_4} Q(x,y)\,dy - \int_{C_3} Q(x,y)\,dy$$

$$= \int_c^d Q[\phi_4(u),u]\,du - \int_c^d Q[\phi_3(u),u]\,du.$$

Since u is a dummy variable, it can be replaced by y. Comparison with $(*)$ gives the result.

19. $\oint_{C_1} = \oint_{C_2} + \oint_{C_3}$

21. $\dfrac{\partial P}{\partial y} = \dfrac{-2xy}{(x^2+y^2)^2} = \dfrac{\partial Q}{\partial x}$ except at $(0,0)$

(a) If C does not enclose the origin, and Ω is the region enclosed by C, then

$$\oint_C \frac{x}{x^2+y^2}\,dx + \frac{y}{x^2+y^2}\,dy = \iint_\Omega 0\,dxdy = 0.$$

(b) If C does enclose the origin, then

$$\oint_C = \oint_{C_a}$$

where $C_a : r(u) = a\cos u\,i + a\sin u\,j, \quad u \in [0,2\pi]$ is a small circle in the inner region of C.

In this case

$$\oint_C = \int_0^{2\pi}\left[\frac{a\cos u}{a^2}(-a\sin u) + \frac{a\sin u}{a^2}(a\cos u)\right]du = \int_0^{2\pi} 0\,du = 0.$$

The integral is still 0.

23. If Ω is the region enclosed by C, then

$$\oint_C v\cdot dr = \oint_C \frac{\partial\phi}{\partial x}\,dx + \frac{\partial\phi}{\partial y}\,dy = \iint_\Omega \left\{\frac{\partial}{\partial x}\left(\frac{\partial\phi}{\partial y}\right) - \frac{\partial}{\partial y}\left(\frac{\partial\phi}{\partial x}\right)\right\}dxdy$$

$$= \iint_\Omega 0\,dxdy = 0.$$

equality of mixed partials ⌐

SECTION 18.6

1. $4[(u^2 - v^2)\mathbf{i} - (u^2 + v^2)\mathbf{j} + 2uv\mathbf{k}]$ **3.** $2(\mathbf{j} - \mathbf{i})$

5. The surface consists of all points of the form $(x, g(x, z), z)$ with $(x, z) \in \Omega$. This set of points is given by

$$\mathbf{r}(x, z) = x\mathbf{i} + g(x, z)\mathbf{j} + z\mathbf{k}, \quad (x, z) \in \Omega.$$

7. $x^2/a^2 + y^2/b^2 + z^2/c^2 = 1$;   ellipsoid

9. $x^2/a^2 - y^2/b^2 = z$;   hyperbolic paraboloid

11. For each $v \in [a, b]$, the points on the surface at level $z = f(v)$ form a circle of radius v. That circle can be parametrized

$$\mathbf{R}(u) = v \cos u\, \mathbf{i} + v \sin u\, \mathbf{j} + f(v)\mathbf{k}, \quad u \in [0, 2\pi].$$

Letting v range over $[a, b]$, we obtain the entire surface:

$$\mathbf{r}(u, v) = v \cos u\, \mathbf{i} + v \sin u\, \mathbf{j} + f(v)\mathbf{k}; \quad 0 \le u \le 2\pi, \quad a \le v \le b.$$

13. Since γ is the angle between p and the xy-plane, γ is the angle between the upper normal to p and $\mathbf{k}$. (Draw a figure.) Therefore, by 18.6.5,

$$\text{area of } \Gamma = \iint_\Omega \sec \gamma\, dx\, dy = (\sec \gamma) A_\Omega = A_\Omega \sec \gamma.$$

$$\llcorner\ \gamma \text{ is constant}$$

15. The surface is the graph of the function

$$f(x, y) = c\left(1 - \frac{x}{a} - \frac{y}{b}\right) = \frac{c}{ab}(ab - bx - ay)$$

defined over the triangle $\Omega : 0 \le x \le a, \quad 0 \le y \le b(1 - x/a)$. Note that Ω has area $\frac{1}{2}ab$.

$$A = \iint_\Omega \sqrt{[f'_x(x, y)]^2 + [f'_y(x, y)]^2 + 1}\ dx\, dy$$

$$= \iint_\Omega \sqrt{c^2/a^2 + c^2/b^2 + 1}\ dx\, dy$$

$$= \frac{1}{ab}\sqrt{a^2 b^2 + a^2 c^2 + b^2 c^2} \iint_\Omega dx\, dy = \frac{1}{2}\sqrt{a^2 b^2 + a^2 c^2 + b^2 c^2}.$$

17. $f(x,y) = x^2 + y^2, \quad \Omega : 0 \le x^2 + y^2 \le 4$

$$A = \iint\limits_{\Omega} \sqrt{4x^2 + 4y^2 + 1}\; dx dy \qquad \text{[change to polar coordinates]}$$

$$= \int_0^{2\pi} \int_0^2 \sqrt{4r^2 + 1}\, r\, dr\, d\theta$$

$$= 2\pi \left[\tfrac{1}{12}(4r^2 + 1)^{3/2}\right]_0^2 = \tfrac{1}{6}\pi(17\sqrt{17} - 1)$$

19. $f(x,y) = a^2 - (x^2 + y^2), \quad \Omega : \tfrac{1}{4}a^2 \le x^2 + y^2 \le a^2$

$$A = \iint\limits_{\Omega} \sqrt{4x^2 + 4y^2 + 1}\; dx dy \qquad \text{[change to polar coordinates]}$$

$$= \int_0^{2\pi} \int_{a/2}^{a} r\sqrt{4r^2 + 1}\; dr\, d\theta = 2\pi \left[\frac{1}{12}(4r^2 + 1)^{3/2}\right]_{a/2}^{a}$$

$$= \frac{\pi}{6}\left[(4a^2 + 1)^{3/2} - (a^2 + 1)^{3/2}\right]$$

21. $f(x,y) = \tfrac{1}{3}(x^{3/2} + y^{3/2}), \quad \Omega : 0 \le x \le 1, \quad 0 \le y \le x$

$$A = \iint\limits_{\Omega} \frac{1}{2}\sqrt{x + y + 4}\; dx dy$$

$$= \int_0^1 \int_0^x \frac{1}{2}\sqrt{x + y + 4}\; dy\, dx = \int_0^1 \left[\frac{1}{3}(x + y + 4)^{3/2}\right]_0^x dx$$

$$= \int_0^1 \frac{1}{3}\left[(2x + 4)^{3/2} - (x + 4)^{3/2}\right] dx = \frac{1}{3}\left[\frac{1}{5}(2x + 4)^{5/2} - \frac{2}{5}(x + 4)^{5/2}\right]_0^1$$

$$= \tfrac{1}{15}(36\sqrt{6} - 50\sqrt{5} + 32)$$

23. The surface $x^2 + y^2 + z^2 - 4z = 0$ is a sphere of radius 2 centered at $(0,0,2)$:

$$x^2 + y^2 + z^2 - 4z = 0 \quad \Longleftrightarrow \quad x^2 + y^2 + (z - 2)^2 = 4.$$

The quadric cone $z^2 = 3(x^2 + y^2)$ intersects the sphere at height $z = 3$:

$$\left. \begin{array}{r} x^2 + y^2 + z^2 - 4z = 0 \\ z^2 = 3(x^2 + y^2) \end{array} \right\} \quad \Longrightarrow \quad \begin{array}{r} 3(x^2 + y^2) + 3z^2 - 12z = 0 \\ 4z^2 - 12z = 0 \\ z = 3. \quad (\text{since } z \ge 2) \end{array}$$

The surface of which we are asked to find the area is a spherical segment of width 1 (from $z = 3$ to $z = 4$) in a sphere of radius 2. The area of the segment is 4π. (Exercise 25, Section 10.10.)

A more conventional solution. The spherical segment is the graph of the function

$$f(x,y) = 2 + \sqrt{4 - (x^2 + y^2)}, \quad \Omega : 0 \le x^2 + y^2 \le 3.$$

Therefore

$$A = \iint_{\Omega} \sqrt{\left(\frac{-x}{\sqrt{4 - x^2 - y^2}}\right)^2 + \left(\frac{-y}{\sqrt{4 - x^2 - y^2}}\right)^2 + 1} \; dxdy$$

$$= \iint_{\Omega} \frac{2}{\sqrt{4 - (x^2 + y^2)}} \; dxdy$$

$$= \int_0^{2\pi} \int_0^{\sqrt{3}} \frac{2r}{\sqrt{4 - r^2}} \; dr \, d\theta \qquad [\text{changed to polar coordinates}]$$

$$= 2\pi \left[-2\sqrt{4 - r^2}\right]_0^{\sqrt{3}} = 4\pi$$

25. (a) $$\iint_{\Omega} \sqrt{\left[\frac{\partial g}{\partial y}(y, z)\right]^2 + \left[\frac{\partial g}{\partial z}(y, z)\right]^2 + 1} \; dydz = \iint_{\Omega} \sec\left[\alpha(y, z)\right] \, dydz$$

where α is the angle between the unit normal with positive $\mathbf{i}$ component and the positive x-axis

(b) $$\iint_{\Omega} \sqrt{\left[\frac{\partial h}{\partial x}(x, z)\right]^2 + \left[\frac{\partial h}{\partial z}(x, z)\right]^2 + 1} \; dxdz = \iint_{\Omega} \sec\left[\beta(x, z)\right] \, dxdz$$

where β is the angle between the unit normal with positive $\mathbf{j}$ component and the positive y-axis

27. (a) $\mathbf{N}(u, v) = v \cos u \sin \alpha \cos \alpha \, \mathbf{i} + v \sin u \sin \alpha \cos \alpha \, \mathbf{j} - v \sin^2 \alpha \, \mathbf{k}$

(b) $$A = \iint_{\Omega} \|\mathbf{N}(u, v)\| \, dudv = \iint_{\Omega} v \sin \alpha \, dudv$$

$$= \int_0^{2\pi} \int_0^s v \sin \alpha \, dv \, du = \pi s^2 \sin \alpha$$

29. $A = \sqrt{A_1^2 + A_2^2 + A_3^2}$;   the unit normal to the plane of Ω is a vector of the form

$$\cos \gamma_1 \, \mathbf{i} + \cos \gamma_2 \, \mathbf{j} + \cos \gamma_3 \, \mathbf{k}.$$

Note that

$$A_1 = A \cos \gamma_1, \quad A_2 = A \cos \gamma_2, \quad A_3 = A \cos \gamma_3.$$

Therefore

$$A_1^2 + A_2^2 + A_3^2 = A^2[\cos^2 \gamma_1 + \cos^2 \gamma_2 + \cos^2 \gamma_3] = A^2.$$

31. (a) (We use Exercise 30.) $f(r, \theta) = r + \theta; \quad \Omega : 0 \leq r \leq 1, \quad 0 \leq \theta \pi$

$$A = \iint\limits_{\Omega} \sqrt{r^2 [f_r'(r, \theta)]^2 + [f_\theta'(r, \theta)]^2 + r^2} \; drd\theta = \iint\limits_{\Omega} \sqrt{2r^2 + 1} \; drd\theta$$

$$= \int_0^\pi \int_0^1 \sqrt{2r^2 + 1} \; dr \; d\theta = \frac{1}{4}\sqrt{2}\pi \left[\sqrt{6} + \ln \left(\sqrt{2} + \sqrt{3} \right) \right]$$

(b) $f(r, \theta) = re^\theta; \quad \Omega : 0 \leq r \leq a, \quad 0 \leq \theta \leq 2\pi$

$$A = \iint\limits_{\Omega} r\sqrt{2e^{2\theta} + 1} \; drd\theta = \left(\int_0^{2\pi} \sqrt{2e^{2\theta} + 1} \; d\theta \right) \left(\int_0^a r \; dr \right)$$

$$= \tfrac{1}{2}a^2 [\sqrt{2e^{4\pi} + 1} - \sqrt{3} + \ln \left(1 + \sqrt{3} \right) - \ln \left(1 + \sqrt{2e^{4\pi} + 1} \right)]$$

SECTION 18.7

For Exercises 1–6 we have $\sec [\gamma(x, y)] = \sqrt{y^2 + 1}$.

1. $\displaystyle\iint\limits_{S} d\sigma = \int_0^1 \int_0^1 \sqrt{y^2 + 1} \; dx \, dy = \int_0^1 \sqrt{y^2 + 1} \; dy = \frac{1}{2}[\sqrt{2} + \ln (1 + \sqrt{2})]$

3. $\displaystyle\iint\limits_{S} 3y \, d\sigma = \int_0^1 \int_0^1 3y\sqrt{y^2 + 1} \; dy \, dx = \int_0^1 3y\sqrt{y^2 + 1} \; dy = \left[(y^2 + 1)^{3/2} \right]_0^1 = 2\sqrt{2} - 1$

5. $\displaystyle\iint\limits_{S} \sqrt{2z} \, d\sigma = \iint\limits_{S} y \, d\sigma = \frac{1}{3}(2\sqrt{2} - 1)$ (Exercise 3)

For Exercises 7–10 the surface S is given by

$$f(x, y) = a - x - y; \quad 0 \leq x \leq a, \; 0 \leq y \leq a - x.$$

Then $\sec [\gamma (x, y)] = \sqrt{3}$.

7. $\displaystyle M = \iint\limits_{S} \lambda(x, y, x) \, d\sigma = \int_0^a \int_0^{a-x} k\sqrt{3} \; dy \, dx = \int_0^a k\sqrt{3} \, (a - x) \, dx = \frac{1}{2}a^2 k\sqrt{3}$

9. $\displaystyle M = \iint\limits_{S} \lambda(x, y, z) \, d\sigma = \int_0^a \int_0^{a-x} kx^2\sqrt{3} \, dy \, dx = \int_0^a k\sqrt{3}x^2(a - x) \, dx = \frac{1}{12}a^4 k\sqrt{3}$

11. $S : \mathbf{r}(u, v) = a \cos u \cos v \, \mathbf{i} + a \sin u \cos v \, \mathbf{j} + a \sin v \, \mathbf{k}$ with $0 \leq u \leq 2\pi, \; 0 \leq v \leq \frac{1}{2}\pi$. By a previous calculation $\|\mathbf{N}(u, v)\| = a^2 \cos v$.

$\bar{x} = 0, \quad \bar{y} = 0$ (by symmetry)

$$\bar{z}A = \iint_S z \, d\sigma = \iint_\Omega z(u,v) \, \|\mathbf{N}(u,v)\| \, dudv = \int_0^{2\pi} \int_0^{\pi/2} a^3 \sin v \cos v \, dv \, du = \pi a^3$$

$\bar{z} = \tfrac{1}{2}a$ since $A = 2\pi a^2$

13. $\mathbf{N}(u,v) = (\mathbf{i}+\mathbf{j}+2\mathbf{k}) \cdot (\mathbf{i}-\mathbf{j}) = 2\mathbf{i}+2\mathbf{j}-2\mathbf{k}$

$$\text{flux in the direction of } \mathbf{N} = \iint_S \left(\mathbf{v} \cdot \frac{\mathbf{N}}{\|\mathbf{N}\|}\right) d\sigma = \iint_\Omega [\mathbf{v}(x(u), y(u), z(u)) \cdot \mathbf{N}(u,v)] \, dudv$$

$$= \iint_\Omega [(u+v)\mathbf{i} - (u-v)\mathbf{j}] \cdot [2\mathbf{i}+2\mathbf{j}-2\mathbf{k}] \, dudv.$$

$$= \iint_\Omega 4v \, dudv = 4 \int_0^1 \int_0^1 v \, dv \, du = 2$$

For Exercises 15–17 $\mathbf{n} = \dfrac{1}{a}(x\mathbf{i} + y\mathbf{j} + z\mathbf{k})$

$S : \mathbf{r}(u,v) = a \cos u \cos v \, \mathbf{i} + a \sin u \cos v \, \mathbf{j} + a \sin v \, \mathbf{k}$ with $0 \le u \le 2\pi, \quad -\tfrac{1}{2}\pi \le v \le \tfrac{1}{2}\pi$

$\|\mathbf{N}(u,v)\| = a^2 \cos v$

15.   with $\mathbf{v} = z\mathbf{k}$

$$\text{flux} = \iint_S (\mathbf{v} \cdot \mathbf{n}) \, d\sigma = \frac{1}{a} \iint_S z^2 \, d\sigma = \frac{1}{a} \iint_\Omega (a^2 \sin^2 v)(a^2 \cos v) \, dudv$$

$$= a^3 \int_0^{2\pi} \int_{-\pi/2}^{\pi/2} (\sin^2 v \cos v) \, d\sigma = \frac{4}{3}\pi a^3$$

17.   with $\mathbf{v} = y\mathbf{i} - x\mathbf{j}$

$$\text{flux} = \iint_S (\mathbf{v} \cdot \mathbf{n}) \, d\sigma = \frac{1}{a} \iint_S \underbrace{(yx - xy)}_{0} \, d\sigma = 0$$

For Exercises 19–21 the triangle S is the graph of the function

$$f(x,y) = a - x - y \quad \text{on} \quad \Omega : 0 \le x \le a, \quad 0 \le y \le a - x.$$

The triangle has area $A = \tfrac{1}{2}\sqrt{3}a^2$.

19.   with $\mathbf{v} = x\mathbf{i} + y\mathbf{j} + z\mathbf{k}$

$$\text{flux} = \iint_S (\mathbf{v} \cdot \mathbf{n}) \, d\sigma = \iint_\Omega (-v_1 f_x' - v_2 f_y' + v_3) \, dxdy$$

$$= \iint_\Omega [-x(-1) - y(-1) + (a - x - y)] \, dxdy = a \iint_\Omega dxdy = aA = \frac{1}{2}\sqrt{3}a^3$$

21. with $v = x^2\mathbf{i} - y^2\mathbf{j}$

$$\text{flux} = \iint_S (\mathbf{v} \cdot \mathbf{n})\, d\sigma = \iint_\Omega (-v_1 f_x' - v_2 f_y' + v_3)\, dx\, dy$$

$$= \iint_\Omega [-x^2(-1) - (-y^2)(-1) + 0]\, dx\, dy = \int_0^a \int_0^{a-x} (x^2 - y^2)\, dy\, dx$$

$$= \int_0^a \left[ax^2 - x^3 - \frac{1}{3}(a-x)^3 \right] dx = \left[\frac{1}{3}ax^3 - \frac{1}{4}x^4 + \frac{1}{12}(a-x)^4 \right]_0^a = 0$$

23.

$$\text{flux} = \iint_S (\mathbf{v} \cdot \mathbf{n})\, d\sigma = \iint_\Omega (-v_1 f_x' - v_2 f_y' + v_3)\, dx\, dy$$

$$= \iint_\Omega (-x^3 y - xy)\, dx\, dy = \int_0^1 \int_0^2 (-x^3 y - xy)\, dy\, dx$$

$$= \int_0^1 -2(x^3 + x)\, dx = -\frac{3}{2}$$

25. $\mathbf{n} = \dfrac{1}{a}(x\mathbf{i} + y\mathbf{j})$

$$\text{flux} = \iint_S (\mathbf{v} \cdot \mathbf{n})\, d\sigma = \frac{1}{a} \iint_S [(x\mathbf{i} + y\mathbf{j} + z\mathbf{k}) \cdot (x\mathbf{i} + y\mathbf{j})]\, d\sigma$$

$$= \frac{1}{a} \iint_S (x^2 + y^2)\, d\sigma = a \iint_S d\sigma = a\,(\text{area of } S) = a\,(2\pi a l) = 2\pi a^2 l$$

27.

$$\text{flux} = \iint_S (\mathbf{v} \cdot \mathbf{n})\, d\sigma = \iint_\Omega (-v_1 f_x' - v_2 f_y' + v_3)\, dx\, dy = \iint_\Omega 2y^{3/2}\, dx\, dy$$

$$= \int_0^1 \int_0^{1-x} 2y^{3/2}\, dy\, dx = \int_0^1 \frac{4}{5}(1-x)^{5/2}\, dx = \frac{8}{35}$$

29.

$$\text{flux} = \iint_S (\mathbf{v} \cdot \mathbf{n})\, d\sigma = \iint_\Omega (-v_1 f_x' - v_2 f_y' + v_3)\, dx\, dy = \iint_\Omega -y^{5/2}\, d\sigma$$

$$= \int_0^1 \int_0^{1-x} -y^{5/2}\, dy\, dx = \int_0^1 -\frac{2}{7}(1-x)^{7/2}\, dx = -\frac{4}{63}$$

31. $\bar{x} = 0, \quad \bar{y} = 0 \qquad$ by symmetry

verify that $\quad \|\mathbf{N}(u,v)\| = v \sin \alpha$

$$\bar{z} A = \iint_S z\, d\sigma = \iint_\Omega (v \cos \alpha)(v \sin \alpha)\, du\, dv = \sin \alpha \cos \alpha \int_0^{2\pi} \int_0^s v^2\, dv\, du = \frac{2}{3}\pi \sin \alpha \cos \alpha\, s^3$$

$\bar{z} = \frac{2}{3}s \cos \alpha \quad$ since $\quad A = \pi s^2 \sin \alpha$

33. $f(x,y) = \sqrt{x^2 + y^2}$ on $\Omega : 0 \le x^2 + y^2 \le 1$; $\lambda(x,y,z) = k\sqrt{x^2 + y^2}$

$x_M = 0,$ $y_M = 0$ (by symmetry)

$$z_M M = \iint_S z\lambda(x,y,z)\, d\sigma = \iint_\Omega k(x^2 + y^2)\sec\left[\gamma(x,y)\right] dx\, dy$$

$$= k\sqrt{2} \iint_\Omega (x^2 + y^2)\, dx\, dy$$

$$= k\sqrt{2} \int_0^{2\pi} \int_0^1 r^3\, dr\, d\theta = \frac{1}{2}\sqrt{2}\pi k$$

$z_M = \frac{3}{4}$ since $M = \frac{2}{3}\sqrt{2}\pi k$ (Exercise 32)

35. no answer required

37.
$$x_M M = \iint_S x\lambda(x,y,z)\, d\sigma = \iint_S kx(x^2 + y^2)\, d\sigma$$

$$= 2\sqrt{3}k \iint_\Omega (u+v)\left[(u-v)^2 + 4u^2\right]\, du\, dv$$

$$= 2\sqrt{3}k \int_0^1 \int_0^1 (5u^3 - 2u^2 v + uv^2 + 5u^2 v - 2uv^2 + v^3)\, dv\, du$$

$$= 2\sqrt{3} \int_0^1 \left(5u^3 - u^2 + \frac{1}{3}u + \frac{5}{2}u^2 - \frac{2}{3}u + \frac{1}{4}\right) du = \frac{11}{3}\sqrt{3}k$$

$x_M = \frac{11}{9}$ since $M = 3\sqrt{3}k$ (Exercise 36)

39. Total flux out of the solid is 0. It is clear from a diagram that the outer unit normal to the cylindrical side of the solid is given by $\mathbf{n} = x\mathbf{i} + y\mathbf{j}$ in which case $\mathbf{v}\cdot\mathbf{n} = 0$. The outer unit normals to the top and bottom of the solid are $\mathbf{k}$ and $-\mathbf{k}$ respectively. So, here as well, $\mathbf{v}\cdot\mathbf{n} = 0$ and the total flux is 0.

41. The surface $z = \sqrt{2 - (x^2 + y^2)}$ is the upper half of the sphere $x^2 + y^2 + z^2 = 2$. The surface intersects the surface $z = x^2 + y^2$ in a circle of radius 1 at height $z = 1$. Thus the upper boundary of the solid, call it S_1, is a segment of width $\sqrt{2} - 1$ on a sphere of radius $\sqrt{2}$. The area of S_1 is therefore $2\pi\sqrt{2}(\sqrt{2} - 1)$. (Exercise 25, Section 10.10.) The upper unit normal to S_1 is the vector

$$\mathbf{n} = \frac{1}{\sqrt{2}}(x\mathbf{i} + y\mathbf{j} + z\mathbf{k}).$$

Therefore

$$\text{flux through } S_1 = \iint_{S_1} (\mathbf{v}\cdot\mathbf{n})\, d\sigma = \frac{1}{\sqrt{2}} \iint_{S_1} \overbrace{(x^2 + y^2 + z^2)}^{2}\, d\sigma$$

$$= \sqrt{2} \iint_{S_1} d\sigma = \sqrt{2}\,(\text{area of } S_1) = 4\pi(\sqrt{2} - 1).$$

The lower boundary of the solid, call it S_2, is the graph of the function

$$f(x,y) = x^2 + y^2 \quad \text{on} \quad \Omega : 0 \le x^2 + y^2 \le 1.$$

Taking **n** as the lower unit normal, we have

$$\text{flux through } S_2 = \iint_{S_2} (\mathbf{v} \cdot \mathbf{n})\, d\sigma = \iint_{\Omega} \left(v_1 f_x' + v_2 f_y' - v^3 \right)\, dx\,dy$$

$$= \iint_{\Omega} (x^2 + y^2)\, dx\,dy = \int_0^{2\pi} \int_0^1 r^3\, dr\, d\theta = \frac{1}{2}\pi.$$

The total flux out of the solid is $4\pi(\sqrt{2} - 1) + \frac{1}{2}\pi = (4\sqrt{2} - \frac{7}{2})\pi.$

SECTION 18.8

1. $\nabla \cdot \mathbf{v} = 2,$ $\nabla \times \mathbf{v} = \mathbf{0}$

3. $\nabla \cdot \mathbf{v} = 0,$ $\nabla \times \mathbf{v} = \mathbf{0}$

5. $\nabla \cdot \mathbf{v} = 6,$ $\nabla \times \mathbf{v} = \mathbf{0}$

7. $\nabla \cdot \mathbf{v} = yz + 1,$ $\nabla \times \mathbf{v} = -xi + xy\mathbf{j} + (1 - x)z\mathbf{k}$

9. $\nabla \cdot \mathbf{v} = 1/r^2,$ $\nabla \times \mathbf{v} = \mathbf{0}$

11. $\nabla \cdot \mathbf{v} = 2(x + y + z)e^{r^2},$ $\nabla \times \mathbf{v} = 2e^{r^2}\left[(y - z)\mathbf{i} - (x - z)\mathbf{j} + (x - y)\mathbf{k}\right]$

13. $\nabla \cdot \mathbf{v} = f'(x),$ $\nabla \times \mathbf{v} = \mathbf{0}$

15. use components

17. $\nabla^2 f = 12(x^2 + y^2 + z^2)$

19. $\nabla^2 f = 2y^3 z^4 + 6x^2 yz^4 + 12x^2 y^3 z^2$

21. $\nabla^2 f = e^r(1 + 2r^{-1})$

23. (a) $2r^2$ (b) $-1/r$

25.
$$\nabla^2 f = \nabla^2 g(r) = \nabla \cdot (\nabla g(r)) = \nabla \cdot \left(g'(r)r^{-1}\mathbf{r} \right)$$

$$= \left[(\nabla g'(r)) \cdot r^{-1}\mathbf{r} \right] + g'(r)\left(\nabla \cdot r^{-1}\mathbf{r} \right)$$

$$= \left\{ [g''(r)r^{-1}\mathbf{r}] \cdot r^{-1}\mathbf{r} \right\} + g'(r)(2r^{-1})$$

$$= g''(r) + 2r^{-1}g'(r)$$

27. $n = -1$

SECTION 18.9

1. $\displaystyle\iint_S (\mathbf{v} \cdot \mathbf{n})\, d\sigma = \iiint_T (\nabla \cdot \mathbf{v})\, dx\,dy\,dz = \iiint_T 3\, dx\,dy\,dz = 3V = 4\pi$

3. $\displaystyle\iint_S (\mathbf{v} \cdot \mathbf{n})\, d\sigma = \iiint_T (\nabla \cdot \mathbf{v})\, dx\,dy\,dz = \iiint_T 2(x + y + z)\, dx\,dy\,dz.$

The flux is zero since the function $f(x, y, z) = 2(x + y + z)$ satisfies the relation $f(-x, -y, -z) = -f(x, y, z)$ and T is symmetric about the origin.

5.

face	n	v · n	flux	
$x = 0$	$-\mathbf{i}$	0	0	
$x = 1$	$\mathbf{i}$	1	1	
$y = 0$	$-\mathbf{j}$	0	0	total flux $= 3$
$y = 1$	$\mathbf{j}$	1	1	
$z = 0$	$-\mathbf{k}$	0	0	
$z = 1$	$\mathbf{k}$	1	1	

$$\iiint_T (\nabla \cdot \mathbf{v}) \, dx\,dy\,dz = \iiint_T 3 \, dx\,dy\,dz = 3V = 3$$

7.

face	n	v · n	flux
$x = 0$	$-\mathbf{i}$	0	0
$x = 1$	$\mathbf{i}$	1	1
$y = 0$	$-\mathbf{j}$	xz	fluxes add up to 0 total flux $= 2$
$y = 1$	$\mathbf{j}$	$-xz$	
$z = 0$	$-\mathbf{k}$	0	0
$z = 1$	$\mathbf{k}$	1	1

$$\iiint_T (\nabla \cdot \mathbf{v}) \, dx\,dy\,dz = \iiint_T 2\,(x + z) \, dx\,dy\,dz = 2\,(\bar{x} + \bar{z})V = 2\,(\tfrac{1}{2} + \tfrac{1}{2})1 = 2$$

9. $\text{flux} = \displaystyle\iiint_T (1 + 4y + 6z) \, dx\,dy\,dz = (1 + 4\bar{y} + 6\bar{z})V = (1 + 0 + 3)\,9\pi = 36\pi$

11. $\text{flux} = \displaystyle\iiint_T (2y + 2y + 3y) \, dx\,dy\,dz = 7\bar{y}V = 0$

13. $\text{flux} = \displaystyle\iiint_T (A + B + C) \, dx\,dy\,dz = (A + B + C)V$

15. Let T be the solid enclosed by S and set $\mathbf{n} = n_1\mathbf{i} + n_2\mathbf{j} + n_3\mathbf{k}$.

$$\iint_S n_1 \, d\sigma = \iint_S (\mathbf{i} \cdot \mathbf{n}) \, d\sigma = \iiint_T (\nabla \cdot \mathbf{i}) \, dxdydz = \iiint_T 0 \, dxdydz = 0.$$

Similarly

$$\iint_S n_2 \, d\sigma = 0 \quad \text{and} \quad \iint_S n_3 \, d\sigma = 0.$$

17. A routine computation shows that $\nabla \cdot (\nabla f \times \nabla g) = 0$. Therefore

$$\iint_S [(\nabla f \times \nabla g) \cdot \mathbf{n}] \, d\sigma = \iiint_T [\nabla \cdot (\nabla f \times \nabla g)] \, dxdydz = 0.$$

19. Set $\mathbf{F} = F_1\mathbf{i} + F_2\mathbf{j} + F_3\mathbf{k}$.

$$F_1 = \iint_S [\rho(z-c)\mathbf{i} \cdot \mathbf{n}] \, d\sigma = \iiint_T [\nabla \cdot \rho(z-c)\mathbf{i}] \, dxdydz$$

$$= \iiint_T \underbrace{\frac{\partial}{\partial x}[\rho(z-c)]}_{0} \, dxdydz = 0.$$

Similarly $F_2 = 0$.

$$F_3 = \iint_S [\rho(z-c)\mathbf{k} \cdot \mathbf{n}] \, d\sigma = \iiint_T [\nabla \cdot \rho(z-c)\mathbf{k}] \, dxdydz$$

$$= \iiint_T \frac{\partial}{\partial z}[\rho(z-c)] \, dxdydz$$

$$= \iiint_T \rho \, dxdydz = W.$$

SECTION 18.10

For Exercises 1–4: $\mathbf{n} = x\mathbf{i} + y\mathbf{j} + z\mathbf{k}$ and $C : \mathbf{r}(u) = \cos u\,\mathbf{i} + \sin u\,\mathbf{j}, \quad u \in [0, 2\pi]$.

1. (a) $\displaystyle\iint_S [(\nabla \times \mathbf{v}) \cdot \mathbf{n}] \, d\sigma = \iint_S (\mathbf{0} \cdot \mathbf{n}) \, d\sigma = 0$

(b) S is bounded by the unit circle $\quad C : \mathbf{r}(u) = \cos u\,\mathbf{i} + \sin u\,\mathbf{j}, \quad u \in [0, 2\pi]$.

$$\oint_C \mathbf{v}(\mathbf{r}) \cdot d\mathbf{r} = 0 \quad \text{since } \mathbf{v} \text{ is a gradient.}$$

3.   (a) $\displaystyle\iint_S [(\nabla \times \mathbf{v}) \cdot \mathbf{n}]\, d\sigma = \iint_S [(-3y^2\mathbf{i} + 2z\mathbf{j} + 2\mathbf{k}) \cdot \mathbf{n}]\, d\sigma$

$\displaystyle\qquad = \iint_S (-3xy^2 + 2yz + 2z)\, d\sigma$

$\displaystyle\qquad = \underbrace{\iint_S (-3xy^2)\, d\sigma}_{0} + \underbrace{\iint_S 2yz\, d\sigma}_{0} + 2\iint_S z\, d\sigma = 2\bar{z}V = 2(\tfrac{1}{2})2\pi = 2\pi$

Exercise 11, Section 18.7

(b) $\displaystyle\oint_C \mathbf{v}(\mathbf{r}) \cdot d\mathbf{r} = \oint_C z^2\, dx + 2x\, dy = \oint_C 2x\, dy = \int_0^{2\pi} 2\cos^2 u\, du = 2\pi$

For Exercises 5–7 take $S: z = 2 - x - y$ with $0 \le x \le 2,\ 0 \le y \le 2 - x$
and C as the triangle $(2,0,0)$, $(0,2,0)$, $(0,0,2)$.   Then $C = C_1 \cup C_2 \cup C_3$ with

$$C_1: \mathbf{r}_1(u) = 2(1-u)\mathbf{i} + 2u\mathbf{j}, \quad u \in [0,1],$$

$$C_2: \mathbf{r}_2(u) = 2(1-u)\mathbf{j} + 2u\mathbf{k}, \quad u \in [0,1],$$

$$C_3: \mathbf{r}_3(u) = 2(1-u)\mathbf{k} + 2u\mathbf{i}, \quad u \in [0,1].$$

$\mathbf{n} = \tfrac{1}{3}\sqrt{3}(\mathbf{i}+\mathbf{j}+\mathbf{k})$ area of S: $A = 2\sqrt{3}$ centroid: $(\tfrac{2}{3}, \tfrac{2}{3}, \tfrac{2}{3})$

5.   (a) $\displaystyle\iint_S [(\nabla \times \mathbf{v}) \cdot \mathbf{n}]\, d\sigma = \iint_S \tfrac{1}{3}\sqrt{3}\, d\sigma = \tfrac{1}{3}\sqrt{3}A = 2$

(b) $\displaystyle\oint_C \mathbf{v}(\mathbf{r}) \cdot d\mathbf{r} = \left(\int_{C_1} + \int_{C_2} + \int_{C_3}\right) \mathbf{v}(\mathbf{r}) \cdot d\mathbf{r} = -2 + 2 + 2 = 2$

7.   (a) $\displaystyle\iint_S [(\nabla \times \mathbf{v}) \cdot \mathbf{n}]\, d\sigma = \iint_S (y\mathbf{k} \cdot \mathbf{n})\, d\sigma = \tfrac{1}{3}\sqrt{3}\iint_S y\, d\sigma = \tfrac{1}{3}\sqrt{3}\,\bar{y}A = \tfrac{4}{3}$

(b) $\displaystyle\oint_C \mathbf{v}(\mathbf{r}) \cdot d\mathbf{r} = \left(\int_{C_1} + \int_{C_2} + \int_{C_3}\right) \mathbf{v}(\mathbf{r}) \cdot d\mathbf{r} = \left(\tfrac{4}{3} - \tfrac{32}{5}\right) + \tfrac{32}{5} + 0 = \tfrac{4}{3}$

9.   The bounding curve is the set of all (x, y, z) with

$$x^2 + y^2 = 4 \quad \text{and} \quad z = 4.$$

Traversed in the positive sense with respect to $\mathbf{n}$, it is the curve $-C$ where

$$C : \mathbf{r}(u) = 2\cos u\,\mathbf{i} + 2\sin u\,\mathbf{j} + 4\mathbf{k}, \qquad u \in [0, 2\pi].$$

By Stokes's theorem the flux we want is

$$-\int_C \mathbf{v}(\mathbf{r}) \cdot d\mathbf{r} = -\int_C y\,dx + z\,dy + x^2 z^2\,dz$$

$$= -\int_0^{2\pi} \left(-4\sin^2 u + 8\cos u\right)\,du = 4\pi.$$

11. The bounding curve C for S is the bounding curve of the elliptical region $\Omega : \frac{1}{4}x^2 + \frac{1}{9}y^2 = 1$. Since

$$\nabla \times \mathbf{v} = 2x^2 yz^2 \mathbf{i} - 2xy^2 z^2 \mathbf{j}$$

is zero on the xy-plane, the flux of $\nabla \times \mathbf{v}$ through Ω is zero, the circulation of $\mathbf{v}$ about C is zero, and therefore the flux of $\nabla \times \mathbf{v}$ through S is zero.

13. C bounds the surface

$$S: z = \sqrt{1 - \tfrac{1}{2}(x^2 + y^2)}, \qquad (x, y) \in \Omega$$

with $\Omega : x^2 + (y - \frac{1}{2})^2 \leq \frac{1}{4}$. Routine calculation shows that $\nabla \times \mathbf{v} = y\mathbf{k}$. The circulation of $\mathbf{v}$ with respect to the upper unit normal $\mathbf{n}$ is given by

$$\iint_S (y\mathbf{k} \cdot \mathbf{n})\,d\sigma = \iint_\Omega y\,dx\,dy = \bar{y}A = \frac{1}{2}\left(\frac{\pi}{4}\right) = \frac{1}{8}\pi.$$
$$\underset{(18.7.8)}{\Big\uparrow}$$

If $-\mathbf{n}$ is used, the circulation is $-\frac{1}{8}\pi$.   Answer: $\pm\frac{1}{8}\pi$.

15. $\nabla \times \mathbf{v} = \mathbf{i} + 2\mathbf{j} + \mathbf{k}$. The paraboloid intersects the plane in a curve C that bounds a flat surface S that projects onto the disc $x^2 + (y - \frac{1}{2})^2 = \frac{1}{4}$ in the xy-plane. The upper unit normal to S is the vector $\mathbf{n} = \frac{1}{2}\sqrt{2}(-\mathbf{j} + \mathbf{k})$. The area of the base disc is $\frac{1}{4}\pi$. Letting γ be the angle between $\mathbf{n}$ and $\mathbf{k}$, we have $\cos \gamma = \mathbf{n} \cdot \mathbf{k} = \frac{1}{2}\sqrt{2}$ and $\sec \gamma = \sqrt{2}$. Therefore the area of S is $\frac{1}{4}\sqrt{2}\pi$. The circulation of $\mathbf{v}$ with respect to $\mathbf{n}$ is given by

$$\iint_S [(\nabla \times \mathbf{v}) \cdot \mathbf{n}]\,d\sigma = \iint_S -\frac{1}{2}\sqrt{2}\,d\sigma = \left(-\frac{1}{2}\sqrt{2}\right)(\text{area of } S) = -\frac{1}{4}\pi.$$

If $-\mathbf{n}$ is used, the circulation is $\frac{1}{4}\pi$.   Answer: $\pm\frac{1}{4}\pi$.

17. Straightforward calculation shows that

$$\nabla \times (\mathbf{a} \times \mathbf{r}) = \nabla \times [(a_2 z - a_3 y)\mathbf{i} + (a_3 x - a_1 z)\mathbf{j} + (a_1 y - a_2 x)\mathbf{k}] = 2\mathbf{a}.$$

19. In the plane of C, the curve C bounds some Jordan region that we call Ω. The surface $S \cup \Omega$ is a piecewise–smooth surface that bounds a solid T. Note that $\nabla \times \mathbf{v}$ is continuously differentiable on T. Thus, by the divergence theorem,

$$\iiint\limits_{T} [\nabla \cdot (\nabla \times \mathbf{v})] \, dx\,dy\,dz = \iint\limits_{S \cup \Omega} [(\nabla \times \mathbf{v}) \cdot \mathbf{n}] \, d\sigma$$

where $\mathbf{n}$ is the outer unit normal. Since the divergence of a curl is identically zero, we have

$$\iint\limits_{S \cup \Omega} [(\nabla \times \mathbf{v}) \cdot \mathbf{n}] \, d\sigma = 0.$$

Now $\mathbf{n}$ is $\mathbf{n}_1$ on S and $\mathbf{n}_2$ on Ω. Thus

$$\iint\limits_{S} [(\nabla \times \mathbf{v}) \cdot \mathbf{n}_1] \, d\sigma + \iint\limits_{\Omega} [(\nabla \times \mathbf{v}) \cdot \mathbf{n}_2] \, d\sigma = 0.$$

This gives

$$\iint\limits_{S} [(\nabla \times \mathbf{v}) \cdot \mathbf{n}_1] \, d\sigma = \iint\limits_{\Omega} [(\nabla \times \mathbf{v}) \cdot (-\mathbf{n}_2)] \, d\sigma = \oint_{C} \mathbf{v}(\mathbf{r}) \cdot d\mathbf{r}$$

where C is traversed in a positive sense with respect to $-\mathbf{n}_2$ and therefore in a positive sense with respect to $\mathbf{n}_1$. ($-\mathbf{n}_2$ points toward S.)

APPENDIX A

SECTION A.1

1. $\{-1,0,1,2\}$

3. $\{0\}$

5. $\{-1,0,1,2,4,6,8,\ldots\}$

7. $\emptyset$

9. $\{1,2,3,4,6,8,\ldots\}$

11. $\{(0,-1),(0,0),(0,1),(2,-1),(2,0),(2,1)\}$

13. $\{(-1,0),(0,0),(1,0),(-1,2),(0,2),(1,2)\}$

15. $\left\{\begin{array}{l}(0,0,-1),(0,0,0),(0,0,1),(0,2,-1),(0,2,0),(0,2,1)\\(2,0,-1),(2,0,0),(2,0,1),(2,2,-1),(2,2,0),(2,2,1)\end{array}\right\}$

17. $\{2\}$

19. $\{0,2,4\}$

21. the set of real numbers

23. $\{x : 2 < x \le 4\}$

25. A

27. the set of real numbers

29. (a) B

31. $\{0\},\{1\},\{2\},\{0,1\},\{0,2\},\{1,2\},\{0,1,2\}$

(b) A

SECTION A.2

1. $\frac{1}{6}\pi$ radians

3. $\frac{3}{2}\pi$ radians

5. $\frac{1}{18}\pi$ radians

7. $\frac{5}{4}\pi$ radians

9. 45°

11. 105°

13. 60°

15. 150°

17. (a) 90°

(b) 10π radians

19. 0.3 radians

21. $\frac{10}{7}\pi$ in.

23. $30/\pi$ in.

25. $\frac{1}{2}\sqrt{2}$

27. $\frac{2}{3}\sqrt{3}$

29. $\frac{1}{2}$

31. 1

33. $\sqrt{2}$

35. $\frac{1}{3}\sqrt{3}$

SECTION A.3

1. Let S be the set of integers for which the statement is true. Since $2(1) \le 2^1$, S contains 1. Assume now that $k \in S$. This tells us that $2k \le 2^k$, and thus

$$2(k+1) = 2k + 2 \le 2^k + 2 \le 2^k + 2^k = 2(2^k) = 2^{k+1}.$$
$$(k \ge 1)$$

This places $k + 1$ in S.

We have shown that

$$1 \in S \quad \text{and that} \quad k \in S \quad \text{implies} \quad k+1 \in S.$$

It follows that S contains all the positive integers.

3.   Let S be the set of integers for which the statement is true. Since $(1)(2) = 2$ is divisible by $2, 1 \in S$.

Assume now that $k \in S$. This tells us that $k(k+1)$ is divisible by 2 and therefore

$$(k+1)(k+2) = k(k+1) + 2(k+1)$$

is also divisible by 2. This places $k + 1 \in S$.

We have shown that

$$1 \in S \text{ and that } k \in S \text{ implies } k + 1 \in S.$$

It follows that S contains all the positive integers.

5.   Use

$$1 + 2 + \cdots + k + (k+1) = (1 + 2 + \cdots + k) + (k+1)$$

$$= \tfrac{1}{2}k(k+1) + (k+1)$$

$$= (k+1)(\tfrac{1}{2}k + 1) = \tfrac{1}{2}(k+1)(k+2)$$

$$= \tfrac{1}{2}(k+1)[(k+1) + 1].$$

7.   Use

$$1^2 + 2^2 + \cdots + k^2 + (k+1)^2 = \tfrac{1}{6}k(k+1)(2k+1) + (k+1)^2$$

$$= \tfrac{1}{6}(k+1)[k(2k+1) + 6(k+1)]$$

$$= \tfrac{1}{6}(k+1)(2k^2 + 7k + 6)$$

$$= \tfrac{1}{6}(k+1)(k+2)(2k+3)$$

$$= \tfrac{1}{6}(k+1)[(k+1) + 1][2(k+1) + 1].$$

9.   By Exercise 8 and Exercise 5

$$1^3 + 2^3 + \cdots + (n-1)^3 = [\tfrac{1}{2}(n-1)n]^2 = \tfrac{1}{4}(n-1)^2 n^2 < \tfrac{1}{4}n^4$$

and

$$1^3 + 2^3 + \cdots + n^3 = [\tfrac{1}{2}n(n+1)]^2 = \tfrac{1}{4}n^2(n+1)^2 > \tfrac{1}{4}n^4.$$

11.   Use

$$\frac{1}{\sqrt{1}} + \frac{1}{\sqrt{2}} + \frac{1}{\sqrt{3}} + \cdots + \frac{1}{\sqrt{n}} + \frac{1}{\sqrt{n+1}}$$

$$> \sqrt{n} + \frac{1}{\sqrt{n+1} + \sqrt{n}}\left(\frac{\sqrt{n+1} - \sqrt{n}}{\sqrt{n+1} - \sqrt{n}}\right) = \sqrt{n+1}.$$

13.   Let S be the set of integers for which the statement is true. Since

$$3^{2(1)+1} + 2^{1+2} = 27 + 8 = 35$$

is divisible by 7, we see that $1 \in S$.

Assume now that $k \in S$. This tells us that

$$3^{2k+1} + 2^{k+2} \text{ is divisible by 7.}$$

It follows that

$$3^{2(k+1)+1} + 2^{(k+1)+2} = 3^2 \cdot 3^{2k+1} + 2 \cdot 2^{k+2}$$
$$= 9 \cdot 3^{2k+1} + 2 \cdot 2^{k+2}$$
$$= 7 \cdot 3^{2k+1} + 2(3^{2k+1} + 2^{k+2})$$

is also divisible by 7. This places $k + 1 \in S$.

We have also shown that

$$1 \in S \qquad \text{and that} \qquad k \in S \quad \text{implies} \quad k + 1 \in S.$$

follows that S contains all the positive integers.

15. For all positive integers $n \geq 2$,

$$\left(1 - \frac{1}{2}\right)\left(1 - \frac{1}{3}\right) \cdots \left(1 - \frac{1}{n}\right) = \frac{1}{n}.$$

To see this, let S be the set of integers n for which the formula holds. Since $1 - \frac{1}{2} = \frac{1}{2}$, $\; 2 \in S$. Suppose now that $k \in S$. This tells us that

$$\left(1 - \frac{1}{2}\right)\left(1 - \frac{1}{3}\right) \cdots \left(1 - \frac{1}{k}\right) = \frac{1}{k}$$

and therefore that

$$\left(1 - \frac{1}{2}\right)\left(1 - \frac{1}{3}\right) \cdots \left(1 - \frac{1}{k}\right)\left(1 - \frac{1}{k+1}\right) = \frac{1}{k}\left(1 - \frac{1}{k+1}\right) = \frac{1}{k}\left(\frac{k}{k+1}\right) = \frac{1}{k+1}.$$

This places $k + 1 \in S$ and verifies the formula for $n \geq 2$.

17. From the figure, observe that adding a vertex V_{N+1} to an N-sided polygon increases the number of diagonals by $(N - 2) + 1 = N - 1$.   Then use the identity

$$\tfrac{1}{2}N(N - 3) + (N - 1) = \tfrac{1}{2}(N + 1)(N + 1 - 3).$$

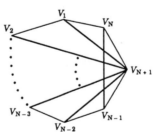

19. To go from k to $k+1$, take $A = \{a_1, \cdots, a_{k+1}\}$ and $B = \{a_1, \cdots, a_k\}$. Assume that B has 2^k subsets: $B_1, B_2, \cdots B_{2^k}$. The subsets of A are then $B_1, B_2, \cdots, B_{2^k}$ together with

$$B_1 \cup \{a_{k+1}\}, \; B_2 \cup \{a_{k+1}\}, \cdots, B_{2^k} \cup \{a_{k+1}\}.$$

This gives $2(2^k) = 2^{k+1}$ subsets for A.